Sacha Kagan
Art and Sustainability

Image | Volume 25

Sacha Kagan is research associate at Leuphana University Lueneburg and founding coordinator of the international network Cultura21. He works in the trans-disciplinary field of arts and (un-)sustainability.

Sacha Kagan

Art and Sustainability

Connecting Patterns for a Culture of Complexity

[transcript]

Printed with the support of Leuphana University Lüneburg.
At the same time dissertation for Leuphana University Lüneburg, 2011.

Bibliographic information published by the Deutsche Nationalbibliothek
The Deutsche Nationalbibliothek lists this publication in the Deutsche Nationalbibliografie; detailed bibliographic data are available in the Internet at http://dnb.d-nb.de

2nd emended edition, 2013

Cover Design: Kordula Röckenhaus, Bielefeld
Cover illustration: Jean-Claude Kagan, Eirtemys, 2010
Typeset & Proofread by Sacha Kagan
Printed by Majuskel Medienproduktion GmbH, Wetzlar
ISBN 978-3-8376-1803-7

Contents

Acknowledgements

The following pages are the result of a research process marked by many exchanges, learnings and bifurcations thanks to encounters with inspiring and generously supportive people.

First of all, my overall research was possible thanks to the active support of my PhD supervisor, and director of our research institute (the Institute of Cultural Theory, Research, and the Arts, at Leuphana University Lüneburg): Volker Kirchberg, who since hiring me in mid-2005, allowed me to spend much of my working time focusing on the research theme of arts, culture and sustainability. Not only did Volker provide this research space-time, but he also actively fostered the vitality of my research theme, not only as doctoral supervisor but also for example as organizer of the 2007 conference of the arts research network of the European Sociological Association (which included sustainability as one of its three key foci).

Equally fundamental was the inspiration and critical support of my second PhD supervisor Hans Dieleman, whose exploration of the roles of artists for sustainability were the yeast that allowed my own interest in this theme to grow, ever since I attended his seminars as an exchange student at the Erasmus Universiteit Rotterdam, back in 2001. As PhD supervisor, Hans brought me much appreciated critical feedback which allowed me to improve and clarify my analysis and my formulations.

Of utmost importance were the insightful exchanges with fellow organizers and members of the international network Cultura21, and of the German organization Cultura21, including: Davide Brocchi, Oleg Koefoed, and Francesca Cozzolino.

Further exchanges with colleagues at the Leuphana University (including Ullrich Günther, Christoph Behnke, Ulf Wuggenig, Anke Haarmann, Nicolas Dierks, Katina Kuhn and the colleagues at the INFU) and at other organizations (including Katelijn Verstraete and her co-workers at ASEF, Guy Gypens at the Kaai Theater, David Knaute, Hatto Fischer, Jennifer Gidley) are also to be thanked, as well as the discussions with social scientists (including Tasos Zembylas, Hans Abbing, Rudi Laermans, Federica Vigano, Arild Bergh, Julien Knebusch, Nelly van der Geest, Jan van Boeckel) artists (including Aviva Rahmani, David Haley, the collective

Wochenklausur, Insa Winkler, HMJokinen, Shelley Sacks, Rosemarie Horn) and curators/writers (including Bettina Steinbrügge, Caffyn Kelley, Hildegard Kurt, Georg Engeli). I also thank Jacques Leenhardt and Werner Faulstich for their stimulating questions and comments as members of the jury at my PhD defense.

By this very process of naming a few persons, I am doing injustice to the many other individuals and groups not mentioned, whose paths I crossed and learned from. As the usual expression goes in 'acknowledgements', "you know who you are" / "you will recognize yourself." I also want to thank the many authors with whom I did not directly interact but whose writings inspired me or stimulated my critical reactions.

I thank the students at the seminars I taught at the Leuphana University Lüneburg and elsewhere (e.g. at the CCC postgraduate program), and my friends, as well as my family and my partner Nino Ruschmeyer, for countless stimulating conversations.

The original version of the manuscript of this book was entirely written with Open Office, under Ubuntu Linux. I thank the FLOSS (Free Libre Open Source Software) community for their great efforts and dedication in providing end-users like me, with alternatives to monopolistic proprietary software. Unfortunately, because an Open Office compatible layout template was not available yet at the publishing house, I had to resort to proprietary software in the final preparation of the current format of the book.

Finally, I thank the team at the publishing house transcript Verlag, who were kindly supportive and responsive, and especially Stefanie Hanneken for her corrections and advices during my copy-editing/proofreading.

Introduction

SUSTAINABLE DEVELOPMENT AND SUSTAINABILITY

The expressions "sustainable development" and "sustainability", which constitute the moving horizon of my research perspective, acquired a widespread popularity across the world within the past two decades, as a response to perceived threats to humanity's continued existence.

The general understanding of "sustainable development" is most often described as a triptych of social justice, ecological integrity and economic well-being, and was popularized by the Brundtland report (1987). It aims at a development of human societies that would achieve "the reconciliation of social justice, ecological integrity and the well being of all living systems on the planet. The goal is to create an ecologically and socially just world within the means of nature without compromising future generations" (Moore 2005, p. 78).

The term 'sustainability' (used in 198 1 by Lester Brown), has an older genealogy in the German-speaking world, as the term 'Nachhaltigkeit' was coined (first in 1713 by Hans Carl von Carlowitz) to characterize a management of forests that would not deplete resources on the long term, but allow the renewable natural resources to regenerate and thus ensure its exploitation over the long-term.[1] I will however not delve further into the specifically German discourses and debates around the term of 'Nachhaltigkeit', for two reasons: On the one hand, my research is focusing primarily on English-speaking literature, and thus on English-language discourses (with a number of insights from French-speaking literature). On the other hand, I consider the contemporary usages of the term 'sustainability' to be rooted mostly in its relatively recent history in English language, i.e. from the 1980's onwards, as a heritage from the notions of "eco-development" as

1 However, this genealogy of the term in German forestry has been reconstructed *a posteriori* by German researchers, when the English term "sustainability" imposed itself in recent years. There is no obvious direct link between the German old expression from forestry, and the contemporary usage of the English term as it emerged in the 1970's.

discussed at the United Nations and other international organizations in the 1970's (for example, the Club of Rome's famous 1972 report *Limits to Growth* uses the adjective 'sustainable' to characterize the desired model of development) as well as from the history of ecology and environmentalist movements across the 20th century.[2]

The use of the term 'sustainability', rather than 'sustainable development', by many authors in this context, reflects "a concern, on the part of NGO and academic environmentalists, that development is seen as synonymous with growth, and therefore that sustainable development means ameliorating, but not challenging, continued economic growth" (Robinson 2004, p. 370). More generally, besides the discussion-point of economic growth, the use of the term 'sustainability' suggests a different priority in framing the future of humanity in terms of its balanced evolution, linking social and ecological issues, rather than framing it in terms of a linear development-course with the economy as its main focus. This is the sense in which I will favor the term 'sustainability' in the rest of this volume, and connect it to the constitution of a complex, 'systemic' perspective on reality.

The expressions 'sustainable development' and 'sustainability' established themselves across a pre-existing division of the environmentalist movement in two divergent discourses known as the 'preservationist' and the 'conservationist' discourses. Simons and Warfield (2007, p. 6) characterized this division as rooted in the divergence between "those who see nature as an art form [with an own spiritual dimension and who] prioritize the non-human" i.e. the preservationists, and "those who prize utility [of nature for mankind and] favour the human" i.e. the conservationists.

The preservationists, such as John Muir and later Aldo Leopold, inspired by the Romanticism of e.g. Henri David Thoreau, consider the whole living world as the community in which we live. A radical form of preservationist discourse is found in deep ecology (with e.g. Arne Naess), which attributes inherent rights to the natural environment. The conservationists, starting with Gifford Pinchot (who coined a 'resource conservation ethic' aiming at minimizing environmental damage so as to ensure continued benefits for human beings), inspired the contemporary mainstream environmental movement, as well as the more radical 'social ecology' movement of Murray Bookchin (which advocates for a participatory direct democracy rooted in anarchism and connects it to an ecological discourse).

Traditionally, preservationists aimed at "the preservation of natural areas in what was perceived to be a pristine (i.e. undeveloped) form" (Robinson 2004, p. 371) and called forward a spiritual awakening to the sacredness of nature, whereas the conservationists heralded the protection of natural areas as a pragmatic goal, i.e. "a form of enlightened self-interest" (ibid.).

---

2 On the histories of the English and German terms sustainability and Nachhaltigkeit, see Grober 2002 and Tremmel 2003. See also Grabe 2010.

Table 1: Biocentric vs. Culture-centric perspectives

Biocentric (from preservationist roots)	**Culture-centric (from conservationist roots)**
Oriented to local, place-specific, closed system	Open System
Culture as nature: e.g. aboriginals as "noble savages"	Nature as culture, e.g. discursive
Stability (peaceful)	Instability (conflicted)
Finite human possibility	Infinite possibility
Independent of other systems	Interacts with and is changed by other systems
Pessimistic: problems result from immutable limits to growth	Optimistic: resourcefulness can solve problems
Culture as art, beauty	Culture as industry, utility
Absolute/fixed/normative (centralized, hierarchical, ranked)	Relative (equity, broad participation, decentralized)
Values secondary	Values primary
Empiricism	Humanism
Maintain/preserve	Expand/grow

Source: Simon and Warfield 2007, p. 10.

Over time, an additional distinction emerged between those who insisted on "questions related to values and fundamental changes in individual attitudes towards nature (the sustainability argument) [and those who, on] the other side take what they believe to be a more pragmatic and collective approach, oriented towards efficiency gains and improvements in technology (i.e. sustainable development)" (ibid.). The distinction has been labeled as "ecocentric" vs. "technocentric" (Pepper 1996 cited in Robinson 2004) or as "biocentric" vs. "culture-centric" (Simons and Warfield 2007). Robinson (2004) discusses further how the Brundtland Commission report constituted a "curious combination" of these two perspectives. But is the discussion of sustainability bound to be divided along such a binary line (with opposed sets of assumptions, as suggested by Simons and Warfield 2007 – cf. table 1)? Or can a complex understanding of sustainability be constructed, which would overcome the problematic assumptions of both approaches – namely, the traditional preservationist assumption of a fixed and pristine natural order, and the conservationist 'optimism' assuming an infinity of possibility thanks to human creativity (i.e. denying – or promising to overcome – hard biophysical limits to growth)?

The arguments to be found in the present volume will tend to argue for the latter, i.e. that a complex understanding of sustainability is possible, and will elaborate a description of such an understanding as the emergence of an alternative worldview. Such a proposal is not an entirely new idea, and it echoes for example the 'interdependency thesis' of Michaelidou et al. (2002) according to which "neither environmental nor community sustainability are possible without the other" (Simons and Warfield 2007, p. 9). It also echoes, at a more fundamental level, a complex idea of nature, which incorporates both nature as inherent beauty and nature as human utility. Edgar Morin's definition of nature already sets the stage for the whole discussion of sustainability and culture that will further unfold itself in the present volume: "Nature is not only *physis*, chaos and cosmos together. Nature is what binds, articulates, makes the anthropological communicate in depth with the biological and the physical" (Morin 1992, p. 382).

But why call it then an 'alternative' worldview? Robinson (2004) echoed the criticisms of sustainable development as "cosmetic environmentalism", both evading the question of "limits to growth" with a faith vested entirely in an imminent technological revolution, and evading the hard political and social challenges of Western industrial societies (as argued e.g. by Naomi Klein 2000). Instead, he argues for an open usage of the term "sustainability", not as a unitary concept, but as "a kind of discursive playing field in which [conflicting views] can be debated", as "a political act, not a scientific concept" in the traditional restricted sense (i.e. supposedly value-neutral), and as "the emergent property of a conversation about what kind of world we collectively want to live in now and in the future" (Robinson 2004, p. 382). Within this playing field and this conversation, my focus, which will lie especially on the cultural dimension of sustainability, will also oppose these mainstream discourses on sustainability that perpetuate techno-optimism and evade political challenges. Laville and Leenhardt (1996, pp. 132-142) already argued that sustainable development requires a "spirit of utopia", in order to reach beyond technocratic superficiality. Dieleman (2001, p. 14) defined "sustainable development" as "challenging existing definitions of reality in virtually all respects" and "working towards sustainable development" as "question[ing] current values and worldviews [...] seek[ing] creativity to develop new and yet unexplored products and practices [and] creat[ing] fear, anxiety, uncertainty and alienation" to be catalyzed and transformed into "empowerment". These three authors also pointed most especially at the roles of culture and the arts in this context.

THE 'CULTURAL DIMENSION' OF SUSTAINABILITY

Moacir Gadotti (the president of the deliberative council of the Instituto Paulo Freire, in Brazil) states: "To us, sustainability is the dream of living well; sustainability is a dynamic balance with others and the environment, it is the harmony among differences" (Gadotti 2009, pp. 13-14). As a shared dream, vision and worldview (as pointed out by Gadotti), as well as a conversation (as described by Robinson), sustainability reveals itself as a cultural phenomenon, if 'culture' is understood as value system and set of signifiers framing social identities and dispositions to act and to believe. Culture is a collective memory of a plurality of knowledge, know-hows and rules/conventions. Cultures, together with the individual minds, are also the ecosystems of ideas and mental images, i.e. of the 'noological sphere' of human imagination and creativity. The noological sphere is capable both of generativity and of regulation (through a relative capacity to resist to external disturbances).[3] Its mythical constructs are not only a superstructure as old-school Marxists used to claim, but a fundamental generative part of the fabric of human societies. As argued by Edgar Morin, an author on whom I will extensively draw in the present volume, "imagination is at the active and organizational heart of social and political reality" (Morin 1977, p. 341, following Castoriadis 1975).[4]

The inclusion of a cultural dimension in the discourses on sustainability has developed gradually in the past decade, at different levels of discourse. In 1996, the exhibition "Villette-Amazone" was organized in Paris at the "Grande Halle de la Villette" (the largest science center in Paris) by the 'Comité 21', French committee for environment and sustainable development. It combined ecological urbanism, architecture, technologies and artistic projects (by artists such as 'the Harrisons' about whom I will talk longer in chapter 5, and other artists).[5] It also was accompanied by a publication, a "manifesto for the environment in the 21st century" written by Bettina Laville (environmental adviser of the French President François Mitterand) and Jacques Leenhardt (philosopher and sociologist). In it, the authors aim to "engage a reflection on the role of culture in the transformations of our relationships with the environment" whereby to "rethink man and his knowledges in the very midst of nature and no longer outside of it" (Laville

3 The meaning of "generativity" will be introduced in the first section of chapter 3.

4 Exact page numbers in quotes from Morin 1977, refer to the 1981 pocket book reprint.

5 The other artists included in that exhibition (whose work I will however not discuss in the present volume) included Frans Krajcberg, Chen Zen, Gilles Clément, Jean-Michel Alberola, Bernard Pagès, Giuseppe Penone, Jaume Plensa, Erik Samakh, Dominique Bailly and Josef Koudelka, among others. The selection of artistic projects in that exhibition focused especially on the activity of landscaping.

and Leenhardt 1996, p. 15). They call forward the need for a "new epistemology" (ibid., p. 12), pointing out that "precaution and contingency [...] consciously break away from violence, *hybris* [...] This here is a matter of culture"(ibid., p. 13). They further discuss how some artists in the 20th century "constructed, maybe without exactly realizing it, the conditions of possibility of a new outlook on nature" (ibid., p. 58) and the emergence of an ecological art (which I will discuss in chapter 5). They denounce the "divorce between the worlds of science and of culture" (ibid., p. 89) and "the civilization of *ex-situ*" (ibid., p. 90) to which they opposed the duty of "environmental sciences" to "replace man in the earthly ecosystem" (ibid., p. 93). Arguing that "ecologists have unfortunately renounced imagination" (ibid., p. 100), they call forward a "moral of transformation" (ibid., p. 102), a "new humanism" founded on the "end of anthropocentrism" (ibid., p. 97) and a "mobilization of the imaginary" (ibid., p. 116) with the participation of artists. Laville and Leenhardt, in their manifesto, suggested that "sustainable development is maybe the new challenge of the 21st century, the sesame [i.e. the key to success] of a cultural revolution [...] on no account the trivial fruit of a technocratic consensus" (ibid., pp. 134-135). Unfortunately, not much happened in France in the decade that followed *Villette-Amazone*, in terms of integration of culture, art and sustainability, but significant developments emerged elsewhere around the world.

The UNESCO Summit on Culture and Sustainable Development in 1998, under the title "The Power of Culture", proclaimed in general terms the interdependence of culture and sustainable (mainly economic) development.[6] In Australia, Jon Hawkes' four-pillar model of sustainability (Hawkes 2001) included cultural vitality as a major dimension of sustainability, pointing at the inherent value of cultural diversity and of a vibrant cultural life of human communities. In parallel, in Germany, a number of cultural actors and of personalities involved in the field of sustainability, called for a greater attention to the role of the arts and culture in the search for sustainability. They organized in January 2002 at the Art Academy of Berlin, a conference of the German Society for Political Culture (Institut für Kulturpolitik der Kulturpolitische Gesellschaft) where they supported the "Tutzinger Manifest", a call for the integration of culture in Agenda 21 processes, which had been launched in April 2001 at the conference "Aesthetics of Sustainability" held at the Protestant Academy Tutzing. The "Agenda 21" action program launched at the 1992 Earth Summit of the United Nations in Rio de Janeiro, and the local agendas 21 that followed at the level of local policies, aimed to work concretely towards sustainable development, but did not directly include, especially in its first decade of existence, a cultural dimension (neither in a wide understanding of culture, nor

6 A further elaboration for cultural policies based on UNESCO's 1998 summit, was developed by Colin Mercer (2002).

in the more restricted sense of specific cultural-artistic activities).[7] The German "Tutzinger Manifest" denounced the dominance of technical, social and political economy discourses in the Agenda 21, and the absence of a discussion of the cultural background and significance of a search for sustainable development. It also denounced the absence of interest for questions of sustainability in cultural policy and among a majority of the art worlds. And it advocated (in the context of the Johannesburg World Summit on Sustainable Development of 2002) for the integration of culture in strategies for sustainability, and for the building of networks in order to overcome this gap. (One of the lead organizers of the two conferences and "Tutzinger Manifest", Hildegard Kurt, will come back under my focus in chapter 5.) The Manifest explicitly stated: "If sustainability is to be attractive and fascinating, if it is to appeal to the senses and convey a meaning, then beauty becomes an elementary component of a future that has a future, a way of life to which all people are entitled. For the Agenda 21 to be successful it is critical to integrate participants with the ability to bring ideas, visions and existential experiences alive in socially recognisable symbols, rituals and practices."[8]

In the meanwhile, a coalition of local governments across the world have launched the "Agenda 21 for Culture", approved by the 4th Forum of Local Authorities for Social Inclusion, in Barcelona in May 2004.[9] This policy document lists principles relating culture and the themes of human rights (which should include cultural rights), of diversity (i.e. the role of cultural diversity for human development), of local sustainable development (pointing at interculturality, at the importance of a "cultural ecosystem" and at the prospects for economic wealth), of social inclusion (cultural activities as part of citizenship) and of the economic role of cultural industries in the knowledge economy. Lately, concerns for ecology, sustainability and sustainable development also gained attention in the field of cultural policy and among prominent arts organizations, as indicated for example by the organization of a conference in December 2009 ("Culture|Futures", alongside the COP15 UN Climate conference in Copenhagen) devoted to the roles of arts and culture for the "transition to an ecological age by 2050".[10] Among the co-organizers of that conference were organizations such as the International Federation of Arts Councils and Cultural Agencies (IFACCA), the network of the European Union National Institutes for Culture (EUNIC), the Asia-Europe Foundation (ASEF), The Danish Cultural Institute, The British Council, the Goethe Institut, The Royal Society for the Arts (in the UK), the

7 For an introduction to the Agenda21, cf. Keating 1993.

8 The Tutzinger Manifest, in English translation, is available online at: http://www.kupoge.de/ifk/tutzinger-manifest/pdf/tuma-gb.pdf

9 The Agenda 21 for Culture and accompanying documents and reports can be found online at http://agenda21culture.net

10 See the conference online documentation at: http://www.culturefutures.org/

Association of Performing Arts Presenters (in the USA) and the European Cultural Foundation.[11] These organizations have not only gathered in that one-off event, but also started engaging, in recent years, in different reflections and programs aiming to integrate their work within the context of sustainability (and particularly of climate change).

At the level of academic research on sustainability, the inclusion of culture and the arts also emerged throughout the past decade. For example, in a 1999 publication about education for sustainability, Stoltenberg and Michelsen pointed at the cultural dimension of sustainable development (next to its ecological, social and economic dimensions) as including the "handling of time", traditional knowledge, a "holistic perception of nature", lifestyles and ethics (cited in Altner 2006, p. 74). At the Erasmus Universiteit Rotterdam, Hans Dieleman launched in 2000 a research program on "art & sustainability" aiming to "explore the various roles artists can play in change processes towards sustainability" and linking the "systems characteristics of sustainable development" with the "beyond rationality" characteristics of art (Dieleman 2001).

I will argue in this volume that the cultural dimension of sustainability involves not only the inclusion of the value of culture and of the arts in the discussion of local sustainable development and of sustainable communities, with an understanding of the contribution of culture (as in 'cultural expressions' of a community, cultural activities and the arts) to economic, social and cultural capital (as argued e.g. by Duxbury, Gillette and Pepper 2007), but also involves an understanding of 'culture(s) of sustainability', i.e. set(s) of norms and values, social conventions and institutions, informing the transition to more sustainable practices. At a more abstract level, the question of culture(s) of sustainability also touches upon transformations in world-views and paradigmatic bases for the knowledge of the world around oneself, i.e. epistemological issues. Under this perspective, my concern will be to understand how the arts are related to such a question of culture(s) of sustainability. My focus, in the present volume, will be lying at the level of ideas, discourses and practices of different social actors including especially artists and intellectuals, rather than at the level of the institutional orientations of (cultural) policies as reflected in the Agenda 21 for Culture.

OVERVIEW OF THE FOLLOWING CHAPTERS

I will now give a brief overview of the progression of the chapters in the present volume. However, this should not be mistaken for an 'executive

11 One of the co-organizers of the "Culture|Futures" conference was the small NGO Cultura21 Nordic, which is a chapter of the Cultura21 international network that I founded and coordinate since 2007, building upon the German NGO Cultura21 which emerged in 2005 and was officially founded in 2006 by Davide Brocchi.

summary'. The main developments and arguments that I will unfold, have to be found in the chapters themselves, and will not be presented in a placated premature version in the following few paragraphs.

The literacy of sustainability, which at a general level can be considered as a common culture (but as a 'uniplural' rather than simply unified culture, as I will argue later), departs from a currently dominant culture based on modes of knowing inherited from modernity. This argument will be developed in the first chapter of this volume, on 'the culture of unsustainability'.

Based on disjunctive thinking, simplification by reductionism and atomization of knowledge and experience, the culture of unsustainability developed an advanced form of cultivated autism in the relationships of modern societies and modern individuals with their environments. This development will be described in the first section of chapter 1, retracing the historical roots as well as the different dimensions and the significance of a culture of unsustainability in the organization of modern to late-modern societies.

The second section of chapter 1 will explore how the social organization and the values of modern and postmodern art have been contributing to the wider culture of unsustainability. This critical overview will focus primarily on the so-called 'high', 'autonomous', 'fine' arts, rather than on popular cultural expressions. I choose to focus especially on 'high art' because of its peculiar contrast between the ambitious claims and hopes often elaborated within art worlds with regards to art's transformative potential, and the challenges and limits set against such ambitions by art's embeddedness in a wider culture of unsustainability. This limitation of my focus to the high arts will also allow me to further explore an alternative development within the scope of the present volume (in chapter 5 on ecological art). More generally, the critical analysis of the culture and the art of unsustainability, as conducted in chapter 1, will serve as a baseline against which the further chapters will unfold the exploration of a culture/cultures of sustainability (chapters 2 and 3) and will more specifically analyze its significance in terms of aesthetic and artistic experience (chapter 4).

As already stated after Robinson (2004), sustainability cannot be summed up in a master plan or a single concept. It requires a reform of thought toward complexity which can help put the pieces together, but without superimposing on reality a new simplifying template replacing the disjunctive literacy of modernity. The exploration of such a difficult undertaking, will require two chapters:

- Chapter 2 will introduce 'systems thinking' as a first step toward a culture of sustainability, introducing a non-linear understanding of life and society and laying the foundations for an 'ecological literacy'; but the limits and pitfalls of systems thinking will also be assessed (most especially with regard to the risk of simplification by holism);

- Chapter 3 will move further into theories of complexity and into the framework of transdisciplinarity, as correction to the simplification of systemic holism and as the basis for a literacy of complexity.

The culture of complexity thus articulated, will be further detailed in its aesthetic dimension in chapter 4 on 'aesthetics of sustainability'. The core insight of this chapter, which constitutes the cornerstone of the present volume, will be based on Gregory Bateson's understanding of aesthetics as the "sensibility to the pattern which connects" as well as on theories of complexity and phenomenological considerations of a sensibility to the "more-than-human".

This sensibility has been most especially at the core of the development of numerous ecological-artistic practices in the past 40 years, which are introduced in chapter 5. This chapter offers an overview of the field of so-called "eco-art", within the art world of contemporary visual art, as an exemplary and illustrative account of a possible development of 'aesthetics of sustainability' in the arts (but by no means an exhaustive mapping of all modern and contemporary artistic and cultural practices which may contribute at different degrees and levels, to a culture of sustainability).

Sustainability and ecology have in the meanwhile turned into a focus of attention for a diversity of artists and other cultural actors (e.g. curators, art critics) involved in the art world of contemporary visual art, most especially in the decade 2000-2010. I will conduct a short overview of some of these developments in chapter 6, assessing how far these artistic discourses and interests may be relevant to the culture and aesthetics of sustainability as I discuss them in this volume, and how far they may contribute relevant insights and sensibilities in this context. Here again, my analysis in no way claims to be exhaustive, but merely to offer an exploratory overview.

The question remains, of the relevance of such artistic practices and experiences for wider social transformations toward sustainability. I will finally address this question in chapter 7, considering art as a field of experimentation where a 'double entrepreneurship in conventions' may allow transformations across social fields. The question of art and social change will also raise political issues related both to the relative window of opportunity for such social change, and to the very question of the political dimension of social sustainability. This later question will also be addressed in chapter 7, with a focus kept on its declination in the worlds of the arts.

Given the width of the area covered in this book, not all possibly relevant approaches can be introduced, and many specific topics will be only shortly evoked. I will focus on the discourses and arguments that are most especially relevant to a discussion of culture, art and sustainability under the perspective of complexity.

METHODOLOGICAL CONSIDERATIONS

My work in the current volume consists in a theoretical exploration, without a properly empirical element, but with especially two chapters (5 and 6) delving into a historical overview of a specific movement in art and a discussion of some contemporary developments in the field of contemporary visual arts which are especially relevant to my analysis.

The choice not to include an empirical study in this work, but to focus my analysis at the theoretical level, stems from the methodological and epistemological challenge of my research theme: The search for sustainability calls forward a movement away from the usual confines of disciplinary, or even inter-disciplinary work, and towards transdisciplinarity.[12] This movement, which I aimed to retrace as part of the description of a culture of sustainability, was also the movement in which I had to engage myself in the making of the current research, and thus I was engaged in a recursive process by which my very theoretical work had to move towards complexity and transdisciplinarity. This imposed that I engage into grounds on which I am not familiar, such as the natural sciences as opposed to the social sciences from which I come. To the reflexive researcher, such a wide horizon as transdisciplinarity finishes to shatter the illusion of a full knowledge of all relevant discourses, which is usually maintained in small (inter-)disciplinary communities of 'peers'.

The preceding couple of self-descriptive sentences are not innocently laid bare to the reader. They reflect an epistemological attention to complexity:

So-called 'Human Sciences' and 'Natural Sciences' (or 'soft' and 'hard' sciences) need to base themselves on each other. In the epistemology of Edgar Morin (whose "method" I will discuss in the first section of chapter 3, not just as a mere methodology but as the basis for no less than a paradigm shift in scientific work), the nature of reality loses a stable primary ontological basis (in a linear view), and establishes itself through the complex circular relation between the level of the social construction of reality (i.e. the social sciences) and the level of natural facts, i.e. the physical and biological substrata for the existence of any human (socially constructed) reality. Thereby, Morin's epistemology aims at a "complex knowledge carrying its own reflexivity" (Morin 1977, p. 19).

This reflexivity also concerns the individual researcher's subjectivity: "The subject who disappears from his discourse in fact takes over the Control Tower. By pretending to give way to the Copernician Sun, he reconstitutes a Ptolemaic system whose center is his spirit" (Morin 1992, p. 19). Consequently, the researcher's subjectivity should be acknowledged and

12 The meaning and significance of transdisciplinarity is not introduced here, but it will be in the second section of chapter 3, as part of my exploration of a culture of sustainability.

made visible, without turning the discourse into an apology for subjectivism. The researcher's culture too, i.e. the social and historical conditions, influence the conceptual work, and therefore a sociology of knowledge is always required as part of the reflexive process.[13] But Morin also warns against a purely social-constructivist discourse which would be equally naive in denying the physical base of any human, 'socially constructed' reality.[14] He calls forward a cyclic, rotational method: "This rotation leads us to physicalize our notions, then to socialize them, then to rephysicalize them, then to resocialize them, and so on *ad infinitum* [so that such a method be] not a vicious circle, but a productive praxis" (ibid., p. 288).

My epistemological position, after Morin's, should therefore not be mistaken for a post-modernist radical constructivism, which it is not. As argued in more details by Smith and Jenks (2006), theories of complexity, in their attachments to emergence, evolution and to the recognition of the reality of 'attractors,' reject an absolutist notion of contingence: In short, not everything in reality is socially constructed, and social constructions and conventions are not floating in an ocean of chaos (in the common speech understanding of the term 'chaos') but are embedded/entangled with attractors, i.e. intervening variables at physical, chemical, biological and ecological levels of reality.

The recognition of subjectivity and the researcher's reflexivity allow science's social praxis to emerge: "All knowledge is a physical praxis which is at the same time an anthropo-social praxis" (Morin 1977, p. 383). There exist only normative knowledge and normative science. Physics is also a human science and, as will be further articulated in chapters 2 and 3, the social sciences are also becoming (for a large part) an emergence rooted in physics. And science, rather than being a disinterested activity, is a historical emergence of the Western civilization.

On complexity

The notion of complexity, which will be properly defined and articulated in chapter 3 with the account of Edgar Morin's theoretical work, plays a key role in the understanding of the search-field of sustainability and in the sort of *Second Enlightenment* that is required in order to achieve such an understanding. To prevent misunderstandings, a few cautionary remarks are warranted from the start: Complexity is not 'complication'.

13 If the present volume may partly fail with regard to Morin's epistemological requirements, it would be by a lack of space devoted to myself, in my written text, rather than by an excess of subjectivity. The kind of self-reflexivity, at a personal level, practiced by Morin in his texts, is something which I am only now barely starting to practice.

14 Cf. Morin 1977, pp. 273-275.

The complicated can be relatively easily reduced to some relatively basic principles (like e.g. the 'complications' of a Swiss watch). But complexity disarms the explanations based on single empirical bases, single and linear logics: Complexity requires multiple logics that are neither separated from each other into neat boxes, nor integrating neatly with each other, but enter into ambivalent relations and tensions. In the universe, in Nature, in our reality, as will be argued in chapter 3 after Morin, *everything is complex.* "The simple is but an arbitrary moment of abstraction torn from complexity, an effective instrument of manipulation flattening complexity" (ibid., p. 386).

The science of complexity as Morin sees it, shall be en-cyclo-pedic, not in the sense of an encyclopedic accumulation of knowledge areas, nor in the sense of a pretense to engulf the whole of knowledge (and be 'exhaustive'), but in the sense of "agkuklios paidea, training which puts knowledge in cycle [...] that is to say of learning to articulate the disjointed points of view of knowledge into an active cycle" (ibid., p. 14). Morin's encyclopedic challenge is concerned with the organization and articulation of knowledge, not with its accumulation.

Furthermore, an understanding of complexity means that *everything shall be ecologized* and that *everything shall be seen in meta-perspective* and with loops. It further means that the Western logical tradition of the disjunctive, i.e. the excluded third, shall be modified in order to also consider the included third and the existence of several levels of reality (and more generally of emergence), as Nicolescu (2002) argued. These notions will all be further introduced in the upcoming chapters. A proper definition and understanding of complexity cannot be laid out in details here and now; it can only be evoked, with the aid of further complex "macro-concepts" that themselves require considerable developments; this understanding will thus emerge in due time: first partly in chapter 2, and then more fully throughout chapter 3.

The rotational approach suggested by Morin, which opposes both radical social constructivism and naturalism, may indirectly remind my German colleagues of the 'Positivismusstreit' (positivism dispute) which famously opposed Theodor Adorno and Karl Popper in 1961, concerning methodology in the social sciences, and which further opposed especially Jürgen Habermas (for the Frankfurt School) and Hans Albert (for the school of critical rationalism) in the following decade. I will not discuss the details of this dispute (which is replete with arcane terminological clashes), but shortly mention one point: In the course of the exchange, Adorno expressed why the complexity of social reality calls forward a dialectical treatment, insofar as "the cognitive ideal of the consistent, preferably simple, mathematically elegant explanation falls down where reality itself, society, is neither consistent, nor simple, nor neutrally left to the discretion of categorial formulation. [...]. Society is full of contradictions and yet determinable; rational and irrational in one, a system and yet fragmented; blind nature and yet mediated by consciousness. The sociological mode of procedure must bow to this. Otherwise, out of puristic zeal to avoid contradiction, it will fall into the most fatal contradiction of all, namely, that existing between its own structure and that of its object. [...] The conception of the contradictory nature of societal reality does not, however, sabotage knowledge of it and expose it to the merely fortuitous. Such knowledge is guaranteed by the possibility of grasping the contradiction as necessary and thus extending rationality to it" (Adorno 1976).

Whereas the Frankfurt school further developed and promoted dialectics as what they considered to be the proper methodological approach to the complexity of society, Morin elaborates a "dia-logic" which claims relevance, transversally, to the natural sciences as well as the social sciences, i.e. to the complexity of the physical and biological as well as the social world (Morin's dialogic will be introduced in chapter 3).

These epistemological and methodological questions are only very lightly touched at this point. Following Nietzsche's claim that "the methods come at the end" (in his *Anti-christ*), it is only much later, in the conclusion of chapter 3, that will emerge a methodical framework for a culture of sustainability.

1. The Culture and Art of Unsustainability

> The kind of culture we inherited from our fathers and grandfathers is beginning to challenge our ability to survive on this planet.
> LASZLO 1996, P. 75[1]

> Comment n'avait-on pas compris que l'ordre pur est la pire folie qui soit, celle de l'abstraction, et la pire mort qui soit, celle qui n'a jamais connu la vie?
> MORIN 1977, P. 62[2]

> La pensée simplifiante est devenue la barbarie de la science. C'est la barbarie spécifique de notre civilisation.
> MORIN 1977, P. 387[3]

INTRODUCTION

The search for 'sustainability' has gained an impressive strength and popularity, because it is posed as an alternative to a civilizational model, which, more or less confusely or precisely, is considered by a great diversity of social actors, as a system in crisis. Sustainability represents the search for a way out of 'unsustainability'. Most analyses of the situation focus on the

1 Exact page numbers refer to the 4th reprint (2002).

2 Or in English: "How could we not have understood that pure order is the worst folly which exists, that of abstraction, and the worst death which exists, that which never knew life" (Morin 1992, p. 58)? The exact page numbers in references to Morin 1977 correspond to the 1981 pocket book reprint.

3 Or in English: "Thought which simplifies has become the barbarity of science. It is the specific barbarity of our civilization" (Morin 1992, p. 397).

economic, ecological and social dimensions of this crisis.[4] But some authors have also explored the cultural dimension of the contemporary crisis of unsustainability, as well as its cultural roots.

This chapter offers an overview of these explorations of our 'culture of unsustainability', first with a wide focus on culture in its anthropological sense and with a long-term historical background (section 1), and then with a closer focus on the roles of the arts as an integral element of a wider culture of unsustainability (section 2).

The following pages will therefore concentrate a great deal of critical, negative impressions. However, the reader should not be thereby discouraged: This critical stage is the precondition, and stepping-stone, which will then allow me, in the following chapters, to identify an alternative culture, conceptualize an alternative understanding of aesthetics and look at alternative artistic practices, as my contribution to the transdisciplinary search for sustainability.

SECTION 1: THE CULTURE OF UNSUSTAINABILITY

The question addressed in the following pages is how modern Western culture constitutes 'the culture of unsustainability' that is preventing our development model from achieving sustainability. The focus will thus be here mainly on the cultural dimension of unsustainability, leaving aside the discussion of other important dimensions, such as for example the economic dimension of the Western development model.[5] The cultural dimension is pointing at a number of social conventions and institutions, norms and values, world-views and epistemological/paradigmatic premises, culture being understood here as a value system and set of signifiers framing social identities and dispositions to act and to believe.

The focus is kept on 'the' culture of unsustainability that concerns especially European societies today, but only as 'a' culture of unsustainability among many possible others, without any claim to define it as 'the' one and only culture of unsustainability in world history.[6] Furthermore, I do not claim to be exhaustive in my exploration of the contemporary culture of unsustainability in the globalized 'West'. An exhaustive exploration would require a study of its own, whereas my goal here is to point at key features against which I will then explore an alternative cultural development, in the following chapters. In the following pages, the critique is oriented against

4 See for example Hossay 2006.

5 The culture of the science of economics requires a specific critical treatment, which I started exploring in Kagan 2004.

6 In history, several forms of human civilizations have self-destroyed, offering numerous examples of unsustainable cultures: See Diamond 2004, and for a critical, more complex account: Eds. McAnany and Yoffee 2009.

several fundamental dimensions of 'modernity'. However, I am not thereby claiming to cover exhaustively all the dimensions of modernity. (For example, I do not discuss directly the whole process of secularization.) Neither do I oppose all aspects of modernity, such as e.g. "the idea of the world as open to transformation, by human intervention" (Giddens 1998, p. 94). Rather, my critique addresses specific ways in which modernity organizes the anthropic transformation of the world, as well as, more generally, the drift from a mere belief in anthropic transformation, to a belief in human control over reality.

First of all, an overview of some critiques of Western culture will be given, focusing most especially on the analyses of authors whose insights will be discussed in the following chapters, in the exploration of cultures of sustainability, i.e. the systems thinkers Fritjof Capra and Ervin Laszlo, the theoretician of complexity Edgar Morin, and the advocate of transdisciplinarity and physicist Basarab Nicolęscu. They are pointing at a crisis in the Western mode of knowing inherited from the classical-modern scientific method, and at consumer culture.

The historical roots of this culture of unsustainability will then be explored, with a discussion of Descartes and Bacon's influences, and further back, of the Western alphabetic culture (considering its indictment by phenomenologist and ecological author David Abram).

Finally, a closer look will be given at specific dimensions of the more recent historical developments in the constitution of a Western culture of unsustainability:

- The constitution of a technological system (cf. Jacques Ellul) following the shift from enlightenment to positivism (as discussed by Horkheimer and Adorno),
- and the constitution of autopoïetic social systems under modern society's functional differentiation, as analyzed by Niklas Luhmann.

A crisis of the Western worldview and mode of knowing

Several authors advocating for sustainability, talk of Western culture, in very general terms, as of a cultural model in crisis, especially concerning its mode of knowing reality.

Ervin Laszlo is one of those who made this point the most straightforwardly. Laszlo is one of the promoters of 'systems thinking' (which will be further introduced in chapter 2), and has been working with the Club of Rome in the 1970's, before creating his own 'Club of Budapest' in the late 1980's, a network of people committed to a 'holistic' intellectual movement (also known as 'Integralism').[7]

7 I will conduct a critical discussion of Laszlo's holistic views in chapter 2.

Laszlo opposes a 'classical' view which we are inheriting from modernity, to a new 'systemic' view:

"The classical worldview was atomistic and individualistic; it viewed objects as separate from their environments and people as separate from each other and from their surroundings. The systems view perceives connections and communications between people, and between people and nature, and emphasizes community and integrity in both the natural and the human world.
The classical worldview was materialistic, viewing all things as distinct and measurable material entities. The systems view gives a new meaning to the notion of matter, as a configuration of energies that flow and interact, and allows for probabilistic processes, for self-creativity, as well as for unpredictability. [...]
The classical worldview was Eurocentric, taking Western industrialized societies as the paradigms of progress and development. The holistic vision takes in the diversity of human cultures and societies and sees all of them as equally valid, ranking them only in regard to sustainability and the satisfaction they provide for their members" (Laszlo 1996, p.11).

Laszlo also denounces a modern, "atomistic" view of the human being, as based in modern scientific method, and which he opposes to the alternative, systemic view he is promoting:

"The classical scientific method led to a vast number of highly accomplished theories concerning man's behavior, dispositions, and even his subconscious. But it also led to the fragmentation of our understanding of human beings. In the midst of all the complex special theories, we have gained little real insight into human nature itself. In fact, some theories would deny that there is any such thing, preferring instead to think of humans as a black box which correlates stimuli with responses. Opposed to atomism and behaviorism, the systems view links the human being again with the world (s)he lives in, for he or she is seen as emerging in that world and reflecting its general character [...] He/she is a natural entity, and an inhabitant of several interrelated worlds. By origin (s)he is a biological organism. By work and play (s)he is a social role carrier. And by conscious personality (s)he is a Janus-faced link integrating and coordinating the biological and the social worlds" (Laszlo 1996, p. 60).

The discussion of the different alternatives to the classical/atomistic worldview denounced by Laszlo, will be conducted at length in chapter 2 and 3. For now, I will rather focus on the characterization of this problematic heritage of modernity: Other authors have further developed critical discourses complementing Laszlo's arguments.

Basarab Nicolescu, quantum physicist and promoter of transdisciplinarity, articulated more clearly, what Laszlo indicted as an "atomistic view" inherited from modern scientific method. Nicolescu describes it as a paradigm of simplicity, which he describes for example in his own scientific discipline (i.e. in classical physics) as imposing assumptions such as:

- Continuity: "one cannot pass from one point to another in space or time without passing through all the intermediary points" (Nicolescu 2002, p. 10);
- Local causality: "each cause at a given point corresponds to a nearby effect, and each effect at a given point corresponds to a nearby cause[and] there is no need at all for any direct action from a distance" (ibid., p. 10-11);
- Determinism : "Physical states are functions of positions and speeds; it follows that if one specifies the initial conditions (the physical state at a given moment of time), one can completely predict the physical state at any other moment of time" (ibid., p. 11).

But the crisis of Western knowing is not only rooted in the paradigm of simplicity, in Nicolescu's analysis: It also finds its root in the "fragmentation of our understanding" as Laszlo called it, and which is caused according to Nicolescu, by the binary character of classical logic: "Classical binary logic confers its patent on either a scientific or nonscientific discipline. Thanks to its rigid norms of truth, a discipline can pretend to entirely contain all knowledge within its own field" (Nicolescu 2002, p. 33).[8] But with the development of disciplinary knowledge, "a multischizoid, complex reality appears to have replaced the simple, one-dimensional reality of classical thought" (ibid., p. 34). This realization has prepared the ground for a consideration of complexity (which I will introduce in chapter 3), but in the meanwhile, a Babelization of knowledge has taken hold of modern societies.

> "This process of 'Babelization' cannot continue without putting our own existence into jeopardy, because a decision maker becomes increasingly more incompetent regardless of his or her intention [...] it is obvious that even a group comprised of the best specialists from all the various disciplines would only be able to develop a generalized incompetence, for the simple reason that the sum total of competencies is not competence: on the technical level, the intersection between different domains of knowledge is an empty ensemble" (ibid., p. 42).

Nicolescu also points at the ravages of the classical-modernist praise of objectivity, turning into a regression of human intelligence: "All knowledge other than scientific is thus cast into the inferno of subjectivity, tolerated at most as a meaningless embellishment or rejected with contempt as a fantasy, an illusion, a regression or a product of the imagination" (ibid., p. 13).

Not only does human knowing suffer from the rigid praise of objectivity: The human subject is also endangered, according to Nicolescu.

8 Classical logic, and Nicolescu's alternative to it, are explained only later, in my section on transdisciplinarity in chapter 3.

"Objectivity, set up as the supreme criterion of truth, has one inevitable consequence: the transformation of the subject into an object. The death of the subject, which heralded all other deaths, is the price we pay for objective knowledge. The human being became an object – an object of the exploitation of man by man, an object of the experiments of ideologies that are proclaimed scientific, an object of scientific studies to be dissected, formalized and manipulated. The Man-God has become a man-object, of which the only result can be self-destruction" (ibid.).

One of the consequences of the development of modern science was the split between science and culture (in the sense of 'humanities'), as part of the fragmentation of human knowing. Nicolescu recalls that up until after the Renaissance, "science and culture were inseparable" (ibid., p. 95). In the 17th century, only "the germ of the split" was present, "when the methodology of modern science was formulated, but it did not become fully blown until the nineteenth century" (ibid., p. 96). Then did the subject get eliminated from the realm of 'hard sciences'. "The 'death of man' coincides with the complete separation between science and culture" (or hard sciences and humanities). This separation entrenched itself in academic institutions in the 20th century. "The two cultures are perceived as antagonists. The split between the two cultures is first of all a split between values" (ibid., p. 97).

Nicolescu even goes as far as arguing that the application of the classical-modern scientific method to the realm of the social sciences, e.g. by Marx and Engels (with "the correspondences between economic, social, and historical laws and the laws of Nature"), advancing the idea that "by assigning certain initial socially specific conditions, one can infallibly predict the future of humanity" (ibid., p. 13), is at the root of 20th century totalitarianisms. "The dogmas and ideologies that have ravaged the twentieth century were derived from classical thought, founded on the concepts of classical physics" (ibid., p. 15).

Overall, for Nicolescu, the main limitation of the paradigm of simplicity of modern scientific culture, lies in its limitation of human intelligence to a simple, linear logic: "scientism has bequeathed us one persistent and deeply rooted idea: that of the existence of a single level of reality" (ibid., p. 14).[9]

A comparable, probably even more accomplished and articulate, critique of classical science, as disjunctive simplification, is conducted by Edgar Morin, a French sociologist and philosopher whose theoretical insights on complexity will be discussed at length in the first section of chapter 3. From the outset of the first volume of his *Method*, Morin argues that the very structures of modern sciences and western knowledge, their paradigmatic baseline, are preventing us from acquiring a proper understanding of the pressing issues of the late 20th century (i.e. issues of unsustainability).

---

9 To move beyond this limitation, Nicolescu advocates a move towards 'transdisciplinarity', which I will introduce in the second section of the third chapter.

"[O]ur principles of knowledge hide what is hereon in vital to know [...] the concepts [...] are mutilated and lead to actions which are inevitably mutilating" (Morin 1992, p. 3).[10]

Morin denounces the mutilation of human understanding by the fragmentation and "dissociation" of the inter-related dimensions of reality, e.g. our human identity as "individual-society-species" being dissociated in three separate terms (Morin 1977, p.10).

Morin articulates how modern science is limited by an "almighty principle of disjunction" (ibid., p.11) and is incapable of confronting "a triple wall: the encyclopaedic wall, the epistemological wall, the logical wall" (ibid., p.12), i.e. of approaching the entire field of knowledge, of learning by linking (and not just learning by isolating) and of overcoming the logical absurdity of circular (instead of linear) causality. Modern science traditionally renounced to face these challenges and kept within the borders of the disjunctive paradigm. "It is precisely this renouncement that the University teaches us" (Morin 1992, p. 6).

On the one hand, the human being is split into pieces: "Man has come apart" (ibid., p.8); on the other hand, the world is also torn into pieces, and even physics is divided into quantum physics, cosmology, and in between, 'classical physics', incapable of communicating with one another.

While science specialized itself to a point of absurdity, losing the possibility to achieve an intelligibility of reality, Morin asks, what is the point of such a science then, i.e. an overall ignorant accumulation of isolated islands of knowledges? As Morin points out, modern science failed to define itself:

> "But then, what is science? *Here we must realize that this question does not have a scientific answer*: science does not know itself scientifically and has no means of knowing itself scientifically. [...] There are epistemological tribunals which, *a posteriori* and from the exterior, pretends to judge and to gauge scientific theories [...] How is [science] incapable, not only of controlling, but of conceiving its power of manipulation and its manipulation by power" (ibid., pp. 8-9)?

A "science of science" is lacking, which should be embedded in the scientific theories and methodologies at their very basis.

Morin denounces three basic modes of simplifying thought: "*to idealize* (to believe that reality can be reabsorbed in the idea, that the intelligible alone is real); *to rationalize* (to want to enclose reality in the order and the coherence of a system, to forbid it all overflow outside the system [...]); *to normalize* (that is to say to eliminate the strange, the irreducible, the mysterious)" (ibid., p. 16).

---

10 For most of the long quotes from Morin's first volume of *La méthode*, I am using the translation from the English-translated edition (Morin 1992). However, for some short quotes, I am doing my own translations and referring to the French original text (Morin 1977).

Modern science tends to forget that simplicity is but an abstraction. "The simple is only an arbitrary moment of abstraction,a means of manipulation torn from complexity" (ibid., p. 149).

> "Simplification reifies, that is to say hides the relativity of the notions [and it] dissolves organization and system. It is certainly necessary to know the *simple* laws of interaction whence flow innumerable, rich and complex combinations. [...] But, to be content with this type of explanation is to hide complexity at departure (the play of order/disorder/interaction) and complexity at arrival: the complex organization of such combinations into systems of systems" (ibid., p. 143).[11]

Classical science stressed the primacy of order, i.e. attributed a mechanical order to the universe. At the astronomical level, classical science pictured a "clockwork universe" of matter and energy (Morin 1977, p. 33). Even the theory of evolution, in biology, was often stripped of its chaotic force and turned into an ordering principle, with an impoverished understanding of "adaptation" supported by a belief in a linear progress: a belief "[t]hat life, obeying natural laws of adaptation and selection, has developed itself to reach this rational order that the name *homo sapiens* symbolizes" (Morin 1992, p. 30). Only starting with thermodynamics did this primacy and linear progression of order start crumbling down in the late 19th century, but only partially and slowly, at least until the appearance of chaos theory in the late 20th century.

Looking back at modern science before catastrophe and chaos theories, Morin wonders how the western belief in the primacy of order could have lasted so long and and remained so strong. "How could we not have understood that pure order is the worst folly which exists, that of abstraction, and the worst death which exists, that which never knew life" (ibid., p. 58)?

Morin's analysis also complements Nicolescu's critique of classical physics. Classical physics is 'reductionist', as it looks for simple elements and general laws. It posits objectivity, i.e. "a universe constituted of isolated *objects* subjected to *objectively* universal laws" (Morin 1977, p. 96). The classically conceived physical object is substantial and self-sufficient i.e. it ontologically exists independently of its environment and of the human observer.

This misconception of the object also relates to the practice of scientific experimentation: "experimentation, which surgically tears the object from its environment and its adherences and, in so doing, manipulates and enslaves [...] the disciplines have closed down upon mutilated objects" (Morin 1992, p. 205). The closed conception of object corresponds to a unidimensional worldview, Morin argues after Maruyama (1974). "Things became objective: inert, fixed, unorganized objects, *bodies* always moved by exterior laws. Such objects, deprived of form, organization, singularity, are, at

11 Morin's notions mentioned here, will be introduced in chapter 3.

that degree of abstraction, terribly unreal; but one has hold of them, by measurement and experiment, and that action is terribly real" (Morin 1992, p. 374).

Furthermore, the experimenter in empirical sciences ignores the very consequences of his observational praxis which is also a transformational praxis allowing the development of a technological system (this aspect will be further discussed below).

"We used to believe, we continue to believe that scientific experiment knows its object by isolating it [...] It certainly provokes a relative isolation by inhibiting certain interactions, but *it provokes, by and of itself, new types of interactions* [...] between the phenomenon and this time the experimenter, interactions of which he remains unconscious [and it] produces transformable information which allows the development of this universal manipulation which we call technique" (ibid., pp. 363-364).

Overall, Western Physics "has not only disenchanted the universe, but also desolated it", argues Morin. It lost the intuitive insights of an animistic universe, an "enchanted vision [which] recognized -mythologically- the presence of generativity [...] and implied a communicating loop between the sphere of physis, the sphere of life and the anthropo-social sphere: anthropomorphism-zoomorphism-cosmomorphism" (Morin 1977, p. 365).

Morin's last argument echoes Gregory Bateson's praise of totemism and animism vs. modern knowledge as a degeneration of these earlier forms of human knowing:

"Anthropologically, it would seem from what we know of the early material, that man in society took clues from the natural world around him and applied those clues in a sort of metaphoric way to the society in which he lived. That is, he identified with or empathized with the natural world around him and took that empathy as a guide for his own social organization and his own theories of his own psychology. This was what is called 'totemism'.

In a way, it was all nonsense, but it made more sense than most of what we do today, because the natural world around us really has this general systemic structure and therefore is an appropriate source of metaphor to enable man to understand himself in his social organization.

The next step, seemingly, was to reverse the process and to take clues from himself and apply these to the natural world around him. This was 'animism', extending the notion of personality or mind to mountains, rivers, forests, and such things. This was still not a bad idea in many ways. But the next step was to separate the notion of mind from the natural world, and then you get the notion of gods.

But when you separate mind from the structure in which it is immanent, such as human relationship, the human society, or the ecosystem, you thereby embark, I believe, on fundamental error, which in the end will surely hurt you.

Struggle may be good for your soul up to the moment when to win the battle is easy. When you have an effective enough technology so that you can really act upon your epistemological errors and can create havoc in the world in which you live, then the error is lethal. Epistemological error is all right, it's fine, up to the point at which you create around yourself a universe in which that error becomes immanent in monstrous changes of the universe that you have created and now try to live in" (Bateson 1973, pp. 460-461).[12]

This last point of Morin's analysis also echoes the phenomenological critique of modernity: The founder of phenomenology, Edmund Husserl, also saw (in his last book: cf. Husserl 1970) a crisis in Western civilization in the increasing overrunning of, and estrangement from, life-world experience ("Lebenswelt", i.e. the subjective and inter-subjective, pre-conceptual lived experience of human subjects) by the qualitatively less nuanced understandings of reality imposed by Western sciences and technologies.

But the contemporary culture of unsustainability can also be analyzed at other levels. Another often-denounced cultural dimension of unsustainability in contemporary late-modern societies, which affects behavior directly and extensively, is consumerism (or in other words, the culture of hyper-consumption).

A critique of consumerism as rooted in patriarchal cultures

Hyper-consumption constitutes one striking dimension of the contemporary culture of unsustainability in late-modern societies.[13]

Fritjof Capra notices that consumerism, over-consumption, is still growing at the turn of the 21st century (even in an 'information' or 'service' economy): "One of the greatest obstacles on the road towards sustainability is the continuing increase in material consumption. In spite of all the emphasis in our new economy on information processing, knowledge generation, and other intangibles, the main goal of these innovations is to increase productivity, which ultimately increases the flow of material goods" (Capra 2002, pp. 262-263).[14] Consumption is considered as an imperative, through "the delusion that the accumulation of material goods is the royal road to happiness [...] and material acquisition is portrayed as a basic human right, increasingly even as an obligation" (ibid., p. 263).

---

12 Exact page numbers refer to the 1978 reprint of the 1973 edition by Granada Publishing.

13 The reader may refer to Veblen 1899, Baudrillard 1970, eds Schor and Holt 2000, eds. Princen et al. 2002 and Kasser 2002 for an analysis of consumer culture, and to Ray and Anderson 2000 and Verdi 2008, for a discussion of claimed transformations toward sustainability with more 'responsible consumers' and 'sustainable lifestyles'. Parallels can be drawn between Ray and Anderson's arguments on the "cultural creatives" and some of my own developments in chapter 2, but in comparison, their analysis remains very superficial.

14 Exact page numbers refer to the 2004 edition by Anchor Books.

Nicolescu, on his part, is not optimistic about the postmodern version of consumer society. His central critical argument concerns the dissolution of the unity of individual personality under the influence of the (televised) image and its cult of personalities.

"Continual disharmony between individual and social life produces multiple personalities in one and the same being. The contradictions and conflicts between different personalities inside a person lead to the dissolution of interior being, which is no longer recognizable under these multiple masks. [...] We now live by proxy. We delegate our lives to some authority figure, to some guru, to the image of some singer or athlete" (Nicolescu 2002, p. 92).

This phenomenon is, according to him, the cornerstone of the contemporary hyper-consumption society:

"In general, the growth of consumerism is calculated to be a function of the number of persons who are susceptible to consuming. However, any given person may exhibit multiple personalities, so the number of potential consumers is much larger than the number of persons who consume. Advertisers have long understood this relatively trivial evidence, which, like all evidence, is not very visible. Every day these advertisers stimulate a different desire, and each desire creates a new potential consumer in one and the same person. A human being's material needs are limited, but his desires are unlimited" (ibid.).

Nicolescu poses the self-constitution of the human being in opposition to this multiplication of desires which, he claims, "leads to the consumption of being"; he instead wants to celebrate "our interior experience" and affirms: "The greatest work – the Great Work – is our own life" (ibid., p. 93).[15]

Fritjof Capra views consumerism as being rooted in patriarchalism, and recognizes this as a cultural and cross-cultural issue: According to him:

"This glorification of material consumption has deep ideological roots that go far beyond economics and politics. Its origins seem to lie in the universal association of manhood with material possessions in patriarchal cultures [...] The association of manhood with the accumulation of possessions fits well with other values that are favored and rewarded in patriarchal culture – expansion, competition, and an 'object-centered' consciousness" (Capra 2002, p. 264).

---

15 Another perverse effects of consumerism, according to Nicolescu, is the 'spiritual consumerism' of 'sectarian insularity': "The rapid multiplication of sects is one of the signs of the disappearance of clear guidelines in a consumer society. The escape into the closed life of a sect actually indicates a need to avoid all responsibility in a world of incomprehensible complexity" (ibid., p. 115).

Patriarchal culture is also playing a role in the scientific culture and more generally, the prevalent mode of knowing in Western, modern societies. For example, Nicolescu notices a clear form of machism or at least an expression of patriarchalism in the works of 19th century scientists: "Nature offered herself as a mistress to man, in order to be penetrated to her depths, dominated, conquered. Even without succumbing to the temptation to psychoanalyze scientism, one is forced to note that the writings of nineteenth-century scientists about Nature abound in the most unbridled sexual allusions" (Nicolescu 2002, p. 12).

While the discussion of patriarchalism suggested a historical analysis of the roots of consumerism, the long quote from Gregory Bateson introduced a historical dimension in the critical analysis of the scientific method, which allows to further explore the possible origins of the mutilated Western mode of knowing.

Historical roots of the culture of unsustainability

As trivial as the remark may seem, time does matter: Historical developments in human societies do impose a degree of irreversibility, a so-called path-dependency (which, however, also knows bifurcations) which has to be acknowledged and understood, prior to any attempt at reforming modern modes of knowing and worldviews. As noted by Laszlo, the irreversibility of human purposive consciousness is a cultural phenomenon. "The apple, not only of knowledge but of the many facets of culture, proved to have irreversible effects [...] consciousness, when evolved, took over the direction of our evolution. The means became the end: the self-maintaining biological species was transformed into a cultural species" (Laszlo 1996, p. 74-75). Laszlo evokes "the vast organizational process of differentiation, complexification and association" taking place in human societies across history, but tremendously accelerating in modern western history.[16] "On the whole, the same branching evolutionary development took place here as in the organic realm. But here it took place within the setting of cultures. And cultures can quicken the pace of development or slow it down" (ibid., p. 75) – or rather, alter the pathways according to emergent logics at a cultural level.

The estrangement of nature

The historical path of modern culture developed an estrangement of the 'environment' from the human subject. Bateson's long quote above, already evoked how the human vision of especially the natural environment went through a number of historical shifts. Eventually, nature came to be seen in Western modern culture as the 'other', the enemy, opposed in a binary fash-

16 The specific issues associated with differentiation in modern society will be analyzed further down in this chapter, with Niklas Luhmann.

ion to human culture, that is, until very recently.[17] Nicolescu describes how the Western vision of nature in Modernity turned away from "magic Nature" and introduced "Nature as machine", resulting in "the death of Nature", and how the 'quantum revolution' prepared "the resurrection of Nature":

"The magical viewpoint depicts Nature as a living organism, endowed with intelligence and consciousness. The fundamental postulate of such thought is one of universal interdependence: Nature cannot be conceived outside of its relationship to us. [...] At the other extreme, the mechanistic thought of the eighteenth and above all the nineteenth centuries (which, by the way, still predominates today) conceives Nature not as an organism, but as a machine: One only has to disassemble this machine piece by piece in order to possess it entirely. The fundamental postulate of mechanistic thought is that Nature can be known and conquered by scientific methodology, defined in a way that is completely independent of human beings and separate from us. [...] The logical end result of the mechanistic viewpoint was the death of Nature – the very disappearance of the concept of Nature from the scientific field [...] there was no more need for a coherent whole, for a living organism, or even for a machine that still kept the musty odor of finality" (Nicolescu 2002, pp. 58-60).

Nicolescu notes that Jacob Boehme and the German 'Naturphilosophie' of Novalis, Goethe and others, revolted against the mechanistic vision of Nature, and did bring some limited insights even into the sciences (leading, according to Nicolescu, to cellular theory and electromagnetism). But, he argues, they lacked the insights that were yet to come from quantum physics.

Indeed, in the 20th century, "Nature was dead, but complexity remained [and the] death of Nature is incompatible with a coherent interpretation of the results of contemporary science. [...] Nature is dead only for a certain viewpoint of the world: the viewpoint of classical thought. The rigid objectivity of such classical thought is not viable in the quantum world" (ibid., p. 60), among other things, because of the non-separability between observer and reality, and because "in the quantum vacuum, all is vibration, everything is a fluctuation between being and nonbeing. The quantum vacuum is full, filled with all potentialities" (ibid., p. 61). Nicolescu further points at

"three theses formulated by the physicist Walter Thirring: (1) The laws of any inferior level are not completely determined by the laws of a superior level. [...] (2) The laws of an inferior level depend more on the circumstances of their emergence than do the laws of a superior level. The laws of a certain level depend essentially on the local configuration to which these laws refer. [...] (3) The hierarchy of laws evolved at the same time as the universe itself. In other words, the birth of laws occurs simultaneously with the evolution of the universe" (ibid., p. 63).

17 Cf. Erzen 2004, p. 23. See also Brocchi 2008.

The insights of quantum physics for a new vision of Nature and a new mode of knowing, are integrated into Nicolescu's transdisciplinary model of Reality which will be introduced in the second section of chapter 3.

The inspiration of Descartes and Bacon

Most authors, when searching for the roots of the modern worldview as based on the scientific method, point their fingers towards the heritage of René Descartes and Francis Bacon. For example, the phenomenologist and ecological author David Abram argues that Descartes' *Méditations* (1641) influentially determined material reality as "a strictly mechanical realm, as a determinate structure whose laws of operations could be discerned only via mathematical analysis" (Abram 1996, p. 32).[18] "Descartes' radical separation of the immaterial human mind from the wholly mechanical world of nature did much to fill th[e] need [for a legitimization of the increasing exploitation of nonhuman nature], providing a splendid rationalization for the vivisection experiments that soon began to proliferate, as well as for the steady plundering and despoilment of nonhuman nature in the New World and the other European colonies" (Abram 1996, p. 78).

René Descartes is often quoted in this context for his claim that human beings are to become "like masters and owners of nature" (Descartes, *Discours de la méthode*, VI). This affirmation, besides its Biblical lineage, is based on the universalistic assumption that there is "no difference between the machines made by the craftsmen and the different bodies that nature alone composes" (Descartes, *Principes de la Philosophie,* IV, §203). Techniques are no longer contingent and subordinated to nature (as in the Aristotelian tradition), but can apprehend, transform and manipulate a universal natural realm. The knowledge of (linear) causes "allows the mastery of all possible effects" (Bourg 1997, p. 10). In the *Discours de la méthode*, Descartes, probably inspired by Bacon, conceives of progress as an accumulation of knowledge allowing the betterment of human civilization.

Francis Bacon can be considered as the father of the modern technological utopia, based on the belief in techno-scientific 'progress' as a temporal process (a notion which Bacon introduced, based on the Judeo-Christian notion of a linear time).[19] In his *New Atlantis*, Bacon even affirms that humanity can recreate Eden thanks to scientific and technological progress. An ideal, utopian city managed by a collegium of wiser men, Bacon's New Atlantis divides work functionally and uses scientific method. Its goal is to realize everything that is possible. In the *Novum Organum*, Bacon argues that the mastery of humanity over nature is a God-given attribute (in Eden) which must be recovered by the human genus. Overall, Bacon's influential works inspired a long-lasted belief that the expansion of the scientific method and derived techniques would bring universal human happiness.

---

18 Exact page numbers refer to the 1997 reprint by Vintage Books.

19 Cf. Delvaille 1910.

The estrangement of Western culture from the lifeworld experience, observed by Husserl, however would find its roots much deeper and earlier than the modern scientific method, if we follow the analysis of David Abram.

Abram's indictment of phonetic-alphabetic culture

> You must forgive me, dear friend. I am a lover of learning, and trees and open country won't teach me anything, whereas men in the town do.
>
> SOCRATES, IN PLATO'S *PHAEDRUS*, 230D

If classical Greece did not know the notion of progress as defined by Francis Bacon, they may have constituted some of the early roots of a Western culture of unsustainability, according to the analysis of David Abram. For Edgar Morin too, the roots of Western dissociative thought plunge back into classical Greek mythology and philosophy which (with the exception of Heraclitus) opposed *ubris* and *dike* (i.e. chaos and "démesure") vs. balance and order.[20] But Abram offers a more thorough and thought-provoking thesis: That Western phonetic-alphabetic culture would be at the root of our estrangement from the lifeworld.[21]

David Abram observes that the Western civilization[22] has developed an acute autism towards the natural environment, in starch contrast to numerous indigenous cultures which have retained an ability to perceive and respond more subtly and qualitatively more effectively to their natural environments.

Abram inquires into the origin of this divorce between the Western civilization and its natural environment, which seems to oddly contradict the very basically animistic character of human perception, as analyzed by the phenomenologist Maurice Merleau-Ponty.[23] Abram observes that, in our civilization, "language functions largely to *deny* reciprocity with nature – by defining the rest of nature as inert, mechanical, and determinate – [so far as

20 Cf. Morin 1977, p. 57.

21 Abram is aware of leaving-out in his analysis, other possible, and complementary historical roots for our estrangement from the natural environment, among which is "the emergence of agriculture at the dawn of the Neolithic era [...] the spread of agricultural techniques radically transformed the experienced relation between humans and other species" (Abram 1996, pp. 263-264).

22 Keeping with the North-American tradition, David Abram speaks of "civilization" where European authors in the tradition of Max Weber, would speak of "culture". In the following paragraphs, I will repeatedly use the term "civilization", keeping in line with Abram, who is under the focus of the coming couple of pages.

23 This phenomenological animism will be introduced in chapter 4, section 5.

to render] our sensorial participation with the land around us [...] mute, inchoate, and in most cases wholly unconscious" (Abram 1996, p. 71). On the contrary, in "indigenous, oral cultures [...] language seems to encourage and augment the participatory life of the senses, while in the Western civilization language seems to deny or deaden that life, promoting a massive distrust of sensorial experience while valorizing an abstract realm of ideas hidden behind or beyond the sensory appearances" (ibid., pp. 71-72).

Abram also quotes Merleau-Ponty's concern that the very view of language as a purely formal, arbitrary, conventional system, reflects a practice of language that has become so much conventionalized and uncreative that it "demand[s] from us no real effort of expression and [...] demand[s] from our listeners no real effort of comprehension" (Merleau-Ponty 1962, p. 184) and, Abram adds, where "meaning has become impoverished" (Abram 1996, p. 77). By contrast to this state of affairs, Abram, borrowing again from Merleau-Ponty, explains how a phenomenology of language highlights the sensory basis of language as practiced by corporeal beings, and therefore, its not-exclusively-human character (or in Merleau-Ponty's words, language "is the very voice of the trees, the waves, and the forests" - Merleau-Ponty 1968, p. 155). "The complex interchange that we call 'language' is rooted in the non-verbal always already going on between our flesh and the flesh of the world" (Abram 1996, p. 90).

However, western philosophy usually claims the opposite, i.e. that language establishes humankind's specialness and superiority over other species. How did such a claim come to be so strongly established in mainstream Western culture? And how did it become a reality for Westerners, argues Abram, by contrast most especially to Indigenous peoples? "How could we have become so *deaf* to these other voices that nonhuman nature now seems to stand mute and dumb, devoid of any meaning besides that which we choose to give it" (ibid., p. 91)?

The monotheist god and the world of pure ideas

Abram recalls two streams of explanations that were proposed in past decades, about Western culture's neglect of the natural world and belief in an abstract source of reality "assumed to exist entirely beyond, or outside of, the bodily world" (ibid., p. 94).

The first stream was initiated by historians (e.g. Lynn White Jr. 1967) analyzing the 'otherworldly' monotheist god of the new testament, as instituting a relationship of domination and exploitation of the natural environment (as exemplified in the Genesic injunction toward mankind, to multiply itself and master all other living things on Earth).

The second stream was initiated by philosophers (starting with Friedrich Nietzsche), and

"turned toward the Greek origin of our philosophical tradition, in the Athens of Socrates and Plato, in their quest for the roots of our nature-disdain. [These critical philosophers] attempted to demonstrate that Plato's philosophical derogation of the sensible and changing forms of the world [as] mere simulacra of eternal and pure ideas existing in a nonsensorial realm beyond the apparent world – contributed profoundly to civilization's distrust of bodily and sensorial experience, and to our consequent estrangement from the earthly world around us" (ibid.).

Abram himself looks for the common feature among the Hellenic and Hebraic traditions, long before they merged together into Christianity, and bases his own critical development on this common feature: the invention and use of phonetic writing, i.e. the 'alphabet'.

The alphabet as the autopoïetic[24] turn of western civilization

While Merleau-Ponty's phenomenology established the animistic and synaesthetic character of human perception, it seems as if this character had been lost in Western civilization. However, Abram argues, it may have not been lost (falsifying Merleau-Ponty's analysis), but have been "transferred from the depths of the surrounding life-world to the visible letters of the alphabet. Only by concentrating the synaesthetic magic of the senses upon the written letter could these letters begin to come alive and to speak", so that written letters unavoidable "speak" to our eyes whenever we see them. "For our senses are now coupled, synaesthetically, to these printed shapes as profoundly as they were once wedded to cedar trees, ravens, and the moon" (ibid., p. 138). "This is a form of animism that we take for granted, but it is animism nonetheless – as mysterious as a talking stone" (ibid., p. 131). To spell, is also to cast a spell, as the Jewish kabbalists believed. However, this is a "self-reflexive mode of animism" (ibid., p. 132).

But why did these printed letters come to turn us away from other dimensions of perception? Abram describes at length how the phonetic Aleph-Beth of the Jews, and even more so its transfer into the Greek alphabet, marked a separation of human sensory participation from the non-human environment.[25] Indeed, if pictographic, ideographic and rebus-like writing systems (e.g. Chinese or Egyptian), are at least still partly based on a reference to the "enveloping natural field" this is no longer the case for the Aleph-Beth. Chinese ideograms have natural counterparts (i.e. they associate certain qualities or characters to depicted visible entities). For example, the Chinese word for the very act of writing, 'wen', "signifies a conglomeration of marks [and] applies to the veins in stones and woods, to constella-

---

24 Although Abram himself does not use the concept of autopoïesis, his analysis of the closure of referentiality and the subsequent autistic tendency of Western civilization, contributes to an understanding of the historical genesis of what was analyzed in its modern form by Niklas Luhmann, i.e. modern society as an autopoïetic communication system. Luhmann's analysis is introduced later in the current chapter.

25 Cf. Abram 1996, pp. 95-131.

tions, represented by the strokes connecting the stars, to the tracks of birds and quadrapeds on the ground [and] has designated, by extension, literature" (J. Gernet, quoted in Derrida 1976, p. 123). However, the principle of the phonetic alphabet is to refer solely to sounds to be made by the human mouth, i.e. human utterances. This principle came to its full completion with the Greek alphabet, which lost the indirect references to natural elements which was still remnant in the Hebraic Aleph-Beth. But the innovation appeared first among Semitic scribes around 1500 B.C.E.[26] As I just suggested, the letters of the Jewish Aleph-Beth still retained an indirect link to the natural world. For example, the first letter 'Aleph' also was the word for "ox", and indeed the letter's form did evoke an ox's head with horn (and so on, for the other letters). But these links were only "vestigial holdovers from the old", Abram argues, pointing out that, unlike many oral cultures which claim that language was given by animals to humans, the Hebraic holy text claims that the verb was godly, given to humans, and that Adam gave names to animals (and not the other way around).

The phonetic alphabet, with its reference to the sounds of the human voice, operated a radical shift towards an autopoïesis in human communication. Abram describes this shift very clearly:

"To be sure, pictographic and ideographic writing already involved a displacement of our sensory participation from the depths of the animate environment to the flat surface of our walls, of clay tablets, of the sheet of papyrus [but] the glyph or character still referred, implicitly, to the animate phenomenon of which it was the static image; it was that wordly phenomenon, in turn, that provoked from us the sound of its name. *The sensible phenomenon and its spoken name, were, in a sense, still participant with one another* [...] With the phonetic *aleph-beth*, however, the written character no longer refers to any sensible phenomenon out in the world, or even to the name of such a phenomenon [...] but solely to a gesture to be made by the human mouth. There is a concerted shift of attention away from any outward or worldly reference [...] to the shape of the utterance itself [...] *for the first time completely bypassing the thing pictured [...] the larger, more-than-human life-world is no longer a part of the semiotic, no longer a necessary part of the system*" (Abram 1996, pp. 100-101).

"Each image now came to have a strictly *human* referent [i.e. a gesture/sound of the mouth]. Such images could no longer function as windows opening on to a more-than-human field of powers, but solely as mirrors reflecting the human form back upon itself. The senses that engaged or participated with this new writing found themselves within a discourse that had become exclusively human" (ibid., p. 138).

26 Cf. Diringer 1948, p. 159.

The development of alphabetic culture, from Socrates, to Gutenberg, to the industrial age

If the Homeric *Iliad* and *Odyssey* illustrate the very beginning of the alphabetic culture in a so-far oral Greek culture, around the 7th century B.C.E., where Gods do speak to humans through natural elements, although they are already characterized by strongly anthropomorphic characters, the Socratic/Platonic dialogues, in the 4th century B.C.E., mark the beginning of the imposition of the alphabetic culture upon a whole society, as "they inscribe for the first time many of the mental patterns or thought styles that today we of literate culture take for granted" (ibid., p. 104).

The spread of alphabetic writing in Athenian society, into Plato's times, allowed the development and dissemination of an autopoïetic reflexivity, i.e. an observation of one's own autopoïetic language. "A new power of reflexivity was thus coming into existence, borne by the relation between the scribe and his scripted text" (ibid., p. 107). This reflexivity inspired, Abram argues after Eric Havelock (1986), the reflexivity at work in the 'Socratic dialectic', which "was primarily a method for disrupting the mimetic thought patterns of oral culture" (ibid., p. 109). Socrates forced his interlocutors to depart from contextual, situated understandings of human virtues "as living occurrences, as *events*", and to search for abstract definition of virtues, independent of their contexts, in the way the written word 'virtue', "once written down [...] was seen to have an unchanging, visible form independent of the speaker" and of the virtue-enacting person (ibid., p. 110). Thus, Abram argues, the Socratic method witnesses and illustrates a "shift in the perceptual field". This shift enabled the development of a belief in pure ideas, in truths eternal and uncorrupted (at the root of Descartes' assumption of universal laws of nature). Plato further digs into this direction with his concept of "eidos" (a greek term for Plato's pure ideas, but which also meant "visible shape or form", Abram notes, pointing at the fixed and abstract forms of alphabetic letters used by the Greek).[27]

Phaedrus or Plato's ambiguous warning

The relationship of Socrates and Plato to writing was not fully unambiguous. In Phaedrus, Plato reports Socrates' criticism of writing. In this critical warning, is conveyed the realization that reading can complement, but not replace direct, interpersonal interactions as the prime locus of learning. But even though this note of caution, embedded as a warning within Plato's own written work, illustrates a reflexive caution towards alphabetic culture, Abram notes that Plato's Phaedrus fails to

27 Abram further points at Derrida in analyzing out how "letters of the alphabet, like the Platonic Ideas, do not exist in the world of ordinary vision" but dissolve their visibility into speech, i.e. sound (Abram 1996, p. 112).

identify "the pervasive influence of letters upon patterns of perception and contemplation in general" (ibid., p. 115).

Note-worthily, Phaedrus is also the only text involving Socrates, where he sits under a tree outside a city, to then claim that nothing can be learned from "trees and open country" (cf. the quote of Socrates quoted at the beginning of my discussion of Abram). In the same text, Socrates praises love for its role in remembering the original world of pure ideas and their transcendental superiority. The autopoïetic order of Platonician ideas is therefore still ruling over the relatively ambiguous cautionary notes that the Phaedrus instills regarding alphabetic culture.

As Abram notes, this "highly anthropocentric [...] mode of experience endemic to alphabetic culture spread throughout Europe in the course of two millennia, receiving [...] a major thrust from the invention of movable type by Johann Gutenberg [and the] printing press", and then by the further innovations accompanying the industrial revolutions, which "have carried alphabetic awareness throughout the globe" (ibid., pp. 138-139).

The abstraction of space and time

Not only does alphabetic culture displace the synaesthetic participation between one's senses and one's immediate, natural environment, as described above. The written stories themselves are being displaced, as "writing down the ancestral stories disengages them from particular places" (ibid., p. 185).

In oral cultures, meaning is place-specific.[28] However: "Writing down oral stories renders them separable, for the first time, from the actual places where the events in those stories occurred" (ibid., p. 183). The change is not inconsequential. "The human sense, intercepted by the written word, are no longer gripped and fascinated by the expressive shapes and sounds of particular places. The spirits fall silent. Gradually, the felt primacy of place is forgotten, superseded by a new, abstract notion of 'space' as a homogeneous and placeless void" (ibid., p. 184).

Furthermore, writing not only imposes a certain notion of 'space', but also insinuates a certain notion of 'time'. Unlike the understanding of time in oral cultures, which is rather cyclical (cf. Eliade 1959), our modern time is conceived as linear (and this is a prerequisite for the Baconian faith in progress). For example, the 'Dreamtime' of Aboriginal Australians does not point at a finished event but at "an ongoing process – the perpetual emerging of the world from an incipient, indeterminate state into full, waking reality" (Abram 1996, p. 169), comparable to the coexistence of the subconscious with our consciousness. On the contrary, writing the mythic stories displaces the notion of time:

28 This dimension will be further explored later: Cf. chapter 4, section 5.

"Recording mythic events in writing establishes, as well, a new experience of the permanence, fixity, and unrepeatable quality of those events. Once fixed on the written surface, mythic events are no longer able to shift their form to fit current situations [while] contemporary events acquire a naked specificity and uniqueness hitherto unknown [...] and so assume their singular place within the slowly accreting sequence of recorded events" (ibid., p. 188).

This new notion of time is of central importance in the Hebrew Bible. "[F]or the first time, we find affirmed, and increasingly accepted, the idea that historical events have a value in themselves, insofar as they are determined by the will of God" (Eliade 1959, p. 104).

The very distinction of time vs. space itself, is also an outcome of written culture. Indeed, "[u]nlike linear time, time considered as cyclical cannot be readily abstracted from the spatial phenomena that exemplify it – from, for instance, the circular trajectories of the sun, the moon and the stars. Unlike a straight line, moreover, a circle demarcates and encloses a spatial field" bordered by its horizon (Abram 1996, pp. 188-189). In oral cultures, space is already temporal, and "spatial phenomena [are] animate, emergent processes, and [...] space itself [has] a kind of dynamism, a continual unfolding" (ibid., p. 190). For example, one can find among "both the Uto-Aztecan and the Athapaskan language groups, a subtle differentiation between manifested and unmanifested spatiality – that is, a sense of space as a continual emergence from implicit to explicit existence, and of human intention as participant with this encompassing emergence" (ibid., p. 193). Even the ancient Hebrews themselves did not yet completely demarcate time from space; "the Greek thinkers were the first to begin to objectify space and time as entirely distinct and separable dimensions" (ibid., p. 197). Aristotle defined time as "the number of a motion with respect to the prior and the posterior", an infinite linear series of 'now' points (Aristotle, *Physics*, book IV, quoted in Abram 1996, p. 198), while Euclid defined in his *Elements* the three-dimensional, homogeneous continuum that we now commonly know as space. Modernity saw the final, objectified separation of time and space by Isaac Newton in his *Principia Mathematica* (1687):

- "Absolute, true and mathematical time, of itself and from its own nature, flows equitably without regard to anything external";
- "Absolute space, in its own nature, without regard to anything external, remains always similar and immovable" (Newton 1687, quoted in Abram 1996, p. 200).

The loss of the presence of the world

The abstraction of space and time take part in the general displacement from natural reality of an increasingly self-referential alphabetic culture.

A deleterious consequence of Western civilization's abstraction of time and of its conception as an infinite line of 'now' points, is the loss of a "living present", in Abram's words (or the loss of a feeling of "*durée*" in Henri Bergson's words).[29] The now is ungraspable, when it is but an instant that either was coming from the future, or will be already lost in the past. "As a result of all these past and future concerns, everyone appear[s] to be strangely unaware of happenings unfolding all around them *in the present* [and] oblivious to the sensuous presence of the world. The present, for them, seem[s] nothing more than a point, an infinitesimal now separating 'the past' from the 'future'" (Abram 1996, pp. 201-202).

Unlike an indigenous or phenomenological perception of space-time, which bases itself on the sensuous world of the earth's phenomenological presence (as will be introduced in chapter 4),

> "the conceptual separation of time and space – the literate distinction between a linear, progressive time and a homogeneous, featureless space – functions to *eclipse* the enveloping earth from human awareness. As long as we structure our lives according to assumed parameters of a static space and a rectilinear time, we will be able to ignore, or overlook, our thorough dependence upon the earth around us" (ibid., p. 217).

Furthermore, Western alphabetic culture inaugurated the separation of the human 'psyche' from the air of the whole earth's atmosphere (previously conceived as one and the same field, from indigenous people to the ancient Hebrews, as will be introduced in chapter 4). "For by using visible characters to represent the sounded breath [i.e. vowels, for the first time], the Greek scribes effectively desacralized the breath and the air" (ibid., p. 252). They engaged the disconnection of the ecopoïetic force of the world's winds from our awareness of our own awareness, as well as the disconnection of our breath-soul's autoecopoïetic participation to the world's winds. "The psychê, that is, was no longer an invisible yet tangible power continually participant, by virtue of the breath, with the enveloping atmosphere, but a thoroughly abstract phenomenon now enclosed within the physical body as in a prison" (ibid., p. 253). Such a disconnection facilitated the rise of the Platonic pure soul toward the realm of pure Ideas, and later infiltrated the Christian doctrines (as well as, lately, even Jewish religion itself).[30] The spread of the technology of alphabetic writing across the world in the Modern age, further pushed away the ecopoïetic dimension of the soul-breath-wind, even among non-Western populations. "It was not enough to preach the Christian faith. [...] Only by training the senses to participate with the written word could one hope to break their spontaneous participation with the animate terrain" (ibid., p. 254).

"[H]uman language became a largely self-referential system closed off from the large world that once engendered it. And the 'I', the speaking self,

29 Cf. Bergson 1889, 1907.

30 On the influence of Platonism on Christian theology, Abram points to Tarna (1991).

was hermetically sealed within this new interior" (ibid., p. 257). The 'respiration' between consciousness and subconscious, between civilization and wilderness, was severed. A new order of sensibility could thus appear, which would animate the "psychic system" (as Luhmann names it) as a primarily (and apparently exclusively) autopoïetic system.

"[A] new, apparently autonomous, sensibility emerges into experience, a new self that can enter into relation with its own verbal traces [and] reflexively interact with itself in isolation from other persons and from the surrounding, animate earth. This new sensibility seems independent of the body – seems, indeed, of another order entirely – since it is borne by the letters and texts whose changeless quality contrasts vividly with the shifting life of the body and the flux of organic nature. That this new sensibility comes to view itself as an isolated intelligence located 'inside' the material body can only be understood in relation to the forgetting of the air [i.e.] this sensuous but unseen medium that continually flows in and out of the breathing body, binding the subtle depths within us to the fathomless depths that surround us" (ibid., p. 255).

Recent historical developments

If the historical analysis of the phonetic-alphabetic culture, and of the modern development toward a disjunctive/simplifying/atomizing culture, do reveal already in depth,a major dimension of the unsustainability of Western modern culture, such an analysis insufficiently accounts for the specificities of the contemporary culture of unsustainability. I have already shortly mentioned the aspect of over-consumption/consumerism. But several other recent historical developments in the constitution of a Western culture of unsustainability, help to better understand the contemporary relevance of a technicized and disjunctive worldview in its late-modern form:

- The constitution of a technological system (cf. Jacques Ellul) following the shift from enlightenment to positivism (as discussed by Horkheimer and Adorno),
- and the constitution of autopoïetic social systems under modern society's functional differentiation, as analyzed by Niklas Luhmann.

'Technique' is the ensemble of man-made means designed and developed in order to allow human societies to free themselves from the direct grip of nature. The development of technique through human prehistory and history is a major dimension of the history of culture. But, according to several authors discussed in the following pages, the instruments of liberation have also become a destructive force threatening the sustainability of human societies. Technique may have become the institutional infrastructure supporting a contemporary culture of unsustainability.

The dialectic of enlightenment

In the 1969 preface to their *Dialectic of Enlightenment*, Max Horkheimer and Theodor Adorno overall confirmed their analysis from 1944 (date of first publication) as a prognosis and criticism of the "lapse from enlightenment into positivism", with the goal to "preserve and disseminate freedom rather than to accelerate, however indirectly, the advance toward the administered world" (Horheimer and Adorno 2002, p. xii). In their 1962 Italian preface, they summed up their argument with the claim that the "book demonstrates tendencies which turn cultural progress into its opposite" (ibid., p. xiii)

The drift of rationality into positivism denounced by Horkheimer and Adorno is, they argue, a threat to human intelligence. "In the belief that without strict limitation to the observation of facts and the calculation of probabilities the cognitive mind would be overreceptive to charlatanism and superstition, that system is preparing arid ground for the greedy acceptance of charlatanism and superstition" (ibid., pp. xv-xvi).[31]

The main concern of Horkheimer and Adorno is attached to the modern notion of human freedom, as both brought to life and endangered by "the self-destruction of enlightenment". "Freedom in society is inseparable from enlightenment thinking" but the later already contains "germs of the regression" because of the extreme development of "rationality" (ibid., pp. xvi-xviii). Against this development, they call for the "necessity for enlightenment to reflect on itself" (ibid., pp. xvii).

Horkheimer and Adorno, defining enlightenment at first very widely (and vaguely) as "the advance of thought", consider it as primarily "aimed at liberating human beings from fear and installing them as masters. [...] Enlightenment's program was the disenchantment of the world. It wanted to dispel myths, to overthrow fantasy with knowledge" (ibid., p. 1). They further point at Bacon's methods and ideas, developing man's mastery over nature as a "patriarchal" relation: "The mind, conquering superstition, is to rule over disenchanted nature. Knowledge, which is power, knows no limits" (ibid., p. 2). The core of this power of knowledge lies in the development of technology. "Kings control technology no more directly than do merchants: It is as democratic as the economic system which with it evolved. Technology is the essence of this knowledge." Technology aims not to produce understanding, but "method" and "capital".[32] Quoting Bacon's own terms on this technological knowledge, they notice that "[i]ts

31 That observation will be echoed by Ellul's analysis of the new forms of irrationality and superstition bred by the 'Technological System' in the second half of the 20th century (as discussed below).

32 By insisting on (economic) capital, Horkheimer and Adorno retain a Marxian attention to the supposed primacy of the economic infrastructure, whereas Ellul will describe technology in the 20th century as primarily self-developing and replacing the economy as the infrastructural basis of social order.

concern is not 'satisfaction, which men call truth', but 'operation,' the effective procedure" (ibid.). This observation already announces the ontological loss generated by this process (which will appear more fully in the discussion of Ellul, p. 39). "There shall be no mystery nor any desire to reveal mystery." Clarity, straight-mindedness are to rule (i.e. linearity). "On their way toward modern science human beings have discarded meaning. The concept is replaced by the formula, the cause by rules and probability" (ibid., p. 3). They further argue that "immanence" and even "Platonic ideas" – as offspring of mythical thought, became suspicious for enlightenment, so that "anything which does not conform to the standard of calculability and utility must be viewed with suspicion" (ibid.). With such a logic, any intellectual opposition, seen as myth, is treated accordingly, so as to strengthen enlightenment, and thereby "enlightenment is totalitarian" (ibid., p. 4). Not only clarity is demanded but also continuity, to ensure Bacon's "una scientia universalis", and discontinuity is intolerable. Positivism implies universalism: There must be only one common logic. Having different orders of reality is dismissed as illusion. "For enlightenment, anything which cannot be resolved into numbers, and ultimately into one, is illusion" (ibid.). They also describe how enlightenment assimilates the differences to the same, with the repetition of established laws.[33]

This development is at the root of the further development of technology into a 'technological system' (TS) in the 20th century (Ellul 1977), whereby technology imposes itself as the only universal mediator (as will be discussed right after these few paragraphs on the *Dialectic of Enlightenment*).

Horkheimer and Adorno further explain how in Western European thought (and already since classical Greek philosophy), "being is split between *logos* [...] and the mass of things and creatures in the external world. The single distinction between man's own existence and reality swallows up all others. [...] The awakening of the subject is bought with the recognition of power as the principle of all relationships" (ibid., p. 5). The knowledge system of enlightenment, while it claims to have purged anthropomorphism, is utterly anthropocentric, and anthropocentrically manipulative, unlike the ecological world of spirits in shamanic magic (which will be further introduced in chapter 4 with David Abram). This causes an 'estrangement' from the surrounding, natural world. "Enlightenment stands in the same relationship to things as the dictator to human beings. He knows them to the extent that he can manipulate them. [...] Their 'in itself' becomes 'for him.' In their transformation the essence of things is revealed as always the same, a substrate of domination" (ibid. p. 6).

In the process, "representation", i.e. symbolic thought, is crushed. "The manifold affinities between existing things are supplanted by the single relationship between the subject who confers meaning and the meaningless object, between rational significance and its accidental bearer" (ibid., p. 7). In

33 Cf. Horkheimer and Adorno 2002, p. 8.

this perspective, other ways of experiencing reality, such as "self-immersion in immediate natural existence," are anathema (ibid., p. 22). Horkheimer and Adorno also notice the importance of the radical separation between "sign" and "image" in Western culture (whereas they coincided, e.g. in "the hieroglyphs"). The parallel is clear here with Abram's critique of the rise of phonetic-alphabetic culture, as already discussed above, and with his discussion of the renewal of a self-immersion in immediate natural existence, pointed at by phenomenology (discussed in chapter 4, section 5).

They further explain how, in this way, loss of meaning progresses, in an "endless process of enlightenment by which, with ineluctable necessity, every definite theoretical view is subjected to the annihilating criticism that it is only a belief, until even the concepts of mind, truth, and, indeed, enlightenment itself have been reduced to animistic magic" (ibid., p. 7).

The dis-empowerment of symbolic thought bears consequences for art under the 'technological system', as argued by Ellul in his discussion of modern and contemporary art (which will be presented further down). Horkheimer and Adorno note that "[t]he impartiality of scientific language deprived what was powerless of the strength to make itself heard and merely provided the existing order with a neutral sign for itself" (ibid., p. 17).

Another consequence, not only for art but for culture in its wide, anthropological sense, is the threat of a possible impoverishment of human experience: "The more complex and sensitive the social, economic and scientific mechanism, to the operation of which the system of production has long since attuned the body, the more impoverished are the experiences of which the body is capable" (ibid., p. 28). While the authors here consider this development as a fact, I will rather consider it as a threat and a possibility. Under the autism of restricted formal rationality, human intelligence, as a fully embodied cognition, is at risk of degenerating further and further.

Ontology recedes further and further, they argue. "Enlightenment finally devoured not only symbols but also their successors, universal concepts, and left nothing of metaphysics behind except the abstract fear" (ibid.). Even thought itself is turned into a means (as part of the universe of means of the TS that Ellul describes). "Mathematical procedure became a kind of ritual thought [...] Mathematics made thought into a thing – a tool, to use its own terms" (ibid., p. 19). This bears consequences for the subject/object relationship. "The equation of mind and world is finally resolved, but only in the sense that both sides cancel out. The reduction of though to a mathematical apparatus condemns the world to be its own measure [...] thought makes itself mere tautology" (ibid., p. 20).

Horkheimer and Adorno point out how the obsessions of positivism are governed by a radical, rigid fear of otherness.

"Humans believe themselves free of fear when there is no longer anything unknown. This has determined the path of demythologization, of enlightenment, which equates the living with the nonliving as myth had equated the nonliving with the living.[34] Enlightenment is mythical fear radicalized. The pure immanence of positivism, its ultimate product, is nothing other than a form of universal taboo. Nothing is allowed to remain outside, since the mere idea of the 'outside' is the real source of fear" (ibid., p. 11).

This is at the root of a radically autopoïetic drive, that will be discussed with Luhmann below.

But, ironically again, they note that the modern societies then turn to fearing themselves (or rather, the increasingly uncontrollable TS growing in their midst) more than the otherness of nature. "The noonday panic fear in which nature suddenly appeared to humans as an all-encompassing power has found its counterpart in the panic which is ready to break out at any moment today: human beings expect the world, which is without issue, to be set ablaze by a universal power which they themselves are and over which they are powerless" (ibid., p. 22).

The *Dialectic of Enlightenment* prepares the rise of a 'technological system' (TS) in the 20th century, which I will discuss next. This development appears quite clearly in the following argument from he authors:

"The technical process, to which the subject has been reified after the eradication of that process from consciousness, is as free from the ambiguous meanings of mythical thought as from meaning altogether, since reason itself has become merely an aid to the all-encompassing economic apparatus. Reason serves as a universal tool for the fabrication of all other tools [...] Reason's old ambition to be purely an instrument of purposes has finally been fulfilled" (ibid., p. 23).

But with Ellul's TS, the focus of attention shifts from the 'economic apparatus' to the phenomenon of technology itself. The main difference between Horkheimer and Adorno, on the one hand, and Ellul on the other hand, is that they stick more closely to Marx's critique of commodification (and, relatively, to the economy as infrastructure of society) while Ellul in his analysis claims that technology has replaced economy as the infrastructure, and focuses on 'instrumentalization' (the turning into means) rather than commodification (the turning into things).[35] So when Horkheimer and Adorno write that "industrialism makes souls into things" (ibid., p. 21), Ellul would write that 'technique/technology makes souls into means'.

---

34 Against this simplistic, linear, disjunctive and reductive form of materialism imposed by enlightenment, a new understanding of sciences under the paradigm of complexity asks to consider both the continuities and the discontinuities between the nonliving and the living: This different episteme will be introduced in chapter 3, with thc discussion of Edgar Morin.

35 Commodification can be considered as one among different possible forms of 'instrumentalization'.

The technological system

A major paradigm unifying contemporary societies has been described by the French sociologist Jacques Ellul as the "Système Technicien" (most often translated as 'Technological System', which I will, here-after label as 'TS'). For him, Technique became in the 20th century the most determinant factor in our society – replacing Marx's Capital in its role of 'infrastructure'.[36]

Originally, technique dealt with a certain way to do things, a how-to, a set of procedures. With the division of labour and the multiplication of techniques (and of machines), technique (or technology in English) came to deal no longer with just single operations but with larger sets of operations and with inter-related systems of efficiencies under a logic of formal rationality: In the second half of the 20th century, a 'Technological System' emerged, both fragmenting society under specialization (i.e. 'differentiation' in sociological terms) and reunifying it under the self-justifying goal of efficiency. The technological environment pushes people to believe that every problem is a technical problem: Every issue, every question should be approachable through a technique. The effective growth of technical problems induces contemporary humans to infer that every problem is technical. This in turn contributes to the locking up of society in the technological environment. Society is still full of short-circuits, of chaos, of not-yet-technicized areas, of unpredictability, of incoherence and of human relationships. The TS is not transforming society into a machine, but it is increasingly installing itself within the existing reality.

To understand the TS in a few paragraphs, it is helpful to review its main properties: an environment characterized by its autonomy, unity, universality and totalization; an evolving system characterized by auto-growth and an accelerating causal progression of means losing sight of ends; an ambivalent force through which regimes of expertise flourish.[37]

By the second half of the 20th century, Technique has reached a state where it no longer needs to be justified by superior ends, argues Ellul. Technique has become autonomous. Technique is not just a means; it is no longer a function of something else. "It is a *complete system of means*, coordinated to each other and constituting a world of powers/forces, and environment fully substituted for the former one. An environment of means instead of an environment of living beings […] What constitutes the universe of technique is the correlation of means" (Ellul 1980, p. 68).

---

36 I will use here the term 'technique', following Ellul's original text in French closely. Although English speakers would probably use the term 'technology' to describe what is referred to as 'technique' here, in Ellul (as in the French speaking world in general), the term 'technology' is more specifically pointing at an organized discourse on technique.

37 The following paragraphs are building on the insights from Ellul 1977, and other specific writings of Ellul when mentioned in the text.

In the TS, the key question for each technique is the question of efficiency, from a formal-rational, means-rational point of view (cf. below, the link with Max Weber). In the TS, social agents get used to the superiority and priority of the "most efficient method", and believe this conventional rule to be self-evident. They hardly change such taken-for-granted priorities and declare that, for example "the most harmonious method" is to be sought instead of the most efficient. And if some do, they are not taken seriously (e.g. they can be categorized as either 'artists', 'dreamers' or backwards unintelligent folks).

This universe of means is organized towards efficiency in the use of techniques in combination with other techniques. With the growing integration of techniques allowed by computers, the different techniques can increasingly be understood as forming a system. Computer technology allowed linking almost all other techniques together, and interconnecting most socioeconomic sectors.

If techniques (and technical expertise – cf. below) are fragmented, the TS is universal and unified.

This system developed itself rapidly over the second half of the 20th century, both in capitalist and communist States. In both political regimes of the cold war era, the TS reinforced (and was in turn reinforced by) the State. Overall, and although the growth of means is accelerated (giving the impression of a diversified society, .e.g at the level of "consumer choice"), technique made contemporary society more uniform. The TS was a necessary precondition for the rise of the contemporary process of globalization (primarily characterized by its planetary network of computerized financial markets).

When means for the sake of means develop, the legitimacy of ends (and ontology in general), declines. Also, relations, structures and processes are more important than the things related. Technique as a system grows itself (auto-growth): The growth of techniques is 'automatic' because existing techniques tend to combine, reassemble and generate new techniques. Under the TS, the growth of means is destabilizing and explosive. A technicized society undergoes, at the level of practice, many continuous and fast-paced changes, and the forms of human action are in almost constant renewal. In most cases, these changes in techniques and in the technicized societies have become out of reach of democratic control.

Technique is ambivalent (rather than ambiguous): It creates problems that it promises to solve (fully, or most often only partly) with the help of new techniques to come – but which will create new problems, etc. This self-reinforcing loop, i.e. this accelerating cycle, is however covered up ideologically: The concept of progress, very much ingrained in a technicized society, largely covers up the ambivalence of technique, and develops the belief that the existing evolution of the TS corresponds necessarily to a linear 'progress', which necessarily benefits the human species as a whole and furthermore, works as a quasi-fatality.[38] Thereby, Technique can appear as a cure-all-remedy, inheriting the Messianic values of science and progress already in currency since Bacon (as already discussed).

The eclipse of ambivalence in the TS is insured through a pervasive dogma of progress, i.e. not just purely technical progress, but overall progress as a catch-all notion. Society is progressing, economic growth is a progress per se, workers unions' struggle for social justice is characterized as 'progressive', etc. Progress is not only pervasive as a traditionally modernist notion, but also in a postmodern version giving way to the feeling that anything new is a 'progress' of some kind (from fashion to 'innovations' in cosmetics and novelties on the art markets).

Technique deprives politicians of much of their power, Ellul argues in *L'Illusion Politique*.[39] It even works towards reorganizing social classes, so that top-level technicians (so-called technocrats) succeed to the old dominance of the Bourgeois.[40] A regime of experts follows, but this does not mean that experts enjoy the stable authority of a straightforward technocracy. People, including experts, must continuously follow the fast pace of technical change if they intend not to become obsolete in their expertise, and are thus continuously at risk, and summoned to update themselves. Furthermore, the regime of expertise does not institute simply a single new overarching elite. The experts are increasingly fragmented into specialized areas. The overall management of society under the TS is increasingly complex while the overall understanding of reality is decreasingly comprehensive, given the fragmentation of expertise.

Nicolescu also sees in technology the main driver of the atomization of scientific disciplines: "the disciplinary big bang is the response to the demands of a technoscience without brakes, without values, without any end other than utilitarianism" (Nicolescu 2002, p. 34). Like Ellul (but without referring to him), Nicolescu sees the in-humanity of the logic of the Technological System: Efficacy for efficacy's sake works "in the name of an efficiency whose rationality totally escapes even those who are its unconditional servants" (ibid., p. 143).

Besides, Nicolescu even sees technoscience as a threat to the world's cultures: With the increasing fragmentation of all forms of knowledge, the "overwhelming advance of technoscience has served only to deepen the abyss between cultures" (ibid., p. 102). Furthermore, "the spread of the Western lifestyle throughout our planet is associated with the destabilization of traditional cultures. On account of its economic strength, the West has a great responsibility: to avoid the cultural disintegration resulting from the

38 The most convincing sledgehammer argument for this mainstream position, is a quantitative measure: The age expectancy has (so far) risen considerably in modern societies.

39 Cf. Ellul 1965.

40 Cf. Ellul 1967.

unbridled development of technoscience" (ibid., p. 103).[41] But an opposite movement also results from the contemporary globalization of modernity:

> "In spite of its chaotic appearance, modernity leads to a rapprochement between cultures. [...] The potential for the birth of a culture of hope is precisely equivalent to the potential for self-destruction that is engendered by the abyss of non-meaning. The multicultural shows that the dialogue between different cultures is enriching, even if its goal is not real communication between cultures" (ibid., pp. 103-104).[42]

Ellul's analysis is also echoing Max Weber's and other authors' earlier concerns about the rising dominance of formal rationality in modern societies. In Weber's terms, technique/technology emphasizes means rationality (Zweckrationalität) rather than axiomatic rationality (Wertrationalität). Weber analyzed how formal rationality supplanted substantive rationality (which unlike formal rationality, is developing a critical reflection in view of certain end-values, such as for example justice, thereby generating meaning for human action).[43]

Other authors made relatively comparable analyses, although they used different terms (which might confuse the readers). A contemporary of Weber, Karl Mannheim, also came to oppose merely 'functional' or 'instrumental' rationality to a 'substantial' rationality which would add an individual-biographical normative framework refraining and reframing pure functionality/instrumentality. As to Horkheimer, he warned against the "subjective reason" of the technician (corresponding to Weber's formal/instrumental rationality), which he opposed to what he labelled as the "objective reason" which allows to determine what is ethical and right among different individual ends, thanks to a wider value-system beyond individual interests.[44] As to Marcuse, he came closest to Ellul's terminology, opposing an "individualistic rationality" (partly comparable to Horkheimer's "objective reason") which he defined as a "critical and oppositional attitude that derived freedom of action from the unrestricted liberty of thought and conscience", to a "technological rationality" which took over and whereby "individuals are stripped of their individuality, not by external compulsion, but by the very rationality under which they live," under the imperative of "efficacy" (Marcuse 1941, pp. 421, 433)

---

41 Nicolescu's claim opens a whole area of potential discussion on cultural diversity. On the threats posed by Western cultural industries on cultural diversity worldwide, see Smiers (2003), and see Cowen (2002) for a provocative contradictory argumentation.

42 Nicolescu's discussion of the multi-, inter- and transcultural will be further discussed in chapter 3.

43 On Weber's types of rationality and their historical significance, see Kalberg 1980.

44 Cf. Horkheimer 1947, p. 4.

Nicolescu also perceives the issue of narrowly-rational reflexivity as especially acute in the domain of education, and links it to the general threat of 'efficacy for efficacy's sake': "At present, education privileges the intellect over the emotions or the body. This was certainly necessary in the previous era, in order to permit the explosion of knowledge. But this privileging, if it continues, sweeps us away in the mad logic of efficacy for efficacy's sake, which can only lead to our self-destruction" (Nicolescu 2002, p. 137). But the introduction of a wider reflexivity, for example in the domain of education, will not be achieved by the mere addition of some 'cultural added-value' in the form of artistic patches and band-aids to the existing curricula. "Of course it is not a question of limiting or increasing the number of hours provided for artistic or athletic activities. This would be like trying to obtain a living tree by juxtaposing roots, trunk, branches, and leaves" (ibid.). What is required is no less than a "revolution in intelligence".[45]

When ends are lacking, means fill the void and technical rationality also leads its followers to new kinds of religious-like irrational behavior, as observed by Ellul (1973), and which can take multiple forms, among which the commodification of the self in a "Société du Spectacle" and the "Re-Enchantment" of life through hyper-consumption.[46] Furthermore, as a religion would, Technique does not let itself be judged. The irrationality of the Technician System is also perceptible in the lack of an end – one which people could perceive and debate about – apart from the delivery of new techniques. In its blindness, the Technician System leads technicized societies towards an unpredictable and alienating construction of their future, according to Ellul.

Bateson's critique of technology and purposive consciousness

In parallel to Ellul's radical critique of the TS, Gregory Bateson also warned against "a growing mystical faith in technology's power of salvation: a perilous faith in the *primacy of conscious purpose*", which he connected to other processes together forming the unsustainability of the Western modern development model according to him. Bateson argued:

"That all of the many current threats to man's survival are traceable to three root causes:

(a) technological progress

(b) population increase

(c) certain errors in the thinking and attitudes of Occidental culture. Our 'values' are wrong.

45 What Nicolescu suggests by "revolution in intelligence", will be introduced in chapter 3.

46 Cf. Debord 1967, Ritzer 1999.

We believe that all three of these fundamental factors are necessary conditions for the destruction of our world. In other words, we optimistically believe that the correction of any one of them would save us.
That these fundamental factors certainly interact. The increase of population spurs technological progress and creates that anxiety which sets us against our environment as an enemy; while technology both facilitates increase of population and reinforces our arrogance, or 'hubris', vis-à-vis the natural environment.
The attached diagram [figure 1] illustrates the interconnections. It will be noted that in this diagram each corner is clockwise, denoting that each is by itself a self-promoting (or, as the scientists say, 'autocatalytic') phenomenon: the bigger the population, the faster it grows; the more technology we have, the faster the rate of new invention; and the more we believe in our 'power' over an enemy environment, the more 'power' we seem to have and the more spiteful the environment seems to be.
Similarly the pairs of corners are clockwise connected to make three self-promoting sub-systems.
The problem facing the world [...] is simply how to introduce more anticlockwise processes into this system" (Bateson 1973, p. 466).

Furthermore, Bateson's critique of purposive consciousness pointed especially at modern technology as an aggravating factor:

"If you allow purpose to organize that which comes under your conscious inspection, what you will get is a bag of tricks [...] Consciousness operates in the same way as [modern] medicine in its sampling of the events and processes of the body and of what goes on in the total mind. It is organized in terms of purpose. It is a short-cut device to enable you to get quickly at what you want; not to act with maximum wisdom in order to live, but to follow the shortest logical or causal path to get what you next want [...] But what worries me is the addition of modern technology to the old system. Today the purposes of consciousness are implemented by more and more effective machinery, transportation systems, aeroplanes, weaponry, medicine, pesticides and so forth. Conscious purpose is now empowered to upset the balances of the body, of society, and of the biological world around us. A pathology — a loss of balance — is threatened. [... Bateson deplores] the curious twist in the systemic nature of the individual man whereby consciousness is, almost of necessity, blinded to the systemic nature of the man himself. Purposive consciousness pulls out, from the total mind, sequences which do not have the loop structure which is characteristic of the whole systemic structure" (Bateson 1973, pp. 409-410).

Figure 1: Bateson's "dynamics of the ecological crisis"

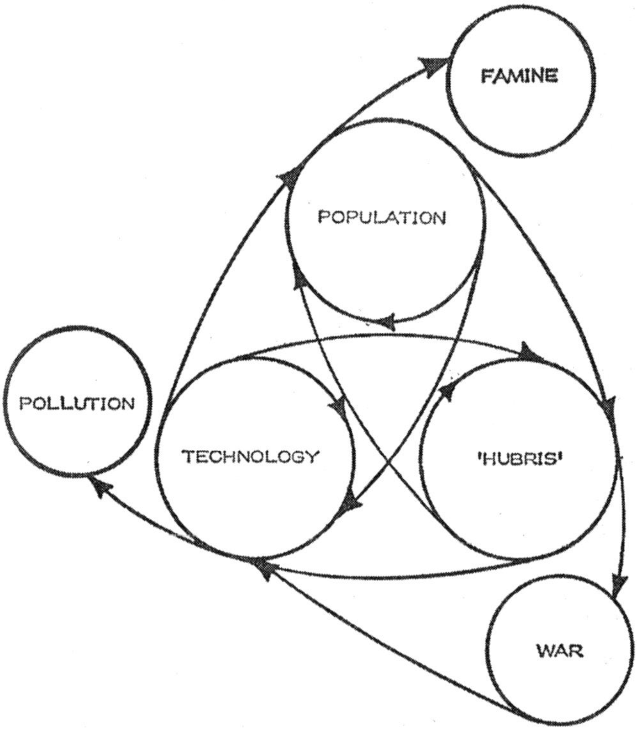

Source: Bateson 1973.

Bateson's version of the fall from Eden

To illustrate his critique of purposive consciousness, Bateson sketched a satirical version of the Biblical fall from Eden:

"Let me offer you a myth. There was once a Garden. It contained many hundreds of species — probably in the sub-tropics — living in great fertility and balance, with plenty of humus, and so on. In that garden, there were two anthropoids who were more intelligent than the other animals. On one of the trees there was a fruit, very high up, which the two apes were unable to reach. So they began to think. That was the mistake. They began to think purposively. By and by, the he ape, whose name was Adam, went and got an empty box and put it under the tree and stepped on it, but he found he still couldn't reach the fruit. So he got another box and put it on top of the first. Then he climbed up on the two boxes and finally he got that apple. Adam and Eve then became almost drunk with excitement. This was the way to do things: Make a plan, ABC and you get D. They then began to specialize in doing things the planned way. In effect, they cast out from the Garden the concept of their own total systemic nature and of its total systemic nature. After they had cast God out of the Garden, they really went to work on this purposive business, and pretty soon the topsoil disappeared. After that, several species of plants became 'weeds' and some of the animals became 'pests' [...] Adam went on pursuing his purposes and finally invented the free-enterprise system. Eve was not, for a long time, allowed to participate in this because she was a woman. But she joined a bridge club and there found an outlet to lie her hate" (Bateson 1973, pp. 410-411).

Disjunctive thinking and knowing, together with the estrangement from nature and the lifeworld, culminated in the shift of enlightenment into positivism and into the growth of formal rationality and a pervasive technoscience. The fragmentation of reality under late modernity is however not only related to techno-scientific specialization under the TS, but affects more generally the whole social system: The modern process of social differentiation, an old tune of sociological theory, has found an especially insightful elaboration in Niklas Luhmann's system theory.

Luhmann's system theory

German sociologist Niklas Luhmann's "system theory" offers an elaborate diagnosis of the unsustainability of modern social differentiation, highlighting the development of self-referentiality and "autopoïesis" in social systems, which has reached a point where, according to Luhmann's analysis, the ecological crisis is probably unresolvable. The interest of Luhmann's theory lies both in its diagnostic value and in its self-intoxication, i.e. its lack of opening to a less unsustainable organization of social systems. The following pages will thus have to take a double-task, both presenting insights from Luhmann's depiction of autopoïesis in modern social systems, and highlighting where the theory itself is intoxicated by a sort of forcefully stubborn autism (as well as points where the subtleties of Luhmann's analysis complicate the accusation of autism).

Luhmann, who bases his concept of autopoïesis on Maturana and Varela's (they developed a theoretical approach to the self-generation and self-organization, i.e. autopoïesis of the living cell – which I will discuss in chapter 2, section 2), defines three systems: the living system (which reproduces life), the psychic system (which reproduces consciousness) and the social system (which reproduces communication). Each of these three systems is "operationally autonomous", and in this way the three systems are hermetically closed to each other, although they can be "irritated" by each other, which can allow "structural couplings" (i.e. reciprocal irritations which are repeated and considered by each system as a permanent characteristic of its environment): In short, in Luhmann's theory each system treats everything else (including other systems) as an environment from which the self-organizing system can select which "irritations" it will perceive and act upon. But in his view, there is no self-organizing ecological system (ecosystem), only ecological relationships in the environment ("Umwelt") of the living, psychic and social systems.

Luhmann's general theoretical framework

In Luhmann's theory, each living organism is a living system. Luhmann speaks only little about the living systems (and the interest of his analysis lies rather in his reflections on social systems). He refers to Humberto Maturana who defines them as dynamic systems and autonomous units, where there is a circularity of molecular production (molecules are contributing to producing each other, in a circular causality). Self-production is especially characteristic of the living cell, and by extension, also relevant to multi-cellular organisms. Thanks to this autopoïesis, a living system is not only able to (cybernetically) regulate itself (self-organization) but is also able to generate itself (self-re-production).[47]

An autopoïetic system "means that the elements of the system are produced within the network of the system's elements, that is, through recursions" (Luhmann 2000, p. 49). Another important characteristic of autopoïesis in Luhmann's strict interpretation is that "[t]he environment cannot participate in the reproduction of systems; the manner in which it affects the system's reproduction is never instructive, only destructive" (ibid., p. 50). What can be instructive, but to a very limited extent, is the "structural couplings" that may occur between autopoïetic systems, as described in the following paragraphs.[48]

The psychic system is the operationally closed system of consciousness, which operations are thoughts. The exterior world only offers stimuli that consciousness autonomously elaborates and structures according to its own rules. For Luhmann, there is no direct communication between a psychic system and any exterior system (e.g. another psychic system), but a mutual opacity. The exterior environment, including e.g. the action of another psychic system, is experienced by the psychic system but only as a stimulus. There can only be a synchrony of mutual irritations between two psychic systems.[49]

In Luhmann's theoretical framework, there is no communication between psychic systems, there is no information that is passed on from one

47 For Maturana and Varela's original theory, see Maturana and Varela 1980.

48 Luhmann thereby follows a strictly Darwinian understanding of evolutionary processes. In chapters 2 and 3, I will follow authors who, unlike Luhmann, reinterpret evolution differently, considering how the environment (including other autopoïetic living systems) can participate in a system's reproduction and transformations, more than merely through "structural couplings". This will eventually lead me to replace Luhmann's concept of autopoïesis with a new concept of "autoecopoïesis".

49 Cf. Luhmann 1990a, p. 218.

person to another. Communication is an emergent phenomenon operating at the level of the social system and its subsystems, not at the level of psychic systems.

Consciousness in the psychic system, constitutes meaning and involves self-observation. And when consciousness represents itself in itself as a unity, the psychic system experiences what is called "reflexion" or reflexivity.[50]

The social system reproduces its basic operational units which are communications. Everything that does not relate to communication is in the environment of the social system (including the individual human being as "psychic system"). The receiver of a communicative action is another communicative action. Consciousness functions at another level of reality (within the psychic system) which is necessary for the emergence of communication at the level of the social system, but not directly interacting with it. There are only massive mutual irritations between the two, which shape a "structural coupling".[51] (The blind spot in Luhmann's theory is how the social and psychic systems can "meet" at all, given that irritations are only internal to each system.[52])

The social system is, in modern societies, subdivided in subsystems (which for ease of language, are called "social systems"): The process of social differentiation that characterizes modern society worldwide, is a process of functional differentiation, where cultural differences are only secondary, according to Luhmann. This process of differentiation led to the constitution of opaque social sub-systems, perceiving each other as part of the environment (e.g. the economic system, the political, the religious, the education system, the scientific, the juridic, the art system, the mass media, family), but integrating the presence and functional differentiation of the other social systems within their own constitution. "The specific system type suggests what kind of other systems can be expected on the other, external side of the form: [i.e. in the case of modern society] other functional systems if the differentiated system specializes itself along functional lines" (Luhmann 2000, p. 136).

Each of the three systems (living, psychic, social), and in modern society, each of the social systems, is closed on itself, "operationally", and autopoïetic: it produces itself what composes its self and organizes itself.

50 Cf. Luhmann 1984, pp. 346-376.

51 Cf. Luhmann 1995, pp. 51ff, 117ff.

52 Cf. Teubner 2001, p. 40. To be noted, Habermas also criticized the "artificial separation of psychic and social systems" in Luhmann (Habermas, quoted in Ferrarese 2007, p. 63) – but I will not myself follow Habermas either in my further developments. As far as I can grasp Luhmann's theoretical elaboration, what I am missing in it is a sufficient attention to the process of emergence, which operates across different systems and generates new systems at new levels of reality. I will come back to emergence in chapter 3, with Edgar Morin.

The system is closed by its own operations and it determines itself how and how much its environment may affect it. The self-closure of the system (setting its borders, its identity) is not done by a set of elements in the system, but by the very chain of operations of the system. The single elements are then no longer to be conceived in themselves but as effects of the system.

A system does not define itself in itself (in terms of a fixed identity), but exists only through distinguishing itself from its environment. But how can the system "draw a distinction" if it cannot communicate with its environment? Luhmann (who needs to elaborate a cumbersome theoretical framework, to keep the autopoïetic autism from communicating with its outside) speaks of "re-entry", i.e. a simulation, inside the system, of the system's hermetic closure from its environment.

Luhmann's perspective allows an understanding of reality in terms of non-hierarchical networks of observation, observing each other but which cannot be brought together under one total observation. Given their relative opacity to each other, the social systems cannot acquire complete knowledge of each other (and therefore none has a superior view, also not the scientific system). But this theoretical perspective also posits a radical self-determination by the system, i.e. a radical non-determination by the environment, which poses problems in terms of evolution: Luhmann's systems have an ability to develop themselves up to a certain extent, insofar as they identify and respond to irritations from the environment. But they cannot evolve by letting in, unexpected elements that may unpredictably recode the system's organization.

That such a view is problematic, can be seen already at the level of the living system: Some phenomena are ignored in such a theory, especially the fact that a retrovirus can 'forcefully' introduce new DNA into a living cell. The role of retroviruses is increasingly considered as crucial in the evolution of new species, i.e. the diversification of life.[53] (The human genome is composed of about 8% of genetic material from retroviruses.[54])

Social systems: Autopoïesis as autism?

Social systems in modern society have reached autopoïesis themselves thanks to social differentiation, which is a functional differentiation, allowing an increasing degree of social complexity with differentiated problem-resolutions in specific social systems.[55] The different social systems observe reality (i.e. also the other social systems in the environment) according to their own semantics and own rules of distinctions exclusively. According to this logic, the rules of one social system may generate irritations but cannot

53 Cf. Luis Villarreal, Professor of Molecular Biology and Biochemistry, University of California - Irvine; website: http://www.faculty.uci.edu/profile.cfm?faculty_id =2705

54 Thierry Heidmann, quoted in Fel 2009, p. 208.

55 Cf. Luhmann 1999, p. 43.

directly interfere within another social system, which responds according to its own rules. Each social system is not just "differenziert", but "ausdifferenziert": Luhmann does not speak of "Differenzierung" (like Max Weber), but of "Ausdifferenzierung", marking out more clearly the process of detachment of each social system from its social environment.[56]

Each social system constructs a description of the whole society (social system as a whole) from its own perspective. From this necessarily biased perspective comes a generalization fueling the illusion that the system's specific operations bear a high relevance for the whole society. But, for Luhmann, there exists, in modern societies (unlike in pre-modern societies), no general view that would sit above the functionality of differentiated social systems and would harmonize their descriptions.[57] As a critical author formulated it:

> "On the level of the social system 'society', this means that society can observe and reflect its own unity only as a difference between its function systems, and since each system perceives this difference in a system-relative way, society can observe its own unity only as a difference between differences, without any possibility of establishing a higher-order perspective as some kind of a collective 'reason' *(Vernunft)* which could overcome (in the dialectical sense of Hegel's '*aufheben*') those differences and thus provide for a co-ordination and harmonization of specific functional operations" (Miller 1994, pp. 108-109).

Nor does the individual harmonize his or her social roles, as roles are fragmented according to social systems, and the very notion of "individual" has lost its social relevance, in Luhmann's view.[58] (However, the human individual/subject is still relevant as the self-distinction of the psychic system, which as Umwelt, irritates the social systems.[59])

The description by Luhmann of what he analyses as the evolution of social systems, is actually, a perfect description of 'development'. For Luhmann, evolution is not the history of the adaptation of society to its environment: "Not only does the system not adapt to its environment, but it chooses or alters its environment to adapt it to what the system itself prefers" (Luhmann 1990b, p. 552). The "primary goal of autopoïetic systems is the continuation

56 Cf. Ferrarese 2007, pp. 46-47.

57 Cf. Luhmann 1984, p. 645. Luhmann himself argues also against the illusion that the scientific system would hold a privileged view (cf. Luhmann 1990b, p. 714). In Luhmann's view, there can also be no real "participant observation" by a social researcher, and his epistemological position does not allow an empirical practice of transdisciplinarity (I will come back to transdisciplinarity in chapter 3).

58 Cf. Luhmann 1984, p. 286.

59 Cf. Gunther Teubner, quoted in Ferrarese 2007, p. 94.

of autopoïesis and any concerns for the environment is filtered by corresponding system references" (Miller 1994, p. 105). In an autopoïetic system, the next operation is more important than taking the future into account. The social system is then, in Luhmann's view, destined to its eventual collapse through the ecological crisis.[60]

However, in the subsequent chapters, I will not follow Luhmann's characterization, and oppose to his view of social development (or evolution), a notion of 'evolution' based rather on Morin's theoretical insights on complexity, and replacing autopoïesis with "autoecopoïesis", along the suggestion that "non-evolutionary development models try to adapt the environment to the social system while an evolutionary development adapts the social system to the environment" (Brocchi 2008, p. 47).[61]

The autopoïesis of social systems according to Luhmann, reveals a form of autism, i.e. an absence of communications across social systems, resulting in the impossibility of any transversal reflexivity. However, some of Luhmann's reflections about complexity hint at different directions (which I will further explore in chapter 3, after Edgar Morin)[62]: If Luhmann argues that the increase in complexity in a social system is the result of the increased autonomization of the system from its environment, this increase in complexity also raises the sensibility and irritability of the system: the system is able to distinguish more irritations from its environment (i.e. it literally constructs the complexity of its environment), because of "an excess of possibilities that results from the differentiation of a system [...] Self-description does not shield the system from constant irritation by excluded excess possibilities" (Luhmann 2000, p. 249). The increase in contingency is also an increase in range of possibilities. As a result, some of Luhmann's reflections come closer to Brocchi's suggestion, quoted above, than one might have expected. Hence: "In systems that are successful in evolutionary terms, more independence typically amounts to a greater dependency on the environment" (Luhmann 2000, p. 158).

A hopeless scenario?

Luhmann shows a clear "hostility to the ethical illusion", as Jean Clam (1997) observed. Luhmann both affirms that society cannot perceive itself from an outside position and correct its evolution thereby, and contends that values are not helpful in a modern society. He even affirms that the normative views on mankind operating in modern society, are responsible for the

60 Cf. Luhmann 1986, p. 38.

61 Cf. chapter 3.

62 Luhmann also developed an elaborate reflection on self-observations by systems (and after Maturana, on second order observations, i.e. the observation of observations, revealing their functioning and their contingency). On second order observation in art, cf. Luhmann 2000, pp. 54-101.

most negative historical moments of modernity.[63] Furthermore, in his theoretical framework appears no possibility for strategic action by the individual ("action" is only an attribution by a social system).[64]

Luhmann defends, at several points, the social differentiation of modern society, which, he claims, offers a multiplication of possibilities (thanks to the autopoïesis of the different social systems), and also does not determine the (autopoïetic) individual consciousness.[65] But this should not be mistaken for a naive belief in 'progress' through modernity. Luhmann rather values the increase in contingency of the multiplication of possibilities, for its own sake, as offering a sort of freedom.[66] Luhmann's consideration of progress is non-complacent: "Technological progress leads to ecological disasters, which can be avoided however only through more advance technological progress and thus at the price of an ever greater dependency of society to technology" (Luhmann 1999, p. 47). Modern social systems could not prevent, and cannot overcome the ecological crisis, in his view. As already described, the overall social system, and its subsystems in modern society, are incapable of communicating directly with the non-human environment. They can only be 'irritated' by their direct environments. An autopoïetic system is able to select which irritations it will notice and ignore the other ones. As long as the systems are only "autopoïetic", nothing guarantees that they will genuinely evolve, acting upon irritations: They are more likely to interpret the irritations in ways that will eventually lead to their self-annihilation through further mis-consideration of the environment. Luhmann's conclusions on the possibility for social systems to overcome the contemporary ecological crisis are especially pessimistic.[67]

Ecological crises are only a relevant reality for social systems insofar as they are taking the form of "ecological communication" and as such, find resonance within social systems. For this to happen, the ecological problems are translated into each social system. For example, the economic system can only treat costs and prices, and does so according to its own rules.[68] The political system is constrained in other specific ways, i.e. by the short-term framework of electoral agendas and 'public opinion', the spatial framework of nation-states, and by the framework of its medium i.e. power (which cannot achieve a transformation of a society's relationship to its environment). Furthermore, the different social systems cannot be coordinated (except through unpredictable "decentred contextual steering", as described in the next paragraph). Thus the political system may disturb other systems, for reasons linked to ecological communication, but with effects beyond control

63 Cf. Luhmann 1994.

64 Cf. Ferrarese 2007, pp. 74-76.

65 Cf. Ferrarese 2007, pp. 63-66.

66 Cf. Luhmann 1992, p. 95.

67 Cf. Luhmann 1986.

68 Cf. Luhmann 1986, pp. 120-123.

and involving multiple perverse effects related to the autopoïetic responses of the other systems to the perceived irritations. The interdependencies (multiple "structural couplings") between different social systems increase this difficulty. Besides, here as elsewhere, Luhmann rejects the 'ethical illusion' vested in the hopes for the emergence of new ecological values with a new morality, which in his view would only raise impatience and fear, and limit the field of possibilities in social systems, but not bring a transformation of autopoïetic systems.[69]

However, the Luhmannian theory is not describing an absolutely 'hopeless' scenario in terms of social transformations. It allows to conceive of some forms of limited influence across social systems: "decentred contextual steering" ("dezentrierte Kontextsteuerung").[70] This consists in modifying the context in which a given system functions, without directly trying to determine, from the outside, the evolution of the targeted social system, but rather trying to stimulate its autonomous corrections and transformation (e.g. in the case of ecological issues, trying to stimulate e.g. the economic system's response to these issues by formulating them in the language of the economic system).[71] This requires a preliminary observation of the targeted social system by the influencing social system. But even with careful observation, decentred contextual steering can only yield very unpredictable, non-linear results, with no possibility of control.[72] Furthermore, as already mentioned above, in an non-centered society as depicted by the Luhmannian system theory, there is no possibility for a global reflexivity, and thus no possible diagnosis of the overall pathologies in the development of modern society.[73]

Max Miller criticizes Luhmann for not considering the possibility of "intersystemic discourse", a communication across social systems which may be under-developed in contemporary societies but should not be a priori rejected as impossible: Miller observes that certain interdisciplinary bodies (e.g. advisory bodies to governments) attempt at such discourses. However, considering this option requires stepping away from the Luhmannian theoretical framework, which does not allow such processes to happen: "Mutual knowledge cannot arise – not even regarding a co-ordinated dissent – because differences mean different things for different observers" (Miller 1994, p. 117). Miller invokes Wittgenstein and other "language philosophers and linguists" to reject a "so-called 'mutual knowledge paradox' [as] only a theoretical construct" (ibid., p. 118). Miller himself argues for the existence of an intersystemic discourse allowed not by a basic consent but by a coor-

---

69 Cf. Luhmann 1986 (p. XVII in the 1989 English edition).

70 Cf. Wilke 1995, pp. 57ff.

71 Cf. Luhmann 1981, pp. 81-88, Luhmann 1988, pp. 346-347, and Luhmann 1986, pp. 122-123.

72 Cf. Luhmann 1981, p. 10.

73 Cf. Ferrarese 2007, p. 125.

dinated dissent, i.e. "the minimal requirement [...] that a mutual knowledge can be developed regarding the differences between different systems' perspectives concerning the difference between systems' perspectives" (ibid., p. 119). Miller points back at the individual social actors who "are not only able to talk in the 'languages' of different sub-systems; more importantly, they do have argumentation or discourse abilities for jointly identifying and answering disputed questions and for passing joint decisions even if they proceed from basically different points of view" (ibid., p. 114).

Overall, Luhmann's analysis offers no normative horizon, and has been accused, among other things, of "transform[ing] empirical facts into logical necessities" (Miller 1994, p. 112) and of thereby developing a sort of cynical metaphysics out of an insightful description of society's "organized irresponsibility".[74] If Luhmann's theory offers several insightful perspectives for the study of complex social systems, and describes well the unsustainability of strict autopoïesis in social systems, it purposively refuses to escape the syndrome of modernity that it analyzes.

Bateson might even characterize autopoïetic tendencies to 'shape' the outside systems according to the own rules of the own system, as something somehow comparable to the secularization of 'totemism' which he criticized (as already discussed above). Of totemism he says: "The fantasy then becomes morphogenetic; that is, it becomes a determinant of the shape of the society." Totemism has to do with analogy between social world and natural world. But its secularization is its disconnection, the loss of patterns that connect and the closure of references on one side only of the relationship: "What seems to happen in such conventional secularization is a shift of attention away from the relationship to focus on one end, on the objects or persons who were related", losing sight of their non-human counterparts (Bateson 1973, p. 132). For Bateson, this development is tragic and profoundly alienating. He characterizes it as a "pathway of degeneracy" (ibid., p. 133).

Luhmann's autopoïetic social systems echo what Bateson criticized as an "ecology of bad ideas":

> "There is an ecology of bad ideas, just as there is an ecology of weeds, and it is characteristic of the system that basic error propagates itself. It branches out like a rooted parasite through the tissue of life, and everything gets into a rather peculiar mess. When you narrow down your epistemology and act on the premise 'What interests me is me, or my organization, or my species', you chop off consideration of other

---

74 Cf. Beck 1988, pp. 166-174. Other critiques of Luhmann are not mentioned here. I will especially not echo the critiques of Habermas against Luhmann, as I reject the Habermassian communicational reason (acknowledging Luhmann's critique of Habermas against the illusion of one single level of public space – but as chapter 3 will show, I will argue that transdisciplinarity can however allow exchanges across levels of reality – an impossibility in the Luhmannian theory).

loops of the loop structure. You decide that you want to get rid of the by-products of human life and that Lake Erie will be a good place to put them. You forget that the eco-mental system called Lake Erie is a part of *your* wider eco-mental system – and that if Lake Erie is driven insane, its insanity is incorporated in the larger system of *your* thought and experience" (ibid., pp. 459-460).

SECTION 2: THE ART OF UNSUSTAINABILITY

How far is the role of the arts in contemporary Western society, contributing to maintain the culture of unsustainability described in the first section of the current chapter?

I will argue in the following pages that the primary dimension of the arts as order factor rather than change factor, is its 'autopoïesis' that results for a great part from the institutionalization of the 'high arts' as analyzed by art sociologists: Major sociologists and economists of culture (e.g. Pierre Bourdieu 1979, 1992, 1993, Howard Becker 1982, Paul DiMaggio 1992, Hans Abbing 2002) as well as historians (e.g. Lawrence Levine 1988) have described this process in details.

Historical and sociological analyses of high art in the 19th and 20th centuries

The social segregations and exclusions developed in the past two centuries through the institution of a field, system, or institution of high, autonomous art, have already been analyzed at length by art sociologists.

The historian Lawrence Levine described how across the second half of the 19th century in the USA, the audiences of theatre, music and museums were gradually segregated, so as to constitute a separate field of "highbrow" culture which would impose an elitist format on the mediation of artistic works and thereby chase away the "lowbrows" (i.e. members of the audiences from popular classes). For example, Levine (1988) analyses at length how Shakespeare was gradually turned from a shared icon of popular and intellectual culture alike, to an exclusive territory of the elites, with the imposition of conventional constraints on acting and on audience-behavior[75], the segregation of theaters[76], and the economic streamlining of popular theater production by Broadway companies.[77] This segregation of social classes according to judgments of taste was

75 Cf. Levine 1988, pp. 26-30, 46-48.

76 Cf. Levine 1988, pp. 57-65.

77 Cf. Levine 1988, pp. 78-79.

analyzed and theorized especially by Pierre Bourdieu (1979), long after the first analytical insights developed by Thorstein Veblen (1899).[78]

The sociologist Pierre Bourdieu (1992, 1993) analyzed the social nature of artistic production, stressing the power relations in the emergence and reproduction of an "artistic field" (as part of a wider theory of social fields developed by Bourdieu), and observing the structures of the 'rules of the game' in the artistic field and the processes of 'consecration' of artists. Bourdieu's work also demonstrated how the idea of autonomous art emerged as a by-product of bourgeois society.[79]

Complementing the previously mentioned works, sociologist Howard Becker (1982) analyzed how the arts as a specific social 'world', have established their own sets of rules and understandings of artistic work, and how a web of conventions is thereby constraining and orienting both production and consumption in the arts.[80]

This institutionalization process established (across the late 19th and the 20th century) the field of art as an apart social world with separate laws, insiders values and discourses. This world of the 'high arts', while proclaimed as 'autonomous', was put on a pedestal that allowed its distinguished consumers and practitioners to set themselves apart from other social groups and impose upon them a symbolic violence fueled by their cultural capital. The institutionalized autonomy of the high arts increasingly turned into a concern for 'art for art's sake' or more generally for self-referentiality, i.e. the art worlds became much more interested in their own internal history, discourses and overall languages, than in their relationships with their environments (cf. Luhmann's analysis of the social system of art, which is introduced below).

As a consequence, alternative developments (such as community arts) have been often looked down upon and the bulk of new developments in the arts and public subsidies for the arts have been oriented to the 'high autonomous arts' across most of the 20th century, especially in Western Europe and North-America. The arts have also thereby become a hyper-specialized professional field of activity, which as other professions, is con-

78 See also eds. Lamont and Fournier 1992. For an elaborate sociological discussion of different modes and degrees of cultural hierarchizations and differentiation across different societies, see DiMaggio 1987. For a discussion of the relevance of Bourdieu's theory today and a more complex level of analysis, including the sociological analysis of intra-individual, distinction from oneself, see Lahire 2004.

79 See also Alexander 2003, p. 86.

80 Howard Becker's analysis will come back into my focus when I will analyze the political dimension of art, in chapter 7.

tributing to the fragmentation of socially constructed reality under the reign of the formal rationality of the 'technological system' (cf. Jacques Ellul, Max Weber, Martin Heidegger) and of the myths perpetuated by a 'Romantic Order' (cf. Maarten Doorman). The seemingly liberating autonomy of the art worlds offers merely an escapist strategy as long as one contributes to the inwards-looking activities of a social world largely disregarding its environment. In such a case, the provision of this autonomous world of art in our societies functions more as a safety valve for the technological system than as a change factor: Releasing the steam of those forces and values beyond formal rationality that cannot accommodate the mainstream cultural order, the art worlds turn potentially radical change agents into agents of social order.

ROTS: The Romantic Order and the Technological System in the arts

The conventions of art worlds typically involve a certain number of founding beliefs with a paradigmatic hold on the artist's values.[81] These institutionalized beliefs both belong to a 'Romantic Order'[82] (RO) and to a 'Technological System' (TS).[83]

The romantic notion of art is not bound to so-called Romanticism: According to Maarten Doorman (2004), it is the product of a *Romantic Order* (which I will further abbreviate as RO), which appeared in the beginning of the 19th century within a short period of time, and spread like an epidemic disease.[84] The RO is more than just Romanticism. It is, Doorman argues, an order or episteme that characterizes any sphere of life: science, art, politics etc. The main features of the RO, which were relatively unimportant before, are imagination, holistic thinking, attention to feelings, admiration of nature and a focus on originality.[85] More specifically in the arts, expression replaces imitation, and the features of the RO produce the dominant belief that (1) artists are specially gifted people who (2) create works of exceptional beauty and depth which (3) express profound human emotions and cultural values.[86]

---

81 The concept and analysis of conventions (and their relevance to the study of art worlds) will be explained in chapter 7, in the context of the discussion of 'entrepreneurship in conventions'.

82 Cf. Doorman 2004.

83 Cf. Ellul 1977, 1980.

84 Cf. Doorman 2004, p. 15.

85 Cf. Doorman 2004, p. 17.

86 I am here echoing Hans Abbing commenting on Doorman 2004 (Abbing, personal communication, 2007).

A similar process was described by Raymonde Moulin (1978), though starting earlier according to her, i.e. already with the industrial revolution of the 18th century, when people started to define the artist and the art product with their 'sublime purposelessness' in opposition not only to artisans and their products, but even more to industrialists and their mass-produced products.

The influence of the RO is especially visible in the obsession of the artistic genius, including the unrecognized genius. Accordingly, not everybody is a genius or an artist. Even if played with, ironically treated and mobile, the line between art and non-art, remains a reality (irony is also considered as typical of the RO by Doorman). Furthermore, although expression, originality, authenticity and autonomy are being regularly questioned in contemporary art, so far there is no widespread evidence that a new, different attitude is quickly spreading like an epidemic as one might expect in the case of a new episteme/paradigm taking over.

According to Doorman, the RO also involves a fascination with dichotomies, i.e. binary oppositions, such as man vs. machine, self vs. other, male vs. female, art vs. science, subjective vs. objective, charismatic vs. bureaucratic, and other interrelated characteristics such as the gifted, unique, authentic, autonomous, vs. the standard, repetitious, replaceable and dependent; or the spiritual, magical, creative, expressive, emotional, vs. the profane, down to earth, businesslike, conceptual and rational.[87] The characteristics of art become partly defined by their opposite. Art is then everything but rational, calculated and economic. But exactly because of the heightened awareness and fascination for the dichotomies in the RO, there is room for doubt as well, and for exceptions which place themselves on the other end of the dichotomy or develop a 'double-mindedness' regarding the dichotomy.[88]

In the RO, freedom, autonomy and authenticity are associated to art. Non-artists envy the artist because they think that his work is more creative, that he can put more of himself in his work, that he is freer not just because he is self employed, but also because in his work he is (supposedly) more independent than they can be. The assumed freedom of the artist both implies 'freedom to' and 'freedom from', while autonomy refers more specifically to 'freedom from'. The free artist has the power to make the art he wants to make, while the autonomous artist is free from demands and influences coming from others. One speaks of an authentic artist and of authentic art works, if the artist's work could not have been created by someone else. Therefore, the work of a hardly autonomous artist with little artistic freedom can still be authentic because people believe that his personal hand is 'in' it.

Art as an interdependent process rather than independent (an interdependence depicted by Howard Becker – who sociologically described how

87 Cf. Doorman 2004, pp. 45-62.

88 Cf. Doorman 2004, pp. 135-148.

art-making is the work of cooperative networks; and advocated by ecological artists – as will appear in chapter 5) is therefore an uncommon and marginal view, under the RO.

At first sight, several features of the romantic worldview might be perceived as interesting elements for the construction of an ethical search process of sustainability. But, to best understand how the art worlds are constrained by these beliefs, they have to be placed in the wider context of the *Technological System* (TS) to which the Romantic Order paradoxically contributes (although the Romantic Order apparently negates the Descartes-Bacon Modern worldview). Doorman (2004), in his criticism of the Romantic Order, fails to contextualize it as a paradoxical feature of Modernity and a relatively weaker and subordinated structuring structure vs. the stronger structuring structure of the Technological System (probably because Doorman's analysis is situated from within the paradigmatic hold of the TS, and uncritical thereof).

As already seen above, the main features of the TS are its autonomy (as an environment of correlated means), its auto-growth (exponential and mostly beyond control) and its self-evidence i.e. its disconnection from ends (apart from a vague idea of a supposedly inevitable self-realization of the individual).

The context of the TS fosters the development of an escapist realm of supposedly autonomous art where a 'Romantic Order' would supposedly manage to defeat the logic of formal rationality. But that second world is harmless as long as it plays within the virtuality of its institutional autonomy, i.e. as long as it remains in its designated sandbox. By design, it is intended to be a spectacle of inefficiency, at the service of otherwise generalized efficiency.

The *Romantic Order in the context of the Technological System* (RO/TS) is a romantic ghetto, an escapist pressure-valve regulating the tensions created by formal rationality's structural hold on modern and hypermodern societies (that does indeed generate a malaise and a romantic yearning among many contemporaries). It is also an order of denials and heightened hypocrisy, where art and art worlds are supposedly all about the spiritual, the unique, the irrational and the non-marketable, while the majority of non-art worlds are in contrast, supposedly all about commodities, the rational, the calculated, the efficient, the marketable, the economic. This institutionalization of art as other-worldly and the "denial of the economy" in art (as coined by Bourdieu 1993) have contributed to a self-alienation of the art worlds.

Even though some art worlds in the 20th century have attempted to break their institutional boundaries, unfortunately, they have often not escaped the paradigmatic realm of the RO/TS, and all that was achieved by many avant-gardes was an extension of the sandbox for art to play in.[89]

89 Cf. ed. Welchman 2006.

In the Romantic Order (RO), art may present itself as an autonomous environment, a counter-milieu that acts as an antidote to general society. In its claim for autonomy, it may even deny its relationships with the surrounding environment. But art, even if it is granted a special place in society (facilitated by the RO), is not isolated from overall developments in society. As the early art sociologist Pierre Francastel commented: "in a society, artists don't play an isolated role independently from technicians. Instead of the idea of separate histories of different disciplines and different human activities, one shall substitute the global conception of the expressive capacities of a society to shape itself through expression" (Francastel 1956).

Ellul (1980) developed an especially critical analysis of the role of modern and contemporary art within the TS, which offers interesting parallels to Doorman's analysis of the RO. In the 19th century, art turned into a specialized realm of illusion, in parallel to the disenchantment of society[90]. However, if the organization of the arts under a RO was installed during the 19th century, a further development took shape across the 20th century, that further integrated the arts in a society where Technique was becoming the dominant 'infrastructure'. In parallel to the accelerating growth of 'technical progress' (with for example the advent of electronics and computer science facilitating the current process of globalization), the arts did not escape the general integration of society under the new paradigm of the Technician/Technological System (TS).

The TS allows and completes the contradictions of the Romantic Order. For example, *spontaneity* is praised; but the 'spontaneous' often must also bear fruits, as the expression of a technique at its best. According to Ellul (1980), 20th century art unknowingly goes into 2 directions in its relationship to the TS: either it rather directly expresses the TS itself and its structural meaninglessness, or it expresses the disorder of a technicized society – a society torn out between old-time values and habits and the new ones installed under the influence of this new world system. Ellul's argumentation (which I will not entirely follow, insofar as I also see alternative developments in the arts, to be explored in chapters 4 to 6), goes as follows: 20th century art is torn apart and reveals a society that is torn apart. Even when trying to oppose the general society, art reproduces the characteristics of our technicized society. A comparable perspective was evoked by Guy Debord: "Art in its time of dissolution [...] is at once art of change and pure expression of impossible change. The more its requirement is grand, the more its real realization is beyond itself. This art is necessarily avant-garde, and it is not. Its avant-garde is its disappearance" (Debord 1967, p. 185).[91]

Art taken as a whole is diverse and incoherent in a society where the TS taken as a whole, is coherent. The diversity of techniques takes precedence

90 On the disenchantment of modern society, see Weber 2002.

91 The page number refers here to the 1992 pocket edition by Gallimard. In other editions, cf. Debord's paragraph numbering 190.

and atomizes the possibilities for meaning. In this context, the diversity of 20th century art is not so much a freedom as it is a limitation and a factor of art's integration in the TS. According to the pessimistic analysis of Ellul (1980), artists are mostly unaware of being integrated in the paradigm of the TS, and this is one of their greatest limits.

Technique is not figurative and the environment it opens does not facilitate figuration and meaning. Art can hardly perform its traditional role of symbolization when confronted to the TS. Unable to symbolize the system, art cannot master/manage it. Ellul gives three reasons why Technique cannot be symbolized: (1) It has become a universal mediation, a realm of means too powerful to be itself mediated by other means. (2) It offers a realm of communion. (3) It institutes a non-mediated, non-distantiated relationship with mankind. To sum it up in one word, Technique is self-symbolizing ("autosymbolisante"): "it produces by itself and for itself the symbols which could still be required" (Ellul 1980, p. 70). Ellul further argues: "I don't understand why conceptual art is called conceptual, I see no concepts in it. In fact what is called non-figurative art is really the figurative art of the technological society" (ibid.).

The overall 'Order of Art' described by Ellul does not just serve social stratification in contemporary western societies: It plays a peculiar *given* role within the logic of the current social order (i.e. of the TS). The role played by the contemporary Order of Art contributes to the establishment of the TS: "the intellectual or artistic superstructure fills the gap left vacant by religion" (ibid., p. 10).[92] Within the TS, art offers a safe haven of indeterminacy: "the uncertainty of intellectual and artistic norms leaves the minds free to look for satisfactions they no longer find in their society". Indeed, "the subjectivity of the superstructure is a response to the frozen objectivity of a scientific infrastructure" (Bernard Charbonneau, quoted in ibid., pp. 10-11).

However, social agents are conscious that Art is "carefully placed in the realm of the unreal, in the domain of things not serious".[93] It is then commonly accepted that, if the artist may well have "the pretension to put society at large into question", it would be foolish to believe in some 'real power' of art to overthrow the questioned social order. Furthermore, the played-out freedom of the artist is little more than the freedom of the actor on a delimited stage.[94] Once off-the stage, this freedom vanishes and the imperatives of efficiency re-appear. With art in the TS, freedom becomes a spectacle by proxy, and the artist a specialist of this freedom-spectacle. Art

92 For Ellul 1980: Translations from the original French text by the author.

93 The obvious issue on the political and ethical level being that Art then remains only an evasion from reality, and offers merely escapism.

94 Cf. Ellul 1980, pp. 227-247.

under the TS thus offers, according to Ellul, only an illusion of revolt, initiative and freedom.[95]

The commonly felt admiration and envy for art and artists is kept at a superficial level. Meanwhile, at a deeper level, the artist is commonly being considered to yield an unrealistic pretension to question 'reality', which relevance is confined to the art worlds.

If art may have replaced religion in a way, it did not sustain the symbolic importance attributed to religion (and religious art) in the past: In a technicized society, people tend to reject art as a symbolic activity. Art's "essentially symbolic character is rejected [...] the technician environment hardly withstands the symbol [...] If the fictional, symbolic character [of art] is maintained, it remains a scandal in the face of the supremacy, the totality, the efficiency of technician productivity: They cast derision upon the aesthetic fiction" (Ellul 1980, pp. 17-18). Rather than having an important symbolic function (as it had in the times when art was linked to the sacred), art is perceived as just a little more than playful, an elaborate form of game. Art is then perceived more and more as superfluous by a majority of people. It has lost its role of giving meaning to life, of acting upon the mysterious and the unknown. And the playfulness of some artists, from Dada to (deconstructive) postmodernist artists, contributes to this general perception.

RO's figure of the genius artist, and its emphasis on individualism and originality, perfectly serve the TS. The prototype of the individual creator is exemplified by Art (with at its extreme the icon of the outcast, bohemian, romantic artist of the late 19th century, e.g. Van Gogh). The RO, exemplified in Art, offers the perfect social setting for the promotion of capitalistic individualism. The creation (of anything new and valuable) is provided by the genius of some individual(s). Belief in individual originality is needed in order to justify the contemporary organization of innovation and the distribution of the economic wealth it creates: Each innovation must be attributed to a single author, so that property rights can be recognized (and the "protection of artists" is heavily assimilated to the extension of copyright laws). Moreover, belief in the intrinsic goodness of this individual originality justifies and contributes to the self-expansion of innovation.

In a society ruled by the TS (where *technical progress* grows by itself and even for its own sake), it is very difficult to accept that some supposedly 'great' progress to come (e.g. some future nanotechnology) might be prevented or slowed down by some interference from other concerns.[96] In Art under the RO, people are afraid that some new Van Gogh might be lost to the world if society were to interfere in the internal affairs of artists.

---

95 Cf. Ellul 1980, pp. 280-282.

96 Some divergent developments are however present, e.g. with the growing importance of the 'Precaution Principle' in recent years (instigated thanks to the ecological movement – inspired by authors such as Ellul, and now introduced in the French Constitution).

Technique needs to organize social activities into specialized fields of production, and to keep these fields relatively isolated from each other. Art has got caught in that process of isolation (being also maintained within the so-called Romantic Order). Furthermore, in their growing isolation from the general public, many art worlds (or smaller circles of recognition) have developed conventions (about style, creativity, etc.) that are hermetic, esoteric, inaccessible.[97] This development participates to the more general development of highly specialized and inaccessible areas of expertise in our technicized society.

But in parallel, the TS also induces the imperatives of 'making-accessible-to-the-masses', standardizing and reifying. In this context, Art under the RO organizes the caricatural perception of cultural production as separated in two genres: First, some 'low-art/junk' that offers a maximum of ease, leisure, conformism, and is easier to integrate in mass-marketing and mass-producing strategies for cultural industries. And then, some 'high/true art' that offers rare and exceptional images of what people actually can't experience in the 'real world' - pointing them at the practical inaccessibility of what art imagines. This is one of the ways in which art plays a role fit for religion in earlier times: The Catholic Church in Europe drew the picture of an egalitarian and just afterlife, thereby justifying and supporting the acceptability of an unequal and often unjust social order.

Visual art and the TS according to Ellul

The following is an overview of Ellul's analysis focusing more specifically on the history of the visual arts in the 20th century, up to 1980.

From the late 19th century onwards, visual art leaves the 'natural' environment and instead mirrors the new 'technician' environment. Impressionist painters no longer consider the human being as a natural given, but in this they do not break up with their environment: The environment is changing, becoming increasingly technical while 'nature' becomes only a secondary environment. What "the artist must take as model is the new mechanical space" (Mondrian 1921).

The visual arts of the 20th century have gone through a process that is expectable in a society strongly under the influence of the Technician System: New techniques quickly imposed themselves as evidently essential to creation, and indeed new techniques were the most able to overcome the limits of earlier conventions. The acceleration of the history of Modern Art has followed the acceleration of the evolution of techniques. Ellul even claims that "it follows exactly the contradictions of the evolution of techniques and adopts their rhythm" (Ellul 1980, p. 43). Specifically, this history was marked by two periods: the machine (from the late

97 One may think for example of most of conceptual visual art and atonal music.

19th century to 1930) and the electricity (from 1930 onwards). In times of the 'machine', characteristics were explosion, class divisions and division of functions and knowledge (in art, Ellul points at cubism). In times of electricity and electronics, characteristics shifted towards implosion, contraction, a global field of consciousness and a reintegration of previously separated functions (in art, he notices the interpenetration of art forms).

Ellul also remarked (back in 1980) that "art now (involuntarily) attempts to find back its symbolic function through abstraction, but it is no longer [an abstraction] of symbols: it is that of the unreachability" of the Technician System.[98] Some other contemporary artists have been pursuing a conceptual formalism of *art for art's sake* (or anti-art for anti-art's sake), being hyper-technical in the sense that they lose sight of ends and of engagement (this process – of losing ends and pursuing means for the sake of means – being typical of the TS).[99] In parallel, much importance was attributed to 'uncertainty'; this interest, Ellul notes, can be linked to random processes in modern electronics.

The Technician System offered artists the power to overcome many traditional conventions about art production. However, this liberation was made possible not in the sake of the originality and revolutionary power of art, but in the sake of a new alienation... to a technicized society. In the TS, relations between things and the structures and processes of these relations have become more important than the things 'themselves'. And indeed, in contemporary visual art, a common view claims that Art is like the metaphor of a structure. Artists and their circles attach then less importance to things in themselves than to the relations between them. However, while Ellul only notices this critically (and seems to retain a Kantian belief in the 'thing in itself'), I do not share his ontology and I will come to appreciate this quality of the arts as sensitive to relations, from the perspective of complexity theories, in chapter 4.

Contemporary visual arts, interpenetrated and globalized, illustrate the domination of a unitarian technician paradigm. If many art forms have been able to interpenetrate each other, notes Ellul, this development should not be credited to a few individual geniuses, but to the new con-

98 If I did follow Ellul all the way, I would come to the conclusion that the Tech–nician System cannot be grasped by art, although it is mirrored by it. This is however not the way I am heading.

99 Disengaged aesthetics, Ellul remarked, also mirror in a relatively realistic way a society which in parallel to segregating its 'art', wishes to become superficially more 'aesthetic' in other applied areas such as design, cosmetics, hygiene, the food, etc. (cf. Ellul 1980 pp. 52-53).

100 Cf. Ellul 1980, p. 222.

text of technicization. "In societies that have attained a comparable technological level, arts tend to a unity of style" (ibid., p. 40). Ellul speaks both of an increasing unity of style, and of an incoherence in contemporary art (reflecting the atomization and unification of social reality under the TS):

"There are a hundred rockets going in all directions with, as their only common element, the use of the most modern technical processes. This art is incoherent [...] it is an ill-matched ensemble of masks veneered on a much more fundamental reality which *is* coherent but prevents art from becoming [coherent]: the technician reality [...] art has become an epiphenomenon of that system [...] It doesn't constitute itself. It responds to certain uses of the system" (ibid., pp. 59-60).

Abstract and conceptual art of the late 20th century often claimed to be revolutionary, calling forward romantic values, notes Ellul. Claiming to contest *everything* in contemporary society, it hardly did tackle anything. Ellul, criticizing artists of the 60's and 70's (including 'land art'), writes: "They want to be radical through refusing the town, the technique, the product, the consumption, the artificial. One proceeds to a romantic return to wandering and so-called 'natural' nature" (Ellul 1980, p. 41). In my own analysis, in chapter 5, I will however come to contradict Ellul to some degree, showing that at least some of these artists of the 1960's and 1970's, did not merely propagate a romantic view of nature.

Even more worrisome is the process in which some artists are claiming that they are revolutionary just because they are using some new techniques. They take up these brand new techniques and say of them 'this is art', waving the Duchamp flag, and then employ propagandistic techniques to impose them as valuable art.[100] This type of attitude is dangerous in the context of the TS, because these artists then contribute to the widespread belief that new techniques are revolutionary in themselves. A confusion is made between technical innovation and artistic creation, and furthermore a confusion is made between a technical innovation and a clear improvement of one's situation. An aspect of this development lies in the belief that more technical stuff means more possibilities for imagination.

However, Ellul also recognized that artists can sometimes enact a protest against, and critically engage their condition within a technicized society. According to him, the multiplicity of attempts, through the arts, to interpret society in the 20th century, reveals the non-toleration of the rigid and Unitarian structure of society under the TS. In some cases, art seems to directly express and expose the traumas suffered in a technicized society. For example, for Ellul "dissonances in music are telling about our own condition in this environment. With Schoenberg are recorded through music the undisguised movements of the unconscious, at grips with this new unbearable environment. We see and hear the traumas inflicted upon us by the Technician System" (Ellul 1980, p. 76).[101]

Nevertheless, Ellul's view on contemporary art is deeply negative, and leaving little room for art as a socially transformative force. For him, as long as the TS will be strongly in place, "the aesthetic paraphernalia will be left to being only a décor and will be perfectly moveable" (ibid., p. 32). For Ellul, the art of our times is doomed to being unable to effectively fight the Technician System: It can deliver either nonsense or propaganda.[102] Even critical art is merely able to preach the converted. "It is a tautological discourse, where the speaker speaks to himself about what he already knows" (ibid., p. 49). Furthermore, critical discourses in modern art are mostly fighting the ghosts of 19th century Bourgeoisie, and perpetuating what Ellul characterized as the "political illusion", with political contestation failing to address the wider issue of the growth of the TS.[103] "Today artists throw themselves at realities that no longer exist. They fight against a bourgeois-mindedness that has not existed since the nineteenth century and not at all against the triumphant utilitarianism and pragmatism that is omnipresent today."[104]

The development of the RO/TS combination in the arts and generally in late modern societies, even adds to the overall efficiency of the TS in recent decades. As described in details by Boltanski & Chiapello (1999), not only does the "artist critique" of capitalism miss its target, but the artistic (RO-determined) model of intrinsically motivated, flexible, project-oriented, creative work, contributed key elements to a (TS-enhancing) "new spirit of capitalism" marked by creative, 'unconventional' entrepreneurs, flatter hierarchies and flexible workers. In this way, for efficiency's sake, work conditions have become increasingly precarious, with the artists as a good example of this (partly self-inflicted) precariousness.[105] A further critical analysis of the consequences of this "flexible capitalism" and its "flexible social relationships" is conducted by Richard Sennett (1998, 2005).

Back to the Dialectic of Enlightenment?

In the *Dialectic of Enlightenment*, Horkheimer and Adorno did, to some extent, describe some features of art under the RO/TS. Discussing the

101 Not all followers of Schoenberg may agree with Ellul's interpretation.

102 On information as propaganda in a technicized society, cf. Ellul 1962.

103 Cf. Ellul 1980, pp. 110-111, 122.

104 Ellul, interviewed by Patrick Chastenet [website text retrieved in 2006, not available anymore].

105 Cf. Abbing 2002, Menger 2002.

parallels between art and science, they argued:

"The prevailing antithesis between art and science, which rends the two apart as areas of culture in order to make them jointly manageable as areas of culture, finally causes them, through their internal tendencies as exact opposites, to converge. Science, in its neopositivist interpretation, becomes aestheticism, a system of isolated signs devoid of any intention transcending the system; it becomes the game which mathematicians have long since proudly declared their activity to be. Meanwhile, art as integral replication has pledged itself to positivist science, even in its specific techniques. It becomes indeed, the world over again, an ideological doubling, a compliant reproduction" (Horkheimer and Adorno 2002, p. 13).

But this realization fails to bring sufficient suspicion against "autonomous art", in Adorno's discourse (and later in late Marcuse's plea for an autonomous aesthetic "subversive imagination" - as mentioned below). In the Dialectic, Horkheimer and Adorno note that "art has in common with magic the postulation of a special, self-contained sphere removed from the context of profane existence. Within it special laws prevail" (ibid., pp. 13-14). However, they fail to see that, in the context of the TS, the supposed autonomy and meaningfulness of this "special sphere" is defused and trivialized. Autonomous art offers a fake and autistic replacement of magic. I will argue later (in chapter 4 with David Abram) that in many non-Western cultures, magic is not a self-contained sphere that works only autopoïetically, but an eco-poïetic dialogue with a more-than-human environment thanks to an animistic phenomenology.

Horkheimer and Adorno argue that autonomous art is the only social activity that can legitimately offer something else than a material for 'progress'. But, they point out, this is "tolerated" only "[a]s long as art does not insist on being treated as knowledge" and rather keeps to its quarters (ibid., p. 25). But then, what can autonomous art really bring, in terms of social transformation? Not much, as Horkheimer and Adorno themselves have to acknowledge: Such an autonomous art is "neutralized as a mere object of contemplation, as art" (ibid., p. 27).

However, as already mentioned above, I will not stop my analysis at the stage of Ellul's extreme scepticism, but consider Ellul's view as a challenging perspective raising questions on the unsustainability of a large part of contemporary art.

Furthermore, Ellul's analysis is also relevant and even inspiring in its appeal to a transformation in the values of art: What Ellul calls forward (but which he failed to find in the art of his time), is for art to renew its ties with ethics and meaningfulness, with the Good, the Beautiful, the Humane.[106] As I will discuss in chapter 5, such values have also been advocated by several key figures in the movement of ecological art (such as Suzi Gablik). Ellul also insists that the critical value of such art would not just consist in the illusion of merely political contestation within the TS, but should "animate our courage, so that man confronts the enslavement of progress and of consumer-happiness" (Ellul 1980, p. 285).

This call is very different from Herbert Marcuse's (and after Marcuse, Carol Becker's) claim that "the political potential of art lies only in its own aesthetic dimension. [...] The more immediately political the work of art, the more it reduces the power of estrangement and the radical, transcendent goals of change" (Becker 1994, p. 120). Such a claim, if it points at the fragility of the valuable processes of estrangement and detachment in the aesthetic experience,[107] however responds to this risk with a fearful insulation from reality which perpetuates the culture of unsustainability of strictly autopoïetic, autonomous art worlds. A full autonomy of the art world indeed pulls artists into an autistic dynamic of autopoïesis (cf. also the discussion of Luhmann below) that leads to a plea for irresponsibility and apolitical, asocial practice, as a statement from the painter Georg Baselitz perfectly illustrates: "The artist is not responsible to anyone. His social role is asocial; his only responsibility consists in an attitude, an attitude to the work he does... There is no communication with any public whatsoever. The artist can ask no questions, and he makes no statement; he offers no information, message or opinion... It is the end-product which counts, in my case, the picture" (Baselitz, quoted in Gablik 1984, p. 119 and Gablik 1991, p. 61). This quote typically illustrates how an artist, who internalized the principles of a fully autonomous art world, thereby alienated himself from society and from ethical, political or social commitments; such an 'attitude' fails to instigate any significant form of "subversive imagination" as Marcuse would have hoped.

The conception that art is for art's sake only (i.e. belief in the necessary uselessness of art in terms of serving a social function), well maintained in the context of the RO/TS, is still relatively strong in a large part of the art worlds, so that art works mainly self-referentially, 'autopoïetically' (cf. the discussion of Luhmann below), involving a widespread a priori conventional suspicion on potentially transformative action. Late Marcuse's defence of the 'aesthetic dimension' contributed to this belief, with the perverse effects that discourses of autonomy in the arts have brought (as the quote from Baselitz illustrates).

A similar conclusion is drawn by Bourdieu and Wacquant (1992) about the fallacy of autonomous art: "As soon as they want to fulfil a function other than that assigned to them by the artistic field, i.e., the function which

106 Cf. Ellul 1980, p. 273.

107 The political dimension in art will be discussed in chapter 7.

consists in exercising no social function ("art for art's sake"), they rediscover the limits of their autonomy" (Bourdieu and Wacquant 1992, p. 110). And related to this autonomy is the belief in authenticity, which would be lost in the process of interacting with outsiders and of achieving a social function.

The significance of autonomy in modern art as a social system, was further analyzed by Niklas Luhmann, along the theoretical framework of his system theory.

The autopoïesis of the art system according to Luhmann

Within the framework of his system theory, Niklas Luhmann analyzed the specific autopoïesis of the social system of art in modern society. His analysis sheds light on the systemic organization that allows modern and contemporary art to function as it does, developing a certain autopoïetic autism, but also some qualities of reflexivity and emergence (qualities which will be relevant to the discussion of cultures of sustainability in the following chapters).

Luhmann analyzes art under modernity as the emergence of a specific form of language, whereby works of art do not simply carry information by themselves, but also by their relationships to other art works.[108] "A work of art without other works is as impossible as an isolated communication without further communications" (Luhmann 2000, p. 53).This is a sort of language "without argumentation. Instead of using words and grammatical rules, people employ works of art to communicate information in ways that can be understood. [What's more,] the work of art is produced *exclusively* for the purpose of communication [...] Art communicates by *using perceptions contrary to their primary purpose*" (ibid., p. 22). Luhmann illustrates this with the practice of poetry: "A poem's 'message' does not allow for paraphrase, nor can it be summarized in a proposition that can be true or false. Rather, connotations, not denotations, mediate its meaning [and it] aims at disrupting automatization and delaying understanding" (ibid., p. 25).

Any communicating object in the art system is an art object, including the previously ordinary objects (such as Duchamp's "Fountain") which are defining themselves as art within the art communication even if pretending to define themselves as non-art (ordinary objects would not need to define themselves as ordinary things).[109] "This is true even when autonomy is practiced as a renunciation of autonomy – when one seeks to reconcile art and life or attempts to commercialize art to the point where it no longer claims any specifically artistic form, where its artistic quality is reduced to the fact that it *wants* to abandon itself and this is how it articulates itself as art" (ibid., p. 294). The only way to be outside the art system is to articulate the

108 Cf. Luhmann 2000, pp. 19-53.

109 Cf. ibid., pp. 145, 293.

communication according to an altogether different code than art/non-art. The codes of the other social systems have a 'rejection value' for the distinction of the art system from them.

Art makes its own distinctions and sets is own constraints; "a work qualifies as art only when it employs constraints for the sake of increasing the work's freedom in disposing over further constraints" (ibid., p. 35). Furthermore, as an emergent, indeterminate process, the production of art is not directly controlled by the psychic system of the artist.

"An artist can control his production only through observation; he must, so to speak, let the emerging work show him what has been done and what can be done further [...] The making of an artwork cannot be understood – or can be understood only in a manner that remains insufficiently formalized – as a means to an external end that is apparent from the start. It escapes planning and programming" (ibid., p. 38).

The desires and intentions of the author/artist (and of other single participants) do not directly matter. At the risk of being redundant, I remind the reader of Luhmann's general system-theoretical assumptions: In communication through art as in other forms in communication in the Luhmannian theory, people (as psychic systems) do not communicate directly with each other, and

"for communication to come about, it is irrelevant whether or not systems of consciousness are capable of figuring each other out. Communication occurs whenever the utterance of an information is understood – which may result in acceptance or rejection, consensus or dissent. Moreover, communication through is not concerned with automating understanding. Rather, it is inherently ambiguous (semiologists speak of polysémie) independently of whether or not the divergence of observational possibilities was planned in the sense of an 'open work'" (ibid., p. 40).

How the structural coupling occurs between the artist and other participants, through art, is neither totally arbitrary, nor pre-determined, but characterized by "nonlinear structures of coupling" (ibid., p. 44). What this also means (though not phrased by Luhmann exactly in that way) is that the art system, a communication system, allows emergence both at its own level and at the level of reflexive individual psychic systems thanks to structural couplings.[110]

Understanding art as "a special kind of communication" also separates it from aesthetics, Luhmann comments, because "aesthetics was founded upon another distinction, one more clearly related to the idea of the subject: the

110 The notion of emergence, as well as the significance of different levels of emer gence which are interrelated to each other, from a system/complexity-theoretical perspective, will be introduced with Edgar Morin, in chapter 3, section 1.

distinction between aistheta and noeta, sensuous and rational cognition, or aesthetics and logic. Cognition (not communication) served as the master concept" (ibid., p. 15). Luhmann, however, clearly distinguishes art as communication, from aesthetics as perception, in *Art as a Social System*, and focuses on the former.[111]

However, Luhmann also points out the interesting specificity of the structural coupling between the social system of art and the psychic system of consciousness: "Art is [...] capable of intensifying the awareness of communication: consciousness becomes aware of being directed and captivated by communication, experiencing the discrepancy between an external control and its own, unrestricted operative possibilities. The self-awareness induced by art is always the experience of a difference" (ibid., p. 21). Furthermore, art allows a relative freedom from the restraints of verbal communication: Art is "integrating what is in principle incommunicable – namely perception – into the communication network of society. Kant already located the function of art [...] in its capacity to stimulate thinking in ways that exceed verbal or conceptual comprehension" (ibid., p. 141).

Art did not always exist as a specific autopoïetic social system, Luhmann notes. In the Middle Ages, art "engaged all the senses [...] Accordingly, the concept of art (ars) was far more comprehensive than it is today, and it had to bridge fewer internal differentiations [between human senses and between different spheres of social reality]. Once art differentiates itself along the lines of a systems-specific play with forms, the situation changes" (ibid., p. 17). Modern art distantiates itself from its former attachment to perceptions, and then rather

> "renders accessible what is invisible without it. In the wake of this transformation, the social relationship between the artist and his audience becomes more problematic, provoking debates on the social status of an expert culture of connoisseurs and art critics in the eighteenth century and eventually leading not only to the realization that conversing *about* art is different from conversing about other objects, but also to the possibility of communicating *through* art" (ibid.).

From the 18th century onwards, "art and the beautiful were now declared purposeful without purpose" (ibid., p. 23).

Consequently, modern art is self-referent (which is not the same as self-isolating, i.e. 'art for art's sake', Luhmann argues)[112]. "Works of art serve no external purpose; having such a purpose is precisely what disqualifies a work as art. [...] The work displays, one might say, itself and its own self-description" (ibid., pp. 44-45).

As was to be expected given his overall theoretical framework, Luhmann takes as a given fact, the fragmentation and mutual non-(direct-)

111 I will come back to aesthetics in chapter 4.

112 Cf. Luhmann 2000, p. 149.

communicability between social systems, and thus also art's incapacity to communicate (directly) with other social systems. In his own words:

"The issue is not to propose a theory that, if properly understood and applied, would [...] assist the art system in coping with its worries about the future. It follows from the general theory of functional social differentiation that functional systems are incapable of directly influencing one another. [But Luhmann then adds:] At the same time, their coexistence increases their mutual irritability" (ibid., p. 3).

The last point leaves one narrow way open, for inter-systemic "decentred contextual steering" (as already discussed above). However, once again as in his general theory, Luhmann leaves no space for a genuine inter-systemic communication, whether between psychic systems and the art system as mentioned in the following quote, or whether art and other social systems. "Boundaries [...] cannot be observed in their capacity as structural couplings, since neither the perceiving consciousness nor communication can explode their operative closure and reach beyond their own systems into the environment" (ibid., p. 46). More generally, Luhmann's analysis is not focusing on the mutual influences between the art system and the rest of society: "We are not primarily concerned with problems of [...] society's influence on art and of art on society. (Such issues are of secondary importance)" (ibid., p. 134).

To summarize Luhmann's insights on art in modern society as a social system:

"Art makes perception available for communication, and it does so outside the standardized forms of a language [...] Art cannot overcome the separation between psychic & social systems. Both types of systems remain operatively inaccessible to each other. *And this accounts for the significance of art* [...] disparate systems operate simultaneously (are synchronized) and constrain each other's freedom" (ibid., p. 48).

Luhmann's analysis, though it leaves some narrow way open for an intersystemic reality and shows certain qualities in structural coupling, remains very limited in its exploration of these possibilities. However, it offers a relevant framework from which to consider the critiques against modern art from certain members of the art system who denounce its overall unsustainability. An especially outspoken and clear example of this is the art historian Suzi Gablik.

A critique of modern and postmodern art

Art historian Suzi Gablik has been a staunch critique of Modern Art from a perspective that is especially informed by ecological thinking, since her publication of *Has Modernism Failed* (1984). My discussion of her arguments will focus more specifically on her 1991 publication, *The Reen-*

chantment of Art, where she associates to her critical discourse, the search for alternative practices and discourses in the arts (which I will then discuss in due time, in chapter 5, section 4).

Gablik's critique of Modernism in art is based on a general critique of Western modern society, which echoes several of the analyses already introduced in the first section of this chapter.

She criticizes modernity for the dominance of a "mechanistic vision [i.e.] the cheerless, clockwork way we have been conditioned to see the world through the myth of science, as a 'single-tracked' universe". She points at the heritage of Enlightenment as "the hegemony of a technological and materialist world view" (Gablik 1991, p. 46). "The individual of today transfers the engineering ethos of modern technology and bureaucracy to his personal consciousness and emotional life. This ethos, characterized by mechanicalness, reproducibility and measurability, produces in consciousness the traits of abstraction, functional rationality and instrumentality" (Peter Halley, quoted in ibid., pp. 45-46).

Her critique points especially at "Cartesianism – the paradigm of the bipolar subject and object" (Gablik 1991, p. 164), and at "patriarchalism". Both points have already been introduced above, in the first section of this chapter. Gablik especially points out the distancing effects of objectivity as a nefarious development. "Objectivity serves as a distancing device, offering the illusion of impregnable strength, certainty and control" (ibid., p. 178). Ultimately, the Cartesian ego turned into a radical individualism associated to the social-economic organization of the market. "Many people find it difficult to imagine a self that is not shaped by the concept of an isolated individual fending for herself in the marketplace [and] it is taken for granted that economic self-interest is a basic given of human nature" (ibid., p. 176). Gablik also warns against the loss of the sense of magic, in modern society.

> "One of the peculiar developments in our Western world is that we are losing our sense of the divine side of life, of the power of imagination, myth, dream and vision. The particular structure of modern consciousness, centered in a rationalizing, abstracting and controlling ego, determines the world we live in and how we perceive and understand it; without the magical sense of perception, we do not live in a magical world. We no longer have the ability to shift mindsets and thus to perceive other realities" (ibid., p. 42).

To recall Luhmann's insights, the ability to communicate the incommunicable and perceive "potentialities" specialized itself into the art system within modern society, but the art system however does not directly communicate such insights with the other social systems.

The critique of Modernism in art developed by Suzi Gablik, targets especially the aesthetics of detachment and the individualistic and antagonistic autonomy developed in the arts.

"Modernism above all exalted the complete autonomy of art, and the gesture of severing bonds with society. This sovereign specialness and apartness was symbolized by the romantic exile of the artist, and was lived out in modes of rebellion, withdrawal and antagonism. Talk about harmony, or fitting in, was anathema to the alienated self" (ibid., p. 5). Gablik also criticizes the critical tradition in art under modernism for being "a contest of wills" perpetuating "a mode of patriarchal consciousness" and denigrating non-confrontational approaches (ibid., p. 60). One of the first historical avant-gardes, Italian Futurism, was typically the most extreme in its celebration of an aggressive hostility against society (and against one's own audience), but this attitude was echoed "in much art of the modern era" across the 20th century.[113]

The autonomy of art is also related to art's self-referentiality. This aspect was already discussed above with Luhmann. Suzi Gablik also points at this dimension, quoting Kurt Schwitters' claim that "[t]he picture is a self-contained work of art. It refers to nothing outside of itself. Nor can a consistent work of art refer to anything outside of itself without losing its ties to art" (Schwitters quoted in ibid., p. 124).

Gablik sees in modern art an enactment of the extreme form of individualism promoted in contemporary Western societies. "Individualism, freedom and self-expression are the great modernist buzz words. To highly individualistic artists, trained to think in this way, the idea that creative activity might be directed toward answering a collective cultural need rather than a personal desire for self-expression is likely to appear irrelevant, or even presumptuous" (Gablik 1991, p. 6) Furthermore, the idea that creativity may work as a collective process rather than an individual one, is difficult to conceive for many contemporary creative professionals.

> "Most of us, in the capitalist world, have never had an experience of true community. We live so much in an ethos of professionalism, which keeps us bound to individualistic modes of thought and directed toward the making of products, that it is difficult not to marginalize, or subtly discount, achievements that manifest less ego-control, and point to the value of cocreativity" (ibid., p. 114).

Gablik denounces "the rational framework of modern aesthetics" for its "ontology of objectification, permanence and egocentricity, which has seriously undermined art's [in her view] inherent capacity to be communicative and compassionately responsive, or to be seen also as a process, rather than exclusively as fixed forms" (ibid., p. 60). She associates this notion of modern aesthetics to art's autonomy in her critique.

---

113 Cf. Gablik 1991, p. 61. Several (often Italian) researchers have since tried to re habilitate the Italian Futurists. For a clear impression of the extreme apology of violence and of technology by the leader of the Futurists, I advise the reader to refer directly to Filippo Marinetti's *Futurist Manifesto* (Marinetti 1909).

"Art as a closed and isolated system requiring nothing but itself to be itself derives from the objectifying metaphysics of science – the same dualistic model of subject-object cognition that became the prototype for Cartesian thinking in all other disciplines as well. This ideal of static autonomy, of the self against the world [...] disregards relationships; it connotes a radical independence from others [...] set apart from reciprocal or participative interactions" (ibid., p. 62).[114]

Her line of argumentation calls forward a move away from this modernist "aesthetic mode" (and towards an aesthetics of "relationships", a question which I will treat directly in chapter 4): "art [...] has become trapped within a rigid model of insular individuality. To reverse this priority, giving primacy to relationships and interaction, is also to reverse the way that artists see their role, and implies a radical deconstruction of the aesthetic mode itself" (ibid.).

The modern aesthetics also involves "a modality where our gaze is that of the detached observer", Gablik argues after Ortega y Gasset (in his 1925 essay "The Dehumanization of Art"). "The artist is *supposed* to be emotionally distant from the event he is portraying [...] Art evokes aesthetic, not real, emotions. [...] Within traditional aesthetics, it is only the image that counts. Lived reality is repressed by the disembodied eye and transformed into spectacle" (ibid., p. 99). Gablik traces back the "illusions that have generated the impression of separation and detachment" not only in Kant's aesthetics, but earlier on, in the heritage of Descartes.[115] One of the dramatic consequences of aesthetics of detachment is an "impaired feeling function, the passive nonintervention of spectacular modes of consciousness" (ibid., p. 113), against which she calls for "restorative action" reviving empathy. Under this perspective, she quotes Aimé Césaire's warning: "Beware, my body and soul [...] Beware above all of crossing your arms / and assuming the sterile attitude of the spectator / because life is not a spectacle" (Césaire quoted in ibid., p. 113).

As a result of all this autonomy, individualism, detachment and antagonism vs. society, the modern artist is most often unable to construct "a symbiotic or complementary relationship with society [...] In modern society, artists see themselves as quintessential free agents, pursuing their own ends. Our cultural myths support economic advancement and the hard-edged individualist writ large, rather than service, caring attitudes and participation" (Gablik 1991, p. 116).

An example of the typically modernist discourse of artistic autonomy, Gablik argues, can be seen for example in a quote of Georg Baselitz's (which I already quoted above).

---

114 As analyzed by Luhmann and already discussed above, the emergence of self-reference and autopoïesis can be seen as a wider phenomenon affecting the whole of modern society.

115 Cf. Gablik 1991, p. 98.

An example of the practice of the confrontational individualism she denounces, can be found in Richard Serra's *Tilted Arc*. The work is a monumental steel sculpture installed in 1981 at Federal Plaza in downtown Manhattan, which directly confronts its audience in a impressive way, blocking sight and movement across the plaza as a massive wall. Gablik quotes art critic Barbara Rose for praising Serra's sculptures' "heroic rectitude" and "self-sufficien[cy]" (Gablik 1991, p. 63). The sculpture ran into a famous controversy with years of litigations in the courts and finally the work's dismantlement in 1989. Serra bitterly fought against his critics and opponents, and accused the US Senate, the government, the New York Time and the whole US judicial system of "censoring" and "destroying" art.[116] The opponents to the work were from the start, especially people living and working around the plaza, who couldn't stand the permanent presence of a sculpture that they perceived as an aggression. Interestingly, Serra defended his individual rights as artist, but did not really consider the complaints against the daily consequences of his 'creative will' upon others. As Gablik notes, Serra did not perceive the site "as a socially responsive field" (although he claimed to be interested in the response of "the viewer" to the work) and she asks "whether art that is based on notions of pure freedom and radical autonomy – without regard for the relations we have to other people, the community, or *any* other consideration except the pursuit of art – can contribute to the sense of the common good" (ibid., p. 66). She notices how readily artists, but also many others, think of their "freedom and rights" without "thinking of responsibilities and obligations".

Modernity also imposed certain specific frames on what would come to be understood as artistic activities, constraining them into specific social institutions. Gablik's analysis also points at this dimension. An illustration of the constraining order imposed by Modernity on art, is taken in the field of music: Following the composer and writer R. Murray Schafer, she argues that:

"In the modern world, the man-made environment has been pervasively organized around the linearity of goal-oriented, object-oriented activities, and for centuries our art has been produced to conform to these kinds of controlled spaces, which have been carefully neutralized so that nothing unwanted can intrude [...] music too has lost its participatory and ecstatic aspect in the modern era [...] In the Euclidean space of the concert hall, everything is willfully arranged for maximum comfort and high-fidelity sounds, and a costly apparatus [...] ensures that working conditions are infallibly maintained" (ibid., p. 85).[117]

116 For more details on the *Tilted Arc* controversy and legal procedures, see http://www.cfa.arizona.edu/are476/files/tilted_arc.htm

117 Schafer's outdoor alternative to the concert hall, and his concern for issues of sound pollution in public space, will be introduced in chapter 5, section 4. Schafer and Gablik's concerns about music losing its participatory and ecstatic

Concerning the contemporary visual arts, Gablik refers to Brian O'Doherty's essays on the "white cube" exhibition space as "a white, clean, artificial space that makes art exist in a kind of eternity of display, isolated from everything that would detract from its own evaluation of itself as art. The outside world must not come in" (ibid., p. 149) For O'Doherty, "[g]enuine alternatives cannot come from this space" (O'Doherty quoted in ibid., p. 149). For Allan Kaprow, the gallery-framed "[a]rtlike art sends its message on a one-way street, from the artist to us [...] You can't 'talk back' to, and thus change, an artlike artwork" (Kaprow quoted in Gablik 1991, p. 150). Following again Ortega y Gasset, Gablik rejects white-boxed art as monologic, i.e. "rejecting dialogue and interaction", or in the words of Ortega y Gasset: "With the objects of modern pictures no intercourse is possible [...] This new way of life which presupposes the annulment of spontaneous life is precisely what we call understanding and enjoyment of art" (Ortega y Gasset quoted in Gablik 1991, p. 150).

The insights of Institutional Critique

The question of the constraints and issues relating to institutions in art, has been a subject of self-reflexive, critical insights, for an artistic movement known as "Institutional Critique" (IC). This movement emerged in the 1990's, and retrospectively identified its roots in a "1st generation" of institutional critique from the 1970's (with the work of artists like Robert Smithson, Daniel Buren, Hans Haacke and Marcel Broodthaers, who critically addressed the institution of the art museum). The proponents of IC thus saw themselves as a "2nd generation IC". Their work included a (quasi-)sociological analysis of the roles of artists as part of the social institutions of contemporary art.[118]

To prevent any misunderstanding: Even the so-called "1st generation

qualities, should however not be applied indiscriminately to the whole of live performing music: It is a problematic condition which, according to art sociologist Hans Abbing, affects the "high art" music world much more than live popular music concerts and what he announces as a "new art" style of performing music (cf. Abbing 2006, 2009, Abbing and Kagan 2006). In response to my words of caution, Schafer would probably respond that the levels of participatory and ecstatic experience in a pop music concert are not as strong as in the case of "the savage drummer or dancer" in some non-Western cultures (cf. Schafer quoted in Gablik 1991, p. 86). One potential counter-example may be the techno rave parties where participants do enter into a sort of trance.

118 Over several years in the mid-1990's, the Kunstraum of the University of Lüneburg was a hot spot of the activities of several 2nd generation IC artists, including Andrea Fraser, and of like-minded theoreticians: Cf. eds. von Bismarck, Stoller and Wuggenig 1996.

IC" was not the first art movement to critically address the institutions of art and its 'autonomy'. The historical avant-gardes of the 20th century (Futurists, Constructivists, Dadaists, the Bauhaus, Surrealists, and later Situationists and conceptual art) already developed radical self-critiques of modern art.

One of the key figures in the 2nd generation IC, Andrea Fraser, states that this art movement is "recognizing the partial and ideological character of artistic autonomy" (Fraser in ed. Welchman 2006, p. 307). She further articulates the awareness of "the inescapability of institutional determination" as a basic condition on which to work reflexively, taking "responsibility for, or action against, the everyday complicities [...] which are driven by our own interests in the field" (ibid., pp. 131-133).

However, if Fraser is right when arguing that "Institutional Critique has the structure of melancholia" (ibid., p. 307) and if such a condition is structural, preventing the development of an institutional agency for social transformations in and beyond art, then this movement shares the fate of other artistic movements that kept playing the strictly autopoïetic game. This risk is especially relevant because IC typically limits its analysis, and its range of intervention, to the art system.[119] A similar critique has been exercised in recent years against the "2nd generation" of IC by authors claiming to identify a "3rd generation" of this movement.[120]

Furthermore, Gablik echoes critiques against commodified art and public art that can be well-integrated into markets and/or with power displays of donors.[121] For example, she quotes the painter Robert Janz: "The subtext of important art today, saleable art today, is power and prestige. That is the real subtext in a Richard Serra sculpture or a Frank Stella painting: It's Brand Art for a Brand Culture" (Robert Janz quoted in Gablik 1991, p. 91).[122] That

119 See for example Simon Scheikh (2006) who considers IC as obsolete.

120 See http://transform.eipcp.net/correspondence/1173106141#redir; some of the artists identified as taking part in this "3rd generation IC" (e.g. Eduardo Kac, the Critical Art Ensemble, Brandon Ballengée) will be discussed in chapters 5 and 6, but I will not follow this IC-generational categorization of artists, prefer ring to analyze their work from a different perspective.

121 For an introduction to how modern and contemporary art serves displays of power by private donors but also by the State through subsidization, see Abbing 2002, pp. 181-202, 232-258.

122 A critique of "Brand Culture" can also be found, concerning contemporary popular culture, in the discourses of "Culture Jamming" and of "Tactical Me dia", two movements which are operating in between the art worlds, design,

point of critique applies as well to postmodernist discourses and practices in art, which are also a target of Gablik's.

She acknowledges postmodern deconstruction for dismantling the beliefs in the value of art's autonomy: "We now know, thanks to deconstruction, that a work of art is *never* pure, never self-contained, never autonomous. Indirectly, a belief system is being reinforced. Underlying the ethos of radical autonomy, for instance, is the metaphor of the market" (Gablik 1991, p. 148).

However, she reports that according to advocates of post-modernism, such as Peter Halley, "[a]rt may reveal the problematic nature of [late capitalism] by mirroring it, or by transforming it into a hollow parody of itself, but it cannot change anything" (ibid., p. 15). She further quotes philosopher Jean-Paul Lyotard's call to "harden, worsen accelerate decadence" and to "[a]dopt the perspective of active nihilism, exceed the mere recognition [...] of the deconstruction of all values" (Lyotard, quoted in ibid., p. 16). Gablik comments Lyotard's 'incredulity toward metanarratives' as the presentation of life as "an endless accumulation of meaningless spectacles, originating in the loss of any unifying narrative of the world" (Gablik 1991, p. 31).

In consequence, Gablik describes "deconstructive postmodernism" in art as a strategy that "does not ward off the truth [of "a now dysfunctional but apparently immovable dominant social structure"] but tries to come to terms with its inevitability, in what are often ironic or parodic modes that do not criticize but simply declare art's pointlessness openly, and bait us with its indifference" (ibid., p. 19). No critical detachment is claimed by such art discourses, as "postmodern parody does not claim to speak from a position outside the parodied. For these artists [e.g. Jeff Koons], it is important that their works function in total complicity with the context they are confronting, and become indistinguishable from it" (ibid., p. 39).

Gablik denounces such a strategy as perpetuating and reinforcing these very social structures, for the simple reason that "a belief that resistance to the dominant social structure is futile, because the structure is too ruthless or too powerful, will have the effect, if accepted by enough people, of stabilizing the relations of dominance" (ibid., p. 23). She quotes Sylvère Lotringer about Baudrillard's de(con)structive postmodernism: "there is a high price to pay in terms of emptiness and disenchantment" (Lotringer quoted in Gablik 1991, p. 40).

Gablik also points at the postmodernist strategies of artists like Jeff Koons and Haim Steinbach as reinforcing the "ideology of consumerism [whereby] freedom in our society has come to mean primarily the consumer's right to choose" (Gablik 1991, p. 39).

In opposition to modernism and to deconstructive postmodernism, Gablik advocates a reconstructive postmodernism and a connective aesthetics

media/journalism and anti-consumerist/left-wing radical activism (cf. Lasn 1999, Comiotto, Kluitenberg, Garcia and Grootveld 2003).

and practice in the arts.[123] These alternatives will be explored in chapters 4 (on aesthetics) and 5 (on ecological art). But before coming to that point, I will elaborate a definition of cultures of sustainability, in chapters 2 and 3.

CONCLUSION

From the analyses introduced in this chapter, it appears that the culture and the arts of unsustainability, are strong and pervasive, and that the challenge of transforming them into alternative states, is not a light enterprise either, as a whole ensemble of structures of modernity is at stake, involving very deeply rooted building-blocks of our culture.

As Ervin Laszlo argues, the development of human cultures is not an inconsequential game, but a social construction that matters in the long run, for the very survival of human civilizations, and maybe even in the current, globalized context, for the survival of the whole human species:

"Our knowledge has made us increasingly autonomous in nature, and enabled us to create the world of culture. It has freed us from many of the bonds of biological existence and given us licence to determine our own evolution. But the possibility of error is the price we pay for freedom. The cultures we can build for ourselves may be manifold, but they must remain compatible with the structured holarchy of nature. We can build culture beyond these limits only at our immediate peril. Any such error must be rectified by using the same capabilities which originally led to the error: our relative autonomy conferred by our reflective consciousness. Here is where the holistic vision of the systems sciences becomes important" (Laszlo 1996, p. 92).

Although I will not follow Laszlo in his "holistic vision", I will however follow him in focusing my attention on the insights brought by "systems science".[124]

The aporia of an opposition between claims to artistic autonomy that denounce usefulness on the one hand, and claims to immediate engagement and relevance that altogether would oppose artistic explorations, is not the direction in which I am heading in the following chapters. With the RO/TS model, I argued that a complementary complicity exists between the two poles of a romantic order with its oppositional-individual-creative figures on the one hand, and of a technological system imposing formal and purposive rationality on the other hand, which together are maintaining a disjunctive mode of knowing and strictly autopoïetic social systems. Moving beyond the RO/TS, requires to move beyond its binary opposition of meaningful uselessness vs. meaningless usefulness. The term of 'emergence' allows to

123 The distinction of ‚reconstructive vs. deconstructive postmodernism' was not invented by Gablik, but is also to be found e.g. in Charlene Spretnak's writings.

124 My critique of Laszlo's holism is to be found at the end of chapter 2.

think beyond this dichotomy. However, to understand emergence, and the process of autoecopoïesis that is related to it, requires a careful introduction of a different mode of knowing, a different worldview, method, and altogether a different epistemology, as developed in 'systems thinking' and in theories on 'complexity'. This task will require the next two chapters with a long theoretical elaboration, which will then serve as a basis to consider the possibility of "aesthetics of sustainability" (chapter 4) and to interpret the work of ecological artists in recent decades (chapters 5 and 6).

2. Toward Culture(s) of Sustainability, Step One: Systems Thinking and the Limits of Holism

INTRODUCTION

Not only the ecosystems, but also human societies, i.e. social justice, democracy, cultural vitality and the welfare of citizens, are put at risk by the continuing trend of an unsustainable development: The challenge of achieving sustainability, in the face of a complex crisis of civilization combining ecological, social, cultural and economic dimensions, demands integrated understandings and responses. One quote attributed to Einstein has passed into common knowledge and is often invoked in the face of the global ecological crisis: "We can't solve problems by using the same kind of thinking we used when we created them." An alternative version of this attributed quote states: "The significant problems we have cannot be solved at the same level of thinking with which we created them." Whether the author of any of the two quotes was Einstein matters little here. What counts is that the quote, in its various forms, has apparently gained popularity and reveals the intuition by many social agents, that the unsustainability of our current development model has to do with the way our thinking is organized. The depth of its message, especially in the second attributed quote, should compel its readers to a very careful attention and a serious rethinking of some of the fundamental epistemological bases of modernity.

The main thesis developed in this chapter will be that sustainability involves a paradigmatic shift in worldviews, away from a great part of modernity i.e. what I described as a 'culture of unsustainability' (cf. chapter 1), and that the foundations of a culture of sustainability can be based on the developments of 'systems thinking' and 'complexity theory' in the second half of the 20th century.

Systems thinking will be introduced, first with its most basic operational tools, and then with an overview of systems views on life, society and physical reality.

But the limits of this approach will also be explored, both in its "holistic" version as represented by the discourse of Ervin Laszlo, and in the risk of technological reductionism in cybernetics.

An increasingly rich and effective understanding of sustainability will come out of the exploration of systems views on society (especially in the sub-subsection below on "Capra's systemic vision for sustainability in the global age"), and a negative definition already emerged throughout chapter 1 with the outline of the 'culture of unsustainability'. However, the following definitions can be used as a starting-point:

Sustainability is most often described as the triptych of social justice, ecological integrity and economic well-being.[1] I proposed alternatively to point at the triptych of biodiversity, cultural diversity and human well-being, which puts the focus on diversities, an aspect of importance in systems thinking and complexity theory, as will be shown in the current chapter.[2] Both triptychs highlight already that sustainability implies both a normative stand and an overarching view that surpasses disciplinary boundaries in human knowledge, expertise and action.

Another aspect of sustainability is that it requires the confrontation to, and overcoming of a complex poly-crisis, "where challenges of increasingly globalizing economic exchanges as well as cultural exchanges are combining with the challenge of interconnected global and local ecological and social crises" (Kagan 2010, p. 1094). Such complexity cannot be addressed with use of existing disciplinary knowledge, even if combined in so-called 'multi-disciplinary' packages, because of the limitations of reductionism, as addressed in chapter 1. "There is one thing such knowledge cannot tell us, and that is how a number of different things act together when exposed to a number of different influences at the same time. And almost everything we encounter around us contains a large number of different things and is exposed to a number of different influences" (Laszlo 1996, p. 3).[3] The search process for sustainability requires a reform of thought, which allows to move beyond these limitations and acquire a better understanding of complexity.

Unfortunately, this necessity is not yet widely recognized. As seen in chapter 1, the dominant culture resists such a development. Mainstream discourses on sustainability limit themselves to the discussion of political com-

1 Cf. Robinson 2004.

2 Cf. Kagan 2008a, 2010.

3 Actually, Laszlo's point itself is a simplification and confusion between "complication" and "complexity": If complication relates to multi-dependencies and numerous interactions, it can be based on simple laws and principles. However, complexity points at something else, i.e. the poly-logical tensions between several principles and laws. Complexity will be further introduced later, in chapter 3. For now, we will have to be satisfied with Laszlo's simplification, to perform a first step in the introduction of systems thinking.

promises between 'three pillars' (i.e. economic, social and ecological targets) and to the hope in salvation through 'green' technological innovations. Critiques of the Technological System from the 1960's and 1970's (Ellul, and after him Ivan Illich) have been conveniently sidelined by promoters of 'ecological modernization', entrepreneurs of a booming green business, and marketers of eco-responsible lifestyles as a growing niche market. In such a context, a 'culture of sustainability' would involve nothing more than a set of superficial lifestyle changes in consumption modes, and the design of an appealing green style for organic and fair trade products. However, the understanding of sustainability developed in the current chapter points at a more fundamental value-shift, exploring culture in its anthropological sense, as the combination of values, beliefs, symbols, practices and 'scripts' or rationalities that organize a worldview in a society. In this sense, the identification of a possible 'culture of sustainability' is nothing less than the exploration of pathways towards a civilizational evolution.

Gregory Bateson already called forward a radical cultural change:

> "We note that every solution which we can imagine is made difficult or impossible by the thinking and attitudes of occidental culture [...] Other attitudes and premises – other systems of human 'values' – have governed man's relation to his environment and his fellow man in other civilizations and at other times. [...] our way is not the only possible human way. It is conceivably changeable" (Bateson 1973, pp. 468-469).

Bateson defined what he called "a healthy ecology of human civilization" as a "single system of environment combined with high human civilization in which the flexibility of the civilization shall match that of the environment to create an ongoing complex system, open-ended for slow change of even basic (hard-programmed) characteristics" (Bateson 1973, p. 470). After him, I will focus on this notion of the 'complex system', exploring "an approach of systems, i.e. anthropo-systems within ecosystems, across space-scales from the local to the planetary, and across time-scales from the short to the very-long term" (Kagan 2008b, p. 15, following Morin 2007). Systems thinking, thus!

The 'systems view' of reality, which will be exposed in this chapter, emerged from several branches of contemporary sciences. As Ervin Laszlo commented, "many branches of the contemporary sciences became, in Warren Weaver's phrase, 'sciences of organized complexity' - that is, systems sciences" (Laszlo 1996, p.8). The rise of systems thinking after the second world war marked a shift in the history of western science and inaugurated a new scientific culture that gradually gained recognition in spite of sciences' path-dependency to the old Descartes-Bacon tradition of linear-causality.[4]

4 Cf. De Rosnay 1975.

Systems thinking offers a way to gain an overview of complex systems, focusing on the relationships rather than on the details, following a conjunctive logic rather than the disjunctive logic of traditional analytical methods.[5]

SECTION 1: SOME BASICS OF SYSTEMS THINKING

As systems thinking introduces radical changes in the understanding of causality and logic, and opens up a field of high complexity (which is further developed in 'complexity theory': cf. chapter 3), I will start this section with a gentle introduction to some of the basic notions of systems thinking.[6] The relative simplifications and colloquial language to be found in this subsection are however only temporary: They shall serve as a practical 'stool' preparing the more advanced discussion of the rest of the current chapter.

What is a system?

A system is a group of interacting, interdependent components that form a complex and unified whole. What does this imply?

- "If you can take components away from something without affecting its functioning and its relationships, then you have just a collection, not a system" (Anderson and Johnson 1997). If you take a bowl of nuts and add or remove some nuts (even different types of nuts), it doesn't change the functioning of the bowl of nuts.
- A system's parts must be arranged in a specific way, and not at random, for the system to work. Otherwise, in our bowl of nuts, where the different types of nuts are distributed does not change the nature of the bowl of nuts. But, to take a counter-example, if the sheep become meat-eaters and the wolves become vegetarian, the food-chain system is undergoing profound changes.
- Systems exist within other systems: Your body is a system in which are interacting a lymph-blood system, a digestive system, etc. You can go down to the tiniest particle and up to the whole universe: All systems are themselves nodes embedded in a giant network in which everything is connected.
- I will introduce further down the notions of "self-reinforcing" loops and "balancing loops". What should be noted already now is that most systems tend to maintain their stability. How do they do that? Thanks to fluctuations and adjustments to changing realities: The different parts of

5 Cf. Lugan 2005, pp. 105-106.

6 This introduction to basics of systems thinking is largely inspired by Anderson and Johnson 1997.

the system are interdependent with each other and/or with the system's environment, so that there is feedback: To take a simple illustration: When you run and your body temperature starts rising, your body adjusts itself to the new situation and lowers your temperature through sweating.

- Feedback is the transmission and return of information. As Virginia Anderson and Lauren Johnson (1997) have put it: "The most important feature of feedback is that it provides the catalyst for a change in behaviour." Feedback comes from within a system, as I explained with the body temperature, but there's also feedback between different systems. And while some feedback loops take only a few seconds to operate (e.g. when you touch something too hot), some other feedback loops can take a very long time to turn back to the initial object/subject (e.g. when you get cancer from over exposure to the sun). These long term feedback processes are both very important to track (because their impact can be tremendous and durable) and very hard to perceive or even recognize, even after they have taken place.

Events, patterns and structures

To keep things simple for now, systems can be said to be built on structures. The structure is the organization of a system, the overall way in which the parts of the system are inter-related. Structure is the most important in a system, but it is also the least visible at first sight, because these interrelationships are not something that is visible (unlike the parts themselves). What is the most visible is the most superficial level: the events: We focus most of the time on events, responding to events in the short term. But responding to events is a bit like behaving as a horse with blinders, avoiding obstacles: It is not the proper way to change reality, to address the roots of the issues that you are interested in.

I will take as an illustration, a current case of unresolved unsustainable development: the food crisis in Karamoja (a region of North-East Uganda that was recently a focus in my research on sustainability)[7]: The event may be food shortage: The event-focused response is to rush in and bring food. Of course this has to be done, in the short term, but it fails to solve underlying issues.

A first effort to take a deeper look is to move from events to patterns of events over time: Can you find a clear pattern of events? Can you draw this trend, for example on a simple graph? But that is not enough. This step only allows to start looking in the right direction: asking, what could have caused these patterns of events?

Going forward with the food crisis illustration: You may, as the World Food Program does, analyze patterns and start adapting your interventions

7 Cf. Eds. Knaute and Kagan 2009.

to them. This way, you may be able to identify where food distribution needs are likely to be the most recurrent, and thus you can better respond. But you are still not addressing the underlying causes.

Asking the "why" question leads then into exploring the structure of a system. The generally applicable question is: "What structures are in place that are causing these patterns?"

In our illustration, the deep roots of food insecurity have to be explored and investigated, and long term feedback effects to be looked up for. This is what is done in systems thinking. Most especially, systems thinkers look for "high leverage points" in feedback loops, i.e. processes that may seem minor at first sight, but which can lead to significant transformations in the long-run, because of their roles in the feedback loops in which they are embedded.

Self-reinforcing loops and balancing loops

Feedback loops can be of two basic sorts: either they are self-reinforcing or they are balancing: Reinforcing loops are engines of growth or collapse. They indeed reinforce change in one direction, accelerate movement in thc same direction (i.e. in one direction or the opposite direction). When you feel an exponential growth or collapse, you are dealing with a reinforcing loop. Balancing loops basically work like 'thermostats': they keep their elements in a certain balanced relationship to each other. They resist change in one direction by producing change in the other direction. These loops can be seen as 'goal seeking', as they keep a target (e.g. the thermostat that keeps a given temperature as a target). If you see certain conditions always coming back to a certain level or 'norm', or resisting change, if a certain behaviour doesn't get dropped. If you see an oscillation or a flattening curve in the evolution of a phenomenon, you probably have a balancing loop at play.

Basically, the complex behaviour of systems starts with the combination of reinforcing and balancing loops.

Feedback delays

The problems with feedback delays is that humans tend not to perceive them appropriately, and thus tend to mismanage them: Often it takes us too long to perceive the feedback, measure results, decide how to respond and implement the decision. Even worse, we often misperceive reality, because of delayed feedback: We do not realize at all that the delayed feedback exists. Delays can make a system's behavior unpredictable and defeat efforts to control the system.

The ways of a systems thinker[8]

A systems thinker is someone who, in the analysis of reality around him or her, will try to follow some of the following principles:

- Look for the 'big picture': getting away from the immediate problems of the day, and away from the local issues only: The issue that is observed is always part of a larger system. So the systems thinker is going to look for an understanding of how that peculiar issue is embedded in a wider system and connected to other systems and other issues elsewhere.
- Balance the short-term with the long-term: The systems thinker is very cautious about short-term 'fixes that fail' in the long term because of longer-term feedback (i.e. so-called side effects or perverse effects). Of course the short term also matters, but short-termed management is very risky. In the case of the Karamoja crisis, one development infrastructure may offer a good illustration for this: installing boreholes for the Karamojong communities (so that they have access to water) is very helpful in the short term. But in the long term it has several detrimental effects: It contributes greatly to sedentarizing nomads in one location, leading to both local overgrazing and overexploitation of the environment and to a loss of adaptability to climate changes, and on top of that, it puts a higher stress on the groundwater table which invariably tends to sink further and further down. Furthermore, wrongly located water-wells can spark conflicts between communities. Thus in the long term, the 'good idea' of installing boreholes all across Karamoja, and the good intentions of some NGOs, may have catastrophic repercussions.[9]
- The systems thinker always reminds him- or herself that: "Things change all the time, life is messy, and everything is connected" (Anderson and Johnson 1997). The systems thinker is very cautious with linear thinking because the simplifying way is the way towards imagining 'solutions' that generate as many problems as they solve. The worst possible situation, from a systems thinker's perspective, is to remain fully unaware of the systemic consequences of one's actions.
- The systems thinker values varied sources and forms of knowledge (e.g. both quantitative and qualitative data), and is aware of everybody's tendency to see only what we can measure, and of our tendency to leave aside what we can't measure: The systems thinker tries not to ignore what he or she cannot measure, or cannot understand.
- The systems thinker is aware that he/she is part of the system being observed: For example, he/she might have already implemented solutions that have had unintended consequences. He/she is aware he/she might be relatively blindfolded by a belief system, a mental model full of prefab-

8 The following lines are largely inspired by Booth Sweeney and Meadows 1995.

9 Cf. Kagan, Pedersen and Ollech 2009.

ricated assumptions of how the world works (and this is also why a systems thinker loves to work in interdisciplinary teams, not all alone as a hermit). One should especially be cautious about what one learns in over-specialized university studies, because over-specialization can have very damaging effects on world views.

- The systems thinker is not afraid of paradoxes: the complex reality of systems is not as nicely logical as linear theories and models may be.
- The systems thinker is curious about new tools and new languages, searching for a language focusing on the interconnections rather than on the single events; a language focusing on cycles rather than on linear causes and effects; a relativizing language and one that digs for the deeper, broader perspectives; a visual, graphical language that allows to perceive several levels of form at the same time (unlike the words-based, spoken/written language which is structurally linear).[10]

Systems are complex

Most real-life systems include many more loops than just one or two main loops. They often cannot be immediately understood by a human being, and especially not by a human being trained in linear logical thinking. This is why systems are complex.

In a complex system, the dominance of some feedback loops or other loops, shifts over time, and there are all sorts of time horizons with delays of different lengths. Some structures may be contradicting each other, conflicting with each other.

Some recurrent characteristics of complex systems include the following:

- A complex system contains many balancing loops, keeping subparts of the system in balance. This explains why complex systems resist change: Even if some balancing loops fail or are too weak to resist some change, some other balancing loops will still, most of the time (but not always: systems are not 'eternal'), be operating. This lead some system theorists to conceive of complex systems as autopoïetic, closed wholes (e.g. Luhmann), but systems thinking is here critical of a system theory that overemphasizes structural stability. Even complex systems can be changed, but this requires equally complex transformation processes.
- Complex systems appear to be purposeful. In their stability, they seem to follow a certain track. That's why many people will say e.g. of a big organization, that it has a 'mind of its own'.
- Complex systems are able to 'learn' from feedback and to modify their behaviour: They are able to change and to learn also from their envi-

10 Cf. Durand 2006, pp. 54-55.

ronment. But the learning opportunities have to be seized by the agents (e.g. species, or human beings) operating in these complex systems.

- Complex systems can modify their environments. Of course, this involves complex interrelationships with the environments which are themselves systems.
- Complex systems are adaptative and/or evolutive: they can reorganize themselves in response to shocks. They also have the ability to replicate themselves and mutate in old and new environments.
- Although a complex system can learn, it also has many subsystems to coordinate and thus many opportunities for things to go wrong, with e.g. conflicting goals, distorted feedback, loss of predictability, centralization vs. decentralization dilemma (some subsystems can have goals that directly oppose the overall goals of the system). But the 'pseudo-solution' of simplifying or e.g. centralizing the system, would destroy its ability to learn.

Complex systems are characterized by both macrodetermination of the whole and functional autonomy of the parts.

"The inverse side of macrodetermination is functional autonomy. The functional autonomy of parts within a natural system adds up to the macrodetermination of the whole. Functional autonomy does not mean independence. A fully autonomous (independent) set of units would not constitute a system, only a heap. Systemicity is imposed as a set of rules binding the parts among themselves. But these rules do not constrain the parts to act in one way and one way only; they merely prescribe that certain types of functions are carried out in certain sequences. The parts have options; as long as a sufficient number of sufficiently qualified units carries out the prescribed tasks, the requirements of systemic determination are met" (Laszlo 1996, p. 86).

The confrontation with complexity further defines the 'ways of the systems thinker'. Systems thinkers are not rushing towards solutions. The learning may have to take long. It is better to be patient and take the time to find higher leverage solutions, rather than to give in to the pressures of others and give quick fixes. Furthermore, confronting complexity is a continuous learning process, and a systems thinker has to be a continuous learner.

SECTION 2: THE SYSTEMS VIEW OF LIFE

In this section, I will principally follow the insights from Fritjof Capra, a prominent advocate of systems analysis in the natural and social sciences. Capra was originally a physicist (student of Werner Heisenberg), but his work also draws heavily on biology and ecology. He is a founding director of the 'Center for Ecoliteracy' in Berkeley which promotes ecology and systems thinking in education.

Living systems as networks

Networks

Capra's development of a "systems view of life" uses the concept of network as its cornerstone: The network is described as a common pattern of all living systems: "As ecosystems are understood in terms of food webs (networks of organisms), so organisms are viewed as networks of cells, organs and organ systems, and cells as networks of molecules. One of the key insights of the systems approach has been the realization that the network is a pattern that is common to all life" (Capra 2002, p. 9). This insight is claimed to unify the understanding of all life on planet Earth. "A central insight of this unified, systemic understanding of life is that its basic pattern of organization is the network. At all levels of life – from the metabolic networks inside cells to the food webs of ecosystems and the networks of communications in human societies – the components of living systems are interlinked in network fashion" (ibid., p. 261).

These networks are both dynamic, i.e. undergoing change, and stable, i.e. maintaining an overall organizational stability: "living networks continually create, or re-create, themselves by transforming or replacing their components. In this way they undergo continual structural changes while preserving their weblike patterns of organization" (ibid., p. 10).

And these networks of life are establishing inter-dependencies. "Living systems are self-generating networks, which means that their pattern of organization is a network pattern in which each component contributes to the production of other components" (ibid., p. 90).

Life forms are not just determined by "a genetic blueprint" but are emergent properties of those networks, also depending on environmental constraints:

"A key insight of the new understanding of life has been that biological forms and functions are not simply determined by a genetic blueprint but are emergent properties of the entire epigenetic network. To understand their emergence, we need to understand not only the genetic structures and the cell's biochemistry, but also the complex dynamics that unfold when the epigenetic network encounters the physical and chemical constraints of its environment" (ibid., pp. 11-12).

More generally, biological reality does not obey linear determinacy, as Gregory Bateson highlighted in very clear words:

"This multiple determination is characteristic of all biological fields. Characteristically, every feature of the anatomy of an animal or plant and every detail of behaviour is determined by a multitude of interacting factors at both the genetic and physiological levels; and, correspondingly, the processes of any ongoing ecosystem are the outcome of multiple determination. Moreover, it is rather unusual to find that any feature of a biological system is at all directly determined by the need which it fulfills. Eating is governed by appetite, habit, and social convention rather than by hunger, and respiration is governed by CO2 excess rather than by oxygen lack. And so on. In contrast, the products of human planners and engineers are constructed to meet specified needs in a much more direct manner, and are correspondingly less viable. The multiple causes of eating are likely to ensure the performance of this necessary act under a large variety of circumstances and stresses whereas, if eating were controlled only by hypoglycemia, any disturbance of the single pathway of control would result in death" (Bateson 1973, p. 476).

The cell: of metabolism and membranes

Living systems are characterized by 'metabolism', which is a dynamic organization of the living systems:

"The essential characteristic that distinguishes living from nonliving systems – the cellular metabolism – is not a property of matter, nor a special 'vital force'. It is a specific pattern of relationships among chemical processes. Although it involves relationships between processes that produce material components, the network pattern itself is nonmaterial. [...] Thus, life is never divorced from matter, even though its essential characteristics – organization, complexity, processes, and so on – are nonmaterial" (Capra 2002, p. 72).

"Metabolism, the incessant chemistry of self-maintenance, is an essential feature of life [...] Through ceaseless metabolism, through chemical and energy flow, life continuously produces, repairs and perpetuates itself" (Margulis 1998 p. 63, quoted in Capra 2002, p. 9).

Together with metabolism, a defining characteristic of living systems is the self-closure of the living system by a membrane: Capra speaks of a single cell as a "system characterized, first of all, by a boundary (the cell membrane) which discriminates between the system [...] and its environment" (Capra 2002, p. 7). Within its boundaries, "the system sustains itself" thanks to its "network of chemical reactions".

The existence of the 'membrane' is indispensable to life: "In such fluid surroundings [i.e. the primordial ocean], a cell could never persist as a distinct entity without a physical barrier against free diffusion. The existence of membranes is therefore an essential condition for cellular life" (ibid., p. 8).

Capra highlights that these membranes are "semi-permeable": "membranes are always active, opening and closing continuously, keeping certain substances out and letting others in. [...] All these activities help to maintain the cell as a distinct entity" (ibid., p. 9). The membrane "preserves [the] identity" of the cell.

The cell is self-generating, producing itself (with elements from the outside). In this sense, it is 'autopoïetic'. Capra argues that the cell is autonomous, because autopoïetic, but is not independent from its environment: Indeed, its survival depends on a continuous input of energy from the outside.

The cellular metabolism indeed "use[s] a continual flow of energy to restore structures as fast as they are decaying" and thus escapes entropy, i.e. death. The cell is 'an open system' that depends on its environment for its very existence.

"Living systems are organizationally closed—they are autopoietic networks—but materially and energetically open. They need to feed on continual flows of matter and energy from their environment to stay alive. Conversely, cells, like all living organisms, continually produce waste, and this flow-through of matter—food and waste—establishes their place in the food web" (ibid., p. 13).

Emergence

Emergence "has been recognized as the dynamic origin of development, learning and evolution" (ibid., p. 14). From there, the concept of creativity can be understood as a basically biological phenomenon: "Creativity – the generation of new forms – is a key property of all living systems." Life is constantly creative: "And since emergence is an integral part of the dynamics of open systems, we reach the important conclusion that open systems develop and evolve. Life constantly reaches out into novelty" (ibid., p. 14).

Capra highlights that emergence also marks a shift of focus by scientists: "the focus is now shifting from the structures to the process of their emergence" (ibid., p. 14). Most especially, the theory of evolution thus gained attention among systems thinkers, who developed a systemic view on evolution.

A proper understanding of emergence and evolution is crucial for a culture of sustainability, because human societies with advanced technologies can only remain sustainable if they do understand and integrate the complex dynamic of natural processes, in order to be able to co-evolve with them.

The evolution of life

Dissipative structures: evolution before life

The emergence of life networks plunges its roots in specific properties of the physical world: These properties were described by Prigogine as "dissipative structures": physico-chemical structures which are capable to generate and maintain a degree of stability far from thermodynamic equilibrium (but which are not life-forms).[11]

In short, the motivation for Prigogine's exploration was to understand how some aspects of physical reality, and most especially life, can maintain themselves apparently 'against' the classical laws of thermodynamics. Prigogine's research then focused on the phenomenon of heat convection, as

11 The influence of Prigogine on further elaboration of complexity theories is also heavily acknowledged by Smith and Jenks (2006).

an example of physical self-organization whereby, with increasing heat i.e. increasing instability, a point is reached at which a coherent behavior of matter emerges. So, unlike classical thermodynamics where dissipation of energy generates waste, in Prigogine's nonlinear thermodynamics, dissipation can become a source of order.[12]

"The dynamics of these dissipative structures specifically include the spontaneous emergence of new forms of order. When the flow of energy increases, the system may encounter a point of instability, known as a 'bifurcation point', at which it can branch off into an entirely new state where new structures and new forms of order may emerge. This spontaneous emergence of order at critical points of instability is one of the most importance concepts of the new understanding of life" (Capra 2002, pp. 13-14).

Prigogine's theory of dissipative structures implies a less clear-cut separation between pleroma and creatura, i.e. between physical and biological levels of reality, than was previously imagined – even among early systems thinkers such as Bateson (1979).

There exists some dissipative structures, which are not autopoïetic systems but can evolve.[13] This was probably the case in 'prebiotic' evolution, i.e. the premises of life. "[T]here is a 'prebiotic' evolution – an evolution of inanimate matter that must have begun some time before the emergence of living cells" (Capra 2002, p. 15). The build-up of molecular complexity led to the emergence of life. There is not a clear-cut line between life and 'non-life': "It is clear that the process leading to life is a continuum process, and this makes an unequivocal definition of life very difficult" (Luisi 1998, quoted in ibid., p.16).

The systemic character of biological life (i.e. with its loops and cycles) has roots in biochemistry:

"Another interesting discovery is that chemical networks in closed spaces that are subject to continual flows of energy develop processes surprisingly like those of ecosystems. For example, significant features of biological photosynthesis and the ecological carbon cycle have been shown to emerge in laboratory systems. The cycling of matter seems to be a general feature of chemical networks that are kept far from equilibrium by a constant flux of energy" (Capra 2002, p. 27).

Capra describes the idea that membranes (which are determinant in the cell's organization), with the closure they allow, preceded life and maybe

12 For a more detailed account, see Capra 1996, pp. 86-89. (The exact page number here refers to the 1997 reprint by Flamingo.)

13 Morin (as will be discussed in chapter 3) went further, considering evolution as primarily a physical phenomenon, starting with the constitution of the universe as we know it.

even prebiotic structures. This is "the hypothesis that very early on, before the increase of molecular complexity, certain molecules assembled into primitive membranes that spontaneously formed closed bubbles, and that the evolution of molecular complexity took place inside these bubbles, rather than in a structureless chemical soup" (ibid., p. 20). These bubbles provided a closed environment in which new, different (and otherwise impossible) chemical reactions could occur.

Capra also encourages a more ecologically literate research on the origins of life, following Morowitz:

> "Sustained life is a property of an ecological system rather than a single organism or species. Traditional biology has tended to concentrate attention on individual organisms rather than on the biological continuum. The origin of life is thus looked for as a unique event in which an organism arises from the surrounding milieu. A more ecologically balanced point of view would examine the proto-ecological cycles and subsequent chemical systems that must have developed and flourished while objects resembling organisms appeared" (Morowitz 1992, p. 54).

The evolutionary paths of life

In the evolution of life forms, Capra points at "three major avenues of evolutionary creativity – mutation, gene trading and symbiosis" (Capra 2002, p. 32).

Evolution is not a continuous gradual change process over time. "The fossil record shows [...] dramatic transitions." Capra follows Margulis in arguing that especially "symbioses are like flashes of evolutionary lightning" (ibid., pp. 30-31). He invokes symbiogenesis to explain phases of quick transitions. Here a closer focus on bacteria will give a better idea of his focus on symbiogenesis: Capra describes how some bacteria can trade their genes (DNA recombination), and argues that "all bacteria are part of a single microscopic web of life" (ibid., p. 29).

The case of bacterias shows that biological systems can exchange a lot between them, unlike what Luhmann assumed, as they can even directly exchange DNA sequences in the case of bacteria.

"Consequently, they have an astonishing ability to adapt to environmental changes. The speed with which drug resistance spreads among bacterial communities is dramatic proof of the efficiency of their communication networks" (ibid.).

Worldwide bacteria come to form a global system, along the conclusions of the Gaia hypothesis (initially formulated by James Lovelock): "bacteria continually transformed the Earth's surface and atmosphere, and established the global feedback loops for the self-regulation of the Gaia system" (ibid.). As a consequence, the evolutionary role of bacteria supports the claim that "Life is a property of planets rather than of individual organisms." (Morowitz quoted in ibid., p. 6). Capra even proclaims: "The planetary network of bacteria has been the main source of evolutionary creativity" (ibid., p. 29).

Bacteria also contributed to the evolutionary importance of evolution through symbiosis, defined as "the tendency of different organisms to live in close association with one another and often inside one another (like the bacteria in our intestines)" (ibid., p. 30). Symbiosis can also lead to symbiogenesis:

> "Margulis [...] proposed the hypothesis that long term symbioses involving bacteria and other microorganisms living inside larger cells have led and continue to lead to new forms of life [...] In their evolution, these organisms continued to absorb bacteria, incorporating parts of their genomes to synthesize proteins for new structures and new biological functions [...] For example, evidence has been accumulating that the microtubules, which are essential to the architecture of the brain, were originally contributed by the 'corkscrew' bacteria known as spirochetes" (ibid., p. 30 , referring to Margulis 1998, pp. 45ff).[14]

An understanding of the importance of symbiosis as one driving factor of natural evolution, and of the global inter-relations between evolving ecosystems on planet Earth, constitute two important elements in a culture of sustainability, learning to be attentive to the complex interrelations constituting evolutionarily relatively stable ecosystems, as a knowledge-base for the sustainable reform of contemporary societies.

Evolution vs. linear thinking

Systems thinkers' understandings of evolution stand in clear contrast to linear causal logics, whether these may come from a classical scientific perspective or from a Christian 'creationist' discourse.

Capra clearly rejects the teleology of 'purpose' and (intelligent) 'design' in Nature, while he "recognizes the pervasive order, self-organization, and intelligence manifest throughout the living world" (ibid., p. 120). Purpose is associated to "reflective consciousness" which is not a property shared by all living systems. Nature only knows emergent structures, not designed structures (unlike human organizations).[15]

In his discussion of genetic research, Capra further elaborates his approach to evolution. His account of recent research in molecular biology, which highlighted that genetic stability is less self-evident than previously assumed, imposes a careful assessment of the working of genetic mutations: Genetic stability is maintained only thanks to a complex error-corrections process involving different enzymes, and thus is an emerging function of the

---

14 The mapping of the human genome confirmed that many human genes originated from bacteria (Capra 2002, p. 168).

15 I already introduced, in chapter 1, Bateson's critique of "purposive consciousness".

whole cellular metabolism and not only the genes. "There seem to be mechanisms that actively generate copying errors by relaxing some of the monitoring processes. Moreover, it appears that when and where mutation rates are increased in this way depends both on the organism and on the conditions in which the organism finds itself" (ibid., pp. 166-167, referring to Keller 2000, pp. 32ff). This discovery has a tremendous significance for the biological theory of evolution: "The new discoveries in genetics will force biologists to adopt the radically different view that mutations are actively generated and regulated by the cell's epigenetic network, and that evolution is an integral part of the self-organization of living organisms" (ibid., p. 167). In the words of a molecular biologist: "Rather than being restricted to contemplating a slow process depending on random (i.e. blind) genetic variations [...] we are now free to think in realistic molecular ways about rapid genome restructuring guided by biological feedback networks" (Shapiro 1999, quoted in Capra 2002, p. 167).

Furthermore, genetic 'determinism' is much less obvious than some biologists have claimed (e.g. Dawkins 1976), as "recent genetic experiments have shown that knocking out single genes, even when they were thought to be essential, had very little effect on the functioning of the organism" (Capra 2002, p. 174; referring to Keller 2000 pp. 112-113). Genetic researchers now agree that a level of 'redundancy' in the gene-pool and in the metabolism, keeps development on track. This understanding of evolution, and of the roles of genes and cellular networks therein, stresses the importance of diversity: "Genetic and metabolic redundancy may be seen, perhaps, as the equivalent of biodiversity in ecosystems. It seems that life has evolved ample diversity and redundancy at all levels of complexity" (Capra 2002, p. 174). To a systems thinker, this insight comes as no surprise and corresponds to the typical structural stability of nonlinear systems, maintained by multiple pathways.

Diversity and resilience[16]

In systems thinking, the notion of diversity is inseparable from another notion: resilience.

Resilience refers to the capacity to adapt to change from the 'outside'. The term is used in ecology, referring to the limits of a system's capacity to be perturbed; once the limits are reached, the system either collapses or finds a new state of equilibrium.[17] As noted by ecological artist David Haley, "the capacity to withstand disturbance is not just a question of how long the status quo can be maintained, but how we might evolve to dwell in this new world" (Haley 2008, p. 204).

---

16 This 'boxed text' is adapted from Kagan 2010.

17 Cf. Walker et al. 2006, cited in Haley 2008, p. 204.

Resilience necessitates the preservation of diversity: Sustainable systems can only exist as long as diversity is preserved, so that the exogenous shocks of the unexpected may give way to the endogenous responses of resourceful (social or eco-) systems. Less diversity in a system means a lower resilience.

The cybernetic "law of requisite variety" already delivered comparable insights half-a-century ago. In the words of Joël de Rosnay:

> "The more complex a system, the more complex its control system must be in order to provide a 'response' to the multiple disturbances produced by the environment. This is the law of requisite variety proposed by Ross Ashby in 1956. This very general law asserts [...] that the regulation of a system is efficient only when it depends on a system of controls as complex as the system itself. In other words, control actions must have a variety equal to the variety of the system. In ecology, for example, it is the variety of species, the number of ecological niches, the abundance of interactions among species and between community and environment that guarantee the stability and continuance of the community. Variety permits a wider range of response to potential forms of aggression from the environment" (De Rosnay 1975, p. 130),

and as well, variety permits a wider range of response to potential disturbance/aggression coming from within the system.

What does this mean, more concretely? The preservation and the advancement of diversity (i.e. both biodiversity and cultural diversity) is a key normative target for sustainability. And the understanding and furthering of resilience must be at the heart of cultures of sustainability.

However, resilience shall not be understood too strictly. Our societies may also need transformations when some social-ecological systems are resilient but undesirable (e.g. for matters related to social justice). In certain situations, relative collapse may even constitute a necessary step for the construction of new configurations of systems.18

What this all means, in systems thinking terms, is that "evolutionary change [is seen] as a manifestation of life's self-organization [which systems theories consider also as] a cognitive process" (Capra 2002, p. 168). This cognitive aspect will now be explained.

Evolution, cognition and consciousness

Capra refers to Bateson and to Maturana and Varela, highlighting that mind is not a thing but a process. Referring to Maturana and Varela's 'Santiago theory of cognition', he argues: "Cognition is the very process of life. The

18 Cf. Haley 2008.

organizing activity of living systems, at all levels of life, is mental activity. [...] Mind [...] is immanent in matter at all levels of life" (ibid., p. 34). Or in other words: "The entire structure of the organism participates in the process of cognition, whether or not the organism has a brain" (ibid., p. 37).

Consciousness is defined by Capra as "a special kind of cognitive process that emerges when cognition reaches a certain level of complexity" (ibid., p. 38). He also emphasizes consciousness as a process, not a thing, and also refers to William James' notion of "stream of consciousness".

Capra distinguishes several levels or stages of consciousness:

- "Primary consciousness arises when cognitive processes are accompanied by basic perceptual, sensory and emotional experience."
- "High-order consciousness involves self-awareness – a concept of self"; Capra calls it "reflective consciousness" (ibid., p. 39).

Capra sees reflective consciousness as an emergence and an "evolutionary stage" leading to social organization and culture. "Conscious experience is an emergent phenomenon, which means that it cannot be explained in terms of neural mechanisms alone. Experience emerges from the complex non-linear dynamics of neural networks" (ibid., p. 41). Varela's 'neurophenomenology' further explains how, in the brain, consciousness is emergent and integrates "many different brain functions", and how consciousness is embedded in sensation, with his concept of "phase-locking":

> "The specific neural mechanism proposed by Varela for the emergence of transitory experiential states is a resonance phenomenon known as 'phase-locking,' in which different brain regions are interconnected in such a way that their neurons fire in synchrony. Through this synchronization of neural activity, temporary 'cell assemblies' are formed, which may consist of widely dispersed neural circuits. According to Varela's hypothesis, each conscious experience is based on a specific cell assembly, in which many different neural activities—associated with sensory perception, emotions, memory, bodily movements, etc.—are unified into a transient but coherent ensemble of oscillating neurons. The best way to think of this neural process is, perhaps, in musical terms. There are noises; then they come together in synchrony as a melody emerges; then the melody subsides again into cacophony, until another melody arises in the next moment of resonance" (ibid., p. 50).

Varela and other researchers characterized "conscious experience as an emergent property of a transient process of interpretation, or synchronization, of widely distributed groups of neurons" (ibid., p. 52).

Language itself "was originally embodied in gesture and evolved from gesture together with human consciousness" (ibid., pp. 58-60, referring to Fouts 1997). The theory of the 'embodied mind' further argues that "conceptual thought as a whole is embodied physically in the body and brain. [...] human reason does not transcend the body, as much of Western philosophy has held, but is shaped crucially by our physical nature and our bodily experience. [...] The very structure of reason arises from our bodies and brains. [Besides,] most of our thought is unconscious [and] the cognitive unconscious shapes and structures all conscious thought" (ibid., p. 61, referring to Lakoff and Johnson's *Philosophy in the flesh*).

"The same neural and cognitive mechanisms that allow us to perceive and move around also create our conceptual structures and modes of reason" (Lakoff and Johnson 1999, p. 4). Capra follows this theory: "How living organisms categorize depends on their sensory apparatus and their motor systems" (Capra 2002, p. 62).

Human consciousness and human reason also reveal themselves as evolutionary emergences, relativizing humanity's tendency to separate itself too radically from the animal realm:

"In [Capra's] view, the main significance of these advances has been the gradual but consistent healing of the Cartesian split between mind and matter that has plagued Western science and philosophy for more than 300 years. The Santiago Theory has shown that at all levels of life, mind and matter, process and structure, are inseparably connected. Recent research in cognitive science has confirmed and refined this view by showing how the process of cognition evolved into forms of increasing complexity together with the corresponding biological structures. As the ability to control precise hand and tongue movements developed, language, reflective consciousness, and conceptual thought evolved in the early humans as parts of ever more complex processes of communication. [...] The unified, post-Cartesian view of mind, matter, and life also implies a radical reassessment of the relationship between humans and animals. Throughout most of Western philosophy, the capacity to reason was seen as a uniquely human characteristic, distinguishing us from all other animals. The communication studies with chimpanzees have exposed the fallacy of this belief in the most dramatic of ways" (ibid., pp. 65-66).

Here again, Capra's argument joins Lakoff and Johnson's: "Reason, even in its most abstract form, makes use of, rather than transcends, our animal nature. Reason is thus not an essence that separates us from other animals; rather, it places us on a continuum with them" (Lakoff and Johnson 1999, p. 4).

This immanent, bodily substrate for the human mind and for human consciousness, formulates the materialist-immanent ground for the self-understanding of the human being in a culture of sustainability: Humanity is embedded and emerging gradually (via the progressive cognitive developments of hominids) from a complex cognitive ecology, and is not separated from it, nor was it suddenly created *ex nihilo* by an 'intelligent designer' or by a totally unprecedented accident that would radically separate the human species from all other animal species.

Capra reaffirms the systemic importance of his view of human nature, vs. 'reductionist' views:

"The new understanding of life is a systemic understanding, which means that it is based not only on the analysis of molecular structures, but also on the analysis of patterns of relationships among these structures and of the specific processes underlying their formation. As we have seen, the defining characteristic of a living system is not the presence of certain macromolecules, but the presence of a self-generating network of metabolic processes. The processes of life include, most importantly, the spontaneous emergence of new order, which is the basis of life's inherent creativity. Moreover, the life processes are associated with the cognitive dimension of life, and the emergence of new order includes the emergence of language and consciousness" (Capra 2002, p. 67).

The Systems View of Life: Autopoïesis, form, matter and process

Autopoïesis

Capra sees autopoïesis (i.e. 'self-making') as "a clear and powerful criterion for distinguishing between living and non-living systems". For example, this allows to clearly exclude both viruses and robots from the list of living systems. But he also argues that "the behaviour of a living organism is constrained but not determined by outside forces" (ibid., p. 85). Autopoïesis then means autonomy but not independence from the environment. "Living systems are organizationally closed [...] but materially and energetically open" as they need both to feed on energy and to reject waste, both of which are co-constituting the food web of an ecological system (ibid., p. 13). However, with a further analysis of the complexity of ecosystems, which I will conduct in the next chapter after Morin, the notion of autopoïesis, i.e. of clear-cut operational closure, will appear as only one aspect of the self-organization of living systems, and not as the 'whole story'. The rise of cultures of sustainability requires an elaborate understanding of what it means to be embedded in a complex ecology of life, and this is an especially difficult challenge.

The concept of autopoïesis in the Santiago Theory of Cognition

The concept of autopoiësis, on which both Capra and Luhmann (as seen in chapter 1) are relying, was developed by Humberto Maturana and Francisco Varela in the 1970's as part of a theoretical framework explaining the self-organization of living cells (and, in a second step; of multicellular organisms) and interpreting it as a cognitive process, i.e.

identifying life and cognition as one and the same thing (i.e. negating the Cartesian division of mind and matter). This framework is known as the 'Santiago theory of cognition').

Capra notes that "[t]he defining characteristic of an autopoietic system is that it undergoes continual structural changes while preserving its weblike pattern of organization. The components of the network continually produce and transform one another" both by renewing dying cells and by operating structural changes in the own structure of the living organism (Capra 2002, p. 34). How the second kind of change, i.e. the structural, developmental change, occurs according to the Santiago theory, is especially interesting. The living organism "couples to its environment structurally, i.e. through recurrent interactions, each of which triggers structural changes in the system. [So far, so good, but the next bit is the most puzzling and problematic:] The environment only triggers the structural changes, it does not specify of direct them" (ibid., p. 35). In simple words, Bateson's illustration of the difference between kicking a stone and kicking a dog offers a perfect illustration of what is meant here: The stone reacts according to the external laws of physics, but the dog's reaction to your kicking it will depend on its own internal nonlinear organization (whether the response will be squeaking, fleeing, ignoring or biting the attacker).[19]

Capra is well-aware that Maturana and Varela's autopoïesis implies that "the behavior of the living organism is dictated by its structure" (ibid., p. 36) or in other terminology, path-dependent. The insightful lesson from this feature of living systems is that "you can never direct a living system; you can only disturb it. More than that, the living system not only specifies its structural changes; it also specifies *which disturbances from the environment triggers them.* In other words, a living system maintains the freedom to decide what to notice and what will disturb it" (ibid.). But this operational closure imposed by autopoïesis involves issues which Capra overlooks, as he goes on, pointing out that autopoïetic systems are learning systems. The main problem is that the theory obscures rather than enlightens, how the autopoïetic organism can learn from the outside, if the criteria for the very recognition of disturbances from the outside is located inside the system. That is an issue that Luhmann was more aware of (but which he did not try to overcome, preferring to elaborate on it with the notion of re-entry).[20] As already discussed in chapter 1, Luhmann pushed the logic of autopoïesis to its maximum, showing that an autopoïetic system will follow its own logic and ignore certain disturbances from the environment 'whatever happens', until

19 Cf. Bateson 1973, pp. 229, 409, 490.

20 On autopoïesis in Luhmann, please refer back to chapter 1.

eventual self-anhiliation (and damages to the own habitat). This would be a very limited form of 'learning'.[21]

However, the concept of autopoïesis can be revised, following the insights from Morin's complexity theory and giving more space to the role of chaos, as will be discussed in the next chapter (where I suggest the alternative term of autoecopoïesis).[22] Concerning more specifically the evolution of living systems, some recent research in molecular biology tends to stress the chaotic-creative role played by retroviruses which, as external elements, forcefully penetrate the living cell and can forcefully introduce new DNA, provoking an unpredictable but certainly not strictly autopoïetic bifurcation in the self-organizational process of the cell.[23]

A revision of the concept of autopoïesis is thus necessary, and would also better fit with Capra's overall ecological discourse.

Form, matter and process

Capra synthesizes the systems view of life around the triptych of form, matter and process:

"The synthesis is based on the distinction between two perspectives on the nature of living systems, which I have called the 'pattern perspective' and the 'structure perspective,' and on their integration by means of a third perspective, the 'process perspective.' More specifically, I have defined the pattern of organization of a living system as the configuration of relationships among the system's components that determines the system's essential characteristics, the structure of the system as the material embodiment of its pattern of organization, and the life process as the continual process of this embodiment. […] However, in view of the fact that the definition of 'structure' in the social sciences is quite different from that in the natural sciences, I shall now modify my terminology and use the more general concepts of form and matter to accommodate different usages of the term 'structure.' In this more general terminology, the three perspectives on the nature of living systems correspond to the

---

21 For a more detailed critique of Maturana and Varela's, and of Luhmann's autopoïesis, see especially Smith and Jenks 2006, who also come to the conclusion that "eco-auto-organization" (after Edgar Morin and with insights from Prigogine) is more convincing than "autopoïesis". Smith and Jenks (2006) describe how the theory of autopoïesis, with its strict operative closure and its epistemological constructivism, constitute a variation on the theme of Kantian idealism, and thereby fails to become an adequate theory of complexity.

22 See also the appendix on the subconscious.

23 See especially the work and writings of Luis Villarreal, Professor in Molecular Biology and Biochemistry at the University of California - Irvine; faculty webpage: http://www.faculty.uci.edu/profile.cfm?faculty_id=2705

study of form (or pattern of organization), the study of matter (or material structure), and the study of process. When we study living systems from the perspective of form, we find that their pattern of organization is that of a self-generating network. From the perspective of matter, the material structure of a living system is a dissipative structure, i.e. an open system operating far from equilibrium. From the process perspective, finally, living systems are cognitive systems in which the process of cognition is closely linked to the pattern of autopoiesis" (ibid., pp. 70-71).

SECTION 3: THE SYSTEMS VIEW OF SOCIETY

Systems thinking author Fritjof Capra argues for a systems view on human societies. About the traditional approach of sociology, he argues especially that Durkheim's 'social facts', which "should be treated like material objects", unfortunately oriented sociology towards an analogy to classical physics rather than to life sciences, with linear cause-effect mechanisms as a central explanative framework, complemented by a "hidden causative reality" when the linear causality couldn't bring proper explanations (Capra 2002, p. 76). Sociology's first steps towards a systemic approach, first with Talcott Parsons and then with Luhmann, are mentioned by Capra, but he especially focuses his attention on Giddens and Habermas, as integrative social theories, able "to integrate the notions of social structure and human agency with an explicit analysis of meaning" (ibid., p. 77), and on Castells for his analysis of a contemporary 'Network Society'. Indeed, Capra discusses the transferability of systemic insights from biology and ecology to sociology and the specificities of social systems. A systems understanding of society further grounds cultures of sustainability in the embeddedness of humanity in a complex ecology.

As will appear in the following paragraphs, I am ignoring Talcott Parsons' structural functionalism, although it was the first sociological theory to introduce systems analysis in the mid-20th century, inspired by contemporary developments in general systems theory.[24] Surely, Talcott Parsons (1961) did make insightful theoretical steps towards understanding the feedback and integration between social actors interactively constructing and playing out roles, and the emergence of social structures in relative equilibrium states. However, he over-emphasized the smooth operation of social norms and of socialization processes and the tendency towards equilibrium, and his theoretical constructions do not satisfactorily discuss how individual agency works. Furthermore, Talcott Parsons' systemic framework was insufficiently informed by the complexity of biological systems. This situation was to change with Niklas Luhmann (himself a student of Talcott Parsons), whose later "functional structuralism" would be grounded in the biology of

24 by Karl Ludwig von Bertalanffy, Norbert Wiener and William Ross Ashby, among others.

autopoïetic systems. However, as I am taking critical distance from Maturana and Varela's strict autopoïesis and will (in the next chapter, following Edgar Morin) prefer a more ecologically-informed theorization in the direction of eco-auto-organization, I am also approaching Luhmann's views with critical caution.

From biological systems to social systems

Capra's "systemic understanding [of social reality] is based on the assumption that there is a fundamental unity to life, that different living systems exhibit similar patterns of organization. [...] As life evolves, these patterns tend to become more and more elaborate, but they are always variations on the same basic themes" (ibid., p. 81). Consequently, Capra aims at "applying our knowledge of life's basic patterns and principles of organization, and specifically our understanding of living networks, to social reality." But he does not claim a direct analogy from biological structures to social structures, thus "we should not expect to transfer our understanding of the network's material structure from the biological to the social domain."[25]

Among the 'basic themes' that Capra finds back in social systems are some common patterns of organization, some common cybernetic principles and processes, such as nonlinearity, feedback, emergence:

> "A social network, too, is a nonlinear pattern of organization, and concepts developed in complexity theory, such as feedback or emergence, are likely to be relevant in a social context as well, but the nodes and links of the network are not merely biochemical. Social networks are first and foremost networks of communication involving symbolic language, cultural constraints, relationships of power, and so on. To understand the structures of such networks we need to use insights from social theory, philosophy, cognitive science, anthropology, and other disciplines. A unified systemic framework for the understanding of biological and social phenomena will emerge only when the concepts of nonlinear dynamics are combined with insights from these fields of study" (ibid., p. 82).

Living systems and social systems are not to be conceived only in their specificities, but also in their inter-relations: Capra affirms a connection between the developments in consciousness and at the social level: "A unified view [...] is now emerging in which human consciousness is inextricably linked to the social world of interpersonal relationships and culture" (ibid., p. 32). "The understanding of reflective consciousness is inextricably linked to that of language and its social context. [And] the understanding of social reality is inextricably linked to that of reflective consciousness."

25 When discussing Morin's notion of socio-organization in the next chapter, I will also shortly discuss another interdisciplinary stream of research, "sociobiology," which exaggerates the structuring role of genes, over socialization processes.

Capra also stresses the feedback from social structures, to human mind, to physical neurons: "The ideas, values, beliefs, and other forms of knowledge generated by social systems constitute structures of meaning, which I shall call 'semantic structures.' These semantic structures, and thus the network's patterns of organization, are embodied physically to some extent in the brains of the individuals belonging to the network. They may also be embodied in other biological structures through the effects of people's minds on their bodies, as, for example, in stress-related illnesses" (ibid., p. 91). The understanding of these pervasive interconnections is in the background of a 'sensibility to patterns that connect' (that I will introduce in chapter 4).

Meaning in social systems

For the understanding of social systems, the concept of '*meaning*' has to be added to the triad form-structure-process that characterizes living systems in general under Capra's perspective. He stresses the importance of meaning in social reality as a hermeneutic dimension: "human language, being of a symbolic nature, centrally involves the communication of meaning, and [...] human action flows from the meaning that we attribute to our surroundings" (ibid., p. 73).

Meaning is a specifically human feature as, contrarily to Aristotle's claim, meaning and its teleology do not operate within most of the nonhuman reality. "[T]he production of material structures in [human] social networks is quite different from that in biological networks and ecological networks. The structures are created for a purpose, according to some design, and they embody some meaning" (ibid., p. 84). "Meaning itself is a systemic phenomenon: it always has to do with context." And such 'meaning' is necessarily co-produced with help of human conscious purposiveness.[26] Furthermore, meaning is especially important for the modern human belief in self-determination: "Because of our ability to project mental images into the future we act with the conviction, valid or invalid, that our actions are voluntary, intentional, and purposeful" (ibid., p. 85).

Therefore, like Luhmann, Capra focuses his analysis of social systems on communication. Both contexts of meaning and social structures arise from communication:

> "To explore the implications of viewing social systems as networks of communications, it is helpful to remember the dual nature of human communication. Like all communication among living organisms, it involves a continual coordination of behavior, and because it involves conceptual thinking and symbolic language it also generates mental images, thoughts, and meaning. Accordingly, we can expect net-

---

26 With a wider understanding of meaning, as in Bateson's extended concept of the mind, 'meaning' may relatively apply to all phenomena in life i.e. to all living systems, and not only to human purposeful consciousness (cf. Bateson 2002).

works of communications to have a dual effect. They will generate, on the one hand, ideas and contexts of meaning, and on the other hand, rules of behavior or, in the language of social theorists, social structures" (ibid., p. 83).

Giddens' structure and agency

Capra directly refers to Giddens' structuration theory and its approach to the duality of structures and agency: Social structures stem from rules enacted in social practice (on the one hand, interpretative rules and schemes, and on the other hand moral rules and norms), and they interact with human agency in cycles: Human agency requires the social structures as preconditions but also reproduces and transforms social structures. Giddens (1984) speaks of "structuration processes" as much as of structures.

Kirchberg (2007) speaks of an "agency-structure feedback loop" being present in Giddens' theory (as well as in Bourdieu's genetic structuralism, with his habitus-field-practical rationality loop – see figures 2 and 3). Giddens recognizes the agentic capacity to see the reactions of other social agents as a reflexive mirror, and he breaks down the artificial separation of structure and agency by pointing out that social structures are the result of social interactions (as well as structuring them in return).[27]

Figure 2: Giddens' structuration theory

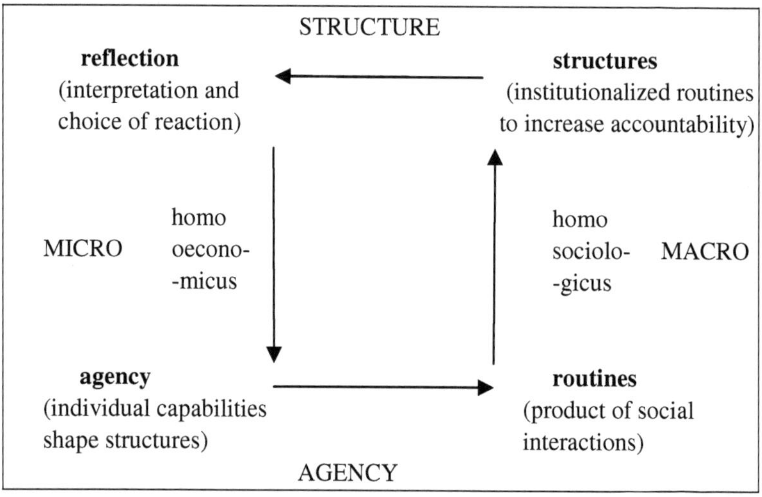

Source: Kirchberg 2007, p. 120

27 Cf. Kirchberg 2007, pp. 118-119.

Figure 3: Bourdieu's genetic structuralism

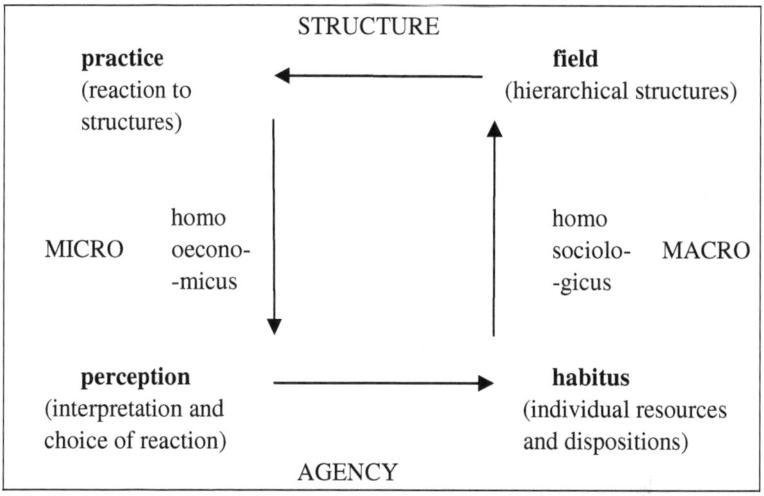

Source: Kirchberg 2007, p. 119

Capra notices that Giddens' "duality of structure" is a form of systems thinking and fits well with Capra's understanding of autopoïesis, even arguing that Giddens "acknowledges the similarity [of his 'duality of structure'] to the circular nature of autopoietic networks in biology" (Capra 2002, p. 79). Capra further observes:

"The conceptual links with the theory of autopoiesis are even more evident when we turn to Giddens' view of human agency. He insists that agency does not consist of discrete acts but is a continuous flow of conduct. Similarly, a living metabolic network embodies an ongoing process of life. And as the components of the living network continually transform or replace other components, so the actions in the flow of human conduct have a 'transformative capacity' in Giddens' theory" (ibid.).

I will come back at a later point, to the role of a human being as social agent in the feedback of social systems, i.e. when introducing the analysis of social conventions and of 'entrepreneurship in conventions' (in chapter 7, section 1). At that point, I will then come back to Giddens, and will also refer to the stream of insights from George Herbert Mead to symbolic interactionism (Harold Garfinkel, Erving Goffman) i.e. sociologists who, before Giddens, already articulated the agency-structure feedback by pointing at the interactive creation of social rules (rituals) by "looking in the mirror through others." (Their insights, as well as those from Talcott Parsons, were acknowledged by Giddens.) It will then become clearer, that a systemic understanding of the structure-agency feedback loop is the prerequisite for strategies of social transformation towards sustainability.

Culture, Technology and Power

As communication and meaning are crucial to a systemic understanding of human societies, Capra's approach gives great importance to culture: "Culture arises from a complex, highly nonlinear dynamic. It is created by a social network involving multiple feedback loops through which values, beliefs and rules of conduct are continually communicated, modified and sustained. It emerges from a network of communications among individuals; as it emerges, it produces constraints on their actions" (ibid., p. 87).

Capra also shows an awareness of the exclusionary and boundary-making properties of culture: "Cultural identity also reinforces the closure of the network by creating a boundary of meaning and expectations that limits the access of people and information to the network. Thus the social network is engaged in communication within a cultural boundary which its members continually re-create and renegociate" (ibid.). Such boundaries can prevent certain communications from the outside to the inside as well as from the inside to the outside. The role of culture is thus central in both social reproduction and social change.

From a systemic perspective, cultural change is inseparable from technological change. "As the culture evolves, so docs its infrastructure – they co-evolve through continual mutual influences. The influences of the material infrastructure on people's behavior and culture are especially significant in the case of technology" (ibid., p. 92). The consequences of technological developments are therefore always more widespread than may appear at first sight. Capra remarks that "defenders of technology do not realize that a specific technology will always shape human nature in specific ways" (ibid.). However, most systems thinkers do not advocate for a freezing of technological developments. "The process cannot be stopped nor the relationship ended; it can only be understood and, hopefully, directed towards goals worthy of [humankind]" (Kranzberg and Purcell 1967, p.11). As Capra stresses, technological developments represent an unavoidable dimension of humanity's history: "Technology is a defining characteristic of human nature: its history encompasses the entire history of human evolution" (Capra 2002, p. 93). However, Capra condemns contemporary technological developments which are based on pre-systemic, insufficiently complex scientific bases and which cater to short-term needs of human societies. A recent example of which is genetic engineering (see boxed text below). I have already discussed, in chapter 1, Ellul's analysis of the Technological System, suggesting a graver situation than what Capra argues. Furthermore, I will point out below (in section 5) the risks associated with a technological reductionism in cybernetics and systems thinking.

What is a sustainable biotechnology?

The integration of systems thinking in the development of new technologies is necessary to the advancement of sustainability. Capra develops this point in the field of contemporary biotechnologies:

Capra strongly condemns genetic engineering as a misconceived and short-viewed practice that intentionally (and, as Capra implicitly concludes, criminally) ignores the recent progress in molecular biology towards an understanding of complexity, and introduces GMOs without having any clue about the consequences thereof.[28] He points out that "gene expression depends on the genetic and cellular environment (the whole epigenetic network) and can change when genes are put into a new environment." Besides, "genes usually have multiple effects and undesirable effects that are suppressed in one species may be expressed when the gene is transferred to another species" (Capra 2002, p. 179). Capra also points out that viruses are used as vectors to introduce foreign genetic elements in species, thereby raising the risk of generating complete new strains of deadly viruses! The reader might wonder why I mention this issue at all. It does also concern artists: I will come back to it when discussing the work of the artist Eduado Kac (in chapter 6).

In contrast to genetic engineering, Capra promotes what he calls another form of biotechnology i.e. biomimicry and ecodesign: New technologies should "understand nature's subtle 'designs' and use them as models [and] integrate ecological knowledge into the design of materials and technological processes, learning from plants, animals and microorganisms how to manufacture fibers, plastics, and chemicals that are nontoxic, completely biodegradable, and subject to continual recycling" (ibid., p. 203). He especially points at industrial ecology and the "cradle to cradle" concept (McDonough and Braungart 2002).

Capra also perceives the importance of the issue of power for social organizations: "In the social realm, the concept of organization takes on an additional meaning. Social organizations [...] are systems whose patterns of organization are designed specifically to distribute power" (Capra 2002, p. 91). However, when it comes to power and political concepts, his analysis shows a severe degree of candidness or naivety, especially when describing power as vested in authority.[29]

Social organizations and social change

A systemic understanding of the structure-agency feedback loop in society, as the prerequisite for strategies of social transformation towards sustain-

---

28 Cf. chapter 6 in Capra 2002.

29 Cf. ibid., pp. 88-89. I will address the political dimension of cultures of (un)sustainability more closely in chapter 7.

ability, also requires an analysis of the meso-social level of the 'organization'.

In contrast to the natural world, Capra observes, change in human social organizations take on different forms and meets specific obstacles. In addressing the difficulties that social organizations have to change, Capra points at the duality of formal and informal structures, serving a duality of purposes: The causes

> "lie in the dual nature of human organizations. On the one hand, they are social institutions designed for specific purposes, such as making money for their shareholders, managing the distribution of political power, transmitting knowledge, or spreading religious faith. At the same time, organizations are communities of people who interact with one another to build relationships, help each other, and make their daily activities meaningful at a personal level" (ibid., p. 99).

This duality engenders a clash between design and evolution in social organizations, i.e. designed formal structures vs. the evolving informal structures: "the designed structure always intersects with the organization's living individuals and communities, for whom change cannot be designed" (ibid.).

Capra's reflection points at formal vs. informal realities in social institutions, which have been one of the foci of attention of organization theory for several decades.[30] W. Richard Scott (2003) categorized organization theories into three perspectives: the "rational system perspective" where formal social structures achieve rational goals (i.e. the oldest and most linear-modernist perspective, unaware of informal processes), the "natural system perspective" where an organization's shared goals are also supported at the informal level by members of the organization, and the "open systems perspective" where "[o]rganizations are congeries of interdependent flows and activities linking shifting coalitions of participants embedded in wider material-resource and institutional environments" (Scott 2003, p. 29) and where formal and informal aspects are not clearly separated, but rather interdependent. The rational system bears no interest at this step of my discussion and belongs to the simplistic discourses denounced in chapter 1 as contributing to a culture of unsustainability. The natural system perspective as defined by Scott, which includes Talcott Parson's structural functionalism as well as Philip Selznick's institutional approach, stresses an organization's drive for survival and the emergence and importance of informal structures. It however shares with the rational system perspective "a substantialist conception of organizations" (ibid., p. 79) and lacks the more elaborate attention to complexity of the open systems perspective. The latter benefits from the insights of cybernetics, and more specifically from the argument that "social organizations, in contrast with physical or mechanical structures, are

30 Cf. Scott 2003, Jaffee 2001.

complex and loosely coupled systems" (ibid., p. 83, pointing at William Ross Ashby and at Walter Buckley), and from Kenneth Boulding's 1956 typology of levels of complexity in systems.[31]

More specifically, the institutional theory of organizations points at how organizations as social institutions are "driven by emotions and traditions" and how subsequently, rational strategies are struck with "unintended consequences" (Jaffee 2001, p. 227, pointing at the work of Selznick in the late 1940's and 1950's). Contributing to bridging the natural system and open systems perspectives, the later development of the institutional theory of organizations further analyzes how the institutional environment of an organization enforces 'rationalized myths' towards which the organization tends to be 'isomorphic' i.e. to adapt (ibid., p. 229, pointing at John Meyer and Brian Rowan's work in the 1970's). As Jaffee argues, the open systems perspective reveals how an organization is "heavily influenced by the environmental forces" (ibid., p. 209). However, that perspective, as discussed by Scott and by Jaffee, ignores autopoïesis (both authors, in their syntheses, do not mention it, nor do they refer to Maturana, Varela or Luhmann – not to speak of Edgar Morin).[32] This is of course not the case of Capra.

Capra places the question of social change within the general context of living systems (as understood in the previous section). "To resolve the problem of organizational change, we first need to understand the natural change processes that are embedded in all living systems. [...] Understanding life means understanding its inherent change processes. It seems that organizational change will appear in a new light when we understand clearly to what extent and in what ways human organizations are alive" (Capra 2002, p. 100). Furthermore, this understanding will further strengthen a culture of sustainability placing humanity within the interconnections of a complex ecology of life.

Organizations as living beings

If an organization is a living being, Capra suggests, it holds some autonomy, and "can never be controlled like machines". This alternative perspective on social organizations bears wide-ranging consequences: "To see the [organization] as a living being [...] is to realize that it is capable of generating itself and that it will naturally change and evolve." Therefore: "It is evident why a management style guided by the machine metaphor will have problems with organizational change" (ibid., pp. 104-105). Capra is here referring to dif-

31 Cf. Scott 2003, pp. 83-92.

32 However, to be fair and precise, I have to mention that Scott does mention that the open systems perspective includes an understanding of "organizations as [functionally] hierarchical systems" in which subsystems are more closely coupled within them than in between them. And thereafter, he points at the "relatively autonomous functioning" of work teams or departments within larger organizations (Scott 2003, pp. 91-92).

ferent metaphors guiding the different existing theories of organizations, as analyzed by Gareth Morgan (1997): the organization as a machine and as an organism (and Morgan identified further metaphors: the organizations as a brain, a culture, a political system, a psychic prison, an instrument of domination, and as flux and transformation). To prevent any confusion, it should be noted however that Capra's organization as "living being" is an especially elaborate variation on the conception of organizations as organism, because it is informed by the systems view of life as introduced above.

Capra recalls how modern management is based on the "mechanistic paradigm that was formulated by Descartes and Newton" (ibid., p. 102); and he warns that "this largely unconscious embrace of the mechanistic approach to management is one of the main obstacles to organizational change today" (ibid., p. 103). Under the perspective of organizations as living beings, certain organizational practices seem wildly alienating: "If organizations were truly living communities, buying and selling them would be the equivalent of slavery, and subjecting the lives of their members to predetermined goals would be seen as dehumanizing" (ibid., p. 104).

Actually, Capra refers to De Geus (1997a) in observing that long-lived (i.e. 'resilient', in systemic terms) business companies are those that most resemble living entities in several aspects: De Geus

> "identifies two sets of characteristics. One is a strong sense of community and collective identity around a set of common values; a community in [which] all members know that they will be supported in their endeavors to achieve their own goals. The other set of characteristics is openness to the outside world, tolerance for the entry of new individuals and ideas, and consequently a manifest ability to learn and adapt to new circumstances" (Capra 2002, p. 105).

But how far can social organizations be said to be 'alive'? "Living social systems [...] are self-generating networks of communications [thus] a human organization will be a living system only if it is organized as a network or contains smaller networks within its boundaries" (ibid., p. 106). Capra further refers to Manuel Castells about the rise of "Network Society". "For an organization to be alive, however, the existence of social networks is not sufficient; they need to be networks of a special type [i.e.] self-generating [through the interplay between communication, thought and meaning:] the entire network generates itself, producing a common context of meaning, shared knowledge, rules of conduct, a boundary, and a collective identity for its members" (ibid., pp. 107-108). Capra refers to Etienne Wenger (1998), labeling such networks as "communities of practice":

> "Wenger defines a community of practice as characterized by three features: mutual engagement of its members, a joint enterprise, and, over time, a shared repertoire of routines, tacit rules of conduct, and knowledge. In terms of our conceptual framework, we see that the mutual engagement refers to the dynamics of a self-generating

network of communications, the joint enterprise to the shared purpose and meaning, and the shared repertoire to the resulting coordination of behavior and creation of shared knowledge"(ibid., p. 108).

Capra identifies the communities of practice to the informal networks of the organizations: "Various communities of practice invariably arise and develop within the organization's formal structures. These are informal networks [...] that continually grow, change and adapt to new situations. [...] The organization's aliveness resides in its communities of practice" (ibid., p. 109).

Organizations as learning beings

Capra argues that the 'interplay' between the formal organizational structures and the "informal self-generating networks" has to be understood in order to allow the organization to learn.

In the informal networks, "the shared practice creates flexible boundaries of meaning that are often unspoken. [Furthermore,] when new people join, the entire network may reconfigure itself; when people leave, the network will change again", unlike the formal organization where functions persist. "Formal policies and procedures are always filtered and modified by the informal networks, which allow workers to use their creativity when faced with unexpected and novel situations" (ibid., p. 110).

"Ideally, the formal organization recognizes and supports its informal networks of relationships and incorporates their innovations into its structures [...] The formal parts of the organization may be 'alive' to varying degrees, depending on how closely they are in touch with their informal networks" (ibid., pp. 110-111). Capra advises to "let the formal structures handle the routine work and rely on the informal organization to help with tasks that go beyond the usual routine. [For this, one has to] support and strengthen its communities of practice. The first step in this endeavour will be to provide the social space for informal communications to flourish."

With the living system model, Capra follows one problematic string of thought following Maturana and Varela: autopoïesis. Indeed, a "machine can be controlled; a living system, according to the systemic understanding of life, can only be disturbed" (ibid., p. 112). However, and unlike Luhmann's, Capra's understanding of autopoïesis in social organizations allows for autoecopoïesis too – with the notion of openness to disturbance.[33] "Force or energy are not the issue; the issue is meaning. Meaningful disturbances will get the organization's attention and will trigger structural changes" (ibid.).

Capra also draws the implications for the leadership of social organizations, if organizational learning and resilience are the goals: "Offering impulses and guiding principles rather than strict instructions evidently

33 See below, and Capra 2002, p. 117.

amounts to significant changes in power relations, from domination and control to cooperation and partnerships" (ibid., p. 113).

Capra's analysis meets a whole school of thought in the field of management, especially around the works of Chris Argyris and Donald Schön (1978) and of Peter Senge (1990) on organizational learning, i.e. relying on informal processes for the management of organizations. I will shortly discuss Senge's *Fifth Discipline* which is, explicitly, an elaboration of systems thinking into management.

Senge draws concrete conclusions from the reality of informal networks in organizations. For example, he warns against the delusion of learning from experience: In a system in which each of us is a specialist, "we each have a learning horizon" beyond which we don't experience the consequences of our decisions – we only learn from what we perceive; and our perception is too fragmented (unless a collective learning process, at a systemic level, is undertaken).[34] Thus, we first need to understand how we are the system in order to change it effectively.[35] Senge points at several aspects of organizational learning, such as dealing with mental models, building shared vision, and especially achieving team learning.[36]

About team learning, Senge refers to David Bohm's notion of dialogue. A quantum physicist by profession, Bohm was also exploring philosophical and social questions. For him, *dialogos* is "a free flow of meaning between people, in the sense of a stream that flows between two banks [...] The purpose of dialogue is to reveal the incoherence in our thought" (Bohm quoted in Senge 1990, pp. 240-241). Senge adds: "In dialogue people become observers of their own thinking. [...] If collective thinking is an ongoing stream, 'thoughts' are like leaves floating on the surface that wash up on the banks. We gather in the leaves [and] misperceive [them] as our own, because we fail to see the stream of collective thinking from which they arise" (Senge 1990, p. 242). Bohm identified three basic conditions for dialogue: "All participants must suspend their assumptions, literally to hold them as if suspended before us; all participants must regard one another as colleagues [to suspend the vulnerability and acknowledge the mutual risk that each takes in opening oneself to inquiry]; there must be a facilitator who holds the context of dialogue" (ibid., p. 243). "Great teams are not characterized by an absence of conflict [...] In great teams conflict becomes productive" (ibid., p. 249). A learning team is also able to identify defensive routines and turn them into a signal showing where learning is not occurring for each

34 Cf. Senge (1990), pp. 21-54.

35 Cf. ibid., pp. 43-44.

36 On mental models, cf. ibid., pp. 174-204; on shared vision, cf. ibid., pp. 205-232; on team building, cf. ibid., pp. 233-269. I carried out an unpublished, more detailed summary of Senge's *Fifth Discipline*, which is available online at: http://sachakagan.chez-alice.fr/docs/Senge.pdf

member. "It is not the absence of defensiveness that characterizes learning teams but the way defensiveness is faced" (ibid., p. 257).

Senge points at organizational learning as a creative force: "In everyday use, learning has come to be synonymous with 'taking in information' [... but] through [real] learning we re-create ourselves [...] generative learning [is] learning that enhances our capacity to create" (ibid., pp. 13-14). This conclusion is also shared by Capra, who then links this creativity to the phenomenon of emergence.

Organizational learning, social change and novelty/creativity

In his discussion of organizational learning, Capra points at Polanyi's distinction of explicit vs. tacit knowledge. He stresses that, "although knowledge is always created by individuals, it can be brought to light and expanded by the organization through social interactions in which tacit knowledge is transformed into explicit knowledge [...] For Polanyi, tacit knowledge is always a precondition for explicit knowledge. It provides the context of meaning [...] which arises from a web of cultural conventions. [Tacit knowledge] is generated collectively" (Capra 2002, p. 115). And because knowledge is not "an entity that is independent of people and their social context – a thing that can be replicated, transferred, quantified and traded", Capra advises to "strengthen communities of practice" (ibid., p. 116).

To further understand creativity and novelty in social organizations, Capra turns to the phenomenon of "the spontaneous emergence of new order": "The phenomenon of emergence takes place at critical points of instability that arise from fluctuations in the environment, amplified by feedback loops" (ibid.). Capra describes how such a phenomenon may concretely proceed through a social organization:

> "In a human organization, the event triggering the process of emergence may be an offhand comment, which may not even seem important to the person who made it but is meaningful to some people in a community of practice. Because it is meaningful to them, they choose to be disturbed and circulate the information rapidly through the organization's networks. As it circulates through various feedback loops, the information may get amplified and expanded, even to such an extent that the organization can no longer absorb it in its present state. When that happens, a point of instability has been reached. The system cannot integrate the new information into its existing order; it is forced to abandon some of its structures, behaviors, or beliefs. The result is a state of chaos, confusion, uncertainty, and doubt; and out of that chaotic state a new form of order, organized around new meaning, emerges. The new order was not designed by any individual but emerged as a result of the organization's collective creativity" (ibid., p. 117).

The systemic perspective elucidates why creativity and the emergence of novelty may appear as 'magical': "Since the process of emergence is thoroughly nonlinear, involving multiple feedback loops, it cannot be fully analyzed with our conventional, linear ways of reasoning and hence we tend to experience it with a sense of mystery" (ibid., p. 119).

But such emergence requires from the social organization, a certain openness to disturbance, which is typical of what I called an autoecopoïesis (Kagan 2010): "To begin with, there must be a certain openness within the organization, a willingness to be disturbed, in order to set the process in motion" (Capra 2002, p. 117). Capra argues that this openness is a "basic property of life" because survival depends on a constant flow of incoming resources.

Capra discusses encouraging openness to disturbances in organizations: "Facilitating emergence includes creating that openness – a learning culture in which continual questioning is encouraged and innovation is rewarded. Organizations with such a culture value diversity and, in the words of Arie de Geus, 'tolerate activities in the margin: experiments and eccentricities that stretch their understanding'" (ibid., p. 123 – quoting De Geus 1997). Capra calls forward, in social organizations, a "right balance between design and emergence" (ibid., p. 121), i.e. between the stability provided by formal structures and the flexibility and creativity provided by informal structures. This echoes the analysis of Karl Weick (whom Scott categorizes in the open systems perspective of organization theories), who articulated already in 1969 an analysis of "organizing" processes as evolutionary, iterative processes "directed toward the establishment of a workable level of certainty" (Weick quoted in Scott 2003, p. 98) while maintaining a level of ambiguity because "organizations continue to exist only if they maintain a balance between flexibility and stability" (Weick quoted in Scott 2003, p. 99).

In other words, as argued by David Jaffee (after Karl Weick and Frances Westley who pointed out the contradictory tension between organizing and learning), "the concept of a learning organization invites disorganization" (Jaffee 2001, p. 176). This insight gives a glimpse into the importance of grasping the dynamic tension between organization and disorganization, which I will further discuss in the next chapter, following Edgar Morin.

Capra also discusses the obstacles to such openness in social organizations, most especially in market-oriented, shareholder-satisfying private businesses. He blames the market's "economic pressures" for the inability of social organizations to evolve and deplores "the life-draining nature of our current economic system" (ibid., p. 126). Unlike most standard economists who would value the market drive for 'creative destruction' and competitive innovation in so-called 'niche markets', systems thinking here points out that market pressures rather drive short-sighted developments in businesses that stifle their long-term evolution, offering little if any ground for creativity and emergence: "Being creative means being able to relax into uncertainty and confusion. In most organizations this is becoming increasingly difficult, because things move too fast" (ibid.). Capra also deplores thereafter, the lack of time-space for reflexivity in modern social organizations; an analysis echoing the analysis of sociologists Scott Lash and John Urry on "reflexive losers" in contemporary late-modern societies.[37]

Capra's systemic vision for sustainability in the global age

Ethics of sustainability based on systemic, ecological literacy

Capra's advocated ethics of sustainability for the age of globalization, is of a planetary scale:

"Ethics refers to a standard of human conduct that flows from a sense of belonging. When we belong to a community, we behave accordingly. In the context of globalization, there are two relevant communities to which we all belong. We are all members of humanity, and we all belong to the global biosphere. We are members of *oikos*, the 'Earth household' [...] What is sustained in a sustainable community is not economic growth or development, but the entire web of life on which our long-term survival depends" (ibid., p. 214).

Sustainability vs Sustainable Development

In his advocacy for sustainability, Capra implicitly rejects the misconceptions embedded in the mainstream conception of 'sustainable development', which assumes economic growth to be the normative target for humanity.[38] Sustainable development is indeed a failing concept in at least two ways:

- Behind the self-evidence of economic growth, lies a hyper-materialistic belief in consumer-welfare as synonymous to happiness and 'good life'. However, the supposed link between happiness and development is questionable (Guerra Gonzalez 2009).
- As a new euphemism for the continuation of modernist 'development' discourses and their culture of unsustainability (cf. chapter 1), sustainable development is a misleading concept, the effect of which is to undermine the vision of 'sustainability' and drive the people away from a concept that they rightly consider as a politically correct, empty formula.

As a result, the concept of sustainable development both reinforces mod-

37 Cf. Dieleman 2008, pp. 116-117.

38 On the history and fallacies of the concept of development and of its later 'sustainable' avatar, cf. Latouche 2004 (especially pp. 13-24 and pp. 51-68). However, rather than Latouche, for a systemic understanding of an economy based on ecological insights, see René Passet's "bio-économie" (Passet 1979, and Ed. Harribey 2004 pp. 192-197). Habbibey and his colleagues at ATTAC argue that the concept of development can and should be salvaged, within Passet's bio-economic conception of the Earth as open system nourished by solar energy.

ernist cultures of unsustainability, and generates suspicions against the adjective 'sustainable', which appears as no more than an especially fashionable 'greenwashing' tool.

Capra points out the fundamentally systemic character of the concept of sustainability: "we need to realize that sustainability – in ecosystems as well as in human society – is not an individual property but a property of an entire web of relationships: it involves a whole community" (ibid., p. 215). Understanding sustainability is a matter of systemic, ecological literacy:

"The key [...] is the realization that we do not need to invent sustainable communities from scratch but can model them after nature's ecosystems. [...] Sustainable communities evolve their patterns of living over time in continual interaction with other living systems, both human and non-human. Sustainability does not mean that things do not change: it is a dynamic process of coevolution rather than a static state. [...] The first step in our endeavor to build sustainable communities must be to become 'ecologically literate', i.e. to understand the principles of organization, common to all living systems, that ecosystems have evolved to sustain the web of life" (ibid., p. 230).

This understanding is based on a "systemic understanding of life" (as described in the previous sub-section of this chapter), where all these systems "communicate with one another".

Capra proposes six principles of ecology that are critical to sustaining life, as the basis of ecological literacy (or 'ecoliteracy') and stresses very strongly that "Ecoliteracy [...] should be the most important part of education at all levels" (ibid., p. 232).

Capra's six principles of ecology

"*Networks*

At all scales of nature, we find living systems nesting within other living systems—networks within networks. Their boundaries are not boundaries of separation but boundaries of identity. All living systems communicate with one another and share resources across their boundaries.

Cycles

All living organisms must feed on continual flows of matter and energy from their environment to stay alive, and all living organisms continually produce waste. However, an ecosystem generates no net waste, one species' waste being another species' food. Thus, matter cycles continually through the web of life.

Solar Energy

Solar energy, transformed into chemical energy by the photosynthesis of green plants, drives the ecological cycles.

Partnership

The exchanges of energy and resources in an ecosystem are sustained by perva-

sive cooperation. Life did not take over the planet by combat but by cooperation, partnership, and networking.
Diversity
Ecosystems achieve stability and resilience through the richness and complexity of their ecological webs. The greater their biodiversity, the more resilient they will be.
Dynamic Balance
An ecosystem is a flexible, ever-fluctuating network. Its flexibility is a consequence of multiple feedback loops that keep the system in a state of dynamic balance. No single variable is maximized; all variables fluctuate around their optimal values " (ibid., p. 231)

At the human social level, the systemic ethics that Capra advocates include "the respect of cultural integrity, cultural diversity, and the basic right of communities to self-determination and self-organization" (ibid., p. 215). His defense of self-organization follows from Capra's views on social organizations, i.e. his analysis that informal self-organization is also the most effective form of social organization, in evolutionary terms. His defense of cultural diversity is in line with the value of resilience in nonlinear systems, which requires diversity; as already mentioned above, the structural stability of nonlinear systems is maintained by multiple pathways; or in Laszlo's words: resilient systems don't just 'work' or 'fail', but have a 'plasticity' which allows them to "act as dynamic, self-repairing wholes in regard to any deficiency" (Laszlo 1996, p. 86).

The values of the global civil society

Capra's political positioning is unequivocally supporting the sort of 'global civil society' movement that is often associated to the independent social-political organization ATTAC: Capra is a strong supporter of the 'Seattle Coalition' and World Social Forums, i.e. a global network of organizations criticizing economic globalization and exploring alternative models of civilization.[39] "The new global NGOs have emerged as effective political actors who are independent of traditional national or international institutions" (Capra 2002, p.219). He also links them to the social movements of the 1960's against traditional values of "patriarchy, the domination and control of nature, unlimited economic growth and material consumption" (ibid.; see

39 Relatively similar political positioning can be found among some other systems thinkers (e.g. Laszlo) and thinkers of complexity (e.g. Morin).

also Castells 1997). Looking at the case of the anti-MAI campaign[40], he observes:

> "Whereas the treaty was discussed in financial and economic terms by the OECD delegates, the NGOs used language that highlighted its underlying values. In doing so, they introduced a broad systemic perspective while at the same time adopting a more direct, frank and emotionally charged discourse. This is typical of the new civil society, which not only uses global networks of communication but is also rooted in local communities that derive their identities from shared values. This analysis is consistent with Manuel Castells' assertion that political power in the network society derives from the ability to use symbols and cultural codes effectively for framing the political discourse" (ibid., p. 221).

"As Castells emphasizes, this means that the power battles of the Information Age are cultural battles" (ibid., pp. 156-157 – referring to Castells 1998 p. 348). Castells argued indeed that political struggles, in the age of the "Network Society" that emerged in the last two decades of the 20th century, are increasingly fought on cultural battlefields: It "will be a cultural politics that [...] is predominantly enacted in the space of media and fights with symbols, yet connects to values and issues that spring from people's life experience" (Castells 1998, pp. 352-353).

The new media of the 'information age' are not *per se* unsustainable, in Capra's perspective. Although Capra, following Castells, criticizes the global financial network that emerged from the information revolution of the 1970's (i.e. as consequences of a technological development combined with political deregulation, and as already discussed in chapter 1), he values positively the development of the Internet, which open structure he claims, following Himanen (2001), to stem from the counterculture values of the 1960's. "The irreverent attitudes, strong sense of community and cosmopolitan sophistication of the sixties formed the cultural background of the informal, decentralized, cooperative, and future-oriented working styles that became characteristic of the new information technologies" (Capra 2002, p. 135).

Capra lists the core values of this global civil society movement as building-blocs for politics of sustainability, i.e. subsidiarity, diversity, a certain conception of food production and a defense of global commons:

> "the creation of new rules and structures that favor the local and follow the principle of subsidiarity ('Whenever power can reside at the local level, it should reside there'); the respect of cultural integrity and diversity; a strong emphasis on food security (local self-reliance in food production) and food safety (the right to healthy and

40 The MAI was the "Multilateral Agreement on Investment", which the OECD was negotiating behind closed doors in the late 1990's (cf. Observatoire de la Mondialisation 1998).

safe food); as well as core labor, social, and other human rights [...] the new rules would also concern goods that belong to the global commons, that is, goods that are part of the fundamental building blocks of life or of humanity's common inheritance. Included in this are goods like bulk fresh water, which should not be traded but should be given away to those in need; seeds, plants, and animals that are traded in traditional farming communities but should not be patented for profit; and DNA sequences that should neither be patented nor traded" (ibid., p. 224).

An ecoliterate industrial revolution

Rather than arguing in favour of de-growth (as Serge Latouche does, in the tradition of Illich), Capra prefers to place his hopes and advocacy into an ecologically literate industrial revolution, based on ecodesign and industrial ecology: "We need to apply our ecological knowledge to the fundamental redesign of our technologies and social institutions, so as to bridge the current gap between human design and the ecologically sustainable systems of nature" (ibid., p. 233). Such a shift "introduces an era based not on what we can *extract* from nature, but on what we can *learn* from her" (Benyus 1997, p.2). Capra argues, as do McDonough and Braungart (2002): "The first principle of ecodesign is that "waste equals food." [...] A sustainable business organization would be embedded in an "ecology of organizations," in which the waste of any one organization would be a resource for another" (Capra 2002, p. 234).

An exemplary Case: the ZERI network

Capra pays tribute to one specifically exemplary case of industrial ecology and ecodesign in action: The ZERI network:

"ZERI consists of three types of interconnected networks. One is the ecological cluster of industries, patterned after the food webs in nature's ecosystems. Closely associated is the human network of the local community where the cluster is located. The third, finally, is the international network of scientists who provide the detailed knowledge necessary to design clusters of industries that are compatible with local ecosystems, climatic conditions and cultural settings. Due to the nonlinear nature of these interconnected networks, the solutions they produce are multiple, or 'systemic,' solutions. The combined value created by the whole is always greater than the sum of the values that would be generated by independently operating components. [...] It is not difficult to see that the principles of organization underlying the ZERI concept—the nonlinear network structure, cycling of matter, multiple partnerships, diversity of enterprises, local production and consumption, and the goal of optimizing instead of maximizing – are basic principles of ecology. This is of course, not coincidental. The ZERI clusters are impressive examples of ecoliteracy embodied in ecodesign" (ibid., pp. 239-240).

Capra also welcomes "the emergence of an international 'ecocity' movement, which tries to counteract urban sprawl by using ecodesign principles, to redesign our cities so that they become ecologically healthy" (ibid., p. 246, referring to Register 2001).

An important insight from the case of ecodesign, as highlighted by Capra, is that, if one of the important points about the cultural dimension of sustainability is to correct an excessive belief in technology (and to deconstruct the hold of the 'Technological System' – cf. chapter 1), the technological question should however not be ignored nor turned into a taboo. What Capra demonstrates is that the dimension of 'ecoliteracy', i.e. a cultural (and social-organizational) dimension, matters a lot to an ecological reform of technological change and industrial design. A cultural-ecological movement can also heal human technologies from Modernism (but this will require a structural deconstruction of positivism and of the Technological System, as well as a humble realization of the inherent limitations in human technologies thanks to complexity theories – which I will introduce in the next chapter). However, this does not mean that Capra or Braungart's optimism about the applicability of ecodesign at the global scale in the medium term (i.e. before global collapse would occur), is warranted. Indeed, a naive belief in the ineluctable advent of an ecoliterate industrial revolution may turn societies away from an urgently needed critique and rejection of mass-consumerism and hyper-materialistic development. This is however not the kind of discourse that Capra is holding on green technologies. His discourse indeed also stresses the need for a radical economic and wider cultural shift, insisting "that the transition to a sustainable future is no longer a technical nor a conceptual problem. It is a problem of values and political will" (Capra 2002, p. 257).

A service economy[41]

Capra welcomes the rise of an "economy of service" which is opposed to an economy based on the ownership of purchased objects. "If the concept of technical cycles were fully implemented, it would lead to a fundamental restructuring of economic relationships. After all, what we want from a technical product is not a sense of ownership but the service the product provides" (ibid., p. 241). Capra calls for the establishment of an economy of service and flow:

41 Besides advocating for a service economy, Capra also shortly evokes State intervention in the economy, advocating for ecological tax reform: "The taxes people pay in a given society ultimately reflect that society's value system. Hence, a shift to taxation that encourages the creation of jobs, the revitalization of local communities, the conservation of natural resources, and the elimination of pollution reflects the cores values of human dignity and ecological sustainability" (ibid., p. 259).

"From the perspective of ecodesign, it makes no sense to own these products and to throw them away at the end of their useful lives. It makes much more sense to buy their services, i.e. to lease or rent them. Ownership would be retained by the manufacturer, and when one had finished using a product, or wanted to upgrade to a newer version, the manufacturer would take the old product back, break it down into its basic components—the technical nutrients—and use those in the assembly of new products, or sell them to other businesses. The resulting economy would no longer be based on the ownership of goods but would be an economy of service and flow" (ibid., pp. 241-242).

Capra further believes that his economic model would "align the interests of manufacturer and customers when it comes to product durability" (ibid., p. 243).

Capra is aware that his 'program' is on a collision course with the current 'program' of economic globalization:

"Whereas global capitalism is concerned with electronic networks of financial and informational flows, ecodesign is concerned with ecological networks of energy and material flows. The goal of the global economy is to maximize the wealth and power of its elites; the goal of ecodesign to maximize the sustainability of the web of life. These two scenarios—each involving complex networks and special advanced technologies—are currently on a collision course. We have seen that the current form of global capitalism is ecologically and socially unsustainable" (ibid., p. 262).

A cultural shift

At the wider cultural level, Capra looks for a way out of consumerism, which he associates to patriarchal cultures (as already discussed in chapter 1). As a remedy to the patriarchal dimension of the dominant culture of unsustainability, he promotes feminism, "overcoming our obsession with material consumption", with

"a new understanding of masculinity and personhood that does not need to associate manhood with material possessions. At the deepest level, feminist awareness is based on women's experiential knowledge that all life is connected, that our existence is always embedded in the cyclical processes of nature. Feminist consciousness, accordingly, focuses on finding fulfillment in nurturing relationships rather than in the accumulation of material goods" (ibid., p. 265, referring to Spretnak 1981).

Capra's discourse points repeatedly at the potential of disturbances, creativity and emergence for social transformation towards sustainability; properties which require a certain productive openness to disturbances that I proposed (in Kagan 2010) to call 'autoecopoïesis', in contrast to the limitations of the strict social 'autopoïesis' as theorized by Niklas Luhmann. Capra stresses that "meaningful disturbances can trigger multiple feedback processes that may rapidly lead to the emergence of new order" (Capra

2002, p. 267). But he also is aware that "these points of instability may lead to breakdowns rather than breakthroughs" (ibid.). Capra takes up Vaclav Havel's definition of hope, i.e. "not the conviction that something will turn out well, but the certainty that something makes sense, regardless of how it turns out" (Havel 1990, p. 181).

SECTION 4: ERVIN LASZLO'S "SYSTEMS VIEW OF THE WORLD" AND THE LIMITS OF HOLISM

One of the available versions of systems thinking is championed by systems theorist Ervin Laszlo, who wrote the third report commissioned by the Club of Rome (Laszlo 1977) and who founded the 'Club of Budapest' in the late 1980's, as well as the 'General Evolution Research Group'. Originally a concert pianist, Laszlo left his career as musician to work in academia, across the philosophical and scientific fields, from the 1960's onwards.

In his *Systems View of the World* (1996) Laszlo provides an analysis of systems that partly differs from Capra's. Most especially, Laszlo's analysis does also include physical systems, i.e. not only life forms but also the physical world, not only bacteria but also atoms, opening up a much wider definition of systems. "The systems view gives a new meaning to the notion of matter, as a configuration of energies that flow and interact, and allows for probabilistic processes, for self-creativity, as well as for unpredictability" (Laszlo 1996, p. 11).

However, his views also promote a 'holistic' understanding of systems thinking, which, while claiming to overcome the limits of disjunctive reductionism, introduces another reductionism: the reduction to the whole. Laszlo also over-stresses the mechanisms that maintain systems stable and increase order, over other mechanisms. In this sense, holism is almost as much prone to "simplifying thought" (as defined after Morin in chapter 1) as is the more classical form of reductionism.

Laszlo's analysis focuses on "natural systems", which he differentiates from the realm of the "artificial" as "any system which does not owe its existence to conscious human planning" (Laszlo 1996, p. 19).

Natural systems

Laszlo categorizes natural systems into three classes: suborganic systems (including e.g. atoms), organic systems (i.e. biological systems) and supra-organic systems (i.e. societies).[42] He then makes "4 interrelated propositions, each grasping an organizational invariance" (ibid., p. 25) and that

42 Laszlo 1996, p. 27.

"traverse the spheres of physical, biological and social phenomena" (ibid., p. 24).

The whole is more than the sum of its parts

Laszlo's first proposition is that "natural systems are wholes with irreducible properties" (ibid., p. 25). This property apply as well to atoms as it does to organisms. The structure and the properties of an atom is not just an addition of the properties of its constituents.[43] The same argument goes for biological organisms. "The difference between Caesar and the chimpanzee is not a difference in substance but in the relational structuring of the substance" (ibid., p. 28). And the same applies to social organizations: "what makes a group what it is, is not just its membership, but the mutual *relations* of the members" in the form of communication. "It is in virtue of such communication that social institutions and organizations can act as entities in their own right" (ibid., p. 29).

Maintaining a steady state

Laszlo's second proposition is that "natural systems maintain themselves in a changing environment", i.e. unlike "ordinary objects" like man-made machines, they can "obtain their own requisite energies to run and keep themselves in good repair" (ibid., p. 30). They form a local and temporary exception to the second law of thermodynamics i.e. entropy. "The technical definition of a natural system is 'open system in a steady state'. Openness refers to the energy import activities of the system, which it needs to [...] maintain its own steady-state" (ibid., p. 32).

Laszlo encounters difficulties with his second proposition for some of the suborganic systems he analyses. Most especially, the atom behaves most of the time as a closed system, but according to Laszlo, it sometimes behaves as an open system maintaining a steady-state.[44] As to the organic systems, the steady state is one of their typical characteristics: Organisms "undergo a slow but inevitable exchange of *all* their parts [...] even when conditions change around them" (ibid., p. 34). Not only individual organisms, but also species tend to maintain their steady state, through reproduction processes. "The individual now becomes like the ripple on the surface of a larger wave in the sea: the individual, like the ripple, is local and temporal, while the species, like the wave, is vast and ongoing. Yet all the ripples together define the curvature of the wave itself" (ibid., p. 36). This does not only concern single species, but also ecosystems and the biosphere as a whole: "The whole of nature [...] is something like a vast, self-regulating and recycling system, drawing energy from the sun and running itself without sur-

43 On the atom as system, cf. Laszlo 1996, p. 27. For a specific discussion of the atom, cf. the boxed text in this volume, upcoming in chapter 3.

44 Cf. Laszlo 1996, p. 33; also discussed in chapter 3, in the said boxed text.

pluses and with minimal waste. It is a beautifully balanced mechanism" (ibid., p. 36). Coming to the case of societies with regard to his second proposition, Laszlo points at "rules, regulations, and laws, and even principles by which we stand" (ibid., p. 37) and argues that "social structures adjust and adapt, maintaining themselves in a dynamic steady state" (ibid., pp. 38-39).

Self-creativity and evolution

Laszlo's third proposition is that "natural systems create themselves in response to self-creativity in other systems". He defines self-creativity as "a response to changing conditions which cannot be offset by adjustments based on the existing structure" and see this ability as "a precondition of evolution [...] Natural systems evolve new structures and new functions; they create themselves in time" (ibid., p. 39). This is an "ability of systems to generate the very information which codes their structure and behavior" (ibid., p. 40).

Laszlo points at a general process of co-evolution, which sets the framework for the evolutionary selection of random mutations:

"Assume that there are a number of at least partially organized objects sharing a field, space, or surface—what topologists call a landscape. Each of these objects is open to some influences from the environment and responds to these influences. Hence each object influences all others, directly or indirectly, by communicating with its own environment. Now, as each natural system (for these are the 'objects' we are interested in) receives and responds to inputs from its fellows, it provides new inputs for the others. And so each system constantly challenges the others by responding itself to such challenges. There is interdependence among the systems—as with points along a net, when one is displaced, all others suffer some displacement, corresponding to where they are relative to the moving point. In natural systems, of course, the points themselves move and produce movement affecting the others—active responses, not merely passive effects. In virtue of the connectedness of all points, there is coordination in the behavior of all systems, and an overall pattern sooner or later emerges" (ibid., p. 40).

From the co-evolutionary process emerges a relative order: "The result is an ongoing reduction of chaos and the patient emergence of discernible order all along the network of systems" (ibid., p. 41).

(However, Laszlo overemphasizes the progression of order, as will be argued further below.)

At the suborganic level, Laszlo observes that "there is a generally one-way build-up of elements during the chemical evolution of stars. Starting with the lightest of all elements, hydrogen, transmutation processes fuse the lighter into the progressively heavier nuclei, building up from hydrogen into helium, and into the still heavier elements" (ibid., p. 45). Atoms and molecules become increasingly complex. "Their interaction results in increasingly integrated levels of organization" (ibid., p. 46). Laszlo describes a similar development at the level of organisms and argues that "if evolution is indeed going anywhere, it is going from the simple system to the complex one" (ibid., p. 47) even if the process is not perfectly linear or if certain exceptions occur. But Laszlo avoids here any teleological misinterpretation: "Complexity of structure or function is not a goal of evolution, it is a result of it" (ibid.).

At the social level, Laszlo perceives a similar development: "Social development conduces [...] toward more complex forms of organization" (ibid., p. 51). In modern society, he mentions the trend toward differentiation (already described by Luhmann – cf. chap. 1) and observes that "the changes tend to be in the direction of greater structuration and improved technology" (ibid., p. 52). He concludes: "we see the tremendous increase in interdependence, complexity and differentiation. The basic laws of development hold true in the supraorganic realm, as they do in the organic and suborganic sectors" (ibid., p. 53). But he also observes that "self-stabilizing feedbacks" such as social institutions are "change buffering", i.e. they prevent change until change comes by crisis (ibid., p. 42). This later observation echoes Capra's warnings against the blockages caused by designed formal social structures (vs. informal social structures which are not designed but emergent).

Coordinating interfaces

Laszlo's fourth proposition is that "natural systems are coordinating interfaces in nature's holarchy" (ibid., p. 53). Different systems (which are also systems of systems most of the time) inter-coordinate with each other.

Laszlo insists especially on the cooperative abilities of cells as part of "complex living systems" i.e. of multicellular organisms:

> "Each variety of cell contributes the integrated energies and functions at its disposal to the community, and in turn receives an environment which enables it to survive. The cooperative endeavor of cells is so tightly knit in higher organisms that the community itself acts as one, and is in fact a system in its own right. [...] Particular cells can provide, as can particular organs, fresh inputs to the systems they constitute. Thus they constitute an influence from within" (ibid., p. 55) which may reform the system, or be rejected, or cause e.g. cancer.

More generally, Laszlo then goes on to describe a process that Morin describes as 'emergence from emergence from emergence' (cf. chapter 3). In Laszlo's words (about the level of suborganic systems): "Established structures jointly constitute new pathways and these, becoming established as structures in time, serve as templates for the production of new systems of flows. The patterns become complex" (ibid., p. 62) and more and more inter-coordinations are operating. And "organisms are likewise steady-state patterns imposed on a continuous flow, this time of free energies, substances (rigidly integrated energies), as well as information (coded patterns of energies). The input-output channels of organisms can further solidify into pathways of definite structure, and the nature of these pathways, plus the organic systems themselves, define the supraorganic (ecological or social) community" (ibid., p.63).

The systems view of the world

"The systems view of the world is non-anthropocentric, but it is not non-humanistic for all that. It allows us to understand that we are one species of system in a complex and embracing holarchy of nature, and at the same time it tells us that all systems have value and intrinsic worth. [...] The status of the human being is not lessened by admitting the amoeba as his kin" (ibid., p. 91).

Laszlo argues that "the self-creating universe is our larger self – our primary sacred community" (ibid., p. 90) and that, to realize this situation, "the rational discoveries of science need to be complemented by affective, basically spiritual insights" (ibid., p. 88). He affects this spiritual role to religions and claims: "They would only need to draw on their own humanism and ecumenism to encourage creative thinking in regards to the elaboration and extension of their traditional insights" (ibid., p. 88-89). "By recognizing and celebrating the world's evolutionary self-creation, religions could promote this process of recognition in each individual" (ibid., p. 90). Laszlo seems here to smooth away all the difficulties related to religions.

Laszlo rightly points out that certain simplifications are necessary at certain points for analytical procedures, also in systems thinking: "Whereas traditional reductionism sought to find the communality underlying diversity in reference to a shared *substance*, such as material atoms, contemporary systems theory seeks to find common features in terms of shared aspects of *organization*" (ibid., p.17). Laszlo points at "invariances of organization", i.e. abstracting structures from systemic organization. However, Laszlo's own simplifications and his defense of 'holism' turn into a form of reductionism, which I will now expose in more detail.

A critique of Laszlo: reductionism by 'holism'

Misunderstanding of emergence and of the irreducibility of 'parts'

Laszlo seems not to recognize fully the principle of 'emergence', i.e. the creation of a new logic at the level of a system, that no analysis of the interactions of the parts alone, could ever account for (even if performed by Maxwell's daemon, as Morin would argue). Indeed Laszlo claims: "Of course, it is quite possible that we could fully account for the properties of each whole if we could know the precise characteristics of all the parts and know in addition all existing relationships between them. Then we could reduce the characteristics of the whole to the sum of the characteristics of the parts in interaction" (ibid., p. 5).[45] Indeed, such an operation would be reductionist, and would unavoidably miss the properties of the whole that are emergent.

Laszlo also too easily ignores the irreducibility of individuality, in a typically holistic reductionism:

For example, he claims "that detail is a specification of some more general trait and must be comprehended within the latter as the relevant context" (ibid., p. 20). However, as we will see with Morin's theory of complexity (in chapter 3), "the parts are also more than the whole".

Too much order

Laszlo oversimplifies the phenomenon of complexity in the following claim: "Organization in nature comes to resemble a holarchic pyramid, with many relatively simple systems at the bottom and a few complex systems at the top" (ibid., p. 53). However, as Laszlo himself would have to agree, complexity is already high at the bottom of physical reality (i.e. at the non-systemic, peculiar level of the quanta).

Laszlo's account of evolution unfortunately runs too neatly into the direction of a sort of 'Great Order' gradually constructing itself. Laszlo argues that there is a general evolution "progressing from a state of great multiplicity and little coordination to a state of highly coordinate general forms of order" (ibid., p. 43). This first argument, already mentioned above, makes sense from a certain perspective. But Laszlo goes further in over-generalization:

> "In evolution there is a progression from multiplicity and chaos to oneness and order. There is also progressive development of complex multiple-component individuals, fewer in number but more accomplished in behavior than the previous entities. [...] We cannot see how evolution could fail to push toward organization and integration, complexity and individuation, whatever forms it may choose for realization" (ibid., p. 44).

In this sense, Laszlo's understanding of evolution comes close to theories developed in the New Age movement and among spiritualists (e.g. Teilhard de Chardin, Rudolf Steiner, Kenneth Wilber).[46] They share in common an exaggeration of the phenomenon of order (and neg-entropy) in the evolution of the universe as a whole, ignoring the opposite tendencies that always bring more disorder together with more order (e.g. entropy, disorganization, chaos). Laszlo even uses the image of a "cosmic cathedral of systems" growing in the universe (Laszlo 1996, p. 62). They even share a misunder-

---

45 Laszlo claims this possibility in principle, but affirms its impossibility in practice, due to the overwhelming complexity of the necessary calculations.

46 Cf. for example Wilber 2007 (who is however probably relatively less simplifying than Laszlo, in his discourse).

standing of self-organization for some sort of teleology (Laszlo writes that there are "guidelines" and that "there is purpose"). Here again, Morin's theory of complexity will do better justice to these aspects of evolutionary processes in the universe. We will see that Laszlo's 'order through evolution' is not an entirely false claim, but is an oversimplification of a much more complex process. Holistic reductionism cannot serve as the basis for a culture of sustainability, although it lured many with its harmonious outlook: Indeed, Laszlo's "holistic vision is one of harmony and dynamic balance. Progress is triggered from below without determination from above, and is thus both definite and open-ended" (ibid., p. 58). But the holistic description of Nature's "dynamic balance" fails to account for its necessary dimensions of conflicts, competitions and chaos.

On the contrary, Laszlo ignores these dimensions and imposes a deterministic discourse which claims that chaos is receding:

> "In the world of organized complexity the arrow of time does not determine which pathway is taken by individual systems, only in what direction their paths converge. The general irreversibilities of organization include the progressive differentiation of existing systems, the merging of smaller systems within large unities without loss of individuality, and the increased level of communication among systems on their own hierarchical level. A corollary of these processes is the increase in the level of order in the largest suprasystem. In the language of information theory, this means that the amount of 'noise' is reduced and replaced by 'signals'.[47] In a process that includes any stops and starts and occasional fallbacks, the system itself becomes less open to chance, more regular and lawlike. Randomness is on the wane, determination on the rise" (ibid., p. 83).

He then observes: "There is empirical evidence that such indeed is the pattern of development in the sociocultural sphere. [...] Increasing communication among a finite number of national and multinational systems can only result in greater mutual determination among them. As the ration of noise to signal is reduced through wider channels of effective international communication, the world will become more and more like a single unit" (ibid., pp. 83-84). Laszlo fails to notice that most of the increase in contemporary communication is also an increase in noise.[48] Furthermore, Laszlo also fails to consider the challenge posed by the differentiation of social systems as analyzed by Niklas Luhmann, and thus the fragmentation of modern society that results.[49]

---

47 Unlike Morin (cf. next section), Laszlo fails to criticize the Shannonian information theory.

48 For example, Morin (1977, pp. 338-339) speaks of a "diaspora" and "degradation" of information into noise, in the context of the modern media.

49 See the pages on Luhmann in chapter 1.

Ill-advised objectivity claims on questions of values

Laszlo makes the claim that "objective value norms can be deduced directly from the contemporary systemic world picture" (ibid., p. 78). This apparent oxymoron may only be a poor choice of words, given that the author then explains, very clearly, that there is nothing 'value-free' from a systemic perspective.[50] In his description of supposedly universal values, Laszlo further speaks of "basic natural values" and claims: "We find that our objective basic values are those which we share with all natural systems. Each of us 'must' (in the sense that he or she cannot help but) commit himself to survival, creativity, and mutual adaptation within a society of his peers; the alternative to these is isolation and death" (ibid., p. 80). It seems that Laszlo is equating 'objectivity' with natural, or rather, living systems. Thus, he qualifies the living systems' survival imperative as a set of 'objective values'.[51] The resulting discourse delivers here again a risky oversimplification of the question of ethics.

Laszlo inquires into "intrinsic human norms", refers in passing to the Aristotelian concept of happiness, and settles for a definition of happiness as *self-fulfillment*:

> "[It] is the end of human purposeful behavior. It is the actualization of potentials inherent in all of us. It is the pattern of what *can* be, traced in actuality. Individual fulfillment can be a human value. And it can be specified and analyzed in the systems perspective. [It] means the realization of human potentials for existence as a biological and a sociocultural being. It means bodily, as well as mental health. It means *adaptation* to the environment as a biological organism constituting an irreducible whole of its parts, and as a sociocultural role carrier collaboratively constituting the multiperson systems in his or her society. Fulfillment also means *acting on* the environment, both the internal one of the organism and the external one of the society, and making it compatible with the expression of one's potentials" (ibid., p. 82).

Although Laszlo's defense of self-fulfillment makes some sense within the perspectives of systems thinking and sustainability, labeling such a value statement as an "intrinsic human norm" bears the risk of carrying a harmony-focused, once again oversimplified (and even essentialist) discourse, which misses how the dynamic balance of ecosystems functions (i.e. not only through self-fulfillment and harmony), as well as how rich and how complex human societies and human ethics are. However, Laszlo's definition of self-fulfillment does offer an especially interesting line of tension,

50 But it may also reflect a back-door return of the concept of "natural law as divine law", which would then explain the excessive focus on order and the lack of focus on emergence in Laszlo's discourse.

51 Unlike Morin (1980), Laszlo ignores here the internal confrontation between the individual survival drive and the species survival drive (which translates itself in the urge to reproduce).

along which a dynamic balance is to be found ethically: the tension between adaptation and acting.

Combining social determinism and the ethics of self-fulfillment

Laszlo combines his advocacy for 'self-fulfillment' with his systemic, but over-deterministic analysis of society:

> "Society can exhibit macrodetermination based on the functional autonomy of individual persons. [....] Our humanistic goal is to enhance individual fulfillment in an increasingly deterministic multilevel society composed of greatly differentiated individuals. Fortunately, this is a feasible endeavor. [...] What is needed is a reorientation of our cultural values in reference to the norms of individual fulfillment in a flexible and dynamic yet fully functional social system" (ibid., p. 87).

The combination of two different forms of over-simplifications (order-focused social determinism and reduction of ethics to 'self-fulfillment') that Laszlo here operates does not yield much more conviction than do these two strands of thought considered separately, despite the author's engaging style.

From the "Systems View of the World" to the "Integral Theory of Everything": Drifting further into holistic reductionism

The reader might suggest that my critique of Laszlo, basing itself on only one of his work, may be unfair. However, the later publications by this author drift even further into a simplifying holism. In *Science and the Akashic Field: An Integral Theory of Everything* (2004, 2nd updated edition in 2007), which is considered as his great contribution to 'Integralism', Laszlo claims (as the title suggests) to develop an overall theory for all aspects of reality in our universe. But this is not yet the worst. Most concerning is Laszlo's claim that a unified field of energy (of course located at the quantum level – which allows all forms of wild speculations by New Age writers) is conveying information at the root of all dimensions of reality (including human psychology and spirituality). This results from his will to obtain coherence with the integration of all scientific knowledge, and is reinforced by his claim that evolution is not a random process, but an "informed" process. In its discussion of consciousness, as noted by Jennifer Gidley, Laszlo's theory of everything is also echoing several elements from Rudolf Steiner's anthroposophy, and can be compared to other such theoretical developments by 'Integralist' thinkers.[52]

---

52 Cf. Gidley 2007, pp. 29-31. A fair critique of the "Integral Studies" movement would require looking more closely and extensively into Rudolf Steiner's, as

Laszlo (2007) argues "that complexity is generated, and can be explained, by basic and relatively simple starting conditions" (Laszlo 2007, pp. 11-12).[53] His key argument is that 'in-formation' is a fundamental, coherence-generating, quantum reality linking "everything to everything else" (not to be confused with 'information', a phenomenon emerging with life, as Bateson (1979) argued). "In-formation is a subtle, quasi-instant, non-evanescent and non-energetic connection between things at different locations in space and events at different points in time" (Laszlo 2007, p. 68), i.e. allowing "nonlocal" interactions as are found in quantum physics.[54] Laszlo then metaphorically speaks of this in-formation field as of a 'sea' on which all other things (from asteroids to human minds) would be traveling, informed by the sea-waves and themselves imprinting on the sea-waves' modulations (so that the quantum void "could carry information on the state of the whole universe" everywhere and across time).[55]

My aim is not to refute all of Laszlo's specific claims as necessarily wrong, but to suggest that, far from offering a "theory of everything", his discourse only highlights the aspects of reality that reinforce a holistic worldview, and thereby, that he instigates a simplifying vision of reality. Even if Laszlo's thesis about a quantum-level, in-formation field, were right, and may explain some elements of "coherence" across the physical, biological and the level of consciousness (e.g. according to Laszlo, quite an eclectic ensemble including Prigogine's dissipative structures, the coherence of multicellular living organisms, universal/cross-cultural archetypes in human societies, and potential 'psi' phenomena like telepathy), it would still not be a 'theory of everything', unless one ignores all the other aspects of reality that are not showing only the coherence that fascinates Laszlo.[56]

Interestingly, in his discussion of the organization of life, Laszlo does not discuss the concept of autopoïesis, which is relevant to the evolutionary phenomena he is evoking. More generally, he seems to ignore the possibility to explain the emergence of complexity by self-organization that "beats the odds" of chance. His only way to explain the statistically improbable development of complexity, is that it has to have been pre-'in-formed', i.e. some-

well as Ken Wilber's and other Integralist authors' writings, and into Jennifer Gidley's review of the different streams in Integral Studies.

53 Laszlo refers here to John von Neumann, and his argument is also valid for chaos theory, i.e. a small set of algorithms can generate a process to unfold which is increasingly complex. But Laszlo goes farther in assuming that all of reality (including not only "every physical thing" but also "life, mind, culture and consciousness" - ibid., p. 11) could be understood under a unified set of principles, if not algorithms.

54 I will discuss the phenomenon of non-locality again shortly (after Nicolescu), in chapter 3.

55 Cf. Laszlo 2007, pp. 70-71.

56 On these puzzles of coherence, cf. ibid., pp. 24-58.

how precoded. Laszlo's limit, it seems to me, is his incapacity to conceptualize emergence. This conceptual blind spot appears especially strongly in some of the wildest speculations in *Science and the Akashic Field*, which I do not wish to discuss at length, but will now shortly mention: Laszlo is of the conviction that the degree of complexity of our universe, which allows the evolution of life and of mind, has been in-formed (through the 'Akashic', in-formation field) by the evolution of previous universes within our meta-verse (or multi-verse – following recent cosmological theories according to which the quantum void generates multiple universes). Laszlo thus posits an evolutionary process of universes, with earlier universes which would have been less complex.[57] But even then, in Laszlo's logic, the issue of original in-formation is only pushed backwards, until he eventually comes to the suggestion that the design-for-evolution-toward-ever-more-complexity of the metaverse, has to be an intelligent design, i.e. a godly creation.[58]

One of the wildest points in Laszlo (2007)'s discourse is reached when he argues that some extra-terrestrial civilizations must have already solved the problem of living sustainably on their biosphere with their own advanced technologies, and that we should let ourselves be inspired by them, or rather, in-formed, thanks to telepathic insights (given that all their knowledge is of course available via the Akashic field).[59] Because, as Laszlo states very clearly, in his view "[a]ll that happens in one place happens also in other places; all that happened at one time happens also at all times after that. Nothing is 'local', limited to where and when it is happening. All things are global, indeed cosmic" thanks to the in-formation field (Laszlo 2007, p. 80). Although the appeal to alien telepathy is very imaginative indeed, I personally shall not acknowledge it as an important contribution toward achieving a culture of sustainability on our planet in the contemporary era.

Laszlo's moral world is divided in two, between those scientists who look for progress and order, and those who believe in a "fall from grace", in the form of a direction-less universe. The later "express the negative face of Western civilization", in his view (ibid., p. 3). In this binary alternative, there is no space for a genuine complexity in which the progress of order can co-exist with the expansion of disorder (with tensions and mutual exploitation between the two – as will be proposed in the next chapter). This also helps to understand his bias toward coherence.

Laszlo himself expresses his bias toward simplicity in the autobiographical appendix of his theory of everything, stating after a quote taken from Einstein (who himself was restating the rule of Occam's razor), that "[t]he attempt to 'create the simplest possible scheme of thought that will

57 Cf. ibid., pp. 82-86.

58 Cf. ibid., p. 87.

59 Cf. ibid., p. 99.

bind together the observed facts' [...] defined my intellectual agenda" (ibid., pp. 162-163).[60] However, as Gregory Bateson (1979) argued, the Occam's razor rule of parsimony does not allow to search for the complexity of reality, and provokes a confusion of logical types (or in the words of quantum physicist Basarab Nicolescu, of levels of reality). I will further describe Bateson's arguments against the rule of parsimony in the next chapter.[61]

The type of unifying theory advocated by Laszlo stands in starch contrast to a theorization in terms of complexity which, as will be introduced in the next chapter, requires to think of multiple levels of reality rather than of merely one unifying field.

A more elaborate consideration of the complexity of the multiple systems of reality (and not only of their connectivity) is thus warranted. Such an undertaking, initiated by systems thinkers such as Bateson, Capra and up to a certain point, Laszlo too, was however only brought to existence by an author who does not identify himself as a systems thinker exactly, but rather as a thinker of complexity: Edgar Morin. The following chapter will introduce Morin's insights.

But before this, another problematic development in systems thinking will be identified: the risk of technological reductionism in cybernetics and systems thinking, as it was perceived by Bateson and Morin.

SECTION 5: CYBERNETIC APPARATUSES AND THE RISK OF A TECHNOLOGICAL DRIFT

Some of the limits and risks of cybernetics (which was a major step in the 20th century history of science, allowing the emergence of systems thinking) have been pointed out by one of the founders of cybernetics, Gregory Bateson, as well as by Edgar Morin (whose views on complexity will be introduced in their own right, in chapter 3), regarding the threat of a technocratic drift of this powerful approach. This threat does not concern directly the

60 Another acknowledged motivation for Laszlo's bias toward holistic unification, is a mystical personal experience (which is recounted in the preface to Laszlo 1993). This oriented him towards theories of "psi" i.e. parapsychology. Although I myself was shortly a student of the empirical parapsychologist Dr. D. Bierman at the Universiteit van Amsterdam, and am thus not biased against such explorations, I am aware-enough of the state of research in that peculiar field of research, to think that no single theory can account for all purported ‚psi' phenomena, even if many clues orient towards some form of connectivity beyond current explanatory frameworks. However, I would characterize such connectivity as an intermittent and weak background with probably only limited significance for other levels of reality, and not as a unified 'Akashic field' as argued by Laszlo.

61 Cf. chapter 3, section 1.

systems thinkers I discussed above (Capra, Laszlo), but it should not be overlooked, as it remains at least a latent risk for systems thinking, next to the risk of holistic reductionism.[62]

Bateson's early warnings

Although, or rather because he was one of the early co-developers of cybernetics and systems analysis, Gregory Bateson was also one of the early critiques of the possible risks associated to this form of knowledge and way of thinking. Bateson was especially concerned with the risk of a technocratic degeneration of cybernetics. As a result, his discourse was cautious and open about the further development and the uses of systems analysis:

"I think that cybernetics is the biggest bite out of of the Tree of Knowledge that mankind has taken in the last two thousand years. But most of such bites out of the apple have proved to be rather indigestible — usually for cybernetic reasons. Cybernetics has integrity within itself, to help us not be seduced by it into more lunacy, but we cannot trust it to keep us from sin. For example, the state departments of several nations are today using games theory, backed up by computers, as a way of deciding international policy. [...] The result is a system in which the rules of international interaction become more and more rigid. [...] The question is not the best thing to do within the rules as they are at the moment. The question is how can we get away from the rules within which we have been operating [...] The problem is to change the rules [...] And, of course, there are other dangers latent in cybernetics and many of these are still unidentified. [...] But this much is sure, that there is also latent in cybernetics the means of achieving a new and perhaps more human outlook, a means of changing our philosophy of control and a means of seeing our own follies in wider perspective" (Bateson 1973, pp. 452-453).

Edgar Morin's critique of artificial machines and of cybernetics, vs. nature's "être-machines" (being-machines)

Edgar Morin links his criticism of cybernetics to a critique of the modern, impoverished conception of the 'machine'. For Morin, the machine we usually think of, the artificial machine, is just a "fragment of prosthesis" of the social megamachine, "incapable of regenerating, repairing, reproducing, or organizing itself [unlike] the least bacteria [...] it does not itself possess the least *generativity* [neither] the least *poiesis*, the least creativity" (Morin 1992, p. 167).[63] The artificial machines, even cybernetic ones, are thus very

62 I especially suspect certain artistic discourses dealing with technology, of being techno-optimistic or techno-naive and influenced by the limited insights of the first cybernetics.

63 Generativity will be introduced in chapter 3.

gross and very unsustainable by themselves, owing their existence to the social megamachine in which they are generated and by which they are manipulated and controlled.

The artificial prosthesis "develops the praxis of the anthropo-social megamachine. But [...] while reflecting, expressing and prolonging social creativity, artificial machines, in their poverty and rigidity, reflect, express and prolong the organizational poverty and rigidity of the societies which produce them" (ibid., p. 168).

In his criticism of artificial machines, Morin also introduces a critique of cybernetics: Indeed, although it offered some insights on the functioning of machines, cybernetics tended to assimilate all physical organizations to its models of self-regulated automata. It "imposes a narrowing of the vision [...] *Technocratic simplification and denaturing constitute the durable effect of such a reductionist extrapolation*" (ibid., p. 171).

One shall notice, however, that Morin's critique of cybernetics is not indiscriminate: Morin does not ignore the insights brought by cybernetics, which he himself further develops: With concepts of feedback, regulation, teleonomy, communication, programs, cybernetics offered *"the first general (that is to say, physical) science having organization as object* [i.e.] envisioning a physical system, the machine, not in function of its constitutive elements, but in function of its organizational characters" (ibid., p. 249).[64]

Such a cybernetic view not only ignores the specificities of life (that will be analyzed by Maturana and Varela): it cannot even properly account for the working and spontaneous regulations of what Morin calls "arkhe-machines" (e.g. stars).[65]

The cybernetic view, if it highlights only the dimensions of regulation and programmes, will cast a shadow on what is a-functional, disorderly and on any creativity, all considered as noise. This is "the basic methodological vice of cybernetics, which, by bringing the concept of machine back to artifact, has missed the generativity and the complexity of the machine-being, and therefore could only conceal existence and self" (ibid., p. 233).[66] Cybernetics became thus also a form of simplification, reductionism and repulsion of complexity.

Besides, Morin also criticizes at length the Shanonnian theory of information on which the first cybernetics as well as genetic research based themselves, especially in its incapacity to understand "order from noise", and he builds an alternative theory of information.[67]

Morin further criticizes (especially the first) cybernetics for:

---

64 Morin further celebrates some developments of the "second cybernetics" (with especially Maruyama and von Foerster).

65 Cf. Morin 1977, p. 177.

66 These notions will be further articulated in chapter 3.

67 Cf. Morin 1977, pp. 291-364.

- Subordinating communication to programming through an informational "apparatus"; and in the background, "the principle of Mind commanding Matter, of Man commanding Nature, of Law commanding Citizen, of the State commanding Society became the principle of Information ruling over Organization" (ibid., p. 237);
- Ignoring the issue of power behind the phenomenon of programming (at two levels: both the transformation of information into coercion, and the issue of the power behind the programming i.e. the programmer's power).

The cybernetic 'apparatus' describes this aspect of organization which can turn information into organizational constraint (i.e. program). But, Morin points out, such an internal constraint also contributes to emancipating the organization from external constraints, by opening an extent of "choice" for the organization. However, the more power the programming "servo-mechanism" obtains, the more it "enslaves" the organization of the system as a whole.[68]

The linear "enslaving" in cybernetic automata and technologies stands in contrast with the biological world, in which the "enslaving/enslaved relation" is mutual and complex:

- "plants possess a polycentric network and not a central apparatus";
- "insects have a polycentric ganglionated system";
- and even among the species with a central nervous system, "the more the brain develops in mammals, primates, and hominians, the more it becomes polycentric [...] the more it functions with some 'noise', that is to say disorder, unlike all artificial computers" (ibid., p. 241).

The role of disorder in nature's self-organizing processes, already started to appear in my discussion of Capra and Prigogine, and will be further developed following Morin, in the next chapter.

Furthermore, in nature's "being-machines" ("être-machines"), the whole emerges as a self which puts its servo-mechanisms at its service while being under their "enslavement". Therefore, Morin points at the "radical difference which separates the artificial apparatus which organizes from the genetic and neuro-cerebral apparatuses of living beings [because indeed] they are part of a complete *whole*, whereas the apparatus of the artificial automaton is the instrument of command for the society which manipulates the machines" (ibid., p. 242).

---

68 In the original French text (Morin 1977) the author uses the verb "asservir" which is slightly different from "enslave" but does not have an exact English equivalent. It would rather mean, 'constrain into servicing' (in a less absolute way than "enslavement"). The translator, in Morin 1992, also sometimes uses the verb "to harness" to translate "asservir".

From the informational apparatus to the State apparatus

Moreover, in historical societies, the notion of 'apparatus' also refers to a specific social institution: the 'State apparatus' i.e. the development of State power, which can be seen, at the level of society as a system, as the macro-encephaly of a 'servo-mechanism' that monopolizes power resources, "as the part which controls the whole, and thereby tends to live-off, exploit, enslave both the parts and the whole" (ibid., p. 243). Overall, the State works to "enslave society and organize it into a megamachine" (Morin 1977, p. 247). Morin does not deny the benefits that State structures bring to historical societies, but insists on the price that the social system has to pay, i.e. "irremediable specialization and a strict subordination of the parts [...] aggravation of the enslavement [and especially] duality and profound scission in the unity of the whole" (Morin 1992, pp. 243-244). Here, Morin's analysis (as well as Luhmann's, to some extent) joins a long line of sociologists and philosophers (e.g. Weber, Heidegger, Adorno and Horkheimer, Ellul) who critically observed the processes of "differentiation" especially in Modern Western societies, i.e. the increasing atomization of the unity of social reality.

Morin further warns against the specifically political consequences of the "informational power of the apparatus" that the modern State and/or further social institutions, monopolize: "It is not power over the "means" of production, it is power over the production of production, that is to say social generativity. It is not only the ownership of things, of goods. Mastery is in the mastery of the means of mastery, enslavement of the means of enslavement, control of the means of control" (ibid., p. 353). The modern forms of domination rely on the monopolization of information, knowledge and creativity, i.e. the monopolization of generativity.

Modern societies do not only "enslave" themselves and their 'parts': They also increasingly "enslave" the physis and nature. Once again, Morin insists on the difference with the usual working of ecosystems which should not be confused with the working of modern human societies: "There are enslavements in eco-systems, but eco-systems are not enslavers by themselves: they have no central nervous apparatus, they are organized through the inter-retroactions of the living beings which constitute them [and] the relation command/communication is always complex" (ibid., p. 244).

Human technologies have achieved the "enslavement" of nature's wild 'machines', i.e. fire (most especially when turned into the explosion engine), "eddies and vortices (watermills and windmills)" and finally atomic fission (and maybe within a few decades, atomic fusion i.e. the sun or in Morin's words the great "arhke-machine"). The historical human societies also developed the "enslavement" of the processes and devices of reproduction of plant and animal species, for their domestication, and later on, of the human work-force (under different regimes such as slavery, state bureaucracies or capitalist industries). Morin argues that these developments not only coincide, but also correlate, and that "the harnessing of nature has from the beginning retroacted in a complex way on the beginning of humanity [...] and created the conditions for generalized self-destruction" (ibid., p. 246).

These last observations further raise Morin's concern about the rise and technocratic development of cybernetics: "the harnessing of cybernetic artifacts may be the prelude to a new type of informational harnessing of man by man" (Morin 1977, p. 247).[69] Indeed, the limits of cybernetics suggest such a risk: "Thus, cybernetics gives a skeleton of organization to living beings but takes away their life [...] such a cybernetics is too capable of introducing its absence of life into our individual lives and our social life, whence consequences both debilitating on the theoretical and possibly terrifying on the practical plane" (Morin 1992, p. 251). The reductionism to 'servo-mechanisms' "forces [cybernetics] to conceive biological organization and social organization only as enslavement" (ibid.). Cybernetics also was incapable, Morin argues, to reflect critically on its own roots in the practice of engineering. "By that very fact it becomes the theoretical pseudopod of an enslaving organization of work and of a technocentric, technomorphous, and technocratic practice [and also] like a new offensive of Cartesian mechanistic theory, which this time, not content to limit itself to laying siege anew to the animal, endeavors to annex man and society" (ibid., p. 252). Morin notices that Henri Lefebvre (1967) too, although from a different analytical perspective, warned against the dream of the "cybernanthrope", which nowadays can be found back in those circles that glorify the soon-to-be advent of post-humans thanks to the marvels of nano-bio-technologies.[70]

Such a development of cybernetics poses a political threat: "It can consider society only as a vast machine to be functionalized. Since it is blind to the command of the command and to the reality of apparatuses, it can only serve the dominant social apparatuses, which always pretend to be the Faithful Bearers of Information/Truth, Servants of the Public Good and General Interest" (Morin 1992, p. 253).

In this way, cybernetics, and its possible extension in the discourses of some systems thinkers, is not immune from the risk of technological reductionism. However, a culture of complexity and of transdisciplinarity may allow to overcome this threat, as will be exposed in the following chapter.[71]

---

69 Translation by the author (this text is missing in the English 1992 edition).

70 See for example the case of Victoria Vesna, for an uncritical art-world engagement into posthuman discourses (cf. ed. Hayles 2004).

71 This chapter does not have an own conclusion, but a conclusion for both this and the following chapter will come at the end of chapter 3.

3. Toward Culture(s) of Sustainability, Step Two: From the 'Big Picture' to the Culture of Complexity

INTRODUCTION

Holistic discourses on ecology and on the contemporary crisis of unsustainability, such as Laszlo's system theory, offer only an impoverished version of the sort of 'new enlightenment' that comes from the search for sustainability. A paradigmatic shift far more elaborate is warranted, than what slogans such as "everything is connected" may suggest at first sight. I already mentioned in the previous chapter that the rise of cultures of sustainability requires an elaborate understanding of what it means to be embedded in a complex ecology of life, and that achieving such an understanding is an especially difficult challenge. Meeting this challenge will now require a relatively long theoretical development at a high level of abstraction.

Systems thinking did open up the inquiry into complexity thinking, which aims to overcome two versions of simplification: the simplification of modernist linear thinking, and the simplification of holistic discourses. Complexity theory will now be approached as a further step in the paradigmatic shift, beyond the possible caveats of systems thinking.

I will introduce this inquiry into complexity following closely two authors: First, I will discuss the 'method' of Edgar Morin: Although Morin can be traditionally qualified as a sociologist or philosopher, his work on complexity extends well-beyond these disciplinary limits, and can be considered as a pioneering exploration across most sciences, humanities and philosophy. Morin's insights point at the constant interplay between order and disorder and at the complex evolutionary processes in physics, biology and society, which involve simultaneously complementary, concurrent and antagonistic dimensions, as well as expressions, as well as inhibitions of emergent organizations of reality.

The consideration of complexity will further engage a rethinking of the epistemological fundamentals of human knowledge, which will be carried out by introducing the *Manifesto of transdisciplinarity* issued by Basarab

Nicolescu in the late 1990's. Morin's work preceded the emergence of 'transdisciplinarity' (which is lately knowing a certain level of popularity in German universities). The second half of this chapter will thus elaborate on insights from Nicolescu, a quantum physicist by discipline, and one of the early and foremost proponents of transdisciplinarity. My discusssion of Nicolescu's discourse will also point out the complementarity of his analysis with Morin's.

Finally, in the conclusion of this chapter, the profile of a culture of sustainability, as a new worldview, will be sketched and proposed as the normative basis for the search process of sustainability, synthesizing the insights from this and the previous chapter.

SECTION 1: COMPLEXITY: EDGAR MORIN'S 'METHOD'

Originally a sociologist and philosopher, Edgar Morin developed a work that can, in retrospect, be labeled as 'transdisciplinary', with as his main *oeuvre* and my focus of attention in this section, the development of a *method* that would meet the challenge of complexity across all disciplines of human knowledge. Morin's method was published in overall six volumes, published in between 1977 and 2004: *La nature de la nature* (1977 – English edition 1992), *La vie de la vie* (1980), *La connaissance de la connaissance* (1986), *Les idées* (1991), *L'identité humaine* (2001) and *Ethique* (2004). My focus will limit itself on the first two volumes of *la méthode*, which unfold Morin's reflections on physis and on life, including ecology (and thereby bring a comparability and complementarity with the discourses of Capra and Laszlo discussed in chapter 2). Morin's 'method' establishes the ground not just for a new inter-disciplinary methodology, but for a paradigm change in scientific method.

A warning to the reader: Caution, spiraling thought inside!
An uncommon warning to the reader is required, concerning this first half of the third chapter: The analysis of complexity, following Morin's subtle analysis, will require that I follow him in a spiral argumentation, rather than a linear one. I therefore ask for the reader's patience, because my developments in the following pages will therefore probably not fit the expectations for a linear-discursive progression of arguments. They will be rather meandering through several concepts. Although I am attempting to synthesize here some key insights of Morin's *method*, I cannot reduce it to one or two key concepts of principles from which follow several consequences, derivations, secondary concepts. On the contrary, Morin's approach to complexity posits that there can be no singular Master Concepts ("pas de Maîtres Mots") in an understanding of complexity, but rather macro-concepts made of the complex (i.e. simultane-

ously complementary, concurrent and antagonistic) interactions between a number of concepts. In this specific way, Morin's (a-)method of complexity increases both clarity and confusion (unlike traditional modern theories which aim only at clarity, and unlike some postmodern discourses such as French sociologist Michel Maffesoli's, which aims only at celebrating confusion).[1]

While trying not to maim Morin's complexity, I will guide the reader by marking key words and key expressions in the following pages with ***bold italics***. The progression of the following pages, which has to be linear by the very structure of written language, will develop, with these key words, declinations (rather than derivations) of the principle of complexity, which will nourish each other (i.e. generate multiple feedback) but also emerge and bifurcate from each other. Also in a relatively linear way, I will progressively shift, after Morin, the focus from physical declinations of complexity to biological and ecological declinations.[2]

Morin aims to find back the unity of our human identity as "individual-society-species" (Morin 1977, p.10).[3] But the unity he seeks is not a simple one: It is a unity through diversity. The whole conceptual construction of Morin's method allows to understand what is really meant by "unity through diversity", behind the catchy slogan itself (which is in widespread use in everyday language, e.g. in the German expression 'Einheit in der Vielfalt').

For Morin, the linkages between different dimensions of reality (e.g. physical, biological, cultural) should not just be established with a few bridging disciplines (e.g. chemistry or thermodynamics, between physics and biology); there should be an "organizational" linkage (ibid.). But Morin explicitly and vehemently opposes the idea of a "unitarian theory [because] it always conceals the difficulties of knowledge, that is to say the resistance which the real opposes to the ideal: it is always abstract, poor, 'ideological,' it is always simplifying [...] hanging the entire universe onto a single global

1 For an extreme example of an apology of a method developing toward confusion (while not aiming at irrationality) and amoral relativism, see Maffesoli's *Eloge de la raison sensible* (Maffesoli 2005). Although Maffesoli makes a number of valuable comments on sensibility-based knowing, his discourse lacks the complexity to be found in Morin.

2 However, as the shift from the physical to the biological and ecological level is progressive, I will not divide this section in a clear-cut way into two subsections.

3 References to specific page numbers correspond to the 1981 pocket-book reprint of Morin 1977.

formula" (Morin 1992, p. 9).[4] Not only would a unitary theory be mutilating. It would, by contrast, strengthen the fragmented, disjunctive paradigm: "the poverty of all unitarian attempts [...] confirms the scientific discipline in its resignation to mourning" (ibid., pp. 9-10). The idea that Morin is developing is however, to build and organize the linkages between the isolated islands of human knowledge: It is "the search for a method which might articulate what is separate and link what is disjointed" (ibid., p.10). Morin's goal can thus be understood only if one moves beyond a binary opposition between the idea of a unitary theory (that could encompass everything under one logical roof) and the idea of particular knowledge (that can only reach narrow confines). This is difficult because "minds which live under the empire of the principle of simplification see only the alternative between fragmented research on the one hand, general idea on the other" (ibid., p. 17).

Articulating a culture of complexity as an alternative to the modernist mode of knowing, will require a long development that will involve introducing notions from several scientific disciplines. First of all, I will now introduce Morin's understanding of complexity through the founding notions of chaos, order, disorder and organization, at work in a "tetralogical loop" operating a complex evolution of our universe, and this will allow to more generally introduce, more abstractly, what are 'complex relations' according to Morin.

Complexity

The end of 'pure order'

Morin's quest for transdisciplinarity does not come into a completely hostile scientific field. Besides the development of cybernetics and systems theories as already mentioned, other branches of science brought insights on which Morin's concept of 'complexity' lays its bases: Noticeably, in the 20th century, the physical sciences themselves, departed in several ways from the classical logics of modern science, as shortly described in the boxed text below.

Inspiring developments in physics

- *Thermodynamics*: The 2nd principle of thermodynamics, entropy i.e. the inexorable degradation of energy (Clausius, 1850), was relatively soon reformulated as a principle of the degradation of order (by Boltzmann 1877, Gibbs and Planck), i.e. of an inexorable disorganization of molecules in the universe (towards homogenization). One

4 In Gregory Bateson's terms, a unitary theory would confuse 'logical types' (described below). In Basarab Nicolescu's terms, it would confuse 'levels of reality' (described in the second section of this chapter).

of the insights of entropy is the general improbability of order both within all 'closed systems' and maybe at the level of the universe (and the inexorably high probability of disorder and physical homogeneity). Then, contrary to the assumptions of classical science, any progression of order and organization is something that needs to be explained. The explanation has to focus on the phenomena of "open systems" i.e. systems in which external inputs of energy can flow.[5]

- *Quantum physics*: The discovery of quanta of energy by Max Planck introduced disorder and a wholly different type of logic at the nano-physical level: As Morin points out: "It is not a disorder of degradation and of disorganization. It is a constitutional disorder, which [...] is part [...] of order and organization, while being neither order nor organization! [It] is a disorder which, instead of weakening, creates" (Morin 1992, p. 35). It is not a 'pure disorder' either, as quantum physicist Basarab Nicolescu notes: "Quantum physics and quantum cosmologies show us that the complexity of the universe is not the complexity of a garbage can, without any order" (Nicolescu 2002, p. 38). The insights from quantum physics combine with the insights from thermodynamics: "A second time order and organization are a problem, become an enigma" (Morin 1992, p. 35).
- The cosmological discovery of the *expansion of the universe*: The continuing dispersion of the universe was first observed by Hubble (1930). In the universe as an 'explosive diaspora', the development of stellar, physical organization (and of living organization on earth) and its combination with the omnipresence and progression of disorder, become daunting questions.[6]
- Later developments in thermodynamics, most especially with Ilya Prigogine, allowed to conceive of the generation of *order and organization from disorder* (i.e. Prigogine's "dissipative structures" with instability in entropy generating order).[7] This insight joins Von Foester's *order from noise principle* in self-organizing systems and

5 Cf. Morin 1977, pp. 34-37.

6 Cf. Morin 1977, pp. 39-40.

7 See chapter 2. See also Prigogine 1972 and Morin 1977, p. 42.

8 Cf. Atlan 1970, 1972a, cited in Morin 1977, p. 42.

9 Cf. Morin 1977, pp. 42-45. Nowadays, many competing theories speculate on the earliest moments in the history of our universe, including several theories (brane cosmology, hypothesis of a pre-existing multiverse) in which for example a pre-existing quantum void/field is assumed to already exist before the big bang 'singularity'. Therefore, Morin's doubts from 1977 are still relevant insofar as the mystery of the earliest moment in our universe is not yet solved.

10 Cf. Morin 1977, p. 47.

Atlan's "hasard organisateur".[8] As Prigogine's thermodynamics revealed, organized phenomena can emerge from initial thermodynamic disequilibria (i.e. very small initial variations can sometimes lead to tremendous organizational changes; cf. also chaos theory).

- The *catastrophe theory* of René Thom allowed to understand the structural link between the disintegration of forms on the one hand, and the genesis of new forms (morphogenesis) on the other. Thom's catastrophe is a continuous genesic process, which still occurs today. Morin contrasts this theoretical insight to the 'big bang' theory (which he denounces as simplistic because attempting to reduce the complexity of genesis by rationalizing it into the known laws of our universe, thereby reaching the grandiose irrationality of equating infinitely great energy with an infinitely small spatial point).[9] The catastrophe is the precondition for the organization of physical order, e.g. in the stars where heavy complex atoms are synthesized, and for the constitution of physical matter as we know it.[10]

As becomes increasingly visible through the scientific insights cited above, the condition for cosmogenesis is ***chaos***. Morin stresses: "Chaos is [...] an idea of indistinction, of confusion between destructive power, between order and disorder, between disintegration and organization, between *Hubris* and *Dike* [...] inseparable in the bi-faced phenomenon by which the Universe simultaneously disintegrates and self-organizes, disperses and polynucleates" (Morin 1992, p. 54). The genesic chaos is still at work in the stars, which Morin describes as Heraclitean fire machines, and at the subatomic level, in the "soup of electrons, protons, photons" (Thom 1974 p. 205, quoted in ibid., p. 56).[11] Morin observes that "chaos is permanently below the surface as the infratexture of our *physis*. The atom is the transformation of this chaos into organization" (ibid.). As well, the stars are the chaotic forges out of which simple atoms are transformed into more complex, more organized atoms, and around which planetary systems are constituted. The physis is thus the "daughter of chaos" (Morin 1977, p. 61).[12]

Inequalities and variations in physical reality (even in their very slight forms) are "the matrix of diversity" as Morin insists. "Now, diversity, which

---

11 Morin's concept of "machine" is an extended concept, not limited to the 'artificial' machines as in everyday language, but also including qualities which are not found in artificial, human-made machines.

12 Morin's understanding of chaos should not be confused with postmodernist notions of chaos (as e.g. in Lyotard's *postmodern condition*). Therefore, it does not fall under the critique of Smith and Jenks (2006) whose rejection of "chaos" is pointed at chaos as understood in everyday speech (or in postmodernist discourses) but not at chaos as in 'chaos theory'.

cannot be born outside of the inequality of conditions and processes, that is to say outside of disorder, is absolutely necessary for the birth of organization – which can only be organization of diversity" (Morin 1992, p. 46). He also points out that, unlike the order of classical physics, the genesic disorder of the new physis is not one but plural: "there is not *a* disorder (as there was *an* order) but several disorders: inequality, agitation, turbulence, chance, encounter, rupture, catastrophe, fluctuation, instability, disequilibrium, diffusion, dispersion, positive feedback, runaway, explosion" (ibid.).

Order is born of the singularity and contingency of the universe (i.e. the organization of matter in this universe is locked into a specific format). Besides, ***disorder*** is required for encounters and interactions to occur, giving rise to interrelations, organizations, and order. The emergence of ***organization*** necessitates that interactions become inter-relations. Encounters are random but their effects institute ordered 'laws' of interaction.[13] Once order is established, it has a relative capacity of resistance to disorganization. "*Order and organization, born with the cooperation of disorder, are capable of gaining ground on disorder* [...] they benefit, in sum, from a principle of *natural physical* selection" (ibid., pp. 49-51). This allows to understand how the development of organization, which thermodynamics first revealed to be a general improbability, not only becomes a local probability but also reproduces itself and allows the expansion of a universe made of increasingly complex matter (with e.g. atoms of helium and carbon) at least in some of its regions. "Organization possesses true cosmic power: *the physical principle of natural selection*. In fact, it self-maintains, resists chance, self-develops" (ibid., p. 60). And from the improbable emerge a number of laws (e.g. the law of gravitation) that consolidate the self-selective process.

However, each physical system exists within an environment. And the negative or stationary entropy inside an open system, has to be 'paid' with an increase of entropy in the system's environment. Therefore, order and organization at some places and times also work for disorganization and disorder at other places and times.[14]

The evolution and complexity of complexity

An important methodological consequence of the evolutionary selection of improbability in our universe, is that statistical probability in general should not be taken for granted. "Statistics have no definitive meaning for a universe unique since its origin and in which everything develops uniquely [...] *everything which was constituted as organizer and creator was made outside all statistical probability*. Statistical probability is out of its own depth before all that is innovation, invention, evolution" (ibid., pp. 60-61).

These different concepts can then only be understood in relation to the others: "disorder, order, organization are henceforth bound, *via* interactions,

13 Cf. Morin 1977, p. 51.

14 Cf. ibid., p. 70.

in a solidary loop, where none of these terms can any longer be conceived without reference to the others, and where they are in complex relation, that is to say complementary, concurrent and antagonistic" (ibid., p. 48).

In the universe, complexity is thus both genesic and growing: "The game is therefore, more and more varied, more and more aleatory, more and more rich, more and more complex, more and more organizing" (ibid., p. 52). Morin links the notion of evolution to the growth of organization and systems in the universe: "Systemic time [...] is also the time of evolution. What is evolutive, in the universe, what develops, proliferates, complexifies, is organization [...] it is the interrelations and intercombinations between systems which will be evolutive. [...] This evolution from matter is in fact the evolution of organization" (ibid., pp. 134-135).

Morin talks thus of a 'physical selection principle' whereby emerging qualities are acquired and consolidating organizational relations which were (prior to the emergence of these qualities) improbable. And in this process, "adapting to the fortuities and integrating it into the organization are going to constitute equally a premium of selection" (ibid., p. 135). And at the core of this physical evolution, as a catalyzer, is diversity.

But, as well as complexity, dispersion is also progressing in the universe as a whole: "The second law [of thermodynamics] is universal in what it prohibits: perpetual motion" (Serres 1971, p. 596, quoted in ibid., p. 70).

One shall of course not mistake Morin's notion of physical evolution for a linear progression of complexity! "There is no linear development in complexity; complexity is complex, that is to say unequal and uncertain" (ibid., p. 149). Overall, Morin does not claim to predict the future, and states an incapacity to say whether entropy will eventually win over organization (as many cosmologies suggest), or whether organization will unexpectedly, eventually beat the odds and thus, improbably, prevent the entropic, cold end of the universe.

Morin's evolutionary tale and his genesic logic follow this insight. "From the very beginning a principle of meandering evolution takes shape, by schismo-morphogenesis [and] there develops a dialectic of the improbable and the probable" (ibid., pp. 377-378), where norms emerge from deviance. Life itself, which requires of course a specifically biological understanding, however emerges as a physical phenomenon as analyzed by Prigogine, i.e. as a "polymachine" (see Morin's concept of machine-being further down). The human being, too, is physical, not by its body, but by its active organization as a physical system (see Morin's concept of organization further down).

The physical universe, in its evolution through chaos, is widely different from the design of man-made physical entities -and of the godly 'intelligent design' stemming out of human imagination. Unlike Laszlo's, Morin's understanding of evolution by self-selection, allows to think of cosmological evolution without necessarily having to include an 'intelligent design' (i.e. a godly inception) into the story. And, as will be seen also of natural evolution in the ecosystems of life on Earth, Morin's complexity thinking is at the 'antipodes' of the linear type of modernist, development thinking, thereby constituting the first serious alternative to Modernity's culture of unsustainability (analyzed in chapter 1).

Figure 4: Morin's "meandering evolution"

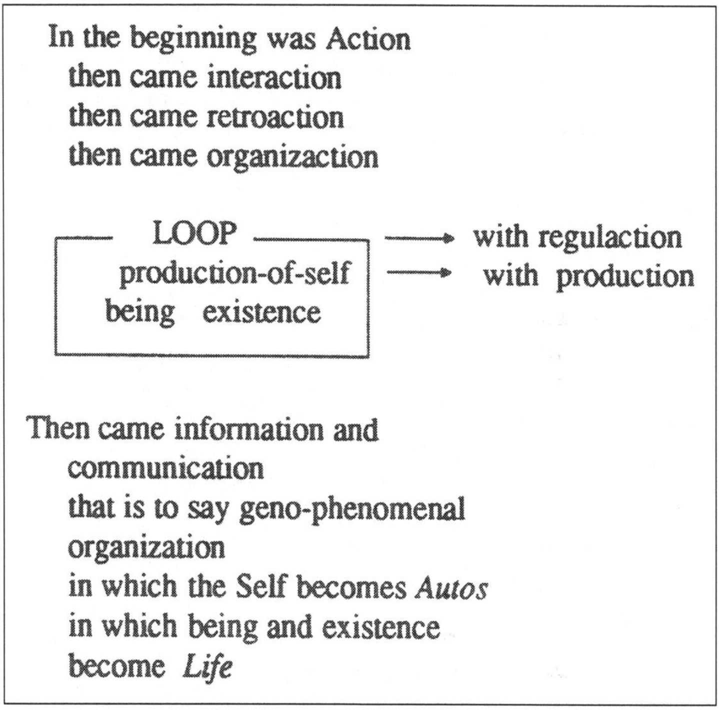

Source: Morin 1992, p. 377

What is complexity?

Repeatedly, across *la méthode*, Morin elaborates how different notions are inter-related in a complex way, i.e. with relations which are (often simultaneously) "***complementary, concurrent and antagonistic***" (the trio of qualifiers that comes back systematically when Morin explicates what is a 'complex relation'). Complex notions, like the notion of chaos, also cannot and should not be conceived as 'clear' notions. Here the 'clear ideas', traditionally sought after by science, should be avoided and mistrusted. "Chaos is beyond our logical intelligibility, it obliges our antagonistic notions to twist towards each other and fasten themselves to each other" (Morin 1992, p. 57).

Based on the principle of complex relations between single notions, Morin elaborates "macro-concepts" that represent complex organizations of reality: The first macro-concept to appear in *la méthode* most especially to explain cosmological evolution, is the "***tetralogical loop***" of the physical universe: "***disorder-interactions-organization-order***" (Morin 1977, p. 56). Apart from the fact that organization and order stem from disorder and interactions (as already noticed), this macro-concept also signifies that "the more organization and order develop, the more they become complex, the more they tolerate, use, indeed necessitate disorder" (Morin 1992, p. 53). The 'tetralogical circuit' of the physical universe constitutes "an immanent principle of transformation and organization" (ibid.).

As Morin insists, in this complex relation, neither order nor disorder is hierarchically superior to the other. As seen above, disorder is also constructive while order is also destructive. Disorder nourishes organization while order nourishes disorganization. Disorder is not one, but plural and ambivalent. Therefore, disorder is not the new king seizing the lost throne of order (after the fall of classical physics). Order is no longer self-evident, but it does gain an ability to evolve itself. It is no longer atemporal and unified, but contextual and plural. Furthermore, order and disorder are interdependently and continuously emerging from chaos, through processes of interactions, transformations and organizations.[15]

Complex relations (complementary, concurrent and antagonistic) institute, no longer a linear logic, but a complex ***dia-logic***: "dialogic signifies the symbiotic unity of two logics, which simultaneously nourish each other, compete against each other, live off each other, oppose and combat each other to death" (ibid., p. 77). A dialogic combines a 'unity' (e.g. the unity of chaos and genesis), a 'complementarity' (e.g. physical organized matter needs disorder to come to existence; neg-entropy also works for entropy), a 'competition' (e.g. dispersion and complexification are competing in writing the history of the universe) and an 'antagonism' (e.g. disorder destroys organizational order while organization dissipates disorders).

Gregory Bateson demonstrated the risks related to Occam's razor, i.e. the rule of parsimony, which enjoins scientists to always prefer the simplest assumptions that will fit the facts, and to base the next predictions on the basis of these assumptions: "All you have is the hope of simplicity, and the next fact may always drive you to the next level of complexity" (Bateson 2002, p. 26). To assume simplicity, Bateson argued, is to take the risk of mistaking one "logical type" (cf. boxed text below) with another, i.e. to risk missing out on the emergence of a higher systemic level where another logic (another logical type) enters into play.

If properly understood, complexity eliminates the very possibility of "universality", even in the universe itself as a whole.[16] Also, rules such as Occam's razor i.e. parsimony and simplicity, no longer make as much sense in the new conception of the universe, full of agitation, dispersion, waste and complementary-contradictions.

---

15 Cf. Morin 1977, pp. 74-78.

16 Cf. ibid., pp. 83-84.

The functionality, rationality and economy of classical rationality then make only limited sense and appears as a "démentielle rationalisation" (Morin 1977, p. 86). "The universe is the game of a child playing chess" (Heraclitus quoted in ibid., p. 85).

Bateson's logical types

Morin's dialogic, introduced in the present section, also bears a relative comparison with Gregory Bateson's developments around "logical types".[17]

Bateson's logical types have to be understood together with his understanding of information and of the mind. To start with, the following preliminary note is thus warranted: a "single thing" (a "Ding an sich") is, in terms of information (in terms of the reality of the 'mind') a non-being. So "it takes at least two somethings to create a difference. To produce news of difference, i.e. information, there must be two entities (real or imagined) such that the difference between them can be immanent in their mutual relationship; and the whole affair must be such that news of their difference can be represented as a difference inside some information-processing entity" (Bateson 1979, p. 64).[18] This is what Bateson calls the "difference that makes a difference". Information, (i.e. reality for the mind) can only emerge out of a triangular relationship, between two things and a third-party – the responder.

Bateson describes "the genesis of information of new logical type out of the juxtaposing of multiple descriptions. In principle, extra 'depth' in some metaphoric sense is to be expected whenever the information for the two descriptions is differently collected or differently coded" (ibid., p. 66). The basic example of that, described by Bateson, is the case of binocular vision, which literally allows to add depth to vision, i.e. to allow three-dimensional vision. 3D vision is of a different logical type than the 2D vision offered by each eye to the brain: "[T]he *difference* between the information provided by the one retina and that provided by the other is itself information of a *different logical type*" (ibid., p. 65). He insists that "the combining of information of different sorts or from different sources results in something more than an addition. The aggregate is greater than the sum of its parts because the combining of the parts is not a simple adding but is of the nature of a multiplication [...] or the creation of a logical product" (ibid., pp. 80-81).

---

17 Bateson borrowed the term from Bertrand Russell.

18 Exact page numbers for Bateson 1979 refer to the 2002 edition at Hampton Press.

Complex relations should also reflexively include the observer-conceiver (i.e. the individual person of the researcher, as well as his social and cultural embededness). For example, the notion of 'disorder' at the cosmological level, also reflects the uncertainty related to the limits of human knowing as well as the uncertainties and disorders in the social realities of the late 20th century.[19] This paradox should not be flattened (as often happens either from a naturalistic or from a constructivist perspectives): The cosmos generates and engulfs the observer that generates and engulfs the cosmos.[20] Complexity thinking, grounding a culture of sustainability, should include such a reflexive dimension.

Overall, Morin does offer an elaborate definition of complexity.

> "Complexity emerges, therefore, at the heart of Oneness simultaneously as relativity, relationality, diversity, alterity, duplicity, ambiguity, uncertainty, antagonism, and in the union of these notions which are each in reference to the others complementary, concurrent, and antagonistic. System is the complex being which is more, less, other than itself. There is no organization without anti-organization. There is no functioning without dysfunction" (Morin 1992, p. 146).

Morin re-asserts the importance of uncertainty all along his work: "complexity will never be totally free from uncertainty, it will never accede to the universe of clear and distinct ideas, since on the contrary it left this universe" (ibid., p. 369). But he also asserts the 'unity in diversity' of complexity, in his search for a 'principle of complexity':

> "It is not sufficient, in order to conceive the principle of complexity, to associate antagonistic notions in a concurrent and complementary way. We must also consider the very character of the association. It is not only a relativization of these terms in reference to each another; it is their integration in the heart of a meta-system *which transforms each of these terms in the process of a retroactive and recursive loop*" (ibid., p. 390).

Morin further warns that this loop should not be 'reified' into a formula but always dynamically move as an invitation to 'generative thinking'.

Systems of Physis

The understanding of complexity, as just introduced, allows Morin to elaborate a number of concepts that explore the complexity of the physical universe as the complexity of an archipelago of systems, where physis is more about organization than about objects, and where the smallest unit of physis, i.e. of organization, is the system. Furthermore, physis as a fundamentally

19 Cf. Morin 1977, pp. 88-92.

20 Cf. ibid., p. 93.

systemic reality, is the theater of operations for the emergence of systems of systems.

The following pages may seem at first to the reader as relatively far away from the discussion of cultures of sustainability. However, a complex understanding of the patterns that connect the living world, have to be grounded in the very systems of physis and in the emergence of life from them, as will emerge in the following pages (and as was already partly discussed with Prigogine in chapter 2).

Physis is organization

First of all, I should explain why, as contemporary physics reveals, physis is about ***organization*** and not about some elementary atom, some basic tiny round object. The new understanding of physical reality is far apart from the view of classical physics. Morin establishes that "everything that ancient physics conceived as simple element is organization" (ibid., p. 91) (atoms, stars, etc.). Indeed, "non-organized particles have hardly any being, any flavor of existence" (ibid.).[21] The quantum particles themselves, which are not proper 'objects', are not even the 'primary element' of physics (which is pushed back to a field, 'bootstrap', 'string' or 'brane' somewhere near the Max-Planck length, none of these later corresponding to the classical notion of an 'object' or 'element'). At the level of the atom, the reductionist, classically 'scientific' approach does not work:

> "Not only is the reductionist explanation no longer suitable for the atom, none of whose characteristics or qualities can be induced starting from the characteristics proper to its particles, but it is the traits and characteristics of the particles which, in the atom, can only be understood in reference to the organization of this system. *The particles have the properties of the system much more than the system has the properties of the particles*" (ibid., p. 95).

Instead, what is required, abandoning the reductionist mode of explanation, is an organizational concept of the atom, which "steps forward as a new object, the organized object or system whose explanation can no longer be found solely in the nature of its elementary components, but is found also in its organizational and systemic nature, which transforms the characteristics of the components" (ibid.). This conceptual revolution, that physicists undertook at the level of the atom, Morin argues, still needs to be expanded into all further levels of scientific inquiry, in order to establish the respective

---

21 In the original French version, Morin speaks not of 'flavor' but of "clignotements d'existence" (Morin 1977, p. 94). Indeed, in the state of "quantum coherence", the quantum particles are said to be in between a state of existence and of non-existence (or in both states), until a "quantum decoherence" emerges, i.e. so to say, reality collapses into existence.

organizational working of different natural systems, from the atoms to the human societies.[22]

Nature is thus to be understood as ***an archipelago of systems*** evolving inside the known cosmos: "Nature [...] is nothing but extraordinary solidarity of mortised systems building one on the other, by the other, with the other, against the other [...] a polysystemic whole" (ibid., p. 97). And each system in this archipelago cannot be understood alone, on its own, but only acquires full significance when put in relation to the other systems: "There exist really only systems of systems, the simple system being only a didactic abstraction" argued Lupasco (1962, p. 186, quoted in ibid.).

But what is a system, actually? Morin argues that the "most interesting" definitions of 'system' link the notion of interrelations between elements and the notion of global unity: e.g. von Bertalanffy (1956): "A system is a set of unities with relationships among them" (quoted in Morin 1992, p. 99); Saussure (1931): "an organized totality, made up of interdependent elements holding together and not able to be defined except one by the other in function of their place in this totality" (quoted in ibid.). Morin adds that the link between the global unity and the interrelations has to be made with a concept of organization. Thereafter, Morin's definition of a system is: "a global unity organized by interrelations between elements, actions or individuals" (ibid.).

Systems are thus especially about organization. But what then is Morin's definition of ***organization*** (i.e. this concept that is omnipresent indeed in his mouth)? Nature does not have an 'initial' principle of organization, he argues (unlike Laszlo), prior to the interactions between elements. A morphogenetic process occurs where the interactions between elements, set by chaos-catastrophe (Thom 1972), generate organization. Morin defines organization on this basis:

> "Organization is the arrangement of relations between components or individuals which produces a complex unity or system, endowed with qualities unknown at the level of components or individuals [...] It assures relative solidarity and solidity to these ties, thus assures the system a certain possibility of duration despite chance perturbations. Organization, therefore: *transforms, produces, binds, maintains*" (Morin 1992, p. 101).

One example from chemistry is that of the isomers (with the same molecular mass and same chemical formula, they have different properties because of the different organization of atoms in the molecules). But organization is

22 Morin observes that several contemporary scientists highlighted in the 1960's-70's the importance of organization: e.g. he refers to Atlan (1974), Koestler's holon (Koestler, 1967) and Chomsky (1967).

also more complex than this first definition suggests (and its further complexity will be appearing throughout the next pages).[23]

System, organization and complexity

From the previous definitions of system and organization, it appears that a system therefore cannot have a simple, substantial identity. It is both homogeneous and heterogeneous; idem for organization, which is both order and variety (and not just a compromise between order and variety). Morin refers to the notion of "***unitas multiplex***" which characterizes quite well the complexity of the physical system.[24] The unity of a system is indivisible and original, but it is also constructed; it can be decomposed but then loses its existence. "[W]e can reduce neither the whole to the parts nor the parts to the whole, neither the one to the multiple nor the multiple to the one" (ibid., p. 102); the parts and the whole have to be understood together in a relationship that is both complementary and antagonistic.

Morin, with the concepts of emergence and of systemic constraints (introduced further down), further defines the system as "*a whole which is formed at the same time as its elements are transformed*" (ibid., p. 112) or in other words a "morphogenesis" indicating a systemic law: "*Everything which forms transforms*" (ibid.). The notion of morphogenesis drawn here shows the temporal (rather than atemporal) character of systems, which also points at the necessary departure from linear logic. Bateson already demonstrated that we cannot understand systems, which operate in time, with the help of simple, timeless logic: "When the sequences of cause and effect become circular (or more complex than circular), then the description or mapping of those sequences onto timeless logic becomes self-contradictory. Paradoxes are generated that pure logic cannot tolerate" (Bateson 1979, p. 54). Timeless, linear logic cannot model causality even in very simple artificial (i.e. man-made) systems, such as a mere "buzzer circuit" as Bateson illustrated.[25]

This leads to the concept of ***emergence***: Indeed, one cannot understand why the system acquires irreducible properties, without the concept of emergence. Earlier in the 20th century, the Gestalt theory stressed the phenome-

23 Furthermore, Morin's concept of organization, is more relevant to complexity than the concept of structure, which is based on a paradigm of order and not a 'dialogic' of order and disorder. But the notion of structure, within a complex conceptual understanding of organization, is useful to describe the stable formal rules of relations and transformations in a system.

24 The expression "Unitas Multiplex" is borrowed by Morin from Angyal 1941 (cf. Morin 1977, p. 105).

25 Cf. Bateson 2002, p. 55

non of globality, which is just one dimension of emergence.[26] Morin identifies emergence at several levels:

- At the global level of the system, feeding back to the elements;
- But also at the level of elements, "micro-emergence" occurs, i.e. "the part becomes more than the part" (Morin 1977, p. 108): some qualities, inherent to the elements, are however absent or only virtual as long as the elements are isolated.

The system has a complex identity in being one and multiple at the same time. But also the parts have a 'double identity': "They have their own identity and they take part in the identity of the whole" (ibid., p. 117).

Emergence allows to see 'the new'. It is discontinuous i.e. does not follow from the logic of the elements when just lined up or lumped together, hence Bateson's insistance on a jump in logical typing, and Morin's point that emergence is "logically nondeductible [and] opens in our understanding the breach through which enters the irreducibility of reality" (ibid., p. 109). Emergence is the outcome of synthesis, but also can retroactively contribute to producing that which produces it.[27] It is both a superstructure (i.e. depending on an infrastructure) and an organizational potential.

The concept of emergence helps to understand 'nature' as an archipelago of systems, because "systems of systems of systems are emergences from emergences from emergences" (ibid., p. 111). Emergence is not only implying logical jumps, but also physical jumps, and thus can help explain why a complex universe can be self-selecting and evolving.

But emergence is not only constructive. It is also destructive. More generally, systems are more complex that described so far. I will thus need now to take a few more pages in order to further expose, after Morin, why systems of physis are complex indeed.

Why systems are really complex

Systems are complex in several ways: They are not just holistically-understood enabling phenomena, but also constraining, inhibiting phenomena (which, however, do not hermetically prevent openings through "eventalities" - this term is explained below). Besides, a system is neither clearly delineated nor clearly objective. Systems are also really complex because they have to manage a qualitative diversity, they have to regulate differences and anti-organization.

Morin warns against "holistic blinding (which sees only the whole)" (Morin 1992, p. 109). Holism ignores that "***the whole is less than the sum***

26 Morin also points at von Foerster's "superadditive composition rule": Cf. von Foerster 1962, pp. 866-867, cited in Morin 1977, p. 106.

27 Cf. Morin 1977, p. 110.

of the parts". Some qualities and properties of the parts are lost when the parts are in the system, i.e. are organized. Organization imposes constraints on the production of possibilities, Morin notes after Ashby (1962). Any association imposes constraints, and systemic order submits elements to multiple constraints, suppressing or inhibiting certain qualities. "Thus, the development of certain systems can come at the expense of a formidable underdevelopment of the possibilities which they hold" (Morin 1992, p. 112).

Especially complex systems, the self-organizing life-forms have elaborate forms of organization that can intermittently constrain or release certain qualities of the parts.[28]

Figure 5: The whole and the parts

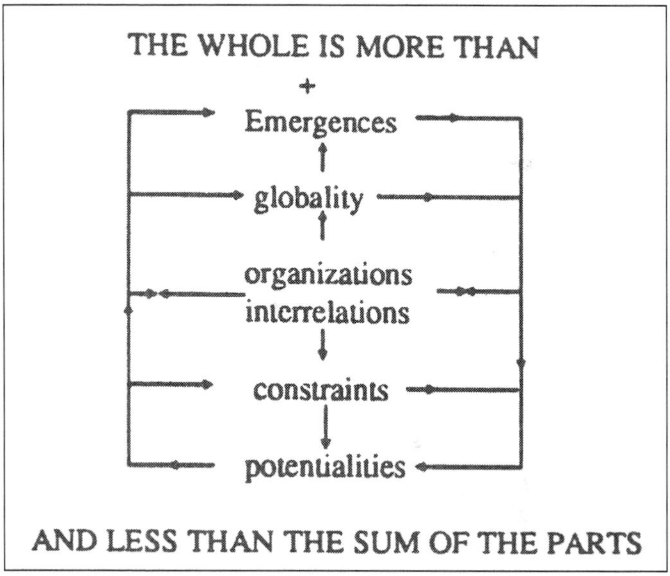

Source: Morin 1992 p. 111.

Thus, a system is often ambiguous and/or ambivalent. "We must henceforth consider in any system, not only the gain in emergences, but also the loss by constraints, subjections, repressions [...] and the impoverishment can be bigger than the enrichment" (ibid.). A system harbors a shadowy zone full of unrealized virtualities/potentialities: "While emergences develop into phenomenal qualities of systems, organizational constraints immerge in a world of silence the characters inhibited, repressed, compressed at the level of parts. Every system includes, thus, its immerged, secret, obscure zone where suppressed virtualities stir" (ibid., p. 124). The duality of the virtualized and the actualized is at the source of tensions and dissociations between

28 Cf. Morin 1977, p. 113; see also further developments in the following pages.

'internal' and 'external' realities, between wholes and parts. They can be interrelating and even solidarian while ignoring each other. This, Morin argues, is for example at the source of the "subconscious" in the case of the human psychic system (although, as Morin warns, this observation is of course not enough to understand the human subconscious).[29] But the subconscious is not always silent.

However, I shall point out, in a spirit of complexity, that the silence is never absolutely complete, and that it exists in varying degrees. To explain this dimension in the complexity of systems, I have to leave Morin shortly, and focus instead on the philosopher Oleg Koefoed (2008), who elaborated the concept of a space of "***eventality***" where the silence of the virtualized, as stressed by Morin above, is relatively lifted and where the possibility-of-becoming-of-an-event is in a space of suspension of virtualities and actualities. This space of suspension (which occurs in the case of the human psychic system, at the moments of intuition) is also, Koefoed argues, a space of "sustension", i.e. of suspended tension between the virtual and the actual. Furthermore, this sustensive space allows to sustain an exploration of potentialities, i.e. suspending the inhibitions/constraints of the system for maybe long enough to glimpse at a possible bifurcation in the system. The space of eventality, as conceptualized by Koefoed, points at a creative potential for emergence, at a moment of emergence-ability. And some of the possible sources of inspiration in this creative potential for emergence, are the inhibited/constrained qualities in the parts that have been, so far, virtualized by the whole.

Learning to value the space of sustension and eventality, is according to Koefoed, a key element in the search process of cultures of sustainability.

The new existence of form, beyond reductionism and beyond holism

The new understanding of reality as advocated by Morin implies that form be no longer an "idea of an essence", but "an idea of existence and organization". Matter is also no longer an ontology captured in a form, but an "organizing *physis*" i.e. an organizing system of elements and processes (Morin 1977, p. 123). Reductionism (à-la Descartes) fails to comprehend this reality (even if reductionism does perceive/construct some elements of reality) and casts a shadowy ignorance on organization. "[T]he simple fact of analyzing an organism starting from its components entails a loss of information about this organism" (Atlan, 1972b, p. 262, quoted in Morin 1992, p. 122).

This divorce from the essentialist view of forms and things is shared

29 I develop a few more (merely sketchy) reflections on the subconscious in appendix 1.

by Laszlo, but Laszlo then turns to a holistic alternative: "Things are emerging from the background of flows like knots tied on a fishing net. These are local actualizations of energies which remain put [...] the particle of matter; Einstein called them 'electromagnetic disturbances' in the space-time matrix [...] they form not isolated units but parts of a continuum" (Laszlo 1996, p. 61). But Morin is opposed to such holistic simplification (as already discussed in the previous chapter): According to Morin, Bertalanffy's systems theory reacted to the simplification of reductionism with the simplification of holism: "holism actually operated a reduction to the whole" that provoked an ignorance of the double-identity of the parts and of organizational complexity (Morin 1977, p. 124). In this context, Morin likes to point at the insight of the French classical philosopher and scientist Pascal (a contemporary of Descartes and Bacon), who argued in his *Pensées*: "I hold it equally impossible to know the parts without knowing the whole, as to know the whole without knowing each term in particular" (Pascal quoted in Morin 1992, p. 122).

More than the knowledge of the whole and of the parts, next to each other, it is towards a "recursive explanation" that the interrelations point.[30] But, at the same time, as Morin explains further, the parts should be conceived both in interrelation and in isolation (otherwise, their irreducibility would be misrecognized).

The observer's subjectivity

The 'whole' of a given system is also a 'part' of a wider system, from another perspective, and can never be conceived as a set 'totality': "we can never truly close, a system among the systems of systems of systems to which it is tied" (Morin 1992, p. 126).

The recognition of a system also depends on the necessarily subjective perception and conception by the observer (or "observer-conceptor", as Morin writes), which are themselves depending on social-historic conditions. Nobody can conceive of systems of physis purely *in abstracto*. Morin refers to Ashby in arguing that the described systems are always abstractions of the human mind, among many other possible and plausible systems in physical reality, that could be perceived in their place.[31] But systems are not either only human constructions, as a radical constructivist perspective would put it: With the notion of emergence, Morin argues, we can point at something that is beyond the logic of the human mind, and thus has a "bottom-up physical phenomenality" ("phénoménalité physique par le bas").[32]

30 Recursivity will be explained further down.

31 Cf. Morin 1977, p. 139, referring to Ashby 1958, p. 274.

32 Cf. Morin 1977, p. 138.

Morin draws two principles out of the subjectivity of systems thinking:

- "a ***principle of uncertainty*** as to the determination of the system in its context and its polysystemic complex" (Morin 1992, p. 139);[33]
- "a ***principle of art*** [...] the art of a skillful butcher who cuts up his beef by following the outline of the joints" (ibid.). What Morin evokes, only shortly, as "principle of art", may be an intuitive opening to the eventality, attentive to the virtualities, i.e. to the subconscious of the system, as well as to its more actualized organizational articulations. It may also relate to what Bateson (1979) discussed as "abduction", i.e. a metaphorical mode of knowing (rather than deductive or inductive) through making analogies, i.e. the comparison of patterns of relationships at different levels, also akin to intuition, guessing, having 'hunches' (as already noted by the philosopher Charles Sanders Peirce, who first called attention to abduction).[34] Finally, this 'art principle' does not necessarily refer to the 'artist', but may refer (as the illustration of the butcher seems to indicate) to the older, wider notion of *ars*, i.e. to the experiential expertise of the "craftsman" as also described lately by Richard Sennett (2008): Sennett characterizes the craftsman as engaging in a playful dialog with materials, finding how to work not against resistant forces but with them, and how to develop an empathic feeling for people and things. Morin's principle of art, as well as Sennett's praise of the craftsman, are pointing at a sensibility to patterns that connect, which I will discuss further in chapter 4 as the basis for aesthetics of sustainability.

Morin also draws out the imperative of self-observation ("auto-observation") because the system is part of the intellect of the observer while the observer is also part of the definition of the observed system.[35]

Systems of physis are also complex because they regulate a ***qualitative diversity***. "The organization of a system is the organization of difference", establishing "organizational complementarity" (Morin 1977, p. 117). Elaborate systems, and especially living systems, have parts which are not only multiple but also diverse. Morin highlights the relationship between diversity, organization and unity: "One of the most fundamental traits of organization is the aptitude to transform diversity into unity, without annulling diversity [...] *and also, to create diversity in and by unity*" (Morin 1992, p. 113). In living systems, in short, both processes occur so that "diversity organizes unity that organizes diversity".

33 I will come back to the 'uncertainty principle' further down.

34 See also the discussion of aesthetics of sustainability as aesthetics of complexity in chapter 4, where I further discuss Morin's as well as Bateson's insights.

35 Cf. ibid., pp. 142-143.

Morin refers to Atlan (1974) in understanding the "trustworthiness of the organization, namely its fitness to survive" (ibid., p. 114). While too much repetitive order stifles emergence, too much diversity dissolves the organization and results in the dispersion of the system. However, this contradiction does not lead to organization finding a "juste milieu" i.e. a mean value, Morin argues, but to a correlative increase of both variety and of the sophistication of organization: "The development of complexity requires, therefore, both a greater richness in diversity and a greater richness in unity (which will be founded for example on inter-communication and not on coercion)" (ibid.).

The organization of difference and diversity also means the maintenance of difference, highlighting the roles of antagonism and entropy. This implies "the existence of forces of exclusion, repulsion, dissociation, without which everything would be confounded" (ibid., p. 116). However, these forces are overcome: "in systemic organization [...] the forces of attraction, affinities, bonds, communications, etc., predominate over the forces of repulsion, exclusion, dissociation; it is necessary that they inhibit, contain, control, in a word *virtualize*" (ibid.). Unlike thermodynamic homogenization, the equilibria of organizations come from antagonistic forces, Morin observes after Lupasco (1962, p. 332).

Furthermore, Morin argues, the complementarity between the parts of a system, their double-identity, and the constraints of the system upon its parts, foster virtual or actual antagonisms. Therefore, "*the complex unity of a system simultaneously creates and represses antagonism*" (Morin 1992, p. 117). The "organization of antagonisms" keeps the system running (as e.g. in a star, with the equilibrium between explosive and implosive processes) thanks to negative feedback loops that are "an antagonistic action which itself actualizes anti-organizational forces" (ibid., p. 118). Morin also calls these negative feedback loops: "***anti-disorganization***". And the evolution of systems' complexity implies more organization, thus more potentiality of disorganization, thus more necessity for such anti-disorganization.[36]

Antagonism is necessary to organization: "In fixed organization, anti-organization is virtual, latent. In active organization, anti-organization becomes active" (ibid., p. 119). Disorganization calls reorganization, but this also means that active systems can know crises because of flaws in their ***regulation***, i.e. in the control of antagonisms. Increases in complexity mean both increasing risks of disorganization and disorder (heading towards entropy), and increasing regulatory capabilities in the system's organization allowing to slow down, halt or reverse the (ultimately inevitable) progress of entropy in the system.

36 The organizational character of antagonism, Morin argues, was not clearly established in the original systems theory, but the importance of antagonism was already noticed: Cf. Morin 1977, p. 121 citing von Bertalanffy 1968, p. 66.

A system's organization remains "an islet of resistance" (ibid., p. 129) both against the ocean of entropy and against self-destruction through the antagonisms of the parts.[37] "Organizational order is a relative order, fragile, perishable, but also [...] evolutive and constructive [where] Disorder [...] is present in a potential and/or active fashion" (ibid., p. 130).

But... 'constructive', 'active', 'evolutive'... how so? I will now focus on these aspects.

Poïesis and Generativity

The analysis conducted so far, allows to think why organization and systems of physis are so complex. However, it does not allow to fully understand a fantastic and fascinating dimension of physical and especially biological reality: How many systems of physis, and most especially living systems, can be actively producing themselves: This is the fascinating reality of ***generativity***, which will be now explored.

Poïesis and the 'machine-being'

For Morin, *"physis* is active*"*. He talks of "active organization", with which wild genesis and organizational processes of production are working together. The term of ***poïesis***, which Morin sometimes uses, gives "a creative connotation to the term production" (ibid., p. 156). And Morin wants to revive the genesic value of the notion of 'production', not letting it framed within its modern technical and economic meaning.

Morin also wants to revive in our understanding of physis, a conception of the "***machine-being***" ("être-machine") which keeps open its genesic force and puts the notion beyond the usual repetitive notion of the man-made machine. Morin's machine-beings act according to organizational properties and are also transformative and poïetic (i.e. they can combine production and creation, as e.g. the sun, an "arkhe-machine" that produces highly organized atoms out of the transformation of simple, light atoms). All machines except the artificial machines, are poïetic (while the poïesis of articificial machines comes from the outside) i.e. they continually produce their own being (e.g. the sun, the aerial whirlwinds, the aquatic swirls, and of course life forms).[38]

About life forms, Morin finds thereafter the notion of a "living machine" at least as interesting as that of a "living system" (mentioned earlier).

37 Organization is then a circular concept: it is also always the organization of organization. Thereby, organization constitutes a self-closure vs the environment. However, Morin very clearly associates the notion of closure to that of opening, in another dialogical step (and thereby differs from Luhmann's conceptualization of systemic closure).

38 Cf. Morin 1977, pp. 161-165 (English version: Morin 1992, pp. 159-163).

"Today we must conceive the machine, not as a mechanism, but as a praxis, production, and *poïesis*. In this sense living beings are auto-poïetic existents (Maturana, Varela, 1972), *formulation where life is not reduced to the idea of machine but includes the idea of machine, in its strongest and richest sense: organization simultaneously productive, reproductive, self-reproductive*" (ibid., p. 163).

The machine as Morin defines it, is a "generic concept" that allows to conceive the physis of all "active organizations", be they biological, social, human, but without a "physical reductionism" (Morin 1977, p. 173). Besides, in the solar system, while they are very diverse, "[w]e could almost consider that all these machines tied together constitute a fabulous polymachine whose center is the sun" (Morin 1992, p. 174). (Morin's idea of "polymachine" is typical of his way to develop a poly-articulated understanding of complexity.)

Culture and society, as seen from their machine-properties
Borrowing the expression from Lewis Mumford (1971), Morin speaks of human society as a "social megamachine" organized in States and class divisions, and borrowing from Noam Chomsky, speaks of a "language-machine" that is capable of both quasi-unlimited poïesis and of quasi-unlimited reproduction and transmission. Culture plays a central role in this social megamachine: "Culture appears. Generative memory, depository of the rules of social organization, culture is a reproductive source of knowledge, of know-how [savoir-faire], of programs of behavior; and conceptual language allows a communication in principle unlimited" (Morin 1992, p. 164).

Self-production, or 'generativity', reorganization and disorganization

Morin articulates how the active organizations (i.e. his 'machines') are self-generating, and thereby builds upon and beyond cybernetics to shape a first physical overview of self-generating systems.

Morin first points at the rotation i.e. loop of aerial whirlwinds and aquatic swirls (which he calls "wild motors"): "This form is born of the encounter of two antagonistic fluxes which, interacting one on the other, are intercombined in a loop retroacting as a whole on each moment and element of the process" (ibid., p. 180). This loop is both genesic and generic, Morin argues, and also organizational and 'generative' i.e. it continuously generates the organization that generates the aerial whirlwinds or aquatic swirls. The loop is retroactive and thus binds together its elements. At a much higher level of complexity, the sun also knows an overall retroactive physical-chemical loop, which, as Morin observes, "assures negative retroaction and regulation without any informational device" (ibid., p. 181): The forms and the emergence come from the retroactions, before the apparition of any "information" (unlike one would assume after cybernetics). The loop introduces a regulation of the whole on itself (hence regulation does not require here information, unlike in cybernetic models and in life-forms). Furthermore, the loops are "***recursive***" (i.e. "process by which an active organization produces the elements and effects which are necessary to its own generation or existence" - ibid., p. 183).

Recursion involves logical consequences: it bears self-production, regeneration and permanent self-re-organization (as Morin points out after von Foerster): "Re" because the system has to fight against degeneration (i.e. entropy) and disorganization, just to maintain its "steady state" (homeostasis, morphostasis). Equilibrium is obtained through permanent disequilibrium, and overall stability is made of unstable elements. "An active system can only be stabilized by action. Change assures constancy. Constancy assures change. [They] mutually co-produce each other" (ibid., p. 185).

Morin insists on the naturally regulative role of this spontaneous poïesis at the physical level. "We are too used to looking for and finding regulation for correcting errors in a device and not in *poïesis* where the play of solidarities and antagonisms makes a loop" (ibid., p. 188). Disorder and disorganization are constitutive of reorganization in active systems: "[E]ach process, taken isolatedly, is disorder or leads to disorder. Together they support the life of organization, namely the loop whose virtue is to combine and transmute disorders into generativity" (ibid., p. 216). Organization tolerates some disorder (in the form of disturbances) and also produces disorganization (in the form of degradation), and more generally permanent reorganization is nourished by permanent disorganization.

Furthermore, ***disorganization*** is the creative drive of active organizations. Positive feedback loops "are the great creators of differences in potential, they are the energy of the world" (de Latil, 1953, p. 187, quoted in ibid., p. 218). Following Maruyama (1963), Morin argues that the disorganization brought by positive feedback loops "plays a genesic role" most especially in the biological and human-social spheres, creating diversity, novelty and complexity. Morphostasis through reorganization is in a complex relation with morphogenesis through disorganization, and "great metamorphoses are always tied to destructurations brought about by positive retroaction" (ibid., p. 222): When deviations turn into tendencies, new organizational forms may emerge and survive if new regulations also co-emerge (Morin calls this "schismo/morphogenesis").

Life itself as self-reproduction ad infinitum, invaded the planet Earth as a positive feedback loop, until it found a self-regulation in the organization of the ecosystems, which will be discussed further down. (A similar corrective self-regulation has not yet been found by human societies in our industrial age characterized by the positive feedback loops of technological and economic development.)

A principle of ***evolution*** comes out of this analysis: "Negative retroaction alone is organization without evolution. Positive retroaction alone is drift and dispersion. Where there is evolution, that is to say becoming, there is a [...] dialogic" (ibid.). The "heroes" of evolution, Morin argues (pointing again at René Thom and Heraclitus), are neither disorganization nor reorganization per se, but the generative potentialities of chaos. Without this omnipresence of chaos, poïesis loses its meaning of a creative-productive force.

Generativity in the living organism

The homeostasis of the living organism is radically different from the regulation of artificial machines, in two ways:

- "the resistance of the living machine is effected by an organizational turnover carrying out the changing and replacement of all the components" (ibid., p. 191);
- "the living machine is devoted to making its own components and to its own reorganization" (ibid., p. 192).

Morin's generalized *physis* gives insights into the biological and anthropological levels, but does not impose a physical reductionism upon reality at the biological or anthropological levels, because it does not deny the cumulative emergence of levels complexity beyond the physical level, and because it actually bases itself on a principle of generativity that calls forth such emergence. Life, with the concept of "machine-being", is a supermachine, while also bringing a level of organization beyond the physical; but life is not metaphysical, if physics is understood as generative organization.[39] In this sense, Morin's approach is maybe one of the most convincingly atheistic streams of thought ever, as it discusses the appearance of life without the need for a godly inceptor, but also without the crude reductionism of classical scientism.

Auto-Eco-Organization

One consequence of complex generativity in active systems of physis and especially in living systems, is that the understanding of active organization has to combine a double dynamic, which is both 'auto' and 'eco'. Morin speaks, alternatively, of an "auto-eco-organization" (in Morin 1977, 1980) and of an "eco-auto-organization" more especially for life on Earth (in Morin 1980). The eco-organization of ecosystems will be introduced in a second step, in the next subsection, but before that, I will now introduce his notion of auto-eco-organization, which bridges on the one hand the active

39 Cf. Morin 1977, pp. 280-281.

organizations of physis (such as whirlwinds and swirls), and on the other hand the active self-organization of living systems.

The open&closed, the auto&eco

Morin argues that the (first) systems theory opposed ***closure and opening*** too much, although its merit was to highlight the working of the opening in 'open systems'. Von Bertalanffy defined living systems as open systems because for them, acquiring energy and other elements from their environment is a matter of survival. Even the properties of supposedly "closed systems", Morin argues, cannot be defined in isolation from their environment: e.g. the mass of an object is a function of gravitational interactions. No absolutely closed system can exist; such a system would be, Morin argues, "a system on which it would be impossible to obtain the least information" (Morin 1992, p. 196); besides, at the quantum level, the absence of any opening to the outside world would mean an absence of "quantum decoherence" i.e. a quantum void where quanta would retain undefined, uncollapsed states.[40] Nor do absolutely open systems exist: they would be indistinguishable from their environment then and thus cease to exist: "Every organization, in the sense that it prevents both the hemorrhaging of the system into the environment and the invasion of the environment into the system, constitutes a phenomenon of closing" (ibid., p. 132). "To conceive of opening, therefore, is to conceive of a corresponding closing" (ibid., pp. 196-197). In the case of so-called open systems, with active organization, Morin argues for a dialogical relation of both closure and opening: "active organizations of systems called open insure the exchanges, the transformations which nourish and effect their own survival: the opening allows them to ceaselessly form and reform themselves; they are reformed by closing [...] Thus, the paradox imposes itself: an open system is opened in order to be closed, but is closed in order to be opened, and is closed once again by opening" (ibid., p. 133). Like the frontier, what closes also opens in the same process. The more complex a system, the more open *and* closed it becomes.

Morin argues that we "have to consider the organizational character of opening" (Morin 1977, p. 198), and he stresses that the opposition between closed and open systems is not the most intelligible framework: "Thus, the decisive cleavage here is not open/close. It is active/non-active [...] the integrity of a non-active system is linked to the absence of exchanges with the outside [...] which prevents hemorrhage but also prevents reprovisioning" (Morin 1992, p. 196), while the integrity of an active system lies in its dialogic of opening and closure. Besides, the same system can be described as a closed or open system, depending on the point of view and reference system of the observer: for example, artificial machines can be seen as functionally open (or passively open as elements of the social megamachine) while living systems are functionally closed but "ontologically and existen-

---

40 On quantum decoherence, cf. Schlosshauer 2005.

tially open [...] The living being can never stop being open, can nowhere escape flux" (ibid., p. 198). Thus, Morin denounces any "rigid opposition" between open and closed systems as "bearing confusion". He instead argues that we should always "situate the problem in a set and a context where opening and closing appear as aspects and moments of a reality simultaneously open and not open"(ibid.).

Different forms of openings

Morin differentiates ecological from thermodynamic opening: If a star, an aquatic swirl and a living being are all 'open' from a thermodynamic view, they are not equally open from an ecological view:

- The star, once formed, does not acquire energy from its environment but from its inner being. The star obtains a fantastic autonomy by "eating its ontological capital until self-annihilation" (Morin 1977, p. 202).
- The aquatic swirl and aerial whirlwinds "are not protected from their environment by the least membrane, they are open on all sides; but this opening on all sides is simultaneously their re-closing on all sides [...] They are absolutely dependent on fluxes, they are incapable of the least chemical transformation, or the least production of object. And yet, they are capable of production-of-self and of permanent reorganization [...] they hold generativity in its pure state [...] in an extreme ecological dependence" (Morin 1992, p. 200).
- The living beings have "an extraordinary autonomy of organization and behavior, which allows them to adapt to the environment, even to adapt the environment to themselves and to harness it. But they are in the same total ecological dependence as eddies, since their provisioning, needed every moment, comes only from that environment" (ibid.).

The case of the atom

The atom is a special case: Although it shows the characters of an active organization and interacts with its environment, the environmental influences (through e.g. radiations) modify the atom, but the atom apparently does not need the exchanges with the environment in order to exist. Also, the atom does not degrade its energy in its own processes (unlike the star). The organizational opening of the atom as system is of a different order than the other active organization: It is an opening "from down under" into the quantum level, as Morin (1977, pp. 231-232) suggests, which is still partly beyond current scientific understanding (and source of much speculations).[41]

If Morin's analysis holds in general terms, there exist certain excep-

41 For a general introduction to these theoretical discussions, cf. Greene 2000.

tional circumstances, according to Laszlo (1996), under which the atom becomes an open system for a very short instant: "The atom's behaviour under conditions of radiations intense enough to penetrate its boundaries but not so high as to smash its nucleus is quite remarkable. Under electron 'bombardment' the atom absorbs the radiant energy and ejects a corresponding quantum of energy from its own structure in the form of one of its electrons" (Laszlo 1996, p. 33).

From Morin's analysis, it becomes clear that the living cell is not a strictly autopoïetic being: the autopoïesis is only a part of a more complex reality, made of the co-organisation of internal and of external interactions. Morin calls this complex reality "***auto-eco-organisation***". Morin explicitly opposes a strict theory of autopoïesis: "The extraordinary perspicacity of Foerster, Maturana, Varela [...] in defining the idea of auto-reference, of autopoïesis" [...] is not for all that legitimized in rejecting the notion of opening, which, while being antagonistic to it, is necessarily complementary to it" (Morin 1992, p. 211).

In auto-eco-organisation, autonomy and dependency are co-constitutive and shape a 'double identity' (own identity / ecological 'belonging'). The auto-eco-organized system is part of its environment which is part of the system. The 'border' between the two has to be reconceptualized: "[T]he word frontier, here, reveals the unity of the double identity, which is both distinction and belonging. The frontier is both opening and closing. It is at the frontier that the distinction and the linking with the environment is effected" (ibid., p. 201). Besides, the environment is more than just a 'milieu': "The environment is not only co-present; it is also co-organizer": This is not only true of living systems (as will be discussed in the next subsection) but also of the aquatic swirl, which is co-organized by the river's flow of water and the bridge's pillar.

The environment is much more than what Luhmann saw in it, who refused to consider the natural environment of human societies as a system in its own right. It is an eco-system of which all living beings are co-constituted, co-constitutive, eco-dependent and eco-autonomous (i.e. an autonomy based on dependence). Morin considers this as "the alpha idea of all ecological thought: the independence of a living being necessitates its dependence with respect to its environment" (ibid., p. 202). The relationship between the two is, once more, complex, i.e. both complementary, competitive and antagonistic. "The environment both feeds and threatens, brings to existence and destroys. The organization itself transforms, pollutes, enriches" its environment and contributes to the environment's organizational complexity (ibid.). The environment's organizational complexity also allows the transformation of the living being's destructive and wasteful productions into food for other living beings. Degeneration and destruction are also integrated into regenerative eco-systemic cycles (also including non-biological, physical-chemical cycles of the biosphere)... at least until a certain point (that the humans' social megamachine seems to have reached and trespassed).[42]

Opening and closure, dependence and autonomy co-evolve in living beings: "the more complex they will be, the more fragile they will be (because they multiply their ecological dependences), the more they will develop the ability to struggle against this fragility" (ibid., p. 203). Besides, Morin points out that death also always threatens from outside and from inside (disorder in self-reorganization) and co-defines life: He recalls Heraclitus' aphorism ("to live from death, to die from life"): "Everything that is open lives *under* the threat of death and *from* the threat of death. In other words, *all existence feeds on what eats it away*" (ibid., p. 204). The negativity not only destroys but also reconstructs (as Hegel already pointed out).

If 'existence' requires opening, 'being' requires closure. The purely closed loop of perpetual movement cannot exist (given the second law of thermodynamics), while the purely open loop cannot be (because it would no longer be a circular movement then but a linear sequence). A loop >opening-existence-closure-being< is in action. "The recursive loop is what ties opening to closing. Opening feeds the loop, which effects closing. [...] The loop is, therefore, both opening and closing. [...] opening produces the organization of closing which produces the organization of opening" (ibid., p. 208). The self ("soi") is an emergence of the loop in the active organization: Generativity, i.e. the production of the self, is "the circuit in which production produces a producer which produces production" (ibid., p. 209). The emerging self both excludes and includes the 'third': the traditional logic of the excluded third has to be associated to a logic of the included third.[43]

A new vision on causality: Generative endo-exo-causality

Morin's developments on auto-eco-organization imply an understanding of causal logic that departs both from a traditional exogenous and linear causality (as articulated e.g. in sociology by James A. Davis in his *Logic of Causal Order*) and from a purely endogenous causality (as implied by Luhmann). The existence of an endo-causality emerged from the study of autopoïetic systems. "Organizational autonomy determines a causal autonomy, namely *creates an endo-causality*, not reducible to the 'normal' play of cause/effects" (ibid., p. 257). But the significance of this endo-causality in auto-eco-organization is also linked to the generative character (cf. above) of auto-eco-organization (hence a generative character of this causality) and to the presence and role of chaos and uncertainty in auto-eco-organization. Also,the overall causal logic of auto-eco-organization points at "the character simultaneously disjointed and associated, complementary and antagonis-

42 See also Lovelock 1979.

43 The logic of the included third will be introduced in the next section.

tic, of exo-causality and endo-causality in a complex of mutual, interrelationed causality" (ibid., following Maruyama 1974).

The internal determinism of an active system prevents certain external causes from exerting their normal effect/influences inside the system. In systemic words, the feedback loops can neutralize external disturbances, up to a certain extent beyond which the external disturbances override certain feedback loops. Morin calls such a process, after Bateson (1967), a "negative causality" or "inverse/antagonistic causality". But Morin also points at the 'positive' i.e. active and productive workings of endo-causality. For Morin, the difference between endo- and exo-causality lies in their nature: Exo-causality is general, it is probabilistic and it does not necessarily result from organized forces. On the opposite, endo-causality is local (specific to the system) and organized. The recursivity of endo-causality allows it to resist exogenous probabilities and to turn the generally improbable into the locally and temporarily probable. Only endo-causality can account for phenomena such as homeostasis.

The self-regulation of active systems called forward the cybernetic conception of 'teleonomy', i.e. the system's own "goal seeking" and finality as an aspect of endo-causality. This conception highlights the autonomy of the active system and its ability to impose its own goals to its environment. But for Morin, this notion of finality is insufficient and also obscures the causal process, as it avoid the question: "where does the finality come from?". As Morin points out, especially in prebiotic stages, before 'life' appeared, there was no such finality. Finality is an emergence, Morin insists: "Finality is a product of self-productive production [...] an emergence born of the complexity of living organization in its communicational/informational characters" (Morin 1977/1992, p. 263).[44] Finality (i.e. ends) is uncertain: ends can appear or disappear, contradict, combine, shift, relativise each other. The notion of teleonomy should be subordinated to the notion of 'generativity', and "there is no "resolution of conflict" between finality and classical determinism; there is a necessary maintaining of conflict within a complex relation [...] between endo- and exo-causality" (ibid., p. 268).

At the level of living systems, "relations between endo and exo attain therein a very high degree of symbiotic complexity and interpenetration" (ibid., p. 269) and Morin speaks of an "auto-eco-causality" whereby both autonomy and ecosystemic dependency are growing: Life forms modify and partly harness their environments, while new environmental dependencies and feedbacks emerge.

As a result, the complex causality of physis and especially of life, generates improbabilities and does not allow to sort out any single "determining

44 Strangely, the translator in Morin 1992 translated here "émergence" into "immergence". I do not follow this probable mistake in translation. (More generally, I took a few liberties with the English translations in Morin 1992, when I found them problematic.)

factors". It implies an '***uncertainty principle***' (Morin 1977, p. 271) and "develops an infinite combinatory dialectic" (Morin 1992, p. 269):

- Similar causes can provoke different and/or divergent effects;
- Different causes can however be led (by teleonomic processes) to result to similar effects;
- Small causes can have dramatic effects (as in systems thinking's points of 'high-leverage' i.e. disturbances which can perturb correction mechanisms at key points/moments);
- Massive causes can have very little effects (if systemic reorganization cancels them out);
- Causes can result into contradictory effects;
- The combination of antagonistic causes leads to uncertain overall effects.

The informational aspect of auto-eco-organization

Although Morin's elaboration on information theories is mostly left outside the focus of my account of Morin's overall theory,[45] a few words are warranted which prove especially insightful about the relationship of information to noise in auto-eco-organization: Indeed, for Morin, in life forms, "information is born from non-information" (ibid., p. 329). Random interactions are at the genesis of a generative, neg-entropic process that creates information (e.g. in the form of genes), and noise i.e. disturbance, is also necessary for the development of information. The genetic information evolves thanks to mutations which are a consequence of disturbances and noise provoking 'errors' in genetic duplication. Noise is not simply ignored (as a Luhmannian systemic interpretation might lead to think) but is a necessary aspect of the information generating complex of "neg-entropic organization/information/interactions/noise" (Morin 1977, p. 324).

The genome represents an organizational competence based on genomic information, as a sort of memory allowing the cell as a whole to elaborate strategies which are not simply, linearly directed by bits of information.[46] On its own, the information makes no sense (as also noted in Bateson 1979). "Solitary information, left to itself, is no longer information but a provisional deformation or a myth of the person processing the data" (Morin 1992, p. 337).

The evolution of auto-eco-organization in life also means a complexification of a "***geno-pheno-eco-communication***": "Organisms more and more evolved, beings more and more cerebralized are going to discern more and

45 Cf. Morin 1977, pp. 291-364.

46 Unlike the cognitive memory of the human mind, the genetic 'memory' is generated by events anterior to the phenomenal life of an individual being, and its "remémoration" (recollection, re-memorizing) process is not imaginary but practical in the cell's reproduction (cf. Morin 1977, p. 329).

more the events of an eco-system more and more diverse, and translate more and more events into information [...] Eco-systems [...] become extraordinarily complex communicational universes" (ibid., p. 340). "In developing, [...] information tends to be *liberalized*, that is to say to lose its force of constraint; it tends to be complexified – to play strategically with ambiguity, error, to make use of ruse; it tends to diversify" (ibid., p. 342). Human communication is highly complex and ambiguous:

> "Every message is in fact multi-connoted and multi-decryptable [and] connotation, and not denotation, can be the true message, then information, redundance, noise are fogged up, lose their clarity and distinction. [...] ambiguity and misunderstanding in communication [...] are one of the principal sources of progress and invention – while remaining sources of error and regression" (ibid., p. 355).

For the noise to be turned into creation, there needs to be a generative apparatus, i.e. an ***auto-eco-poïesis*** (and not a purely autopoïetic i.e. autistic system).[47] "*Thus, information can only be born starting from an interaction between a generative organization and a chance perturbation or noise.* Ergo, *information can only be developed starting from noise* [with] an organizational aptitude of negentropic character [...] which goes beyond itself by transforming the event into novelty, the 'error' into 'truth'" (ibid., pp. 356-357).

The progress of 'generative information' sees a parallel progression of diversity/variety, novelty and complexification and flexibilization of organizational regulation. The evolution of organization requires a balanced progress of novelty and redundancy: "*the new can only be inscribed on the already known and the already organized*; otherwise the new [...] returns to disorder" (ibid., p. 358). Thus evolution can only move step by step, level after level: Once the new has been acknowledged by the autoecopoïetic apparatus, "the durable inscription of the new allows the constitution of a new redundance, which is ready in its turn to receive a new newness" (ibid.).

The eco-organization of ecosystems

Morin (1980) further explores the complexity of life on Earth, articulating the specific complexity of the eco-auto-organization of life. After his insights, I propose to consider the macro-concept of eco-auto organization as the main medium of a new mode of knowing founding cultures of sustain-

47 The notion of autoecopoïesis is a neologism which I am introducing, and which is not found in Morin's work. As far as I could find out, the term was already used once before I started using it myself, but in a Spanish-language publication from 2001 by two Chilean authors (Lavanderos and Malpartida), which I did not access. I will further articulate this notion in the conclusion on the current chapter.

ability, and to be articulated with a practice of autoecopoïesis (which I will describe in the conclusion of this chapter).

The notion of ecosystem, which was introduced by Tansley (1935), constituted the emergence of the awareness that "the interactions between the living, by combining with (and feeding back on) the physical biotope, organize the environment precisely into a system" (Morin 1980, p. 17).[48] Thereafter, ecology "is the science of the combinatory/organizing interactions between each and every of the physical and living constituents of ecosystems" (ibid., pp. 17-18). This science points out that life is not solely based on living organisms (and their autopoïesis) and on specied reproduction (and the process of evolution), but also on an ***eco-organization***.

Eco-organization

The environment is not only an environment or an Um-Welt for living systems, but is itself a system, an 'Unitas Multiplex', an organization that both produces emergence (both at its overall level and at the micro-level of single organisms – cf. the earlier explanation on 'micro-emergence') and produces constraints upon living organisms (suppressing certain potentialities).

Not only is the ecosystem an active and complex organization, Morin observes, it is also an astonishingly temperate and self-regulating organization, given the incredible diversity and disorder of living forms that it hosts . The question fascinating Morin is thus that "eco-organization is a spontaneous organization which, though on the basis of deterministic geophysical carriers and of genetically determined beings, makes its self by itself, without being incited or constrained by a program, without disposing over an autonomous memory and an own computation, without being organized and ordered by a control, regulation, decision or government apparatus" (ibid., p. 21). The eco-organization achieves the complexity of linking and interrelating extreme diversity with unity, disorder with order, and extreme antagonisms with solidarity.

Eco-organization sets up, at a meta-level, an overall complementarity of antagonisms (parasitism, phagism, predation), competitions and complementarities (association, symbiosis, society and other interdependencies) between living organisms and between species. The food chain is also a CO^2/O^2 chain, antagonisms co-organize regulations while solidarities generate antagonisms, parasitism can turn into symbiosis, etc. Morin talks of an "uninterrupted rotative circuit of antagonism and complementarity" which have "a common basis: the existential need for the other" and in between which there is no neat border (ibid., p. 25).

---

48 All the translations of quotes from Morin 1980 into English are of my own making, as unfortunately, no English edition is available of the second (and subsequent) volumes of Morin's *method.* All references to specific page numbers in Morin 1980 refer to the pocket book reprint of 1985.

Overall, eco-organization constitutes a "Pluriboucle", or loop of loops, which integrates itself and the living organisms into the cyclical order of the solar polymachine (climates, seasons, days/nights), and as an 'open system' from an energetic perspective, needs the permanent input of solar energy to ensure its neg-entropic permanent reorganization and counteract the thermodynamic process of entropy. Eco-organization constitutes not just one big unified loop, but a "Boucle uniplurielle" (uniplural loop) constituted of several great cycles (e.g. water cycle, CO^2-O^2 cycle, food cycle) in which "each moment of a cycle simultaneously constitutes the moment of one or several others, where it plays different, or maybe even opposite roles" (ibid., p. 29).

Eco-organization follows the logic of permanent disorganization/reorganization, already articulated above. Reorganization is present in disorganization, any waste becoming also an ingredient reintroduced in the cycles. "Eco-organization is nourished and regenerated not only by life but also by death, and it is regulated by the antagonism between the excesses of both" (ibid., p. 31). Furthermore, the ecosystems evolve through their open fragility:

"Through their extreme opening and sensibility, ecosystems are extremely vulnerable to disorganizing agents. But here again, frailty makes vigor. They are so open to one another that they inter-organize and inter-feed in case of devastations. [And] their extreme sensibility is inseparable from an extreme suppleness which allows them to reorganize themselves in new ways, i.e. to transform themselves and evolve" (ibid., p. 32).

This property of ecosystems allows their resilience.[49]

Eco-evolution

Eco-organization is indeed capable of "***creative eco-evolution***", i.e. of metamorphosing itself in response to irreversible changes:

"The most remarkable eco-organizing quality [...] is to be also able to produce or invent new reorganizations from irreversible transformations [...] the supreme virtue of eco-organization: this is not stability, but the ability to construct new stabilities; this is not the return to balance, but the ability of reorganization to reorganize itself in a new way under the impulse of new disorganizations. In other words, eco-organization is able to evolve under the disturbing burst of the new" (ibid., p. 35).

By 'eco-evolution', Morin conceives together the evolution of eco-systems and the co-evolution of species. Indeed, 'natural selection' does not impose altogether-fixed rules on species, as an "atomized conception of evolution" would have it. These selective 'conditions' are themselves changing, follow-

49 The concept of resilience was already introduced in chapter 2.

ing the evolution of eco-systems.[50] Moreover, "what is selected is not only the species fit to survive in such or such conditions, but everything that facilitates the regulation and reorganization of eco-systems" (ibid., p. 36). What is selected are the feedback loops which stabilize each other and become co-selective. Eco-evolution also is an evolution in the sense that it establishes "an increasingly refined reorganizing complexity [based on] the capability to reorganize the rules of reorganization" (ibid.).

In this context, the 'adaptation' of single species should not be conceived in a linear, mechanical way, but under the perspectives of reorganization and resilience: "One has to detach oneself from the rigid concept of adaptation defined as the perfect match of a species with a given environment. That perfect adaptation becomes unadapted and fatal as soon as occurs a modification of the conditions of adaptation [...] on the contrary, the capability to adapt under different conditions or different environments facilitates survival" (ibid., p. 48). Morin thus prefers the notion of "***adaptativity***", which also includes the ability to re-adapt oneself to changing contexts (in various ways, from warm-blooded animals' homeothermy to the occurrence of random genetic mutations) and the relative ability to adapt one's habitat to one's needs.

In parallel, the notion of "natural selection" should also not be conceived in a linear, mechanical way, but under the light of eco-organization[51]. Noise, uncertainties, relative errors and disorders are not necessarily eliminated by a narrowly rational selection process. "Actually, the more complex a system is, the richest diversity it has, the more supple it gets, the more fuzziness, uncertainty, chance it contains, the more it allows neutral or useless traits to remain" (ibid., p. 52). Imperfect species can support each other in symbioses and other complementarities instituting cycles which are then balanced enough to allow their survival. Integrative interactions give the eco-organizational logic which would appear as an irrationality from the perspective of a linearly-rational selection principle.

---

50 Smith and Jenks (2006) stress the same reality when they point out that "the ecology of the environment evolves as well" (p. 100). Their account is however more sketchy and less well articulated than Morin's. I therefore do not rely on their text in the current chapter, although they do often come close to or complement my own arguments (i.e. not only for eco-evolution but more generally for eco-auto-organization, which they stress repeatedly, explicitly after Morin). But there are also some differences. For example, they repeatedly proclaim that complexity "must develop from simple beginnings" (p. 130), a rather simplistic position that neither Morin nor I do share. (Their discussion and use of Morin is also limited to one single article.)

51 This could not be achieved in the original Darwinian theory (Darwin ignored the concept of ecosystem), and it was still not achieved in neo-Darwinian theories of the 1970's, according to Morin (cf. Morin 1980, pp. 51-52).

Such an eco-organizational selection penalizes not so much the less-well-performing species in a fixed context as a Darwinian model would predict, but rather penalizes those species which do not insert themselves in the cycles of complementarites/antagonisms (e.g. because they have become over-specialized, or because they extinguished all their preys): "those eliminated are the ones who are not solidarian with eco-organization" (ibid., p. 53). Overall, eco-organization is self-selective, but without a fixed selection principle, given that the selection principles themselves evolve together with the ecosystems.[52]

Eco-communication

As already mentioned, eco-organization does not have a computational center or memory of its own, and nevertheless regulates itself in an exquisitely complex and overall balanced way. Morin proposes to characterize eco-organization as a ***poly-centric and a-centric machine-being*** – referring to the concept of a-centric automata by Rosenstiehl and Petitot (1974). Morin further develops the notion of "***eco-communication***", which is based on:

- the numerous communication network relevant exclusively within species (and especially animal societies);
- the forms of communication evolving between preys and predators, which are based, on each side, on the double-aim to extract as much information as possible about the other side, and to give away as little information oneself as possible – or better: to confuse the other side with "a fog of noise and pseudo-informations" (Morin 1980, p. 37); Morin points out that "it is through these antagonisms that non-communicating systems communicate, that information crosses over obstacles" (ibid., p. 38);
- the active extraction of information by a living being from its environment (given that the eco-system does not produce 'information' in itself but events – as Bateson 1979 also pointed out), and the release of information/events by the living being;
- the presence of multiple forms of 'noise' in the environment and of errors and cognitive limitations among living beings, that not only degrade information, but also nourish the complex co-evolutions of species in their communicational dimension.

Altogether, these dimensions constitute a communicational "poly-network" that has no particular center, and which regulation is nourished by chaos: "The living eco-tower of Babel, far from collapsing and dislocating itself, builds and rebuilds itself unceasingly in cacophonia" (ibid., p. 40).

---

52 Cf. Morin 1980, pp. 54-55. For a further discussion of the new understanding of the evolution of life under the influence of complexity theories (with several examples), see Wesson 1991.

The inter-relations of diversity, complexity, organization and spontaneity

The apparition of the eco-organizing loops of the ecosystems(i.e. the great cycles mentioned above) was made possible by the diversification of life forms on earth, and in return allowed further diversification. In eco-organization, diversity and complexity are complementary. But, as Morin notices (after Gauthier et al. 1978, p. 120), not any diversity is stimulated, not an infinitely increasing diversity is fostered, but an "optimum of diversity" rather than a "maximum of diversity" (Morin 1980, p. 41). Such an optimum, at a certain moment for a certain eco-system, involves a dominating feedback-loop involving one or a few species (e.g. herbs and herbivores in prairies), and a great variety of other species organized around them and contributing to maintaining the dynamic balance of the eco-system. The dominant species of an ecosystem is not controlling the eco-organization (or rather was not, until the human species recently increased its relative control over eco-systems). Rather, an ecosystem has "weak spots" (ibid., p. 44) where changes introduced by even a minor species or physical element can have devastating effects, and thus relative control lies at these weak spots.

Diversity also matters at other levels: The genetic diversity among individuals in a species, increases the species' resilience to various threats of annihilation (unlike human-cultivated species, which genetic diversity has been reduced and which are thus much more vulnerable). At another level, ecological diversity means that open-systems are relatively open into one another and certain species can migrate in between them, constituting new feedback loops where existing ones have collapsed.

The complexity of eco-organization also preserves diversity, because highly complex life forms still do need – and inter-relate in various ways with – simpler life forms (parasitism, symbiosis). "Complexity is not the rejection of the less complex by the more complex; on the contrary, it is the integration of the less complex into the diversity" (ibid., p. 43). Cultures of sustainability should keep this aspect in mind, and thereby avoid becoming technocratic cultures of neo-cybernetic optimization.

The spontaneity at play in eco-organization, means that no central memory, no 'genetic code' of the ecosystems is at work. But it does not mean pure spontaneity, insofar as it is built upon a long evolutionary history that allowed the eco-evolution of spontaneous eco-organization, and also because spontaneous eco-organization is built upon the non-spontaneous auto-organization of living species (with their genes and computational abilities).

Indeed, the selfishness of auto-organization is turned into a building-block of eco-organization: "a living being becomes an existential requirement for another one; this requirement creates immediately a solidarity and a de facto complementarity of the other with regard to the self [and] introduces the egocentric being into the interdependences and the polycentric/acentric inter-feedback" (ibid., p. 46).

Eco-Nature

The notion of eco-organization revives the idea of a generative, integrative and organizing nature, which had been pushed aside by Modernity into the realms of Romanticism, and associates it to the insights stemming from the Darwinian conception of nature. Such an eco-nature both shows organizational virtues of integrative, spontaneous reorganization, and shows the barbarity of exploitations, servitudes, competitions, and gigantic 'wastes' of lives i.e. death as a major dimension of eco-organization.[53] Also, to the "Ubris of life" of the masses of seeds and eggs being disseminated responds the "Ubris of death" that destroys most of them right at their first instants of existence: "It is not only in its disorders, its misfirings that nature is barbarian, but also in the construction and the regeneration of its Harmony" (ibid., p. 59). Therefore, the "good" and the "bad" nature are two poles of the same reality which is both coherent and contradictory. "Nature reveals us no unambiguous virtue: it is wise-crazy." (ibid., p. 60). Life and death are nourishing each other.

Of eco-organization and auto-organization, none of the two is unilaterally alienating the other or servicing the other, but both are founded in their autonomy by the alienation and servitude of the other.

> "It is a phenomenon more complex than the 'guile of reason' of Hegel where the particular action, believing itself to be selfish, is unaware of working for the general interest into which it becomes totally integrated. Here, unlike in Hegelian rationalization, we see that it is about a complex relation where the particular interest works for the general interest while continuing to work against the general interest, but thereby, it maintains some antagonism and complexity which, in a sense, are also part of the general interest" (ibid., p. 59 in footnote).

The eco-auto-logical relationship

Now that the eco-organization of ecosystems of life has been introduced in the specificity of its complexity, it is time to reconsider the eco-auto relationship in living systems.

The 'autos' of the living being is both integrated into the 'oikos' and distinguished from it. Morin aims to "conceive the ***eco-auto-logical relationship***", i.e. how the two logics can both oppose and integrate into each other. First of all, neither genetic determinism nor eco-determinism do offer satisfying accounts of the organization of life forms. Rather, "eco-organization is the co-programmer of auto-organization" (ibid., p. 62), as e.g. when the biological cycles of plants' self-organization adapt to the exterior cycles. Furthermore, the uncertainties and ambiguities of the environment are stimulat-

53 Cf. ibid., pp. 58-59.

ing the "development of intelligence" of living beings, and thus "eco-organization is the school of auto-organization" (ibid., p. 63).

The more complex and autonomous the cognitive abilities of mammalians and eventually humans, the more these abilities are turned to the outside: About the mammalian brain, Morin notices: "The internal nature of the animal is less known to it than the outside nature [...] the development of the brain is first oriented towards action upon and knowledge of the outside" (ibid., p. 221). Also, the outside is then integrated into cognition: "among mammals, the slowness in the development of the brain after birth and the scale of non-specified zones in this brain allow specification by outside events" (ibid., p. 63).

Six principles

From these observations, Edgar Morin establishes six "fundamental principles that govern the relation between the self-organization of living beings and the eco-system/organization" (ibid., pp. 65-66):

- The "principle of bio-thanatic inscription" ("principe d'inscription bio-thanatique" from Thanatos, the ancient Greek personification of death) i.e. the mutual nourishment of individual existence and of the eco-organizational cycles and life & death, and the fact that any life is embedded in biocenosis on the basis of its existential need for other lives;
- The "principle of eco-auto-organization", i.e. eco-organization as co-organizer of self-organization;
- The "principle of a mutual and recursive development of the complexity of eco-auto-organization", i.e. the complexity of each reinforces the complexity of the other, and the two complexities are inseparable from each other;
- The "principle of dependence of independence", i.e. the development of autonomy in life forms is a development in complexity which is increasingly dependent on eco-organizational complexity;
- The "principle of dialogical explanation of life phenomena", i.e. the necessity to elaborate a complex understanding of life based on both internal (autopoïetic) and external (eco-organizational) explanations;
- The generalization of these principles to a "generalized ecology" also exploring its insights at the levels of animal and human societies.

The cycle of life and death
Morin's "bio-thanatic" principle echoes certain features of indigenous cultures.[54] Indeed, a comparable principle is embedded in the mythic narratives of indigenous peoples such as the Pueblos peoples of North America: "The Hopis – like the other Pueblos – believe their ancestors to be fertilizing clouds, bringers of rain who will nourish the crops upon which the living subsist. The necessity of death [...] becomes even more accentuated, therefore [...] death feeds life" (Vecsey 1991, p. 45).

The understanding of the eco-auto-logical relationship necessitates an understanding of the complex dynamics of opening and closure in self-organization, as developed earlier, as well as the realization that autos is both a part of a wider whole and a whole in its own right, i.e. what Arthur Koestler (1967) characterized as a "holon". The relationships between autos and oikos are characterized by complementarity, mutual ignorance and uncertainties.

The integration of ecology and human societies

Not only are human societies a part of ecosystems, but eco-systems are also a part of human societies since the development of agriculture and of cities, which instituted a new kind of control on eco-systems, i.e. an anthropological control, and an ***anthropocentric organizational principle*** which complements and conflicts with eco-organization. In the same process, historical societies have become more vulnerable to the eco-relationship, as disasters and famines illustrate.

The intensification and industrialization of agriculture in the modern age lead to a drastic reduction of the variety of cultivated species, and extensive monocultures lead to the degradation of the complexity of entire ecosystems, and to the proliferation of parasites leading to an 'arms race' between those parasites and the techno-chemical poisoning capacities of the agro-industry.

Furthermore, the impoverishment of eco-organization further stimulates contemporary societies to trap themselves into the logic of the Technological System (as described after Ellul in chapter 1), which Morin also deplores as a "course infernale" (Morin 1980, p. 75). Humanity, having become a "hyper-parasite", has thus come to threaten its own survival by the double-dependency to impoverished ecosystems and to technological prostheses.

We are thus dealing now with a vast "***eco-socio-system***", Morin argues. There was never any "pure society" (void of nature), and there is not any-

54 A closer look at the insights offered by indigenous cultures for cultures of sustainability, will be given in chapter 4 especially with the analysis of David Abram.

more any "pure nature". In response, a planetary ecological science and conscience develops, which itself carries the contradictions of the technological society: Hence, Morin observes, the 1972 Meadows Report does mark not only the strength and the promises of this new conscience, but also shows the signs of simplifying thought and of human technocratic arrogance: "Ecological thinking is already parasited by technocratic thinking which is nevertheless its personal enemy" (ibid., p. 77).

The planetary system now has two pilots. The human being has reached transformative capabilities and responsibilities that make him/her the copilot of nature, but do not allow humanity to imagine itself anymore as the single pilot (master, or even shepherd) of nature. A "symbiotic co-development" is required (ibid., pp. 96-97).

The ecologization of thinking

The understanding of the eco-organization and eco-auto-logical relationship allows the development of an ***ecological gaze***, which "perceives any autonomous phenomenon [...] in its relationship with its environment" (ibid., p. 78). The ecological gaze attaches itself to any environment, and not only the eco-system (e.g. the social environment for an individual, or the human intestine as environment for bacteria). The ecological gaze highlights both the complexity of the relationship between the autonomous phenomenon and its environment (cf. above: complementarity, antagonism, etc.), and the active role of the observing subject (i.e. the conceiver of the gaze) who is re-interpreting a certain reality as an 'environment'.

A social ecology allows to perceive the "quasi eco-organizational" dimension of historical societies, a dimension of spontaneous social self-organizing processes, which enters in a dialogical interplay with the top-down social order emanating from the State apparatus (which introduces dimensions unknown to eco-organization, such as a hypertrophy of control allowing levels of domination and exploitation, of both societies and nature, yet unattained in non-human societies).

An urban ecology allows to perceive the social ecosystem of the relationship between cities and their hinterland, and the urban ecosystem of cities where the techno-sphere partly substitutes the ecosystem with an artificial oikos of technical artefacts.

An ecology of ideas, of mind (as initiated by Bateson) and of action also emerges. Morin insists on an ***ecology of action*** which allows to perceive how the action of the human being (or social agent, in the sociological language of Bourdieu and Giddens) "enters into processes that escape the purposefulness, and even the understanding and the awareness of the actor" (ibid., p. 82). An ecology of action thus allows to move beyond the methodological individualism of some sociological and economic models that root action in the 'social actor', revealing the limits of individual strategic opportunities: "the action uproots itself from the actor, either to dampen itself in the negative feedback, or to activate unexpected positive feedback":

here the eco-logic supplants the auto-logic of the actor. The basic principles of an ecology of action, Morin suggests, are thus that:

- The optimum efficiency level of action is situated at the beginning of its developmental stage, before it drifts too far away from the social agent;
- The eventual consequences of action are unpredictable, and inspire an uncertainty principle at the social level;
- The multiple mediations of action (e.g. in the political field) unavoidably prevent linear strategies (i.e. political 'will' alone cannot overcome the complexities of given social situations); strategy is thus also concerned by the challenge of complexity.

Concerning the ecology of ideas, Morin speaks of "noological beings" which take hold of human minds at paradigmatic or ideological levels, and eco-co-organize the meanings of single words, texts and ideas: "The same ideas or theories can have completely different, even opposite meanings, according to the mental and cultural ecology which feeds them. [In this sense] Marxism in a libertarian or open mental ecology lives in a way completely set against the Marxism fed by an authoritarian or dogmatic mental ecology" (ibid., p. 85).[55] The ecological insight on ideas also reveals, reflexively, that a theory of complexity and an "ecologized thinking" require a complex, open "mental ecology". The epistemological framework of transdisciplinarity, explored in the next section of the current chapter, works for the maintenance of such an open context for the life of the human mind. The ecological gaze also reveals that "our culture is the ecosystem of our ideas of nature" (ibid., p. 93), thus a double ecology of nature and culture is revealed.

But in general terms, Morin (1977) considers that ideas "are still much less complex in their organization as theoretical systems than the organization of the least living being. They are still [...] very barbarous in their stiffness, stupidity, coarseness" (Morin 1992, p. 149).[56]

The ecological gaze on human ideas also multiplies the notions of creativity and authorship, allowing to perceive the ecological life of creativity and the presence of multiple eco-coauthors nourishing any single author's work.[57]

Overall, the ecologized gaze overcomes the dichotomy of reductionism to external causality vs. reductionism to auto-causality, and allows the eco-auto-logic to enter "into the description and explanation of anything that is

55 Morin (1991) further explores the ecology of ideas at a greater level of detail.

56 Morin also mentioned in several interviews, that we would be still living in a "planetary iron age", in terms of the complexity of our ideas, and that we should hope to let ideas evolve much further in coming decades and centuries.

57 An interesting development of such an understanding of creativity can be found in the organizational culture of FLOSS (free, libre, open source software).

living, including society, the human being, the mind, ideas and knowledge" (ibid., p. 87). It inaugurates an ecological paradigm, which should evolve into a paradigm of complexity, avoiding the risk of simplification of an "ecologism" that ignores the evolutionary "adventure" (ibid., p. 90) and turns into a restrictive, frigid conservatism (a symptom of which Morin claims to see in the "zero growth" discourse).

The ecologization of thinking, and the new perceptions that it develops, pass through a re-thinking of rationality, and the development of a certain sensibility, which will be explored further in chapter 4.

Socio-organization

In the sake of relative brevity, I will not be able to account for Morin (1980)'s very long developments on the auto-logic of the living subject, starting with the "computo" of the bacterium. However, the further insights given by Morin's notion of "***socio-organization***" cannot be bypassed, as they integrate an understanding of social phenomena with the understanding of eco-organization and auto-organization. Indeed, Morin addresses the question of how far animal societies, as "***third type entities***", are different from the former two (i.e. the first type entities of unicellular organisms with their auto-logic, and the second type entities of multicellular organisms which combine the relative autonomy of cells with the hierarchy of their functional integration in the autonomy of a larger organism).

Acknowledging the researchers who pointed out, along the second half of the 20th century, that societies are an animal phenomenon rather than a merely human one,[58] Morin establishes as precondition for the emergence of societies, the association of second type entities with "a sex and a head", i.e.: "Societies form from communicative/associative interactions between animals endowed with a nervous system and a sexual reproduction system" (ibid., p. 237). Therefore, already, he observes that societies, by definition, require a very high level of individuality of their members. A society emerges when a social system is constituted, i.e. when the interactions between the animals produce a whole which acquires its own emerging qualities and feeds back upon its parts (i.e. the individual animals). A society is also more than just a system: it is an organization in Morin's sense, i.e. it "retroactively organizes and controls the production and reproduction of the interactions that produce it, assuring its homeostasis through the turnover of mortal individuals, and thus, it constitutes a self-producing and self-organizing being-machine" (ibid., p. 238).

58 Cf. Morin 1980, p. 236.

Socio-organization compared to eco-organization and auto-organization

Like eco-organization and multicellular organisms, socio-organization is polycentric, living off the multiple connections between computing beings. However, unlike eco-organization, socio-organization also develops self-reference and ***sociocentrism*** (i.e. a community that fraternally defends its members against the outside world). A "sociocentric computo" is established (ibid., p. 239). But the autonomy of a society is far less developed, Morin argues, than the autonomy of either first type or second type entities (i.e. auto-organizing bacteria or integrated organisms). Even the most integrated animal societies, i.e. some insect societies (bees, ants), do not reach the level of integration and specialization of the parts that one can find in a multicellular organisms: The single bee retains a large autonomy of decisions and behavior, compared to any single body-cell or neuron. And animal societies among vertebrate species are even less strictly integrated than insect societies. Furthermore, no animal society (except the human societies, which will be treated below) has a central control apparatus comparable to a multicellular organism's head.

The evolution of socio-organization follows two paths, Morin argues: On the one hand, a quasi-ecological tendency whereby the spontaneous interactions between individuals are complex (with solidarities, complementarities, competition, antagonisms and disorder); on the other hand, a quasi-organic tendency whereby individuals are integrated like cells in an organism. Among societies of mammalians, the first tendency is predominant, while the second tendency is stronger among insect societies. In the societies of mammalians, the eco-organizational tendency reinforces the individuality of its members. However, solidarity still overweighs antagonisms (which are often euphemized and ritualized) and sociocentrism installs a "trans-subjective unity" among its members. The disorganizing factors (antagonisms, competitions, disorders) also contribute to the socio-organization in bringing it complexity.[59] And as discussed already in the preceding pages, such complexity brings with it a degree of flexibility and of resilience, allowing societies to evolve.

Socio-organization in the human species[60]: the emergence of culture, States and nations

The emergence of human language and with it, the emergence of ***culture as a "social genos"*** (i.e. an informational capital of societies, in comparison to

59 Cf. ibid., pp. 241-243.

60 Morin warns his readers not to assume that his developments, presented in the current paragraph, would "reduce the anthropological to the biological" (ibid., p. 244). The goal here is not to reduce but to root the complexity of the social in the biological.

the genotype as informational capital of first and second type entities), mark the evolutionary jump of human societies.[61] This informational capital covers a very wide range, from the knowledge of the natural world, technical know-hows, social rules, to beliefs, values and worldviews, and even meddles into sexuality (introducing a "cultivated sexuality") - thereby retroactively influencing even the biological genos and Darwinian 'sexual selection'. Culture also reinforces sociocentrism into ethnocentrism.[62]

A further step in the evolution of human societies is reached with the development of historical societies, which institute a "***social megamachine***" (Mumford 1971) with the constitution of States differentiating societies into casts, classes, professions and other specializations, i.e. a hierarchic construction under the control and domination of a central State apparatus. Here the quasi-organic tendency, seen in insect societies, finds a new vigor and developmental potentialities in the historical human societies. Especially in the Modern age, the development of the 'Nation' allows a society to identify itself as an individual and a subject in its own right, transcending individual human beings. As a "brain-machine-being", the modern Nation-State enters in a complex (i.e. complementary and antagonistic) relationship with the individual human beings. From the perspective of the human beings, the development of the Nation-State bears a considerable (and always threatening) risk of engendering a totalitarian socio-organization (an early example of which is found in the totalitarian States of the 20th century), which could eventually wipe out human individuality with the aid of techno-science.[63]

Therefore, the preservation of human individuality and of 'human rights', requires not only a socio-organization, but a '***socio-eco-organization***' where a social eco-organization balances the organic tendency of the State, with the notions of citizenship, participation and polyarchy (which will be further introduced in chapter seven). However, social eco-organization should not be mistaken for an ideal tendency on its own, because it generates not only spontaneous collaborativity but also spontaneous conflicts, competitions and the potential exploitation of certain social groups/classes by other social groups/classes. The term 'socio-eco-organization' thus is inscribed with the complex relationship between the socio- and the eco-organization of the social level (like eco-auto-organization concerning the auto- and eco- logics).

61 The evolutionary jump is however probably not abrupt but gradual, given mounting evidence of "proto-culture" among apes as discussed by primatologist Franz de Waal (2001).

62 Cf. Morin 1980, pp. 244-246.

63 Cf. ibid., pp. 246-254.

Is Morin a sociobiologist?

Readers informed by the interdisciplinary field of 'sociobiology' may wonder whether Morin's socio-organization is a form of sociobiology, as developed by Edward O. Wilson and Richard Dawkins. Sociobiology is a neo-Darwinian theoretical strand which explores the genetic evolution of advantageous social behavior, under the pressure of natural selection (see e.g. Wilson 1978).

In short, Morin's own answer to this question is a resounding "no!" If he plays with turning the notion from "sociobiology" to "bio-sociology," Morin (1980) opposes the organicist tendency among several sociobiologists and their "pan-geneticism" or "pan-ethologism" (Morin 2008, pp. 1091, 2176-2179; see also Morin 1973, pp. 63-105). But a fair answer has to be more carefully weighed: Morin's view affirms the irreducibility of human culture to genetic determinism. However, Morin does not deny that human complexity is based and rooted in biological and ecological complexity, and that the complex relations between these dimensions require interdisciplinary research (to which sociobiology does contribute – and Morin does quote and refer to Wilson's research, although he rejects his "genetic fundamentalism").

So in short: Yes, there are biological and genetic "attractors" as argued repeatedly by Smith and Jenks (2006).[64] This implies constraints indeed, and the rejection of an absolute contingency of human culture (as was uncritically implied by many social scientists until the late 20th century) or of autopoïesis (à-la Luhmann). But this evolutionary basis should not lead us to under-estimate the possibility for culture to impose emergent structures that can diverge from biological attractors in a very extensive way, and/or the high degree of post-natal plasticity allowed within the constraints of biological attractors.

In this sense, Morin's view of human complexity is relatively more compatible with the insights from the evolutionary psychology of John Tooby and Leda Cosmides (see Cosmides and Tooby 1997) than to Wilson's views.

Socio-organization may develop itself (eco-)evolutionarily or non-evolutionarily, although in the latter case the likeliness of a society's

---

64 The concept of "attractor" comes from chaos theory, and points at the topology of the sensibility of complex systems to their initial conditions. To be noted, "initial" conditions should here not be understood as "once and for all" but as related to the constantly present genesic potential of chaos which may actualize itself at birfurcation points and moments of emergence in complex systems of systems.

ecological "collapse" (à-la Jared Diamond) increases in probability in the long-term view.[65] Meanwhile, eco-evolution itself, also in non-human societies, is more complex that neo-Darwinism would assert. As already argued above, selection principles themselves evolve together with the ecosystems, and in the human case, with the social (and technological) system: This imposes a constant re-assessment of what the "selection pressures" are.

The life of life[66] or the Great Tinkering

Overall, Morin reveals the organization of life as a "Great Tinkering" ("Grand Bricolage") in which four logics are operating and interacting in complementary-competing-antagonistic ways: a centralizing-hierarchical logic, a polycentric-polyarchic logic, an anarchic logic and an eco-logic.[67] The dominance of one of these different logics is contextual and temporary, but at the root of all living organization is always the anarchic logic, which is however always relativized by the very organizational processes that it allows.[68]

Overall, the "polylogic" of life is "tinkering" because it is opportunistic, using any available materials, working with disorder and eventually resorting to "détournements" (i.e. diversions, re-routings, re-appropriations). Morin compares it to the 'Palais du facteur Cheval', i.e. a famous chef-d'oeuvre of Art Brut from the late 19th century, a dream palace built by a French countryside postman with stones he found along the roads, iron and concrete of his own invention, and mixing symbolic elements from world cultures as glimpsed by him on postcards.

A 'bricolage' cannot be optimized, and thus the 'lesson' of the complexity of life should be that a "pseudo-rationalization" through optimization is a destruction of the life of complexity. "Any ideal conception of an organization which would be only order, functionality, harmony, coherence is a mad dream of an ideologist or/and a technocrat" (ibid., p. 326). One should not try to eliminate disorder, uncertainty and error. Alas, the mainstream thinking about social organization in the age of modernity, is ignoring this insight and pursues "rationality, efficiency, economy and expediency" through the logic of hierarchy, centralization and specialization, perpetuating a culture

65 Smith and Jenks (2006) for example propose to use a criterion of "vitality" which, although interesting, may lead to over-simplifications. Their proposal is however too vaguely formulated to be criticized on clear grounds.

66 The life of life, or "la vie de la vie", is the overall title Morin gave to the second volume of *la méthode*.

67 Cf. ibid., pp. 303-324.

68 Cf. ibid., pp 322-323.

of unsustainability. This "model conceived according to the rationality and functionality of artificial machines" is fundamentally problematic because, as Morin argues, it should be clearly inconceivable to think of our contemporary society with lesser complexity than we have come to understand life.[69]

SECTION 2: TRANSDISCIPLINARITY

The 'pensée complexe' (complex thinking) that Morin started elaborating, "is the journey in search of a way of thinking which would respect the multidimensionality, the wealth, the mystery of reality, and would know that the intellectual, cultural, social, historic determinations that any thought bears, always co-determine the object of knowledge" (Morin 1980, p. 10).[70]

Beyond the recent boost in popularity of the term in academic circles, transdisciplinarity, as it was defined in the 'Transdisciplinarity Charter' at the First World Congress of Transdisciplinarity (Convento de Arrabida, Portugal, November 1994), offers a methodological framework for the understanding of the world of complexity that Edgar Morin exposed (cf. previous section). This approach allows to reintegrate the observer and his/her values in relation to the object of study as part of the research, opening a new type of knowledge, "*in vivo* knowledge" by contrast to disciplinary "in vitro knowledge", as Basarab Nicolescu, a major figure in the transdisciplinarity approach, recently coined (Nicolescu in ed. Nicolescu 2008, p. 3). Or in the words of Alfonso Montuori: "the knower is not a bystander looking at knowledge in its pristine cognitive state, but an active participant, *a being-in-the-world*" (Montuori in Ed. Nicolescu 2008, p. xi).

> "The project of transdisciplinarity is an emancipatory one. It provides us with a way of thinking and a way of organizing knowledge and informing action that can assist us in tackling the complexity of the world, while at the same time inviting us to come to grips with the role of the inquirer in the process of inquiry. [It] recognizes the role of values in inquiry, rather than attempting to suppress or 'bracket' them [and thus] engages the inquirer as an active, ethical participant in world affairs" (Montuori in Ed. Nicolescu 2008, p. ix).

Nicolescu (2002) proposed a broad outline for the transdisciplinary approach in his *Manifesto of Transdisciplinarity*, which will serve as the basis for the current section.

---

69 Cf. ibid., pp. 324-325.

70 In 1972 already, Edgar Morin concluded an article in the *Nouvel Observateur* (entitled "L'an I de l'ère écologique") with an explicit call for a "generalized ecology" as "a truly transdisciplinary science" (reproduced in Morin 2007, p. 20).

Transdisciplinarity as a new way of thinking

The insights of quantum physics

A theoretical physicist, Nicolescu founds his claim for transdisciplinarity on some of the discoveries of quantum physics, which destabilized the simple order of classical physics (cf. the 'paradigm of simplicity', in chapter 1):

- Max Planck's discovery of discontinuity at quantum level, questioned the notion of local causality in physics.
- Bell's theorem of 'nonseparability', i.e. in very short: "quantum entities continue to interact no matter what their distance from one another" (Nicolescu 2002, p. 17), does not mean (as misrepresented in popular science fiction) that quantum interactions break through the speed of light, but does point at a "new type of causality – global causality, which concerns the system of all physical entities in its entirety" (ibid., p. 18). As a result, the "existence of nonlocal correlations enlarges the field of truth, of Reality."
- Classical determinism also crumbles at the level of quantum physics, with Heisenberg's uncertainty principle:

> "The indeterminism that rules on the quantum level is a constituent, fundamental, irreducible indeterminism that signifies neither chance nor imprecision. 'Quantum randomness' is not 'chance'. [It is] neither chance nor necessity. Quantum randomness is really a constructive gamble, which has a meaning – that of the construction of our own macrophysical world" (ibid., p. 19).

Quantum physics and the macrophysical reality: two different „levels of reality"

Quantum laws are drastically different from macrophysical laws. "The discontinuity that is manifested in the quantum world is also manifested in the structure of the levels of Reality. That does not prevent the two worlds from coexisting" (ibid., p. 21). The coexistence of the quantum and macrophysical levels forces us to accept the coexistence "of what were formerly considered to be pairs of mutually exclusive contradictories (A and non-A): wave and corpuscule, continuity and discontinuity, separability and non-separability, local causality and global causality, symmetry and a break in symmetry, reversibility and irreversibility of time" (ibid., p. 23).

Nicolescu gives a strict definition to his concept of "levels of Reality": They "are radically different from levels of organization as these have been defined in systemic approaches. Levels of organization do not presuppose a break with fundamental concepts: several levels of organization appear at one and the same level of Reality. The levels of organization correspond to different structurings of the same fundamental laws" (ibid., p. 22).

Therefore, what Nicolescu is evoking is a rather radical type of emergence, with a more fundamental restructuring of reality from one to another level. Also compared to Bateson's notion of 'logical types' introduced in the previous section, Nicolescu's understanding is pretty radical. It remains to be seen, at which points one can speak of a jump from a level of reality to another one, in the strict sense adopted by Nicolescu (e.g. in the jump from quantum to macrophysical reality), and where it suffices (and is sufficiently insightful) to speak of different levels of organization. But the situation is often not allowing such a clear-cut distinction. Emergence of new structures also implies logical jumps and the emergence of new laws which are different and possibly contradicting the laws prior to emergence (but maybe only contradicting them partly or intermittently, and/or also complementing them, in a complex relation). Such is the case in the transition from active physis, to dissipative structures, to living autoecopoïetic organisms. Would Nicolescu here speak of different 'levels of organization' (given that the emerging laws do not *systematically* radically overturn the pre-existing laws, at their new level of existence) or of different 'levels of reality' (given that the advent of autopoïesis is a radical restructuring of reality)? In the rest of this volume, I will tend to refer to 'levels of reality' in a less-strict way than Nicolescu might do himself.

The logic of transdisciplinarity: The included Middle

Logic plays a fundamental role in establishing validity in the sciences and philosophy, i.e. constitutes a powerful Cerberus keeping 'non-sense' outside the gates of Reason. "A certain logic is the implicit and hidden agenda that determines all social regulation" (ibid., p. 27). But, with the advent of quantum physics, the idea of a unique, unchangeable logic came to be questioned. Nicolescu points out that "there is a direct relation between logic and the environment [which] changes with time", and the microphysical environment changed dramatically with quantum physics:

The discontinuity between levels of Reality conflicts with

"the interpretive filter of classical logic. This logic is founded on three axioms: (1) The axiom of identity: A is A. (2) The axiom of noncontradiction: A is not non-A. (3) The axiom of the excluded middle [or excluded third]: There exists no third term T ('T' from 'Third') which is at the same time A and non-A. According to the hypothesis of a single level of Reality, the second and third axioms are obviously equivalent" (ibid., p. 26).

But what happens if we consider the existence of several levels of Reality? Then, a new elaboration at the level of logic is warranted: Nicolescu turns here to Stéphane Lupasco who established a "logic of the included middle". He adds that Lupasco's logic makes sense once combined with the concept of different levels of Reality:

"In order to obtain a clear image of the meaning of the include middle, we can represent three terms of the new logic – A, non-A and T – and the dynamics associated with them by a triangle in which one of the vertices is situated at one level of Reality and the other two vertices at another level of reality. The included middle is really an included third. If one remains at a single level of Reality, all manifestation appears as a struggle between two contradictory elements (example: wave A and corpuscle non-A). The third dynamic, that of the T-state, is exercise at another level of Reality, where that which appears to be disunited (wave or corpuscle) is in fact united (quanton), and that which appears contradictory is perceived as noncontradictory. It is the projection of the T-state onto the same single level of Reality that produces the appearance of mutually exclusive, antagonistic pairs (A and non-A). A single level of Reality can only create antagonistic oppositions" (ibid., pp. 28-29).

As Nicolescu explicitly states, his understanding of the 'logic of the included third' respects the axiom of non-contradiction, at a given level of reality.

Nicolescu sees a popular illustration of the logic of the included third, in a famous story of the French stand-up comedian Raymond Devos ('le bout du bout') with "a man who desperately wants to separate the two ends of a stick. He cuts his stick and then sees that instead of having separated two ends, he now has two sticks, both of which have two ends of their own. He goes on cutting his stick, all the while becoming more and more anxious – the sticks multiply ad infinitum, but he finds it impossible to separate the two ends" (ibid., p. 31).

Nicolescu's logical triangle further develops Morin's "dialogique", rather than Hegel's dialectics, as Nicolescu highlights:

"The entire difference between a triad of the included middle and a Hegelian triad is clarified by consideration of the role of time. In a triad of the included middle, the three terms coexist at the same moment in time. This is why the Hegelian triad is incapable of accomplishing the reconciliation of opposites, whereas the triad of the included middle is capable of it. In the logic of the included middle the opposites are, rather, contradictories: the tension between contradictories builds a unity that includes and goes beyond the sum of the two terms" (ibid., pp. 29-30).

Therefore, the logic of the included middle as understood by Nicolescu, "is perhaps the privileged logic of complexity; privileged in the sense that it allows us to cross the different areas of knowledge in a coherent way" (ibid., p. 30). Within more restricted contexts, in "relatively simple situations", the logic of the excluded middle still applies.

Multi-, inter- and transdisciplinarity

Nicolescu proposes the following definitions, to discriminate between multi-, inter- and transdisciplinarity: "Multidisciplinarity concerns studying a research topic not in just one discipline but in several at the same time [...] Any topic in question will ultimately be enriched by incorporating the perspectives of several disciplines" (ibid., pp 42-43).

Interdisciplinarity "concerns the transfer of methods from one discipline to another" (ibid., p. 43). It allows epistemological transfers between disciplines but it also contributes to the "disciplinary big bang" i.e. to the contemporary explosion of disciplines, which poses a challenge that only transdisciplinarity may address:

For Nicolescu, "transdisciplinarity concerns that which is at once between the disciplines, across the different disciplines, and beyond all discipline. Its goal is the understanding of the present world, of which one of the imperatives is the unity of knowledge" (ibid., p. 44). And the space outside disciplines is not empty: "In the presence of several levels of Reality, the space between disciplines and beyond disciplines is full, just as the quantum vacuum is full of all potentialities" (ibid.). But the field of research of transdisciplinarity is neither unified, nor simple, but necessarily complex and discontinuous:

"The discontinuous structure of the levels of Reality determines the discontinuous structure of transdisciplinary space, which in turn explains why transdisciplinary research is radically distinct from disciplinary research, even while being entirely complementary. Disciplinary research concerns, at most, one and the same level of Reality; moreover, in most cases, it only concerns fragments of one level of Reality. In contrast, transdisciplinarity concerns the dynamics engendered by the action of several levels of Reality at once" (ibid., pp. 44-45).

This implies the construction of an integrative, patterning knowledge. Transdisciplinarity develops a "principle of articulation between different forms of knowledge [...] which accepts that an object can pertain to different levels of reality, with attendant contradictions, paradoxes, and conflicts" (Klein 2004). This principle, as already expressed by Morin, implies a "dialogic" allowing "unity through diversity" or "unitas multiplex [i.e.] we cannot reduce the whole to the parts, nor the parts to the whole [...] but conceive the notions of whole and parts, unity and diversity, together, both as complementary and antagonistic" (Morin 1977, p. 105).

Transdisciplinary research does not replace disciplinary research, as a sort of super-discipline. On the contrary, it complements disciplinary research "in a new and fertile way". Nor does it aim to replace multi- or interdisciplinary research, but it should not be confused with them: "Transdisciplinarity is nevertheless radically distinct from multidisciplinarity and interdisciplinarity because of its goal, the understanding of the present world, which cannot be accomplished in the framework of disciplinary research" (Nicolescu 2002, p. 46). Nicolescu advices not to confuse "the different goals of these three new approaches" (ibid., p. 47).

Research can be more or less transdisciplinary, in degrees, according to how much it takes "into account, more or less completely, the three methodological pillars of transdisciplinarity: the levels of Reality, the logic of the included middle, and complexity" (ibid., p. 122). A first degree of transdisciplinarity can be operative from within any discipline: "It is the spirit of a researcher in such and such a discipline that can be transdisciplinary". Furthermore, the said three methodological pillars complement and enrich existing disciplinary methodologies, "by bringing them new and indispensable insights, which cannot be produced by disciplinary methods. The transdisciplinary method could even lead to some true discoveries at the heart of disciplines". A further degree of transdisciplinarity is relating to the paradigmatic and epistemological foundations of disciplinary practices, and "implies not only that which crosses the disciplines but also that which structures them [with] a transdisciplinary attitude that gives the discipline its meaning" (ibid., p. 123).

The logic of the included middle and the "open unity of the world"

Transdisciplinarity opens a vast number of questions: "What is the nature of the theory that can describe the passage from one level of Reality to another? Is there truly a coherence, a unity among all levels of Reality? What is the role of the subject-observer of Reality in the dynamics of this possible unity?" (ibid., p. 49).

Nicolescu bases his exploration of these questions on the logic of the included middle, which he poses as the cornerstone of a transdisciplinary research approach: "Two adjacent levels are connected by the logic of the included middle in the sense that the T-state present at a certain level is connected to a pair of contradictories (A and non-A) at an immediately adjacent level. The T-state allows the unification of contradictories A and non-A, but this unification takes place at a level different from the one on which A and non-A are situated" (ibid., p. 50).

Nicolescu further explicates how the included middle allows to conceive of unity across levels of Reality:

> "In other words, the action of the logic of the included middle on the different levels of Reality induces an open structure of the unity of levels of reality. This structure has considerable consequences for the theory of knowledge because it implies the impossibility of a self-enclosed complete theory. In effect, in accordance with the axiom of noncontradiction, the T-state realizes the unification of a pair of contradictories (A, non-A) but is associated, at the same time, with another pair of contradictories (A^1, non-A^1). This signifies that starting from a certain number of mutually exclusive pairs, one can construct a new theory which eliminates contradictions at a certain level of Reality, but this theory is only temporary because it inevitably leads, under the joint pressure of theory and experience, to the discovery of new pairs of contradictories, situated at new levels of Reality" (ibid., p. 51).

However, if a single unified theory is then unattainable, it is nevertheless possible to "speak of an evolution of knowledge, which encompasses all levels of Reality: knowledge that is forever open" (ibid., p. 52).

To reinforce his argument about the impossibility of a single unified theory of everything, Nicolescu appeals to the incompleteness theorem of logician and mathematician Kurt Gödel, "which states that a sufficiently rich system of axioms inevitably leads to results that are either indecisive or contradictory" (ibid.): Nicolescu argues that the "Gödelian structure of the unity of levels of Reality, associated with the logic of the included middle, implies that it is impossible to construct a complete theory for describing the passage from one level to the other, and, a fortiori, for describing the unity of levels of Reality" (ibid., p. 53).

Nicolescu's argument also preserves a domain for the sacred, which would be at the 'lowest' and 'highest' levels of Reality, beyond "the limitations of our bodies and of our sense organs – limitations that apply regardless of what measuring tools are used to extend these sense organs. [This] corresponds to the sacred – to that which does not submit to any rationalization" (ibid., p. 54).

The transdisciplinary object and the transdisciplinary subject

Nicolescu's transdisciplinarity reorganizes the relationship between the subject and the object of knowledge: "A new Principle of Relativity emerges from the coexistence between complex plurality and open unity: no level of Reality constitutes a privileged place from which one is able to understand all the other levels of Reality. A level of Reality is what it is because all the other levels exist at the same time" (ibid., pp. 54-55). From the perspective of the subject, Nicolescu develops a certain form of relative constructivism: "This Principle of Relativity is what originates a new perspective on religion, politics, art, education, and social life. And when our perspective on the world changes, the world changes. In the transdisciplinary vision, Reality is not only multidimensional, it is also multireferential" (ibid., p. 55).

And in order to access these multiple referentialities, the subject must also develop his/her ability to perceive in different ways: "The different levels of Reality are accessible to human knowledge thanks to the existence of different levels of perception, which are found in a one-to-one correspondence with levels of reality" (ibid.).[71] Nicolescu thus posits a "transdisciplinary subject" (with his/her/its levels of perception) as a counterpart to the "transdisciplinary object" (with its levels of Reality). Consequently, "knowledge is neither exterior nor interior: it is simultaneously exterior and interior".

71 Nicolescu is not a radical social constructivist, i.e. for him, the levels of reality do exist independently/outside of human perception (and outside of the levels of perception). However, they do not exist as information, as long as no perceiving subject (whether human, or another life form) is perceiving them, at one or another level of perception. In that respect, Morin (1977) also rejects radical social constructivism as "conceptual ethylism".

Transvision

As already noticed, changing perceptions of reality followed technological developments in human societies. "With the evolution of tools and measuring devices, we observe a spectacular example of the relationships between the levels of human perception and those of physical Reality" (p. 67). The development of technologies extended the sense organs, most especially the sense of sight with telescopes, microscopes, and indirectly with particle accelerators. With the particle accelerators, which allow to "see" non-localizable particles, the "transgression of the field of vision leads to a transvision: a new level of Reality that can be explored by means of science [...] toward that which is beyond the visible" (ibid., pp. 68-69).

But Nicolescu argues that a perception of the quantum level of Reality, basing a "transvision", cannot rest only on the "exterior perception" allowed by the particle accelerator, but also has to rest on an "interior vision", i.e. a "quantum imagination" which is not hampered by the habitual thoughts based on the macrophysical level of Reality. And taking distance from these habitual thoughts requires a "moment of silence", allowing the unfolding of "an *imaginatio vera*, a foundational, true, creative, visionary imagination" (ibid., p. 71).

A new way of thinking

Montuori (in Ed. Nicolescu 2008, p. xi) summarizes the novelty of transdisciplinarity, as a way of thinking, along four 'requirements':

- "A *focus* that is inquiry driven rather than discipline driven [...] for the purposes of action in the world";
- "A stress on *the construction of knowledge* through an appreciation of the meta-paradigmatic dimension", i.e. questioning and inter-relating the paradigmatic assumptions of different disciplinary knowledges;
- "An understanding of *the organization of knowledge*" (e.g. disjunctive or conjunctive); and
- "*The integration of the knower in the process of inquiry*" reflecting on the process of construction of knowledge.

Kirchberg (2009) further synthesizes the shift from disciplinary to transdisciplinary thinking, as "a knowledge shift:

- from simplicity to complexity
- from clearness to hybridity
- from linearity to non-linearity
- from the universality of the unified to the integration of the differentiated
- from fragmentation to cooperation

- from the hardening of borders to the dissolution of borders[72]
- from short-termed to sustainable research
- from the reduction to the knowing few, to the dialogue of many about systemically networked knowledge",[73]

highlighting a number of keywords (complexity, non-linearity) that were already introduced earlier in this chapter about the insights from systems thinking and complexity theory.

Transdisciplinary research

The transdisciplinary vision of Nature

As earlier described (in chapter 1), Nicolescu argues for a "resurrection of Nature" from the ashes of the "death of Nature" caused by modernity's mechanistic vision of the natural world. "Transdisciplinary Nature has a ternary structure" (Nicolescu 2002, p. 64):

- "Objective Nature, which is connected with the natural properties of the transdisciplinary object [i.e. with levels of Reality] [and] is subject to subjective objectivity" as in the sciences;
- "Subjective Nature" which is connected with the transdisciplinary subject, i.e. with levels of perception, and which finds methodological roots in "the ancient science of being, which is present in the traditions and religions of the world";
- Trans-Nature, which Nicolescu describes as a "communion" between the transdisciplinary object and subject, which "concerns the domain of the sacred" (ibid.).

Nicolescu argues that transdisciplinarity should "elaborate a new philosophy of Nature" on the basis of this ternary structure and of an affirmation of both science and subjectivity. Nature is not an already written book to be read by humans, but "seems more like a book in the process of being written [and] not so much to be read as experienced, as if we are participating in the writing of it" (ibid., p. 65).

An exclusive focus on only one of the three facets of transdisciplinary Nature, nourishes "a dangerous phantasmagoria" that inevitably leads to destructive social-historical developments, according to Nicolescu: "Reality reduced only to the Subject has caused traditional societies to be swept away

72 I would note however that speaking of a "dissolution" of disciplinary borders is misleading, as disciplinary borders do not disappear because of transdisciplinary thinking, nor should they entirely disappear, as Nicolescu's arguments make clear.

73 Translation from the original German text by the author.

by modernity. Reality reduced only to the Object has led to totalitarian systems. Reality reduced to the sacred has led to fanaticism and religious fundamentalism. A viable society joins these three aspects of Reality in a balanced way" (ibid., p. 72).

The transcultural

Transdisciplinarity also involves the emergence of a transcultural communication. Nicolescu values multi- and intercultural exchanges across the world, though longing for a more profound, transcultural process:

"The multicultural helps us discover the face of our own culture in the mirror of another culture [but] the multicultural and the intercultural by themselves do not assume the kind of communication between all cultures that presupposes a universal language founded on shared values, but they certainly constitute important steps toward the act of transcultural communication. The transcultural designates the opening of all cultures to that which crosses through them and transcends them" (ibid., p. 104).

Nicolescu points at one example of a transcultural approach in the arts (according to him), with the case of Peter Brook's theatrical productions such as the *Mahabharata* and *The Tempest* (at the Centre International de Créations Théâtrales).

Other authors also explored the notion of transcultural art. For example the dance education researcher Laurien Saraber (2000) evokes the possibility of "transcultural dance". On multi-, inter- and transculturality in the case of dance, see also ed. Pontbriand (2001), especially for the articles by Richard Shusterman and by André Lepecki who evokes "sensous knowing" as "a sensuality generating cultural identification rather than split" (Lepecki 2001, p. 167) and advocates for transcultural "moments of intense corporeal becoming, of immanent transcen/dance" (ibid., p. 168).

Nicolescu defines the transcultural as an experience across and beyond cultures.

"Each culture is the actualization of a potentiality of the human being, in a specific place on Earth and at a specific moment in history. Different places on Earth and different moments in history actualize different potentialities of the human being, that is, different cultures. It is the open totality of the human being that constitutes the 'place without place' of that which crosses and transcends cultures. The perception of the transcultural is first of all an experience, because it concerns the silence of different actualizations. The space between the levels of perception and the levels of Reality is the space of this silence; it is the equivalent, in interior space, of that which is called the quantum vacuum in exterior space" (Nicolescu 2002, p. 105).

Therefore, the transcultural is in no way aiming to realize the unification of all cultures on one plane. Along the spirit of transdisciplinarity, transculturality "both enables the dialogue between all cultures and prevents their homogenization" (ibid., p. 108).

The transcultural experience grounds the "transdisciplinary attitude" (which will be further described in chapter 4).

"In effect, the complex plurality of cultures and the open unity of the transcultural coexist in the transdisciplinary vision. The transcultural is the spearhead of transdisciplinary culture. Different cultures are the different facets of the human being. The multicultural allows the interpretation of one culture by another culture, the intercultural permits the fertilization of one culture by another, and the transcultural assures the translation of one culture into various other cultures, by deciphering meaning that links them, while simultaneously going beyond them" (ibid., p. 107).

Transdisciplinarity as action-research

Although Nicolescu does not use the term, his approach is clearly advancing a form of action-research (or action & research, action in research and research in action): "Transdisciplinarity is both a body of thought and a lived experience. These two aspects are inseparable. Transdisciplinary language translates the simultaneity of these two aspects into words and actions. Any excessive slipping one way or the other – to the side of discursive thought or to the side of experience – takes us away from the domain of transdisciplinarity" (ibid., p. 119).

Transdisciplinarity is not just about knowledge either, as Montuori noted: "The transdisciplinary approach does not focus exclusively on knowing, but on the inter-relationship between knowing, doing, being and relating" (Montuori in Ed. Nicolescu 2008, p. xi).

The risk of wrongly understood transdisciplinarity: levels of confusion

Nicolescu warns against "levels of confusion" (a notion that he borrows from Philippe Quéau), which "are generated by lack of respect for the unique and singular role that each level of Reality and each level of perception plays within the open unity of the world" (Nicolescu 2002, p. 111). Nicolescu further lists several possible levels of confusion:

"The most elementary confusion consists in forgetting the discontinuity of levels of Reality and levels of perception, replacing them implicitly with continuity. Then an inevitable reduction of all levels of Reality and all levels of perception to one and the same level occurs; complex plurality is reduced to a complexity without any order other than the horizontal one of levels of organization, and the open unity of the world becomes a plural world closed in upon itself, and thus favorable to all ideological and dogmatic mixups" (ibid.).

Another level of confusion could lead to a neoscientism: "One can envisage the arbitrary reduction of all the levels of perception to one and the same level of perception even while recognizing the existence of several levels of Reality. This level of confusion can lead to a new scientism being taken as the intellectual foundation of a badly understood transdisciplinarity" (ibid., p. 112). No single level of perception, however inter-disciplinarily informed it becomes, can ever account for the totality of everything that exists.

The other important risk Nicolescu sees here is political: "The same level of confusion can lead to the absorption (and therefore the destruction) of transdisciplinarity by extremist ideologies [...] looking for renewed virginity" (ibid.).

The inverse confusion can also occur: "Another extreme level of confusion consists in the recognition of the existence of several levels of perception, all the while refusing to admit the existence of several levels of Reality. This wrong turn leads to the annexation of transdisciplinarity by hermetic irrationalism" (ibid., p. 113).

This irrationalism can take the form of "the current infatuation for cheap esotericism". But Nicolescu's warning here also concerns the position of radical constructivism: "the current academic style is to reduce everything to language: as if there were no Reality, in the ontological sense of the term, but merely languages that construct a reality, and no science which explores Nature, but merely a social construction that we call 'science'" (ibid., p. 114). Such a form of radical constructivism indeed degenerates into a new mode of simplification of complexity. Furthermore, radical constructivism and its actualization in the postmodern worldview, lead to a dissolution of normativity which Nicolescu also rejects: "In a world where everything is presented as a social construction; where the relative has become, in fact, the absolute; where violence is the other face of solidarity and exclusion is the other face of well-being [...] it is unthinkable that we can find the true reason for living in such a world" (ibid., p. 141).

Finally, confusion can arise, even if acknowledging the plurality of both levels of Reality and of perception, when one is not "taking into consideration their rigorous correlation". Concerning the 'New Age' movement, Nicolescu does not altogether reject it as improper but sees such a confusion emerging from the New Age movement: the danger associated with the New Age is rooted in its lack of rigor, which leads to unconsciously mixing everything into an amorphous catchball, where it will attempt to include transdisciplinarity as an honorable and more or less exotic component. "Notwithstanding the motivations of one or another of its representatives, the New Age appears like a giant hypermarket of our consumer society, where each and every person can come to search for a little of the Orient and a little of the Occident, in order to regain, fairly cheaply, his peace of mind" (ibid., p. 115).

At a practical level, Nicolescu warns against "the marketing of transdisciplinarity" in "seminars for business leaders", "bestowing new legitimacy to decision makers in distress without doing anything to change their approach" (ibid.).

More generally, "there is an urgent need for the formulation of a transdisciplinary deontology" (ibid., p. 116). Indeed, the special status of transdisciplinarity can very easily be confused. "The place of transdisciplinarity is a place without a place. It is found neither in interior man (either generating a new religion, or a new philosophy, or a new metaphysics), nor in exterior man (generating a new science, which would be the science of sciences)" (ibid., p. 117). Only then can mutilating simplifications be avoided, as Morin also argues. Because all the confusions listed above have one dynamic in common: "At its core, that which links all the wrong turns is the impoverishment of the trans-subjective dimension of being" (ibid.).

Conclusion to Chapters 2 and 3

The wide-ranging overview of systems thinking, complexity theory and transdisciplinarity conducted in the previous and the current chapter, inform the formulation of key notions and principles for an open definition of a "culture of sustainability" and of "cultures of sustainability", as a paradigmatic shift in the context of contemporary societies.

Following Morin's dia-logic which combines unity, complementarity, competition and antagonism in complex relationships, I cannot talk either of one single "culture of sustainability", at one single level, nor of a multiplicity of "cultures of sustainability", but I must speak of both a culture and cultures: Sustainability requires a diversity of cultures of sustainability, but also a common set of characteristics which are common to these cultures, i.e. "a culture-on-a-meta-level, a meta culture with its characteristics of sustainability".[74] Or these key notions and principles may even contribute to a "trans-cultural" dimension, in Nicolescu's sense, rather than to any specific 'culture'.

Yet another alternative would have been to stop altogether talking about culture or cultures, and instead speak of "culturality", as proposed by the philosopher Oleg Koefoed, following sociologist Michel Maffesoli's preference for the term "sociality" over "society" (Koefoed 2008, p. 60). However, I will depart from Maffesoli and Koefoed insofar as Maffesoli focuses too exclusively on localized meanings, i.e. on the aspect of multiplicity, negating unity. (In this, Maffesoli is typically post-modern, whose level of constructivism and hyper-fragmentation are eventually incompatible with complexity theories).

74 Hans Dieleman, personal communication; I am thankful to Dieleman for pointing out this dia-logic of culture and cultures, which I had failed to notice in a recent article on the subject (Kagan 2010).

Notions and principles from systems thinking

Several insightful notions from systems thinking have been introduced in chapter 2. Especially, the notions of *networks*, *non-linearity*, *emergence* and *self-creativity* opened up a first phase in the outlining of a paradigmatic shift, away from linear causality.

Some principles have also come out of the systems perspective, most especially Capra's six principles of ecological literacy, i.e. the *networking principle* of living systems, the *continual cycles* of the web of life, the *openness* of the Earth system *to solar energy*, the importance of *partnership and cooperation* for evolution, the necessity of *diversity for resilience* (and first of all, of biodiversity for ecological resilience), and the search for *dynamic balance* (rather than the attempt to optimize any single value or dimension).

At the social level, the systems perspective highlighted how social organizations may be able to evolve, if understood and managed as *living and learning organizations*. Systems thinking also points out the *importance of culture* for both social reproduction and social change.

Notions and principles from complexity thinking

A second phase in the outlining of a paradigmatic shift was opened with the introduction of Morin's approach to complexity, away from both the simplification of reductionism and the simplification of systemic holism (as seen e.g. in some passages by Laszlo). Morin introduces the possibility to think *unity and diversity* alongside each other, and to think about any pair of terms, with a combination of *unity, complementarity, competition and antagonism*, altogether forming a *complex relationship* and calling forward a *dia-logical* thinking process.

This second phase, with the current chapter's discourse, also introduced a number of notions and principles at the level of what Morin calls "*macro-concepts*", which organize complex relations between single notions (unlike "concepts" which are limited by the classical logical axiom of non-contradiction). One such macro-concept is *eco-auto-organization*, which explores the complex organizational relationships between individual life forms and the ecosystems in which they *co-evolve* and *eco-evolve*. *Eco-organization* highlights the complementarity of diversity, complexity, spontaneity and organization. Auto-eco-organization further encompasses six principles, i.e. the "*principle of bio-thanatic inscription*" (which complements Capra's continuous cycles principle), the continuous role of *eco-organization as co-organizer of self-organization*, the *mutual reinforcement of the complexities of autos and oikos*, the *correlation of increase in autonomy with increase in dependence* on eco-organizational complexity, the *dia-logical explanation of life phenomena* (i.e. the recourse to the eco-auto or auto-eco)[75], and the extension of the former principle into a principle of *generalized ecology* exploring the eco-auto-logic in human societies and in the world of ideas.

At the social level, Morin thus introduces the macro-concept of *socio-eco-organization*, which aims to balance the organicist tendencies with the eco-organizational tendencies in contemporary human societies, in order to prevent the risks of both organicist totalitarianism (e.g. the totalitarian state) and eco-organizational conflictuality (e.g. class struggle under an unbridled market economy). Morin's insights thus also reveal, next to the *centrality of culture in social evolution* already highlighted above with Capra, the *issue of the State* (pointing at the political dimension of sustainability, which I will further explore in chapter 7).

Overall, the planet Earth in its current state is revealed as a wide and complex *eco-socio-system* which is *double-piloted*, i.e. piloted by nature's eco-auto-organization but also co-piloted by humankind, and which requires a *symbiotic co-development* of its two pilots.

Autoecopoïesis

Morin's insights point at the *genesic force of chaos* at the root of any creation, and highlight the continuous presence of chaos (and of order from disorder) in *generative endo-exo-causality* and *de-re-organization*, two macro-concepts that, in turn, are at the root of phenomena of *emergence*, are fueling the *poïesis* of *machine-beings* (i.e. the creative evolution of the universe) and are imposing an *uncertainty principle* which disarms all forms of determinisms (whether they are based on linear causality, on autopoïetic teleonomy or on Laszlo's 'evolution towards more complexity' argument).

Morin points out that poïesis and organization are not having the same meaning and should not be confused with each other. Poïesis is about the creative-constructive-productive, while organization is about the processing-of-structuring-of-processing; so one reality is approached with both (poïesis and organization), but from different perspectives. Although Morin uses the word organization much more widely than poïesis, the word poïesis bears a transformative potential, of especially high relevance to the search process of sustainability in its dimension of cultural change.

Nicolescu points at the contemporary relevance of 'poïesis': "The word *poetics* comes from the Greek word *poïein*, which means 'to do': To do, today, means to reconcile the contradictories" (Nicolescu 2002, p. 90). In the Greek language, the verb poïein relates to a wide range of action. "It applies

---

75 In Morin's work, the order between 'auto-' and 'eco-' is not of utmost importance, as he conceives them as a feedback loop (including in the typography of *la méthode*, where a special typographic device visualizes the loop, inside the body of the text).

to all sorts of operations, from those that model some clay until the highest realizations of the artist or the poet" (Blondel 1949, p. 55).[76]

Poïesis gives "a creative connotation to the term of production" (Morin, vol. 1, p. 158). Nicolescu also points at the productive-creative force of 'poïein': "In the last analysis, 'learning to do' is an apprenticeship in creativity. 'To make' also signifies discovering novelty, creating, bringing to light our creative potentialities" (Nicolescu 2002, p. 134).

Already at the level of physical machine-beings, Morin expresses that poïesis is built upon chaos and implies both a generative organization and the intrusion of noise; the active organization then turns noise into novelty. At the level of living beings and human societies, I propose to use another expression than Luhmann's autopoïesis, more in line with Morin's analyses: *autoecopoïesis*.

As already mentioned earlier, Capra's discourse points repeatedly at the potential of disturbances, creativity and emergence for social transformation towards sustainability; properties which require a certain productive openness to disturbances: Such an openness is the hallmark of autoecopoïesis, by contrast to the rigidity or autism of a strict autopoïesis.

"Strict autopoïesis as a culture of unsustainability in hypermodernity, is not an inescapable trend. Autopoïesis is but only a tendency that, if strong and dominant so far, may be balanced by ecopoïetic tendencies, i.e. tendencies of psychic systems and social systems to construct themselves in open communications with their environments (implying a co-determination and co-evolution of both the system and its environment through the emergence of properties stemming from the open communication between system and environment). [...] Not only 'eco-' is necessary, but also 'auto-' because the capacity for relative autonomy (i.e. a capacity for self-closure) is a pre-requisite for a system's ability to participate in its own (re-)construction" (Kagan 2010, p. 1097).

The inter-... and the trans-...

Working with the notions and principles synthesized above, also involves a shift away from traditional methodologies in all areas of knowledge-creation (from the scientific disciplines to the arts and philosophy), away from reductionism, holism and other forms of simplification of reality. Achieving such a transformation requires that working methodologies be developed at interdisciplinary and transdisciplinary levels, without altogether abandoning the insights gained from disciplinary work.

The "inter-..." of interdisciplinarity and interculturality, "operate[s] most especially at the level of the membranes, of the borders, of the contact areas between different elements or different systems, and [...] foster[s] dialogues across the membranes. The 'inter-...' should not be mistaken for the

76 Translation from the original French text by the author.

'multi-...' nor for the 'integrative'. Neither does the 'inter-...' set differences apart as irreducible, nor does it integrate differences (making them increasingly indistinguishable)" (ibid., p. 1096). Understood as a sensibility that operates at the borders/membranes, the inter-... of interculturality is cultural bridging, fostering mutual learning processes, inter-contextually i.e. across different context-specific knowledges.

The inter-... develops a praxis that allows to work with the insights from systems thinking, which in turn open up a bridge into the "trans-...".

The "trans-..." of transdisciplinarity and transculturality, operates both outside and across different systems: It calls forward the practice of several levels of perceptions attentive to several levels of reality, the exploration of a logic of the included middle across levels of reality. It announces a practice that is neither a science of sciences, nor a philosophy of philosophies, nor a 'theory of everything', but a complex relational methodology that allows to conceive of unity in diversity, and invites everyone to turn Morin's complexity into a praxis.

The significance of autoecopoïesis at the level of cultural production and experience will be further explored in chapter 4, with the definition of "aesthetics of sustainability" alongside Gregory Bateson's notion of a "sensibility to the pattern which connects"; a sensibility that echoes the insights from systems and complexity thinking and the understanding of eco-auto-organization.

4. Aesthetics of Sustainability

> Our civilisation comprehends great variety and complexity, playing upon a refined sensibility, it must produce various and complex results. The poet must become more and more comprehensive, more allusive, more indirect in order to force, to dislocate if necessary language into his meaning... cultivating all the possibilities of words as medium and when the speech of one sense is insufficient to convey the entire meaning (using) the languages of another
>
> T.S. ELIOT (IN DREW 1949)

INTRODUCTION

Edgar Morin recently commented in the French daily newspaper *Le Monde* on political ecology and the Green party in France, pointing out that their political vision is still lacking a focus on "the sense of the quality and poetry of life": "There is also something deeper, which is still found in no political program: the positive necessity of changing our lives, not only in the sense of sobriety, but especially in the sense of the quality and the poetry of life. But this second face is still not developed enough in political ecology." (Morin 2009). Morin regrets that the message of Ivan Illich about "psychic ill-being" in late 20th century affluent societies and his call to "regenerate human relationships in what he called conviviality" (and I would add, of Illich's inspirer, Jacques Ellul) was side-lined in Green politics. Morin asks:

"When will thus politics take into account the immense longing for love of the human species lost in the cosmos? A policy that would integrate ecology in the whole range of the human problem would face the problems posed by the negative effects of the developments of our civilization (which are growing comparatively to the positive effects), among which is the deterioration of solidarities – this would make us understand that the introduction of new solidarities is a key aspect of a policy of civilization" (ibid.)

The necessary change Morin evokes here is not a reform or a revolution but, he claims, a "metamorphosis": "Because, when a system is not capable of handling its vital problems, either it splits apart, or it produces a richer metasystem, capable of handling them" (ibid.).

Morin's most recent interventions in mainstream media point at the urgency of a social transformation, not just at the level of climate and agricultural policies, but at the level of a re-animation of certain sensibilities that Modernity has numbed. In a recent interview, Morin also states creativity and the arts as one of the two things that allow him to retain some hope despite the gravity of the contemporary crisis of civilization and the inertia of political and social structures:

"Two things allow me to keep hoping. The first one is that the unforeseen often arose in the history of humanity [...] and awarenesses can accelerate, increase. [...] The second source of hope lies in the existence of creative capacities which exist among all individuals and in all societies. These capacities are put to sleep, generally, because the individuals are domesticated in ossified societies. At the margins of societies, artists, poets, musicians and inventors keep these creative capacities. A crisis can trigger regression and disasters, but also awaken, wake creation, imagination and invention" (Morin in Tariant 2009).

How can the creative capacities invoked here by Morin, contribute to a cultural evolution towards a better understanding of the complexity of the global crisis, as well as towards strategies of resilience and sustainability? How can the sensibility to the "poetry of life" that Morin evokes, be raised? The current chapter will address this question, and suggest that the answer may lie in the development of an aesthetic sensibility that follows what Gregory Bateson described as a "sensibility to the pattern which connects". But first of all, aesthetics will be introduced as an experiential notion, following John Dewey, as well as some existing uses of the expressions "environmental aesthetics" and "ecological aesthetics".

The common thread across this chapter is experience: The expanded mode of knowing required for cultures of sustainability, is linked to an expanded experience of reality, to be achieved through a heightened sensibility (to patterns that connect, to complexity and to the more-than-human life-world in general). As this chapter will argue (after Bateson), the contribution of art to a culture of sustainability is to be found especially in its potential to create experience, make relationships, expand mental activity beyond the linear confines of purposive consciousness.

SECTION 1: FROM AESTHETICS, ENVIRONMENTAL AESTHETICS AND ECOLOGICAL AESTHETICS TO AESTHETICS OF SUSTAINABILITY

The aesthetic experience according to John Dewey

John Dewey's *Art as Experience* (1934) oriented the understanding of aesthetics towards the "aesthetic experience", beyond the classical focus on the museum-art object, and pointing at personal affectivity in everyday life. The word 'aesthetics' or 'esthetics', derives from the Greek words *aisthetikos* (sensitive) and *aisthanesthai* (to perceive, to feel).[1]

The domain of 'experience' as understood by Dewey, has to do with a human being's overall interrelationship with his/her environment: "Experience is the result, the sign and the reward of that interaction of organism and environment which, when it is carried to the full, is a transformation of interaction into participation and communication" (Dewey 1934, p. 22).[2]

The aesthetic quality of experience unites meanings and values. "Whenever there is a coalescence into an immediately enjoyed qualitative unity of meanings and values drawn from previous experience and present circumstances, life then takes on an aesthetic quality–what Dewey called having '*an* experience'" (Field 2005).

Aesthetic experiences can arise in everyday life, and not only in the confrontation with works of art. However, artistic expressions are especially conducive to aesthetic experience. "What distinguishes artistic creation is the relative stress laid upon the immediate enjoyment of unified qualitative complexity as the rationalizing aim of the activity itself, and the ability of the artist to achieve this aim by marshaling and refining the massive resources of human life, meanings, and values" (ibid.). Artistic creation consists in a highly reflexive activity which processes senses and emotions, not merely directly, but in the manifold attachments that they have to each other and to a variety of meanings. Because artistic creation is consisting in this emotional connection to meanings, the form of a work of art cannot be separated from its matter (and the means and ends cannot be separated either).

Furthermore, the aesthetic experience integrates the subject and the object: "no such distinction of self and object exist in it, since it is esthetic in the degree in which organism and environment cooperate to institute an experience in which the two are so fully integrated that each disappears" (Dewey 1934, p. 259). And as seen in chapter 3, the non-separability of subject and object is also one of the conditions for an understanding of complexity.

---

1 Cf. Douglas Harper, Online Etymology Dictionary, http://www.etymonline.com, accessed September 13 2009.

2 The exact page numbers refer to the 2005 Perigee edition of Dewey 1934.

Aesthetic experiences have as much to do with the activities of 'observers' as with those of 'creators'. Indeed, whereas an artist may realize an "art product", it is only through the dialog between the "art product" and the observer that a "work of art" arises, "for to perceive, a beholder must create his own experience": If the observer is "too lazy, idle, or indurated in convention to perform this work [...] his 'appreciation' will be a mixture of scraps of learning with conformity to norms of conventional admiration and with a confused, even if genuine, emotional excitation" (ibid., p. 56). Therefore, aesthetics has to do with the shared *and* autonomous sensibilities of human beings, producing meaningfulness. In other words: "Aesthetics as a multi-dimensional and multi-sensory opening of one's perceptual faculties to the world meet the 'other' in transgressing the limits of the 'self' or of the 'ego'" (Erzen 2004, p. 23).

Artistic and other creative productions are expressions opening a dialog of experiences. The aesthetic experience is not only an individual, but also a cultural phenomenon. Dewey insisted on the significance of artistic expressions as articulating the significance of life for a certain culture (i.e. for a certain society at a certain historical period). He even characterized aesthetic experience as a "manifestation, a record and celebration of the life of a civilization, a means of promoting its development, and [...] also the ultimate judgment upon the quality of a civilization" (Dewey 1934, p. 339). For Dewey, artistic expressions reflect on what a society considers as a meaningful and satisfying life. He argued that

> "instruction in the arts of life is something other than conveying information about them. It is a matter of communication and participation in values of life by means of the imagination, and works of art are the most intimate and energetic means of aiding individuals to share in the arts of living. Civilization is uncivil because human beings are divided into non-communicating sects, races, nations, classes and cliques" (ibid., p. 350),

i.e. because they are missing the connectedness of aesthetic experience, i.e. the "sense of the including whole" (ibid., p. 201).[3]

Environmental aesthetics

As Dewey already suggested, the aesthetic experience involves the apprehension of the environment. The environment is not merely a setting, but is integrated in human experience and contextualizes experience.[4]

---

3 Furthermore, the disassociation from everyday life of the concerns of what sociologist Howard Becker will later characterize as "art worlds", across the 20th century, was already a source of concern for Dewey (and contributed to a 'culture of unsustainability' – cf. chapter 1, section 2).

4 Cf. Berleant 1992.

Dewey's understanding of aesthetics is comparable to the discourses of some contemporary philosophers of aesthetics focusing on the relationship between aesthetics and ecology: Jale Erzen, for example, defines

> "an aesthetic disposition [...] as the resulting mental and sensual reaction to a corporeal relationship to the environment, which is affected through a multi-dimensional, synaesthetic perception. Perception seen this way means a constant give and take amongst all beings that dwell on the earth. This constant, intimate, direct and plural alliance amongst all things is what makes the physical processes of life on earth possible" (Erzen 2004, p. 22).

Such an understanding of the aesthetic experience calls forward a link to the natural world, and Erzen thus further argues that "aesthetics and ecology have to be understood as related in all domains" and traced back to "the basic kinship with the earth".

The direct experience of the natural environment is however relatively distinct from the experience of a work of art: Natural environments are relating relatively differently to our aesthetic experience than do works of art, insofar as they are (up to a certain extent) not the products of human intentionality, frequently immerse us and engage multiple senses (whereas the work of art often focuses on a selected number of senses) and are often marked by spontaneity and change.[5]

Environmental aesthetics is a branch of aesthetics that studies the aesthetic appreciation of natural (and man-made) environments, and "asserts the importance of aesthetic value to discussions of our relationships with the natural world" (Brady 2003, p. 1). Emily Brady's work in this field highlights that the aesthetic experience is not merely self-oriented and hedonistic (as supposedly opposed to environmental ethics). On the contrary, "the appreciative side of aesthetic experience is active, engaged and even performative"; and in practice, "the importance of imagination and emotion to moral choice" should not be underestimated: e.g. "'sensitivity' and 'attentiveness' suggest a careful kind of perception that is a feature of both moral decision-making and attitudes and the appreciation of aesthetic qualities" (ibid., p. 255). Brady (2003) further articulates how aesthetic skills contribute to moral sensitivity and discusses the relation to ethics in aesthetics. John Dewey also discussed the relationship between aesthetics and ethics, finding their unity in what we experience and understand as being "fair" behavior (Dewey 1932, p. 275).

5 Cf. Brady 2003.

Ecological aesthetics

The expression "ecological aesthetics"[6] has been used and developed among artists who came to be called "ecological artists", after the example of US artists Helen and Newton Harrison, from the 1970's onwards.[7] One of the first typical aspects of the self-imposed 'ecological' constraints of ecological artists was to allow "only what could be created from materials available on site. The shaping process, a necessary characteristic of art, was not allowed to change the materials from the outset and fundamentally – transform them by adding or subtracting energy-, but only to "arrange", in other words organize them in a new way that resisted chance. The hand of man [...] intervenes in the procedure and then leaves what emerges to "natural" events, to "metamorphosis"" (Boberg in eds. Strelow, Prigann and David 2004, p. 8). This example shows how the movement of ecological art developed the notion of 'ecological aesthetics' as an aesthetics that pays attention and respect to the own complex dynamics of natural phenomena in their relationships to human interventions, and that wants to highlight these aspects in the artistic working process. In other words, the 'ecological aesthetics' aims to highlight the form and meaningfulness of natural processes (i.e. complex processes of auto-eco-organization, if we follow Morin's characterization). In her introduction to *Ecological Aesthetics, Art in Environmental Design: Theory and Practice*, Heike Strelow insists on the holistic integration of aesthetics and ethics:

"The perception of aesthetics on which this book is based draws on the ancient idea of aesthetics as the teaching of sensual perception and cognition (aisthesis). At the same time, aligning aesthetics in the sense of a wholistic perception with an integrative understanding of nature and culture leads to a discrete form of 'normative' aesthetics. On the plane of thinking and acting in terms of the environment and human ecology, ecological aesthetics links the integrated experience of the world with ethical criteria defined in terms of the humanistic tradition" (Strelow in eds. Strelow, Prigann and David 2004, pp. 10-11).

Ecological aesthetics also qualifies as a "sensibility to the pattern which connects" (to be discussed below), as it is "inseparably linked with the idea that ultimately everything, nature and culture as well, and thus man and his habitat, are connected in an infinite, diverse systems of relationships" (Stre-

6 The expression "ecological aesthetics" has also been used in other contexts. For example, it is in use in the psychology of perception (cf. ed. Landwehr 1990) but that approach is largely irrelevant to the focus of my current research.

7 I will introduce ecological art in chapter 5. For the moment, I merely approach the notion of "ecological aesthetics".

low in eds. Strelow, Prigann and David 2004, p. 11).[8] This idea emerged together with the ecological movement of the late 20th century, and allowed to move beyond a Romantic dichotomy between a pristine nature and an extra-natural human culture, and the Modern opposition between primitive nature and civilized culture: "In the course of the growing ecological understanding that did not start until the late sixties, man came to perceive himself as an integral part of a set of connected, natural and cultural eco-systems, and thus also part of the nature surrounding him" (ibid.). Strelow locates the emergence of this idea in art in the movement from "Land Art" to "Art in Nature": indeed the latter, unlike the former, "do not just seek stimulus from nature, but build her as a partner, as their fellow creator". Ecological aesthetics points at "the traces of this interpenetration of nature and culture" (ibid., p. 12).

I shall remark, however, that one can find some premises of this awareness that arose in the 1960's, already in earlier writings from individual modern artists: For example, in the early 20th century, Paul Klee noted: "For the artist communication with nature remains the most essential condition. The artist is human; himself nature; part of nature within natural space" (quoted in Haley 2008, p. 202).

Because ecological aesthetics thus deals with multi-layered systems of relationships, a superficial parallel can be made with Nicolas Bourriaud's "relational aesthetics", in the stimulation of visitors participation.[9] Bourriaud also argued that "nothing (no art) will be possible without the deep ecological transformation of subjectivities, without the awareness of the interdependencies on which subjectivity is based" (Bourriaud quoted in Erzen 2004). However, unlike ecological aesthetics, relational aesthetics limits itself to a superficial participative spectacle with no or little reflexivity.

Because culture is part of nature, "within art, an 'ecological aesthetic' would be a reflexive, socially and environmentally shaping activity", argues Herman Prigann (in eds. Strelow, Prigann and David 2004, p. 111), who further characterizes ecological aesthetics as an "aesthetic of integration" by contrast to the former "aesthetic of exclusion towards nature, excluding it as something external, something material, something understood as a resource" (ibid., p. 113). However, this should not be confused with an aesthetics of confusion (i.e. a holistic simplification), as Jacques Leenhardt's response to Prigann suggests.[10]

8 The original text speaks literally of "an infinite [...] systems", leaving the reader wondering whether Strelow speaks here of one system or of plural systems.

9 The parallel is explicitly made by Erzen 2004, p. 24.

10 Alas, instead of warning against an aesthetics of confusion (as I explicitly do here), Leenhardt falls into a binary dichotomy between differentiation-specialization on the one hand, and a Romantic fusion and de-differentiation/de-specialization on the other hand (cf. Leenhardt in eds. Strelow, Prigann and David 2004, p. 113). This is probably due to the fact that Leenhardt is here not us-

Timothy Collins suggests "that the aesthetics of ecology are revealed through the perception and understanding of the physical characteristics of diversity" (Collins in eds. Strelow, Prigann and David 2004, p. 170), with effective change emerging from diversity rather than from "primary authorship or the exclusive manifestos of modernist practice" as was attempted at in artistic avant-gardes. Collins further points at the ecological value of biological and social /cultural diversity as "a complex interrelationship of diversities" (ibid., p. 172), i.e. its value for resilience, which Collins metaphorically understands as an issue of health: "a lack of health can be described in terms of emergent dominant systems that mitigate or constrain diversity".

Strelow mentions "interdisciplinarity and transdisciplinarity" as typical features of art projects based on ecological aesthetics.[11] "Connecting things separate spatially and intellectually, in other words transdisciplinary thinking and acting, are essential for conceiving and realizing these projects" (Strelow in eds. Strelow, Prigann and David 2004, pp. 12-13). The project interdisciplinarity also involves bottom-up political participation by communities, or at least their approval and legitimation: In this sense, Strelow also argues that ecological artists, who are "initiating, promoting and presenting these processes", are thus becoming "catalysts for social transformation processes. They mediate between various pressure groups and disciplines" (ibid., p. 13). This also points at another feature of ecological aesthetics in such projects: It is "aimed at practicality" and founds a concretely transformative, i.e. transform-active artistic practice.

A further feature stemming from the 'expanded concept of art' initiated by Joseph Beuys (and in opposition to historical avant-gardes) is, according to Jacques Leenhardt, that by moving away from museum art, such art projects "involve [themselves] once more in all the dangers emanating from uncertain spaces like the street, from nature, from opaque human situations [and thereby] rediscover less theoretical, more directly human aspects of existence in which the complexity of ideas and gestures finally achieves its full identity, its full density" (Leenhardt in eds. Strelow, Prigann and David 2004, p. 112). As Morin indeed pointed out, the understanding of complexity, in nature and in human society, requires such an openness to uncertainties and to the agitations of disorders outside the organized fields of art

---

ing macro-concepts, and does not clearly envision a dialogical alternative (as e.g. Morin would do: Cf. Morin 1980, pp. 327-328) and thus remains captive of the binary tension he describes.

11 However, even though she uses the word repeatedly, Strelow only succinctly defines transdisciplinarity as "a research approach that defines problems independently of any discipline" and as "a further development of interdisciplinarity" (Strelow in eds. Strelow, Prigann and David 2004, p. 15) - for a more thorough approach to transdisciplinarity, see my earlier developments in chapter 3 and my forthcoming discussion below in this chapter.

worlds. Leenhardt is explicitly pointing at the "ecological idea" for its introduction of "complexity and the interaction of causalities [into] the circle of artistic disciplines, whose unduly confined framework it opens up": In other words, he argues that the ecological idea, as in 'ecological aesthetics', offers to the art worlds the opportunity to leave the orbit of the 'culture of unsustainability' (as discussed in chapter 1).

But this opportunity does not come without challenges: Leenhardt, in his discussion of the insights of the "ecological idea" to art, warns about the consequences of such insights for artistic practices and the kind of aesthetic experiences that are to be expected: These can no longer be limited to merely local objects and relations, but must relate them to wider contexts: "the new interest in complex causalities leads to increased attention to global connections rather than spatially limited situations that cannot carry the real driving forces of the phenomena within them. [...] Objects of ecological aesthetics are not permitted small frames of reference" (ibid.). This requirement further qualifies ecological aesthetics as a "sensibility to the pattern which connects".

Towards aesthetics of sustainability

The notion of a sensibility to connectedness was central in Gregory Bateson's understanding of aesthetics. Based on Bateson's aesthetics, which I will now introduce and discuss, I will further elaborate an understanding of "aesthetics of sustainability". This very expression has already been used by Hildegard Kurt in the past decade, but in a different way, which I will discuss in chapter 6 (I will also discuss in chapter 6, the expression "sustainable art" as introduced by Maja and Reuben Fowkes). But for now, it suffices to say that my use of the expression does not borrow from their precedents.

SECTION 2: FROM BATESON'S SENSIBILITY TO THE PATTERN WHICH CONNECTS, TO A SENSIBILITY TO PATTERNS THAT CONNECT

The anthropologist Gregory Bateson, one of the founders of cybernetics and systems sciences in the decades following the second world war, understood early on that "there is an increasing necessity for an awareness of being part of relational contexts [...]: persons, groups, populations, genders, species" (Foreword by Sergio Manghi in Bateson 1979, p. xii). Bateson perceived that necessity as the need for a major cultural shift. "His interest was in addressing the very way we think about issues. [...] Bateson was after the very principles of organization that informed the thinking of our culture as a whole. [...] Bateson was engaged in what, again following Edgar Morin, we might now call transdisciplinary work, whose nature it is not merely to cross disciplinary boundaries, but to rearrange our mental landscape" (Alfonso Montuori, Series Editor's Introduction, in Bateson 1979, p. xvii-xviii).

Bateson's approach *connects*, in contrast with Luhmann's system theory, which *dis-connects* (and thereby allows to think unsustainability in systemic terms - cf. chapter 1, section 1). Bateson's work highlights the complex living systems as that which connects together the whole ecology of the living. In his late work *Mind and Nature*, he asks himself: "What is the pattern which connects all the living creatures?" (Bateson 1979, p. 7).[12] His thinking contributed to this reunification of nature and culture already mentioned above, away from the romantic/modern dichotomy. The elaborate properties of humanity do not stand out against a primitive natural background: On the contrary, the most elaborate human qualities are expressions of the functioning of living systems.

"It was, rather, the more complex, the aesthetic, the intricate and the elegant aspects of people that reflected nature. It was not my greed, my purposiveness, my so-called 'animal', so-called 'instincts' and so forth that I was recognizing on the other side of that mirror, over there in 'nature'. Rather, I was seeing there the roots of human symmetry, beauty and ugliness, aesthetics, the bodily grace, and even his habit of making beautiful objects are just as 'animal' as his cruelty. After all, the very word 'animal' means 'endowed with mind or spirit *(animus)*'" (ibid., p. 5).

Cybernetic insights

Bateson's cybernetics helps him to think Time through.[13] This points at an important property of the pattern that connects: the patterns are not to be thought of "as fixed affairs [but as] a dance of interacting parts and only secondarily pegged down by various sorts of physical limits and by those limits which organisms characteristically impose" (ibid., p. 12) . Furthermore, time for living systems implies a sort of narrativity. All "minds" in Bateson's sense, i.e. from the tree to the Senate of the United States and the ecosystem of the Amazon forest, think in terms of stories.[14] Stories are that which puts A and B together into a constructed 'context'. Context is to be understood as "pattern through time". Relevance comes in contexts; stories shape relevance. Noel Charlton also pointed at this: "It is too easy for us to understand a concept like pattern as something fixed, like the pattern printed on a piece

12 Exact page numbers refer to the 2002 edition. In *Mind and Nature*, Bateson explicitly restricts his focus to the world of the living, taken altogether and set apart from the world of the nonliving. His analysis does not touch upon physics, unlike Laszlo's or Morin's (as seen in chapters 2 and 3).

13 More generally, systems thinking approaches take time into account at a serious theoretical level (e.g. with the notion of irreversibility), unlike mechanistic theories and their offspring such as e.g. neoclassical economics.

14 Cf. Bateson 1979, pp.12-13.

of cloth. For Bateson, pattern is always dynamic – formed by all the shifting relationships that make-up the ever-changing dance of social, biological and psychological contexts" (Charlton 2008, p. 77).

Also, context is necessary for fixing, providing, assigning meaning. Bateson asserts "that all communication necessitates context, that without context, there is no meaning and that contexts confer meaning because there is a classification of contexts" (Bateson 1979, p. 16).

Furthermore, another basic insight from Batesonian cybernetics is that we cannot properly understand the world around us, by focusing on the things in themselves (das Ding an sich), but that we should instead investigate all the forms of inter-relations between things. Bateson explains that you cannot "define something by what it supposedly *is* in itself" and that the only sensible way to define a thing is "by its relation to other things" (ibid., p. 15). Relationship rather than things should be the starting point of any definition.

The aesthetic according to Bateson

To Bateson, the aesthetic is that which is "responsive to *the pattern which connects*" (ibid., p. 8). He defines the "aesthetic preference" of a mind, as being "able to recognize characteristics similar to their own in other systems they might encounter" (ibid., p. 118). A typically aesthetic question, would be "*How are you related to this creature? What pattern connects you to it?*"

Bateson gives the illustration of a group of art students whom he once asked to explain why a dead crab being displayed, used to be a living thing (the students were asked to find answers by just looking at the dead crab, and to do as if they had never seen a crab before). The students moved from the observation that the crab showed some symmetry between its parts (left/right), to the observation that the symmetry was not absolute (e.g. one claw bigger than the other), to the conclusion that there existed a similar relation between parts, in the case of one crab ("both claws are made of the same parts") as well as in the crab/lobster comparison and (crab-lobster)/human comparison. They "discarded an asymmetry in size in favor of a deeper symmetry in formal relations" (ibid., p. 8).

Bateson calls these patterns within the crab, *first order connections*. The pattern connections between crab and lobster, he calls *second-order connections*, or what biologists call "phylogenetic homology". Finally, he points at the pattern which connects the patterns connecting, on the one hand, the crab and lobster, and on the other hand, the human being and horse. This comparison of comparisons is labeled as *third order connections*. These three levels of connections, and of perception-conceptualization of connections, are pointing at three different "logical types" (to use Bateson's terminology, after Bertrand Russell; or different "levels of organization" to rephrase into Nicolescu's terms), i.e. different levels of functioning of systems within systems.

This movement illustrated by the art students' progression in the example, of pattern recognitions across different levels, is what Bateson proposes as the way to think about "the pattern which connects": "The pattern which connects is a metapattern. It is a pattern of pattern" (ibid., p. 10).

In this sense, a strong aesthetic sense is a heightened responsiveness to the meta-pattern uniting the living world, rather than an arrested perception, stumbling upon the first-order or second-order differences. To prevent a misreading of Bateson here: The differences are indeed what allows the mind to emerge, so that it can perceive the differences, so of course Bateson's argument here is not against the perception of difference, but against a perception of superficial difference that satisfies itself with the fact of superficial difference and hinders the pursuit of the mind's aesthetic probing of the world around itself, i.e. a probing for connections across differences.

Bateson viewed this aesthetic sensibility as rooted in the biological, and not a uniquely human quality. In a presentation to a psychiatric conference in 1958, Bateson suggested "that we are genetically conditioned to like things that share our own formative influences and that we recognize, mainly at unconscious levels of mind, those aspects of the world that are convivial to our systemic survival" (Charlton 2008, p. 98). Besides, Bateson asserted (in a 1967 conference paper on primitive art) that there is, at the biological level already, a redundancy, a shared (meta)pattern in all of life that allows information about a part to also contain some information about the whole and about other parts, allowing a meaningful relational context to emerge.[15] "Responding to beauty (or ugliness) is recognizing a systemically related being that is an extension, an enlargement, of our own systemic sense of self. *This is the link between systems theory and aesthetics*" argues Charlton (ibid., p. 141).

The aesthetic sensibility, which potential lies beyond purposive consciousness, is also not reserved, among humans, to the artists only, although it is most often repressed in non-artistic social systems. For example, Donald Schön pointed out, in *The Reflective Practitioner* (1983), how professionals (and more specifically in his analysis: engineers, architects, managers, psychotherapists, and town planners) do not only apply rational decision-making as they are supposed to do (and claim to do), but use their whole selves, including their intuitions, although they often do not acknowledge it.

For Bateson, the aesthetics of the pattern which connects is that which can provide a sense of aesthetic unity (and an ecological ethics in the same process) that modern societies are critically lacking.[16] This aesthetic lack is

15 This is something that Edgar Morin will also refer to, as a "holographic principle".

16 Bateson made the link between aesthetics and ecological ethics most explicit in his "last lecture" held in October 1979 at the Institute of Contemporary Arts in London: "matters of beauty are really highly formal, very real, and crucial to the

an epistemological lack: "our loss of the sense of aesthetic unity was, quite simply, an epistemological mistake" (Bateson 1979, p. 17). As Charlton notes, the linking of the urgency of the ecological crisis and of aesthetics' possibility of enlightenment, became central to Bateson in the last ten years of his life.[17]

Bateson's understanding of artistic activity as offering an expanded experience

As Noel Charlton observes, from 1954 onwards (with the metalogue "why a swan?"), "Bateson had realized that the *metaphorical functions* of art processes are the central and most important dynamic aspect of the relationship between artworks and those who interact with them" (Charlton 2008, p. 96). In a conference paper on primitive art in 1967, Bateson pointed out that it is "of prime importance to have a conceptual system that will force us to see the "message" [i.e. the artistic creation] as *both* itself internally patterned *and* itself a part of a larger patterned universe – the culture or some part of it" (Bateson quoted in Charlton 2008, p. 102). The metaphorical linking between the internal pattern and the external pattern occurs outside the reach of purposive consciousness, Bateson argued, noting the "complex layering of consciousness and unconsciousness" (ibid., p. 105). The aesthetic reflects a mental capacity which exceeds consciousness, and poetry is not distorted prose, but rather prose is poetry subjected to logic.[18] Art can engage a wider-than-conscious communication. Quoting choreographer Isadora Duncan's claim that "If I could tell you what it meant, there would be no point in dancing it", Bateson proposed to translate her saying as meaning "This is a particular sort of partly unconscious message. Let us engage in this particular sort of partly unconscious communication," or "This is a message about the interface between conscious and unconscious" (ibid., p. 109). In a lecture held in 1970, Bateson re-asserted that art "is concerned with the relations *between* the levels of mental process [...] artistic skill is the combining of many levels of mind [...] to make a statement of their combination" (ibid., p. 121).

Across his works, as Charlton analyzed closely, Bateson increasingly gave attention to this metaphorical process as a corrective to the limits of "purposive consciousness" (which I discussed in chapter 1, section 1). In a lecture also held in 1967, Bateson stated: "I don't know the remedy but there is this: that consciousness can be a little enlarged through the arts, po-

entire political and ethical system in which we live" (quoted in Charlton 2008, p. 129).

17 Cf. Charlton 2008, p. 101.

18 Cf. ibid., p. 106 (citing Bateson).

etry, music and the like" (Bateson quoted in Charlton 2008, p. 100). Such a remedy is required especially

"because our interactions with our surroundings have (until recent times) been constant, they have been consigned to primary process operations [a Freudian term for the subconscious] and so we no longer have the *conscious* capacity to deal with environment wisely. [...] However, artistic engagement [...] provides a root into primary process whereby the buried wisdom, the otherwise inaccessible responsiveness, can be accessed and utilized" (Charlton 2008, p. 107).

Without this wisdom, rational planning guided by purposive consciousness short-circuits the mental process and mistakes its short-circuits for straight lines of causality. Bateson argued vehemently against this situation in his 1967 conference paper on primitive art:

"Purposive rationality unaided by such phenomena as art, religion, dream, and the like, is necessarily pathogenic and destructive of life [...] its virulence springs specifically from the circumstance that life depends on interlocking *circuits* of contingency, while consciousness can see only such shorts arcs of such circuits as human purpose may direct. Unaided consciousness must always tend towards hate; not only because it is good common sense to exterminate the other fellow, but for the more profound reason that, seeing only arcs of circuits, the individual is continually surprised and necessarily angered when his hardheaded policies return to plague the inventor" (Bateson quoted in Charlton 2008, pp. 112-113).

Bateson thus sees art as opening up the human mind to more than the narrow "purposive consciousness". A long quote from Bateson's *Steps to an ecology of mind* (which includes a reprint of the 1967 lecture) sheds light on his understanding of art as a "royal road" to an expanded, aesthetic reflexivity:

"It is, however, possible that the remedy for ills of conscious purpose lies with the individual. There is what Freud called the royal road to the unconscious. He was referring to dreams, but I think we should lump together dreams and the creation of art, or the perception of art, and poetry and such things. I would include with these the best of religion. These are activities in which the whole individual is involved. The artist may have a conscious purpose to sell his picture, even perhaps conscious purpose to make it. But in the making he must necessarily relax that arrogance in favour of a creative experience in which his conscious mind plays only a small part. We might say that in creative art man must experience himself — his total self — as a cybernetic model. [...] What is required is not simply a relaxation of consciousness to let the unconscious material gush out. To do this is merely to exchange one partial view of the self for the other partial view. I suspect that what is needed is the synthesis of the two views and this is more difficult" (Bateson 1973, p. 414).

Aesthetics in general or aesthetics of sustainability?

I will depart from Bateson insofar as he defines aesthetics, in general terms, as that which is "responsive to *the pattern which connects*".

Aesthetics may not always be "connective" to the fullest extent described by Bateson. Indeed, an aesthetic experience can exist, which does not reach the level of "third-order connections" and the generality of the unity of all life forms described by Bateson, and which satisfies itself with a unity of meanings and values (in Dewey's sense) with a narrower scope/at a more limited range. In a Luhmannian sense, the existence of more exclusively autopoïetic aesthetic experiences should be acknowledged. The aesthetics described by Bateson should then be qualified as characteristic of aesthetics of sustainability, rather than of aesthetics in general.

In refusing to cover aesthetics in general terms, I am not advancing an unprecedented argument. Indeed Timothy Collins also discriminates ecological aesthetics as "depart[ing] from the autonomous object of classical aesthetics, defined as unity, regularity, simplicity, proportion, balance, measure and definiteness" (Collins in eds. Strelow, Prigann and David 2004, p. 172). By contrast to classical, object and essence-centered aesthetics (cf. analytical aesthetics)[19], aesthetics of sustainability is to be understood as a subset of aesthetics as understood by Dewey, i.e. a form of relation and process-centered aesthetics, which bases itself on a sensibility to patterns that connect at multiple levels (i.e. first, second and third-order connections as described by Bateson).

Pattern or patterns?

Another departure from Bateson consists in speaking of pattern**s** that connect, rather than of a pattern that connect**s**, in order to avoid a holistic bias simplifying and impoverishing the theoretical understanding of aesthetics of sustainability (which may not be clearly present in Bateson's own work but as a risk of misunderstanding among his readers).[20] As already seen with Morin in chapter 3, the issue is to express unity in diversity, and not unity against diversity, and to be sensitive to complexity. As the next section will show, this requirement will necessitate a further characterization of aesthetics of sustainability as sensibility to complexity, based on Bateson's pattern that connects but also beyond its holistic caveat.

Another option would be to talk of patterning that connects, partly in the way Oleg Koefoed proposes to talk of culturality as a way out of the alterna-

---

19 For an introduction to the difference between analytical aesthetics and Dewey's aesthetics, cf. Shusterman 1992, pp. 3-33.

20 For example, Charlton (2008) repeatedly flirts with the risk of holism, as e.g. p. 141 where he insists on "wholeness" and "oneness".

tive culture/cultures. In this sense, one could then speak of aesthetics of sustainability as a sensibility to patterning that connects. The verbal form 'patterning' stresses the time-dimension, i.e. the process-value of the "dance of interacting parts", as well as the relational value of pattern(s) that connect, which should not be confused with the assumed thing-ishness of purported things-in-themselves and of pure ideas. However, I will prefer to use the expression "patterns that connect", in order to avoid excessive stylistic esotericism.

Topics, processes and values that connect

What does this characterization of aesthetics of sustainability as based on a sensibility to patterns that connect, signify in practice? How can cultural practices, e.g. the arts, express and foster such a sensibility? I argue that they can do so at three interconnected levels:[21]

- The level of a sensibility to topics that connect, i.e. issues characterized by the inter-relatedness of cultural, social, economic, political and ecological processes, of local and global realities and different time frames (from the short-term to the very-long term), and by intercultural linkages.
- The level of a sensibility to processes that connect, i.e. search, research, learning, playing and working processes, insofar as they:[22]
  - involve all-out reflexivity about 'ourselves' in a wide sense (from individual routines to social institutions and polities);
  - develop reflexivity skills of different types, appealing to a diversity of human qualities, beyond the limited types of rationality tapped by most scientific discourses and beyond the limitation of imagination embedded in established rules and routines;[23]
  - develop an ability to work in inter- and transdisciplinary teams on projects (in the arts especially, this implies a shift towards relatively less autonomous, less individualistic, more collaborative and more interactive working processes);
  - develop intercultural and interconventional interactions, tapping into a human capacity for enhanced empathy, beyond sociocentrism and ethnocentrism.[24]
- The level of a sensibility to values that connect, inquiring into the meanings and implications of justices, in a pluralistic way, opening up multiple layers of interpretations, neither assuming universal values nor retreating into a postmodern hyper-atomization of values.

---

21 Cf. Kagan 2008b, pp 17-19; Kagan 2010.

22 The following list is taken from Kagan 2010, pp. 1098-1100.

23 Cf. Dieleman 2008.

24 On intercultural and interconventional interaction, cf. Kagan 2004.

The sensibility to the topics that connect expresses itself fully in transdisciplinarity and in Morin's dia-logic, which were discussed in chapter 3 and which imply the construction of an integrative, patterning knowledge. This sensibility requires indeed a "science and art of discovering bridges between different areas of knowledge and different beings" (Klein 2004). The more the focus of attention is placed on the comparisons, the interrelations, the connectedness between different dimensions or 'levels' of reality, the more one may speak of a sensibility to patterns that connect, in terms of contents.

The sensibility to processes that connect also evokes,at the level of the individual, Richard Sennett's "craftsman", insofar as he is skillfully engaged in a dialogue with materials (Sennett 2008). Lewis Hyde summarizes well this dimension of Sennett's craftsmen:

> "And what is it that such persons know? They know how to negotiate between autonomy and authority (as one must in any workshop); how to work not against resistant forces but with them (as did the engineers who first drilled tunnels beneath the Thames); how to complete their tasks using 'minimum force' (as do all chefs who must chop vegetables); how to meet people and things with sympathetic imagination (as does the glassblower whose 'corporeal anticipation' lets her stay one step ahead of the molten glass); and above all they know how to play, for it is in play that we find 'the origin of the dialogue the craftsman conducts with materials like clay and glass'" (Hyde 2008).

At the level of social interactions, the sensibility to processes that connect also evokes quantum physicist David Bohm's notion of social intelligence as vested in the connectivity of genuine "dialogue",vs. the exclusionary process of "discussion" as mere contest of wills (Bohm 2006). The social intelligence of dialogue is complementary, but not identical with the dialogue with materials of the craftsman. Because "the intelligence that comes from dialogue may make it possible for something new to come into human relations [...] I think that dialogue will liberate a more subtle kind of intelligence than that used in making tools. The intelligence that creates and uses tools is not able to organize society properly so as to take into account the consequences of these tools" (Bohm quoted in Gablik 1991, p. 162).

Last but not least, the sensibility to processes that connect also relates to Arnold Berleant's understanding of aesthetic engagement, which Noel Charlton compared to Bateson's aesthetics (see boxed text below).

Arnold Berleant's aesthetic engagement

Noel Charlton (2008, pp. 145-157) draws parallels between Bateson's and Berleant's notions of aesthetics, which have many common points. Berleant points at an understanding of aesthetics which, he argues, was prevalent in Europe before the eighteenth century and is re-emerging since the 20th century: an engaged, participative understanding of aesthetics, with artistic practices involving observers as participants, as opposed to the disinterested, disengaged aesthetics founding a contemplative approach to art (as developed after Immanuel Kant), separating the art object from the perceiving subject.[25] Berleant's aesthetics, based on experience like Dewey's, also echoes the phenomenological insights, arguing that "[t]he arts bring us closer than any other social form to the immediacy of the human world as we know it" (Berleant 1991, p. 210). Berleant (1993, pp. 199-227) further analyzes how Kantian aesthetics drew strong 'borders' around art objects in a separate special domain (see also chapter 1). He argues that such an aesthetic stance is especially untenable in the case of the appreciation of nature, which involves a more integrated relational experiential process. Even Kant himself developed a separate understanding of the aesthetic appreciation of nature, which he called the "sublime": As noted by Berleant, the sublime is "the capacity of the natural world to act on so monumental a scale as to exceed our powers of framing and control [...] to produce [...] a sense of overwhelming magnitude and awe" (Berleant 1993, p. 234). The sublime is beyond the judgment of taste on beauty and ugliness.[26] Berleant argues that the experience of the sublime invalidates the illusion of separatedness and instead imposes connectedness and "total engagement [...] sensory immersion that reaches the still uncommon experience [...] of unity" (ibid., p. 237). In his view, the sublime can also become an example for more more mundane aesthetic experiences which can be experiences of a participatory aesthetics, not only with nature but also with art.

The sensibility to values that connect favors participatory polyarchic polities, i.e. regimes of authority allowing experimentation with various non-hierarchical configurations of work.[27] This sensibility also conveys a sense of humility towards the non-human, and an openness to the potentiality of moral inspiration from non-human elements (which will be further discussed in section 5, after David Abram's understanding of a phenomenological and animistic sensibility).

The characterization of aesthetics of sustainability is however not thereby completed. As mentioned a few paragraphs above, the insights from complexity theories necessitate a further characterization of aesthetics of sustainability as sensibility to complexity, based on Bateson's pattern that connects but also beyond its potential holistic caveat.

25 Cf. Berleant 1991.

26 For a discussion of Kant's sublime in the context of ecology and aesthetics, see also Fel 2009, pp. 50-53.

27 The topic of participatory polyarchic polities will be introduced in chapter 7.

SECTION 3: THE SENSIBILITY TO COMPLEXITY

> Everything which does not bear the mark of disorder and of subject is insignificant and mutilating.
> MORIN 1992, P. 395

> It is the tendency to reduction that deprives us of the potentialities of understanding.
> MORIN IN ED. NICOLESCU 2008, P. 31

Morin's "art principle"

Bateson's definition of aesthetics as the sensibility to the pattern that connects, is relatively comparable to Morin's "art principle" for systemic sensibility, already mentioned in chapter 3. However, Morin not only points at connectivity but also at complexity, i.e. the distinctions, contradictions, competition as well as the connections, symbioses and harmonies:

> "The systemic sensibility will be like that of the musician's ear which perceives the competitions, symbioses, interferences, overlappings of themes in the same symphonic flow, where the brutish mind will recognize only theme surrounded by noise. [...] The notions of art and science, which oppose each other in the dominant techno-bureaucratic ideology, must be associated here" (Morin 1992, p. 139).

To be fair to Bateson, he too used the symphonic metaphor to evoke the aesthetics of the pattern which connects (while recalling the experience of an art exhibition in a personal letter) because listening to a symphony is at one level a sequence of sounds, and at another level an integrated experience. But Bateson did not explore the metaphor further, in its sensibility to complexity.[28]

And this requires a strong personal engagement and sensibility, Morin adds: It "requires the full use of the personal qualities of the subject in his communication with the object" (ibid.).

Indeed, the insights from complexity theories point not at a holistic sensitivity which would only consider complementarity and symbiosis (as e.g. Capra sometimes tends to do), but:

- a complex sensitivity that considers as much antagonisms and competitions as complementarities and symbiosis, and that transcends the contradictions so as to reveal the complementary tension of antagonism and complementarity;

28 Cf. Charlton 2008, p. 83.

- a sensitivity to wholeness and order that also considers and values disorder, disharmony, as well as uncertainty, and that respects genesic chaos.

Besides, the imperative of self-observation also stressed by Morin, links the sensibility to patterns that connect, to a self-critical reflexivity, i.e. a praxis of critique, that embraces the concrete and the abstract, from the experiential to the epistemological levels: "*this development necessitates not only that the observer observe himself observing systems, but also that he make an effort to know his knowledge*" (Morin 1992, p. 142).

Dewey's aesthetics vs. fixed harmony

Actually, such a sensibility to complexity is more relevant to Dewey's understanding of aesthetics than a solely holistic sensibility fixed on harmony. Indeed, Dewey's characterization of the aesthetic experience as an experience of unity should not be misunderstood as a search for permanent contemplation. Rather, as Shusterman explains, "for Dewey, the permanence of experienced unity is not only impossible, it is aesthetically undesirable; for art requires the challenge of tension and disruptive novelty and the rhythmic struggle of achievement and breakdown of order" (Shusterman 1992, p. 32). Tensions and conflicts are recognized as harboring potentialities for new levels of unity. Dewey's position echoes, at an aesthetic level, with Lupasco's logic of contradiction as applied by Nicolescu across levels of perception and levels of reality (cf. chapter 3). And Dewey clearly rejects holistic simplifying closure from aesthetic grounds:

> "In the process of living, attainment of a period of equilibrium is at the same time the initiation of a new relation to the environment, one that brings with it potency of new adjustments to be made through struggle. The time of consummation is also one of beginning anew. Any attempt to perpetuate beyond its term the enjoyment attending the time of fulfillment and harmony constitutes withdrawal from the world. Hence it marks the lowering and loss of vitality" (Dewey 1934, p. 16).

The sensibility to complexity

The culture of complexity that Morin suggests (as described in chapter 3) "necessitates a reform of understanding" (Morin 1992, p. 143), which suggests how elaborate and complex an aesthetics of sustainability would be:

- "We must first of all be capable of conceiving plurality in oneness" (ibid.), which involves associating different concepts in complex relationships (i.e. both contradictory and converging), into what Morin calls macro-concepts (e.g. "system-organization-interrelation"). An aesthetic experience of complexity shall offer an insight into the unicity of plural experiences, but without erasing their plurality.

- One shall also "mentally conceive oneness and diversity together [...] as two notions which are not only antagonistic or competitive, but also complementary" (ibid., p. 144). For example, this applies to the concept of the 'system': "A system is a unity which comes from diversity, ties in diversity, carries diversity in itself, organizes diversity, produces diversity" (ibid.).
- One shall be capable of conceiving an apparently paradoxical logic (or rather "meta-logic", according to Morin 1991), where for example: "One is both *one* and non-*one* [and] *has become relative in reference to the other*. It cannot be defined only in an intrinsic fashion. To emerge, it needs its environment and its observer. [...] There is, likewise, alterity at the heart of the one [which] has unity and also scission" (Morin 1992, pp. 144-145). One should also conceive the apparently paradoxical "antagonism internal to unity" and the complementarity of antagonisms: Morin sees the seeds of this capacity dispersed among different Western philosophers: First of all Heraclitus (cf. boxed text below on Heraclitean aesthetics), but also Hegel in his dialectics and Marx in understanding the "negativity of constructivity", Nicholas of Cusa in his "coincidentia oppositorum", and Stéphane Lupasco's claim of a physics based on a "principle of antagonism" (Lupasco 1951, 1962). This sensibility was also especially shared by quantum physicists: Anecdotally, as Antoine Faivre notes about Niels Bohr (who theorized the wave/corpuscule complementarity): "In 1947, while receiving the Danish Order, Bohr chose the *Yin* and the *Yang* for his coat of arms, along with the legend: 'Contraria sunt complementa'" (Faivre 1994, p. 289).
- Therefore, one shall abandon clarity, distinction and substantiality and instead "integrate ambiguity" (Morin 1977, p. 148). The aesthetics of complexity is thus also an aesthetics of ambiguity.
- One shall not despise mystery but appreciate it for its contribution to the virtue of humility. "Mystery is not only privative; it liberates us of all delirious rationalization which claims to reduce the real to the idea, and it brings us, in the form of poetry, the message of the inconceivable" (Morin 1992, p. 393). The wonder and awe of mystery are thus valuable elements in an aesthetics of complexity.
- In the new mode of perception and thought that Morin suggests, "complexity is everywhere" (i.e. in the origins of the universe, in the quanta, as much as in the human realm: "at the base of physis there is not simplicity, but complexity itself" (ibid., p. 148). In this context, the most 'basic' unit of thought, the most basic concept, is not anymore an element or a particle, but it is a system: "*System is, therefore, conceived here as the basic complex concept* [...] if we may say so, the most simple

complex concept. In fact, *there are no longer, there will be no longer any basic simple concepts*" (ibid.).[29]

- One shall also develop an ecological understanding of the notion of organizational opening (and closure): "to search as much for relation with the environment as for distinction from it, plus the complex association between dependence and autonomy; as much for opening and closing as for the alternative between these terms; more for reorganization than for organization, more for praxis than for structure [and one shall be set to conceive together with the living system] its ecosystem of inscription and to elaborate a meta-system of reference" (ibid., p. 206). At a logical level, this means "to link the constant and the changing, the moving and the stationary, the autonomous and the dependent" and to introduce "at the heart of the identity principle of the existent, the excluded third part: the environment". This logic "cannot isolate or exclude one by the other, either the internal logic of the system or the external logic of the situation (that is to say the environmental conditions); the explanation must be dialogical and dialectical, linking the inside and the outside processes in a complementary, concurrent and antagonistic way" (ibid., p. 207).[30] Morin's dialogic, understood in this way, develops the sensibility to patterns that connect into a sensibility to eco-auto-organization and autoecopoïesis.[31]
- One shall be open to chaos as source of aesthetic novelty: For Morin, chaos is the source of creativity in a wide sense (i.e. the creativity of poïesis in the physis, in life and in humanity): The "regression towards turbulence and chaos revives along the way the poïetic virtues, which, if they are not submerged, call forth a new genesis, which becomes the source of a new generative loop. Creation is always an irruption of genesis in generativity" (ibid., p. 224). Creativity is a characteristic of all life forms, and not only of human cognition (cf. Capra on bacteria as quoted in chapter 2, section 2). As ecological artist David Haley noted: "If we take the root of the word 'art', we find 'rt', an Indo-Aryan noun/adjective of the *Rg Veda*, meaning the dynamic process by which the whole cosmos continues to be created" (Haley 2008, p. 201). Life's "creative evolution" (cf. Bergson 1907) emerges not from computational capacities alone, but from the ability to deal with disorder and ambiguity as genesic forces. Under the light of complexity, the Romantics' longing for creative chaos finds a renewed legitimation. Also, an aesthetics of sustainability, which is open to the generativity of chaos, implies a sensibility to emergence.

29 This also means that the concept of "system" is itself just a starting-point, for Morin and for us, but not a satisfactory outcome in itself: it is too crude for an understanding of life on earth.

30 Cf. also endo-exo-causality as described in chapter 3.

31 On eco-auto-organization and autoecopoïesis, cf. chapter 3.

- Therefore, one shall not only be open to chaos, but also open to 'Nature'[32]: "We must therefore rediscover Nature in order to rediscover our Nature, as the Romantics had sensed it, genuine guardians of complexity during the century of the great Simplification" (Morin 1992, p. 382).

Heraclitean fragments on aesthetics

To come back to Heraclitus, his fragments relating to aesthetics already give a sense of an aesthetic sensibility to complexity, as documented in Tatarkiewicz et al.'s history of aesthetics:

"This philosopher, who chiefly stressed multiplicity, change and opposition in the world, also saw, however, its unity and harmony. Four of his fragments on harmony have survived [see the direct quotes below]. One says that harmony is most beautiful when it is derived from various sounds. Another even goes as far as to say that harmony arises out of opposing forces. As examples of harmony, Heraclitus cited the bow and the lyre: the greater their tension, that is, the more divergent are the forces acting upon them, the more effectively the bow shoots and the lyre sounds. He concluded that harmony may arise also out of opposing, divergent elements. [...] Finally, the fourth fragment states that out of opposites there arises a symphony not only in nature, but also in art, which thereby imitates nature. [...] The doctrine of harmony issuing out of opposites was the special Heraclitean contribution to aesthetics" (Tatarkiewicz et al. 2006, pp. 84-85).

That which is in opposition is in concert,
and from things that differ comes the most beautiful harmony.
HERACLITUS (ARISTOTLE, ETH. NIC. 1155B 4; FRG. B 8 DIELS)[33]

[People] do not understand how that which differs with itself is in agreement:
Harmony consists of opposing tension, like that of the bow and the lyre.
HERACLITUS (HIPPOLYTUS, REFUT. IX G; B 51 DIELS)

The hidden harmony is stronger [or

32 Morin's definition of nature was mentioned in the introduction.

33 All quotes from Heraclitus in this boxed text, are taken from Tatarkiewicz et al. 2006, pp. 88-89.

> "better"] than the visible.
> HERACLITUS (HIPPOLYTUS, REFUT. IX G; FRG. B 54 DIELS)
>
> But perhaps nature actually has a liking for opposites; perhaps it is from them that she creates harmony, and not from similar things... It seems too that art does this in imitation of nature.
> HERACLITUS (PSEUDO-ARISTOTLE, DE MUNDO, 396B 7)

One shall not either think and perceive the "patterns that connect" or think and perceive distinctions, but think and perceive both and also move them into an active and generative process: "to join and to isolate must be inscribed in a recursive circuit of knowledge which never stops nor is ever reduced to one of these two terms" (Morin 1992, p. 392).

However, Morin's "knowledge of complexity" is not fit as an operational knowledge nourishing techniques and technologies as the manipulation and "asservissement" (enslavement/servocontrol) of Nature (including human beings of course). But it calls forward a different form of action and "calls upon us to enrich and change the meaning of the word action [with] a principle of action which does not order but organizes, does not manipulate but communicates, does not direct but animates" (ibid., p. 397). This may be indeed the complex principle of the action-research of a science-art, vs. the barbarian and functionalist praxis of technosciences.

SECTION 4: THE TRANSDISCIPLINARY SENSIBILITY

The sensibility to complexity is also a transdisciplinary sensibility. In his *Manifesto*, Nicolescu discusses the sujectivity and sensibility of a transdisciplinary researcher. His arguments further complement Morin's insights into the kind of sensibility which would nurture aesthetics of sustainability.

The transdisciplinary attitude

Nicolescu borrows the expression "transdisciplinary attitude" from the Argentinian poet Roberto Juarroz, and characterizes it as "the individual or social capacity to preserve a constant, unchanging direction, no matter what the complexity of the situation or the hazards of life. On the social level, this direction is that of the flow of information crossing the different levels of Reality, whereas on the individual level, it is that of the flow of consciousness crossing the different levels of perception" (Nicolescu 2002, p. 83).

In the hybrid context of transdisciplinarity, Nicolescu here points at the necessity to maintain a certain stability and rootedness while dealing with complexity, and thereby to avoid the rootlessness of a postmodernist attitude, as expressed in a warning about the context for transdisciplinarity in humanities, by Gibbons et al. (and which echoes concerns already articulated in chapter 1):

"The alarming tendencies in modern societies towards narcissism and atomism, the decline in civil participation, the increasing sense that all relations and commitments are revocable, and the growth of increasingly 'instrumental' attitudes towards nature and society are all manifestations of a slide to subjectivism to which modern culture is prone. The social contextualisation of the humanities, now more explicit and insistent than ever, carries with it this danger, although some would describe such contextualisation as reflexivity run amok. In its most radical and theorised form, as postmodernism, it can even be carried to self-contradictory extremes. If all is incoherent and unconnected, playful shadows, how is reflexivity possible?" (Gibbons et al. 1994, pp. 102-103).

Such a transdisciplinary attitude is required in order to contribute to a civilizational shift towards sustainability[34] because it allows effectivity and affectivity:

"To preserve a constant direction across the levels of Reality guarantees a growing effectivity (or effectiveness: ability to make something happen) of our action in the world and in collective life [...] To preserve a constant direction across the levels of perception guarantees a growing affectivity (or feeling; influenced by or resulting from the emotions), which ensures the link between the whole and ourselves" (Nicolescu 2002, p. 84).

Nicolescu advocates for "effectivity and affectivity" as "the slogan of a plan for civilization to meet the challenges of our time".

This transdisciplinary attitude as stressed by Nicolescu, highlights explicitly the normative dimension of his transdisciplinary approach. And this normative positioning opposes the normativity of the 'Technological System' (Ellul) or 'technoscience' (in Nicolescu's words) of modern western civilization (which was described in chapter 1).

"Alas in today's world mere efficacy at any price exists, which is only a caricature of effectivity. Affectivity, on the other hand, has no commercial value: it is therefore scoffed at, ignored, forgotten, even scorned [and] this contempt for affectivity is none other than contempt for the human being, who has been transformed into a commer-

34 Nicolescu (2002) does not employ the term 'sustainability', but refers to Morin's "politics of civilization" and talks of the global crisis, of "the challenges of our time" and of "the threat of self-destruction of our species" (pp. 84-85).

cial object [...] it is the growing disequilibrium between effectivity and affectivity that puts our species in danger" (Nicolescu 2002, pp. 84-85).

Echoing some form of gender perspective, Nicolescu also identifies the transdisciplinary attitude with the achievement of a balance of masculinity and femininity: He associates masculinity with the levels of Reality and effectivity, and femininity with the levels of perception and affectivity. The above-mentioned contempt for affectivity has been "giving preference to the unbridled development of masculinity [in] efficacy for efficacy's sake." Against this trend, achieving balance requires now to "pass through a process of social feminization [...] that can give birth to the social ties that are now so conspicuously absent in the cross-communication of human beings" (Nicolescu 2002, p. 86).[35]

The transdisciplinary attitude implies avoiding the binary oppositions between the 'for' and the 'against' on a single level of Reality, and avoiding the mere compromise in-between them. "The reconciliation between 'for' and 'against' cannot be produced except by placing oneself on another level of Reality, where 'for' and 'against' appear as two contradictory poles of a larger unity, something that signifies being with, meaning that one takes into account everything that is positive and constructive within the 'for' and the 'against'" (Nicolescu 2002, p. 87).

However, such an attitude can also degenerate, if not associated to movements also between different levels of perception:

"If one engages oneself exclusively in the crossing of different levels of Reality, this new attitude, to be with, neither for nor against, but both for and against, traps one in a new dogmatic, and even totalitarian system, even if through thought one changes levels of Reality. It is only through the harmony between levels of Reality and levels of perception, that is to say, through an accord between thought and one's own experience of life, that this trap may be avoided" (ibid.).

Here again, a holistic trap is present which an aesthetics of sustainability must learn to overcome.

Transdisciplinarity and the sensibility to the pattern that connects

Nicolescu's insights join Bateson's on the sensibility to the pattern that connects and asserts its social relevance. He proposes to understand transdisciplinarity "as being both the science and the art of discovering these bridges"

35 Nicolescu is careful to point out that this does not mean a homogenization of society towards some sort of androgyny, which would collapse transdisciplinarity into yet another single-level simplification of the world (cf. Nicolescu 2002, pp. 86-87).

that will allow "to found some new, durable and powerful social ties" (ibid., p. 89). Indeed, as he observes, the prolonged absence of such a sensibility pushes social fragmentation toward mutual autism between social groups/institutions/systems (cf. also Luhmann's analysis of this phenomenon, in chapter 1). "What is social dialogue in the absence of bridges between social partners? A market of fools that only aggravates the social fracture" (ibid.).

Nicolescu further affirms this sensibility as an aesthetics/ethics that has become a political imperative: "The word *revolution* is not devoid of meaning because of the defeat of the social revolution. Today, revolution can only be a revolution in intelligence, which transforms our individual and social life into an aesthetic as well as an ethical act, an act that unveils the poetic dimension of existence. In our time, an effective political will can only be an affective poetic will" (ibid., p. 90).

This 'revolution in intelligence' implies the inter-relations between and across different forms of reflexivity as identified by Dieleman (2008), and grounds "the emergence of a new type of intelligence, founded on an equilibrium between analytic intelligence, feelings, and the body. It is only in this way that the society of the twenty-first century can reconcile effectivity and affectivity" (Nicolescu 2002, p. 138). Here Nicolescu's focus also reminds of Lakoff and Johnson (1999) and the multiple intelligence theory of Howard Gardner (1983).

Transrepresentation and transperception

Nicolescu further characterizes the parallels between scientific and artistic creativity in their sensibilities to the pattern that connects, and introduces the notions of 'transrepresentation' and 'transperception'.

> "The encounter between different levels of Reality and different levels of perception engenders different levels of representation. [...] These levels of representation of the sensible world are [...] connected with the levels of perception of the creator, the scientist, or the artist. True artistic creation arises at the moment of crossing several levels of perception simultaneously, engendering a transperception. True scientific creation arises at the moment of crossing several levels of representation simultaneously, engendering a transrepresentation. Transperception permits a global, non-differentiated understanding of the totality of levels of perception" (Nicolescu 2002, p. 99).[36]

36 Nicolescu here also further points at the psychological research carried out by the mathematician Jacques Hadamard on the creativity of leading physicists and mathematicians: "Essai sur la psychologie de l'invention dans le domaine mathématique" (Hadamard 1954).

Sensibility to deviations and avoidance of confusions

Small deviations and small but stubborn 'anomalies' can bring about historical shifts. Indeed, "the deviance acts as a vision opening toward a level of Reality different from the level at which the system under consideration is situated" (ibid., p. 110). Therefore, a transdisciplinary sensibility is attached to deviations. Or even, as Nicolescu argues, "by nature transdisciplinarity has the status of deviation, not of dissidence (which always ends by being absorbed by the system in place)" (ibid.).

But a transdisciplinary sensibility should also be cautious with "levels of confusion" which are, as already described earlier (in chapter 3), "generated by lack of respect for the unique and singular role that each level of Reality and each level of perception plays within the open unity of the world" (Nicolescu 2002, p. 111). Such confusions, linked to different forms of holistic simplifications, are to be avoided by a discerning transdisciplinary sensibility. The sensibility to patterns that connect shall thus not form a locked-in, homogenized 'art-scientist' perception of the world in its plurality, but inform several, discontinuous levels of perception in an 'open unity'.

Rigor, opening and tolerance

The avoidance of confusion is pointing at some further values of the transdisciplinary attitude: "Rigor, opening and tolerance are the three fundamental characteristics of the transdisciplinary attitude" (ibid., p. 119). Thus the sensibility to patterns that connect is not to be understood as floating purely freely, but is based on rigorous attention to complexity.

Such rigor also applies in communication:

> "Because communication is first of all the correspondence between the right places in myself and in the Other, it is founded on authentic communion, beyond all dream or all fantasy to manipulate the Other. Rigor is therefore also the research for the right place in myself and in the Other at the moment of communication. This rigor is therefore the result of perpetual search, continually nourished by new knowledge and new experiences" (ibid., p. 120).

"Opening brings an acceptance of the unknown, the unexpected and the unpredictable" (ibid.) coming from the other levels – of Reality/perceptions/the sacred. "Transdisciplinary culture is a culture of questioning, one perpetually accompanied by responses accepted as temporary" (ibid., p. 121).

The transdisciplinary sensibility and the artistic perspectives on reality

This transdisciplinary culture is practiced in the arts, argues Rosemary Ross Johnston: "The arts construct and bring together (as a novel, as a painting, a a piece of music, as theatre) multiple sites of looking again, multiple sites of reinterpretation, and multiple ways of comprehending the world". Working as situated "bricolage", artistic practices "are an example in practice of a transdisciplinary research approach" (Ross Johnston 2008, p. 232). I will prefer to qualify Johnston's argument as pointing at a potentiality of artistic practices rather than at a fixed quality of the arts, which can also develop a potential for obtuseness as already discussed (cf. chapter 1, section 2).

The transdisciplinary sensibility, Johnston further argues, "like the arts, gives a central place to imagination" and operates a bit like looking at a picture, and then looking again at it after being told that there is a 3D picture to be seen inside or beyond it:

"Without consciously looking for the something more, without my prior knowledge that tells you that there *is* something more to a picture that looks quite complete the way it is, and without your prior knowledge about what 3D is, you would never see that other image. Transdisciplinarity alerts us to this something more – it invites us to comprehend the world by looking again, and reinterpreting existing knowledge" (Ross Johnston 2008, p. 234).

The feeling of the sacred

As seen earlier (cf. chapter 3, section 2), transdisciplinarity also relates to the notion of the "sacred". Furthermore, Nicolescu's description of the feeling of the sacred comes relatively close to Bateson's idea of aesthetics as the sensibility to the pattern that connects:

"The sacred is that which connects. The sacred links, as indicated by the etymological root of the word *religion* (*religare* – 'to bind together again') [...] The sacred is first of all an experience; it is transmitted by a feeling – of that which links beings and things and, in consequence, induces in the very depths of the human being an absolute respect for the others, to whom he is linked by their all sharing a common life on one and the same Earth" (Nicolescu 2002, pp. 125-126).[37]

Such a feeling is also shared by Morin at the outset of his *Méthode*: Morin states that his motivation to engage into research on complexity also stems from his subjective feeling for "infinite solidarity; what the Tao calls the *Spirit of the Valley* which 'receives all the waters which flow into it'" (Morin 1992, p. 19).

37 Nicolescu also further argues that the sacred, understood from the transdisciplinarity approach, can neither be either immanence or transcendence, but a reconciliation of the two (cf. Nicolescu 2002, pp. 127-128). In the case of Bateson, the sacred points at a monism of immanence (nowhere in his work does Bateson move towards transcendence, as noted by Charlton 2008, pp. 159-161).

If we follow this definition of the sacred, then the sensibility to patterns that connect is also a sensibility that opens up to the sacred, or in other words, it constitutes to some extent a spiritual sensibility. Alfonso Montuori[38] makes the link between Nicolescu's and Bateson's expressions rather explicitly: "we reach Nicolescu's call for transdisciplinarity as [...] one that begins to integrate the sacred – where inquiry, inquirer, and the subject of the inquiry all are part of the larger pattern that connects, a *re-ligio* or re-connection of what has been torn asunder" (Montuori in Ed. Nicolescu 2008, p. xvi).

Such a spiritual sensibility is by far not an unknown or new phenomenon in the arts, but rather a recurrent feature in art history. Nicolescu, aware of this convergence, refers to artistic illustrations in his discussion of the feeling of the sacred (e.g. pointing at Brancusi's *Princess X*)[39], and suggests that his approach would be "very near that which the great Arab poet Adonis called the *mysticism of art* – a movement towards the hidden face of Reality, a living experience, a perpetual travel towards the heart of the world, a unification of contradictories, the infinity and the unknown as aspiration, freedom from any philosophic or religious system" (Nicolescu in Ed. Nicolescu 2008, p. 17).

SECTION 5: THE PHENOMENOLOGICAL AND ANIMISTIC SENSIBILITY TO A MORE-THAN-HUMAN WORLD

"Humans are tuned for relationship. The eyes, the skin, the tongue, ears and nostrils – all are gates where our body receives the nourishment of otherness. [...] The simple premise of this book is that we are human only in contact, and conviviality, with what is not human." (Abram 1996, p. ix).[40]

The philosopher David Abram echoes, poetically, Morin's account of the eco-evolution of the species (cf. chapter 3, section 1), for ourselves as humans: "Our bodies have formed themselves in delicate reciprocity with the manifold textures, sounds and shapes of an animate earth – our eyes have evolved in subtle interaction with *other* eyes, as our ears are attuned by their very structure to the howling of wolves and the honking of geese" (ibid., p. 22), and to cut these connections "is to rob our senses of their integrity, and to rob our minds of their coherence".

Abram's insights also echo Bateson's understanding of aesthetics which, according to Noel Charlton, "restores to us a lost tradition of closeness to the rest of nature and can provide renewed access to our wider-than-

38 (who re-edited and wrote a preface to Bateson's *Mind and Nature* as well as articles about Bateson)

39 Cf. Nicolescu 2002, p. 129.

40 Page numbers refer to the 1997 paperback edition.

conscious mental processes" (Charlton 2008, p. 157). Abram further explores this lost tradition, as an animistic-phenomenological sensibility to the the sensuousness of a more-than-human world.

Abram voices the aesthetic experience of connectedness with Nature as an imperative, especially in contrast to technology-mediated experience. He does not profess a so-called "neo-luddite" perspective or rejection of technology on these grounds, but suggests a distanciation from technologies grounded on the experiential encounter with the non-human. "We need to know the textures, the rhythms and tastes of the bodily world, and to distinguish readily between such tastes and those of our own invention. Direct sensuous reality, in all its more-than-human mystery, remains the solid touchstone for an experiential world now inundated with electronically-generated vistas and engineered pleasures" (Abram 1996, p. x).

The ecological shaman

Abram writes about the example of shamans in non-Western cultures, the ecological dimension (rather than simply "supernatural")[41] of their dealings with Nature, and thereby gives an insight into this potentially shamanic role of the artist which the German artist Joseph Beuys called forward (and aimed to incarnate), elucidating why a shaman may have restorative, healing powers.[42]

"The medicine person's primary allegiance [...] is not to the human community, but to the earthly web of relations in which that community is embedded – it is from this that his or her power to alleviate human illness derives" (ibid., p. 8), as a systemic healer of a community rather than of single individuals.

Furthermore, the shaman needs to shift his human awareness in order to perceive this web of relationships: "The traditional magician cultivates an ability to shift out of his or her common state of consciousness precisely in order to make contact with the other organic forms of sensitivity and awareness with which human existence is entwined. [He or she is] temporarily shedding the accepted perceptual logic of his culture [and] altering the common organization of his senses" (ibid., p. 9). Abram defines the shaman by "the ability to readily slip out of the perceptual boundaries that demarcate his or her particular culture [...] in order to make contact with, and learn from, the other powers in the land". This attitude is based in the belief, or "intuition", that "every form one perceives [...] is an *experiencing* form, an

41 Abram describes how in indigenous cultures, the "spirits" are not, as Westerners would expect, immaterial and supernatural anthropomorphic phantoms, but "those modes of intelligence or awareness that do *not* possess a human form" (Abram 1996, p. 13).

42 A further discussion of Beuys' social sculpture perspective is conducted in chapter 5, sections 2 and 3.

entity with its own predilections and sensations, albeit sensations that are very different from our own" (ibid., p. 10).

An understanding of these forms of non-human "intelligence" (in Abram's terms) is also sought after in indigenous cultures because, "[t]o humankind, these Others are purveyors of secrets, carriers of intelligence that we ourselves often need [about] unseasonable changes in the weather, or [...] imminent eruptions and earthquakes [...] where we may find the ripest berries or the best route to follow back home", etc. (ibid., p. 14). Here, we find an ancient relative to the contemporary development of the idea of biomimicry, that was introduced earlier in chapter 2, section 3.

The necessity of sustained immersion in natural environments

However, this shamanic practice cannot be compared with the contemporary fad for "shamanic methods of personal discovery and revelation" in the Western world, however well-intentioned, because the "role of the indigenous shaman [...] cannot be fulfilled without long and sustained exposure to wild nature, to its patterns and vicissitudes. [...] For the source of stress lies in the relation *between* the human community and the natural landscape" (ibid., p. 21).

Abram's own personal experience confirms the difficulty and vulnerability of the experiential learning of such sensibilities: Indeed, Abram also accounts for his own personal learning of an increasingly sensitive experience of natural phenomena, which allows to discern subtle and shifting differences as well as connections. "I became a student of subtle differences [such as] the way the intensity of the sun's heat expresses itself in the precise rhythm of the crickets [...] and other elements I could only isolate after many days of stopping and listening" (ibid., p. 20). However, Abram also notes from his personal experience, that the "new sensibilities" he gradually acquired through sustained contact with natural environments and indigenous cultures, rapidly subsided when immersed back into his North-American, urban context. "I was indeed reacclimating to my own culture, becoming more attuned to its styles of discourse and interaction, yet my bodily sense seemed to be losing their acuteness, becoming less awake to subtle changes and patterns" (ibid., pp. 25-26).

The level of ecological sensibility witnessed by Abram, and its reflexive as well as aesthetic potential for social transformations towards sustainability, therefore cannot be reaped as long as the human experience is kept within the habitual realm of our contemporary usual everyday practices, embedded in the technological system (cf. chapter 1). Western "culture's assumptions regarding the lack of awareness in other animals and in the land itself was less the product of careful and judicious reasoning than of a strange inability to clearly perceive other animals – a real inability to see, or focus upon, anything outside the realm of human technology, or to hear as meaningful anything other than human speech" (ibid., p. 27).

A phenomenological basis for autoecopoïesis

The recovery of the eco- in an autoecopoïetic sensibility requires an acknowledgment of the participatory nature of perception.

First of all, perception is more than just the activation of pre-programmed receptors to already internally pre-selected forms of information (as e.g. in the Luhmannian conception of the structural coupling of psychic systems). From a phenomenological perspective, perception ensures "the constant thwarting of [narrowly autopoïetic] closure" (ibid., p. 49) and no living body's activity can be reduced to (genetic/systemic) programming. Abram illustrates the dimensions of improvisation and creativity in perception, with the case of a spider weaving its web at the entrance of a cave on the island of Bali (a long quote is here warranted):

"Consider a spider weaving its web, for instance, and the assumption still held by many scientists that the behavior of such a diminutive creature is thoroughly 'programmed in its genes.' Certainly, the spider has received a rich genetic inheritance from its parents and its predecessors. Whatever 'instructions,' however, are enfolded within the living genome, they can hardly predict the specifics of the microterrain within which the spider may find itself at any particular moment. They could hardly have determined in advance the exact distances between the cave wall and the branch that the spider is now employing as an anchorage point for her current web, or the exact strength of the monsoon rains that make web-spinning a bit more difficult on this evening. And so the genome could not explicitly have commanded the order of every flexion and extension of her various limbs as she weaves this web into its place. However complex are the inherited 'programs,' patterns, or predispositions, they must still be adapted to the immediate situation in which the spider finds itself. However determinate one's genetic inheritance, it must still, as it were, be woven into the present, an activity that necessarily involves both a receptivity to the specific shapes and textures of that present and a spontaneous creativity in adjusting oneself (and one's inheritance) to those contours. It is this open activity, this dynamic blend of receptivity and creativity by which every animate organism necessarily orients itself to the world (and orients the world around itself), that we speak of by the term 'perception'" (ibid., p. 50).

Furthermore, perception, seen from a phenomenological perspective, is imaginative: "that which we call *imagination* is from the first an attribute of the senses themselves; imagination is not a separate mental faculty (as we so often assume) but is rather the way the senses themselves have of throwing themselves beyond what is immediately given, in order to make tentative contact with the other sides of things that we do not sense directly" (ibid., p. 58).

The living body's boundaries are not to be seen as closing it off from its environment; "more like membranes than barriers, they define a surface of metamorphosis and exchange" (ibid., p. 46). Perception is an encounter where neither the perceiver nor the perceived is wholly passive. Indeed, the perception of, say a flower, is neither a pure construction of the subject's mind, nor a thing-in-itself imposing a specific perception-of-itself to the perceiving subject. Rather, perception works as "an attunement or synchronization between my own rhythms and the rhythms of the things themselves" (ibid., p. 54), or in Merleau-Ponty's words: "Hardness and softness, roughness and smoothness, moonlight and sunlight, present themselves in our recollection not pre-eminently as sensory contents but as certain kinds of symbioses, certain ways the outside has of invading us and certain ways we have of meeting this invasion" (Merleau-Ponty 1962, p. 317).

Furthermore, Merleau-Ponty pointed out that at the moment of encounter, the perceiving subject ceases to exert complete control upon the perceptive encounter: "I give ear, or look, in the expectation of a sensation, and suddenly the sensible takes possession of my ear or my gaze, and I surrender a part of my body, even my whole body, to this particular manner of vibrating and filling space known as blue or red" (ibid., p. 212).

Abram insists that "such animistic turns of phrase [are not] to be attributed simply to some sort of poetic licence" by Merleau-Ponty, but that Merleau-Ponty properly described spontaneous, pre-conceptual perception as "necessarily a reciprocal encounter" with an other which has a "dynamic presence". By contrast, "[t]o define another being as an inert or passive object is to deny its ability to actively engage us and to provoke our senses; we thus block our perceptual reciprocity with that being" (Abram 1996, p. 56).[43]

This acknowledgment of the participatory nature of perception, is based on the recognition of the whole human body as the subject of experience (and not only an idealized Aristotelician soul, a Cartesian disembodied thinking mind, or a Luhmannian enclosed psychic system). Immaterial emotions and thoughts, even the most abstract, are rooted in the sensuous life of the body. Abram traces back this explicit recognition of the body to Merleau-Ponty's writings, themselves rooted in Husserl's notion of "Lebenswelt", i.e. life-world, or pre-conceptual lived experience.[44] In his late writings, Husserl for example argued that all conceptions of space in humankind (including classical Newtonian physics' conception of space) are basically rooted in the physical bodily experience of the earth. The earth, which sup-

43 Interestingly, sociologist Bruno Latour recently introduced new perspectives in sociology by considering objects as "actants" interacting with social agents [i.e. human subjects], rather than assuming them to be merely passive objects, unworthy of sociological attention: Cf. Latour 2005 (see also Latour 1991).

44 Cf. Husserl 1970.

ports one's body, provides the first experience of space, from which other conceptions can then arise.[45]

Another basis for an autoecopoïetic enactment of the participatory nature of perception, is Abram's extension of Husserl's notion of intersubjectivity to inter-species-subjectivities: Indeed, if "[t]he encounter with other perceivers continually assures me that there is more to any thing, or to the world, than I myself can perceive at any moment", these others are not only human beings: Taking the example of a particular oak tree, "it remains an experience for others – not just for other persons but [...] for other sentient organisms, for the birds that nest in its branches and for the insects that move along its bark, and even, finally, for the sensitive cells and tissues of the oak itself, quietly drinking sunlight through its leaves" (Abram 1996, p. 39).

Perception is also participatory in its primordially synaesthetic character.[46] The "full participation of my sensing body with the sensuous terrain" (ibid., p. 60) integrates together the inputs of the different senses into one experience, as "complementary powers evolved in complex interdependence with one another" (ibid., p. 61). The senses, although they are distinct, are however converging upon the sensible: "My various senses, *diverging* as they do from a single, coherent body, coherently *converge*, as well, in the perceived thing, just as the separate perspectives of my two eyes converge [...] and convene there into a single focus" (ibid., p. 62). This convergence is also what allows the perceiving, bodily self to exist, i.e. to complete itself in the perception of the other. The participation between the senses and the participation with the sensible/perceived are inseparable, and "my body is a sort of open circuit that completes itself only in things, in others, in the encompassing earth".

Furthermore, as Merleau-Ponty suggested in his last, unfinished book *(The visible and the invisible),* the subjective body is co-constituting and co-constituted by "the flesh of the world" (cf. Merleau-Ponty 1968), i.e. in Abram's words: "the perceiver and the perceived are interdependent and in some sense even reversible aspects of a common animate element, or Flesh, that is *at once both sensible and sensitive*" (Abram 1996, p. 67).

Especially the living beings that I encounter (whether animal or plant), are also experiencing subjects probing me as object. More generally, perception can only exist from within the fields of perception. To see is also to feel seen, and to touch is also "to experience one's own tactility, to feel oneself

45 Cf. Husserl 1981. Several contemporary cognitive scientists (e.g. Damasio, Lakoff and Johnson) bring support to the argument that the body founds all cognition. However, cognitive science also falsifies the notion of an unmediated access to reality, found in Merleau-Ponty and in Abram. This is why Smith and Jenks (2006, pp. 142-143) reject Merleau-Ponty's phenomenology as inadequate for a theory of complexity.

46 Cf. Merleau-Ponty 1962, p. 229.

touched *by* the tree" (ibid., p. 68; see also Merleau-Ponty 1968, p. 127). Perception, Abram repeatedly argues, is primarily animistic.

On the perception of human-designed artifacts vs. the perception of evolved forms of nature

With Edgar Morin's distinction of the evolving Being-Machines of nature (e.g. atoms, stars) from the artificial artifacts of human invention, we have already seen how comparatively poorer the later are, as they serve only functionalist, mostly linear logic of functioning, cannot repair themselves, and do not evolve (cf. chapter 2, section 5 & chapter 3, section 1). With Bateson's indictment of the limits and autism of purposive consciousness, we can understand why human technologies, guided by purposive consciousness, have so far been so inferior to natural systems in terms of complexity (cf. chapter 1, section 1). This qualitative difference also bears consequences at the phenomenological level: Indeed, whereas the "patterns on the stream's surface as it ripples over the rocks, or on the bark of an elm tree, or in a cluster of weeds, are all composed of repetitive figures that *never exactly repeat themselves*", but are in constant metamorphosis and draw "our awareness in unexpected and unpredictable directions [... i]n contrast, the mass-produced artifacts of civilization [...] draw our senses into a dance that endlessly reiterates itself *without variation*" (Abram 1996, p. 64).

This general linear functionality, and thus predictability, of artificial artefacts, also explains at least for a part, why the post-industrial society is marked by the accelerating rhythm of both technological innovation and its hyper-consumption: "Once our bodies master these functions [for which the artifacts were built], the machine-made objects commonly teach our sense nothing further; they are unable to surprise us, and so we must continually acquire *new* built objects, new technologies, the latest model of this or that if we wish to stimulate ourselves" (ibid.).

Despite technological innovations, the predominance and loudness of artificial elements in the daily environment of most industrial and post-industrial societies, both:

- tends to practically cover up the sounds and sights of the remaining natural elements,
- and tends to numb and atrophy our perceptive capacities.

Concerning the first phenomenon, Abram contrasts this situation with the example of exceptional instances where the roaring tumult of human artifacts is temporarily silenced by accidents (such as the great hurricane that swept across Long Island, New-York, in the autumn 1985: Cf. ibid., pp. 62-63).

Concerning the second phenomenon, we have also already seen

Abram's wider account of the autopoïetic closure of perception introduced by phonetic alphabetic language (in chapter 1, section 1).

The ethical aesthetics of animistic perception

Abram explicitly suggests that the environmental ethics sought after by environmental philosophers, "will come into existence not primarily through the logical elucidation of new philosophical principles and legislative strictures, but through a renewed attentiveness to this perceptual dimension that underlies all our logics, through a rejuvenation of our carnal, sensorial empathy with the living land that sustains us" (ibid., p. 69).

Such an ethical aesthetics based on animistic perception, which Abram finds in several indigenous cultures (cf. e.g. ibid., pp. 69-71), also applies to "genuine art", according to him:

"Genuine art, we might say, is simply human creation that does not stifle the nonhuman element but, rather, allows whatever is Other in the materials to continue to live and to breathe. Genuine artistry, in this sense, does not impose a wholly external form upon some ostensibly 'inert' material, but rather allows the form to emerge from the participation and reciprocity between the artist and his materials, whether these materials be stones, or pigments, or spoken words. Thus understood, art is really a cooperative endeavor, a work of cocreation in which the dynamism and power of earth-born materials is honored and respected. In return for this respect, these materials contribute their more-than-human resonances to human culture" (ibid., p. 278).

Remembering the participative nature of language

Abram illustrates with the cases of several indigenous peoples around the world, how not only the actions of a single shaman, but the whole of a human language can be contributing to a participatory perception (which is difficult to conceive from our modern vantage point, in-formed by the heritage of phonetic alphabet – cf. chapter 1, section 1). Abram even speaks of "interspecies linguistics", insofar as e.g. human hunters gain a subtle knowledge of "the nuanced meanings of particular calls made by various creatures", and can themselves voice those calls, whether to catch a prey or communicate with another hunter discretely, or can perceive warning signals from other species, that a dangerous predator is nearing.[47]

Also in our modern languages, Abram notes, contrary to Saussure's or Derrida's claims (i.e. the supposed "arbitrariness of the relation between spoken sounds and that which they signify"), some traces of the original in-

47 Cf. Abram 1996, pp. 140-144.

timacy between the language and the land still remain present in "a subtle sort of onomatopoeia [which] is constantly at work in language: certain meanings inevitably gravitate toward certain sounds, and vice versa" (ibid., p. 145, referring to Jacobson and Waugh 1979).

However, our alphabetic culture considers "reading omens [as] a superstitious and utterly irrational activity, prevent[ing] us from recognizing the practical importance, for foraging peoples, of such careful attention to the behavior of the natural surroundings" (Abram 1996, p. 153). We fail to see how the sense of the world requires such a participation in all of the world's meaningful gestures, when meaning is not only constructed and ascribed by autopoïetically human, text-based discourses.

Another important quality emerges, at an epistemological level, from the indigenous attentiveness to a widely meaningful environment: It allows "a worldview that simply has no notion of pure meaninglessness", although some natural elements are deemed less meaningful than others. Such a worldview neither assigns predestination, nor does it posit pure free human will. "No event for the Koyukon [a tribe of Northwestern Alaska] is ever wholly accident or chance, but neither is any event entirely predetermined. Rather [...] the sensuous world is a spontaneous, playful, and dangerous mystery in which we participate" (ibid.).

Furthermore, such an epistemological insight establishes that the meaning of an action depends on its rootedness in a particular location. "A particular place in the land is never, for an oral culture, just a passive or inert setting for the human events that occur there. *It is an active participant in those occurrences*" (ibid., p. 162). In other words, meanings stem from particular ecologies. Abram illustrates this dimension vividly with the case of Aboriginal Australians in their personal relationships to song cycles attached both to 'Dreamtime' and to one's birth place – or more exactly, the place of one's spiritual conception, at a specific point of the mother's pregnancy.[48] Abram concludes that "the synaesthetic association of visible topology with auditory recall – the intertwining of earthly place with linguistic memory – is common to almost all indigenous, oral cultures" (ibid., p. 176).

Worst case scenarios among indigenous peoples

Abram draws repeatedly on inspiring examples from indigenous cultures. However, he does not promote a naively harmonious discourse that would claim some sort of absolute perfection among indigenous peoples' relationships to their environments. Most notably, he points at the examples of catastrophically destructive human practices that often stemmed from the migrations of indigenous peoples to altogether new (for them) natural environments:

48 Cf. ibid., pp. 164-176.

"Of course, the intensely place-centered character of the older, oral cultures was not without its drawbacks [...] there were times when human bands were displaced from their familiar lands [...] and suddenly found themselves in a world where their ritual gestures, their prayers and their stories seemed to lose all meaning [...] where nothing seemed to make sense [...] without an etiquette matched to *this* land and its specific affordances of food, fuel and shelter, the displaced and often frightened newcomers could easily disrupt and even destroy a large part of the biotic community" (ibid., p. 269).

Abram points at the examples of the migrations across the Bering Strait, into Australia, in New Zealand, Hawaii and Madagascar.[49]

Phenomenological space-time and the sensibility to the presence of the world

The contrast between the notions of space-time in indigenous cultures, vs. the abstracted notions of space and time in Western culture, was already discussed in chapter 1, as well as the negative consequences of the Western notions for a sensibility to the presence of the world. As phrased by Abram:

"As long as we structure our lives according to assumed parameters of a static space and a rectilinear time, we will be able to ignore, or overlook, our thorough dependence upon the earth around us. Only when space and time are reconciled into a single, unified field of phenomena does the encompassing earth become evident, once again, in all its power and its depth, as the very ground and horizon of all our knowing" (Abram 1996, p. 217).

To find back the presence of the world, from which the Westerner's past and future are cut-off, Abram moves along an inverse path: "When I allow the past and the future to dissolve, imaginatively, into the immediacy of the present moment, then the 'present' itself expands to become an enveloping field of *presence*. And this presence, vibrant and alive, spontaneously assumes the precise shape and contour of the enveloping sensory landscape" (ibid., pp. 203-204). Abram further grounds his argument, phenomenologically, especially referring to the late Heidegger's understanding of "time-space" as a phenomenological preliminary to the abstracted time and space (cf. Heidegger 1972), and to the late Merleau-Ponty who referred to "this very time that is space, this very space that is time, which I will have rediscovered by my analysis of the visible and the flesh" (Merleau-Ponty 1968, p. 259).

49 I referred to Jared Diamond's *Collapse* in a footnote in chapter 1, section 1.

Abram grounds his distinction of 'the present as presence', from the common Western notion of 'the present', in Heidegger's essay "Time and Being" (in Heidegger 1972). Heidegger argues indeed that "the present in the sense of presence differs so vastly from the present in the sense of the now [... i.e.] in the sense of a succession of a calculable sequence of nows" (Heidegger 1972, pp. 11-12). This present-as-in-presence, is an open present that unfolds itself into a "horizon" (Abram insists on Heidegger's metaphorical use of the term horizon) – or rather, the horizon of this present opens this present to "that which lies beyond it". That which is beyond the horizon is that which "exceeds this open presence". For Abram, this horizon is the very visual horizon of the perceived landscape.

Heidegger further discusses the presence of the 'past' and 'future': Both remain hidden from the open presence of the present, but both also bring about the presence of the present.[50] Both make their absence felt within the present. But each remains hidden in a different way: While the future withholds its presence, the past refuses its presence.[51] Coming back to the sensible experience of the landscape, Abram interprets Heidegger's insights as follows:

- The future is felt in the presence of the present as the landscape's horizon, withholding, lying beyond; and I can walk toward it to disclose its withheld promises.[52]
- The past is felt in the presence of the present as the inside of my body, of another living being's body, or the earth's underground, hidden from visibility but grounding the present.[53]

Or in Abram's own words: "The beyond-the-horizon, by withholding its presence, holds open the perceived landscape, while the under-the-ground, by refusing its presence, supports the perceived landscape" (Abram 1996, p. 214).

Space-time is thus no longer abstract space and abstract time, but (again) one sensuous experience, where "the visible landscape has the other moments of time 'inside itself', precisely in that the past preserves itself under the ground, as well as *inside* every entity that I perceive [and t]he sensorial landscape [...] not only opens onto that distant future waiting beyond the horizon but also onto [...] an immanent field of possibilities waiting behind each tree" (ibid., p. 215). And such an experience, before or beneath abstraction, also reveals that ground and horizon "are granted to us only by the earth. Thus, when we let time and space blend into a unified space-time, we

50 Cf. Heidegger 1972, p. 13.

51 Cf. ibid., pp. 16-17.

52 Cf. Abram 1996, p. 212.

53 Cf. ibid., p. 213.

rediscover the enveloping earth" (ibid., p. 216), which remains "the very ground and horizon of all our knowing" (ibid., p. 217).

The soul is in the air

Abram further follows Heidegger in arguing that there is something of the present that is invisible in its presence, "an enigmatic, hidden dimension at the very heart of the sensible present, into which phenomena may withdraw and out of which they continually emerge. [...] It is the invisibility of the air" (ibid., pp. 222-223).[54] Flowing in and out of the earthly living beings constantly bathed in it, the air "is the soul of the visible landscape [...] As the very mystery of the living present, it is that most intimate absence from whence the present presences, and thus a key to the forgotten presence of the earth" (ibid., p. 226).

Furthermore, the air's ineffability echoes awareness / the mind's own ineffability, and "many indigenous people construe awareness, or 'mind', not as a power that resides inside their head, but rather as a quality that they themselves *are inside of*, along with the other animals and the plants, the mountains and the clouds" (ibid., p. 227).[55] Within such conceptions, one's own "wind" (breath) is "not entirely autonomous" from the whole earth's winds, "but rather momentary articulations within the vast and fathomless body of Air itself" (ibid., p. 235). However, this holistic presence of the air does not imply any strict and linear determinism. If "humans are themselves one of the Wind's dwelling places, one of its multiple centers", they are not just conceived as purely passive receptacles of an allopoïetic force. On the contrary, "our actions and thought affect the Air in turn. The individual [...] participates in *it* [...] and hence can engage and subtly influence events in the surrounding terrain", as e.g. the Navajos aiming to influence the natural environment through "the ritual power of speech and song". Nevertheless, this participation does not grant humans a full autonomy: "by insisting that the 'winds within us' are thoroughly continuous with the Wind at large [...] the Navajo elders suggest that that which we call the 'mind' *is not ours*, is not a human possession", but rather, a property of ecological relationships (ibid., p. 237).

The etymology of our own words witnesses a degree of kinship to such indigenous insights. Indeed, the ancient Greek 'psychê' signified not only the 'soul' or 'mind' but also 'breath' and 'wind', and the verb 'psychein'

---

54 Cf. Heidegger 1967, p. 416 and Heidegger 1972, p. 15.

55 Abram also follows contemporary anthropologists who are now deconstructing an ethnocentric bias of their predecessors, which consisted in attributing the Western notions of mind and soul to indigenous cultures (e.g. the "wind within one" of the Navajo which is not separable from the other winds altogether forming the Holy Wind entering all natural phenomena – cf. Abram 1996, pp. 230-233).

meant 'to breathe', while the Latin 'spiritus' meant 'breath' and 'wind', as did 'anima', which also meant 'soul'.[56] The Sanskrit 'atman' (root of 'atmosphere') meant 'soul', 'air' and 'breath'. Even among the Hebrews, whom had rejected animism, however still remained a strong relation of the wind and breath to the spirit, and the vowels, unlike consonants, were not in the original a*leph-beth*: "The vowels [...] are nothing other than sounded breath. And the breath, for the ancient Semites, was the very mystery of life and awareness, a mystery inseparable from the *ruach* – the holy wind or spirit" (ibid., p. 241).

As we have already seen in chapter 1, it is only with the Greek alphabet that the human soul engaged in its divorce from the environing air. If any human language forms a boundary, a filter around a community of human speakers, between that community and its environment, such a filter and boundary does not always necessarily function as a barrier, as it massively does in Western alphabetic culture. It can also function rather as a membrane. Indigenous languages are such "permeable membranes binding the people to their particular terrains, rather than barriers walling them off from the land" (ibid., p. 256).

And the shaman, already evoked at the beginning of this section, contributes to maintaining this plasticity: "By regularly shedding the sensory constraints induced by a common language, periodically dissolving the perceptual boundary in order to directly encounter, converse, and bargain with various nonhuman intelligences [...] and then rejoining the common discourse, the shaman keeps the human discourse from rigidifying, and keeps the perceptual membrane fluid and porous" (ibid.).

Observing the development of the "Gaia hypothesis" among scientists working on the world's planetary cycles (including atmospheric, geophysico-biological cycles), Abram suggests that the "emergence [of the Gaia hypothesis] provides a striking illustration of the way in which a renewed awareness of the air forces us to recognize, ever more vividly, our interdependence with the countless organisms that surround us, and ultimately encourages us to speak of the encompassing earth in the manner of our oral ancestors, as an animate, living presence" (ibid., p. 302).[57]

Rational information vs. phenomenological sense

As long as we remain autopoïetically sealed, rational information and arguments about the ecological crisis (e.g. climate change, biodiversity collapse, etc.) will not reach us very deeply, because "it does little to alter our intellectual detachment from the sensuous earth" (ibid., p. 259). This concerns especially the phenomenology of the air, which "is that element that we are

56 Cf. Abram 1996, pp. 237-238.

57 The Gaia hypothesis was articulated in James Lovelock's *Gaia: A new look at life on earth*, first published in 1979.

most intimately *in*. As long as we experience the invisible depths that surround us as empty space, we will be able to deny, or repress, our thorough interdependence" with the natural world (ibid., p. 260).

Besides, we need to make sense. Because, for a human sensible being that recovers its sensibility to the presence of the world, "to *explain* is not to present a set of finished reasons, but to tell a story. [...] And making sense must here be understood in its most direct meaning: to make sense is to *enliven the senses*. A story that makes sense is one that stirs the sense from their slumber, one that opens the eyes and the ears to their real surroundings [...] to make the senses wake up to where they are" (ibid., p. 265).

Abram makes a number of further suggestions as to what makes sense, i.e. re-awakens the senses:

- First of all, the sensuous world has to be encountered locally. "In contrast to the apparently unlimited, global character of the technologically mediated world, the sensuous world – the world of our direct, unmediated interactions – is always local. The sensuous world is the particular ground on which we walk, the air we breathe" (ibid. p. 266). "For it is only at the scale of our direct, sensory interactions with the land around us that we can appropriately notice and respond to the immediate needs of the living world" (ibid., p. 268).[58]
- Abram suggests a measure of distanciation from our technologies that "short-circuit the sensorial reciprocity between our breathing bodies and the bodily terrain" (ibid. p. 267), and a careful attention to our senses: "Only as we come close to our sense, and begin to trust, once again, the nuanced intelligence of our sensing bodies, do we begin to notice and respond to the subtle logos of the land" (ibid., p. 268).
- Warning against any Utopian thought-construction, Abram argues that a "genuinely ecological approach does not work to attain a mentally envisioned future, but strives to enter, ever more deeply, into the sensorial present" (ibid., p. 272).
- Finally, Abram advocates for the use of synaesthetic effects in cultural production, including in the medium of alphabetic writing. Here his call echoes that of T.S. Eliot which was quoted at the beginning of the current chapter. For Abram, a good poem awakens, re-members all the senses: "Finding phrases that place us in contact with the trembling neck-muscles of a deer holding its antlers high as it swims towards the mainland, or with the ant dragging a scavenged rice-grain through the grasses" (ibid., p. 274).

While acknowledging (halfheartedly) the value of a global human fraternity, Abram insists on the requirement that it be rooted on local experience, "let-

58 Abram directly points at the contemporary practice of "reinhabitation", i.e. people re-apprenticing themselves to their local ecologies: Cf. Abram 1996, p. 271.

ting the vision of a common world root itself in our direct, participatory engagement with the local and the particular" (ibid., p. 270).

However, a merely locally-based sensibility would also fail to meet the challenges of sustainability (as already argued at the beginning on the present chapter, following Leenhardt in eds. Strelow, Prigann and David 2004, p. 112).

The phenomenological challenge of climate change

In the contemporary context, a phenomenologically understood sensibility for sustainability, would also imply a sensibility for such a global and long-term phenomenon as climate change. This issue was addressed by Julien Knebusch at the occasion of the conference of the arts research network of the European Sociological Association (ESA), that our team at the University of Lüneburg organized in 2007.[59] Knebusch's analysis, published in the following-up publication from the research stream 'sustainability', forms the basis of the following considerations.[60]

Knebusch shares with Abram, concerns about "a deterioration of sensitivity to meteorology and climate. Today, very few professions still reveal a direct relationship to climate; our intuitive understanding of it is thus weakened. [...] Thus the divide is growing in our consumer societies between our extended knowledge of climate change and our relatively limited experience of the phenomenon" (Knebusch 2008, p. 242).

However, Knebusch also argues that the phenomenological basis of individual human perception does not allow a perception of the global and long-term reality of climate change. Whereas, at the social level, the "social infrastructure" of history, science and education is "allowing a long-term perception and helping to educate populations to consider and to take larger temporalities and geographical spaces into account", at the individual level (phenomenologically understood) "generally speaking we do not experience climate change" (ibid., p. 251). Not only is the social, abstracted and rational perception of climate change lacking the level of personal sensibilities, but such an individual level cannot arise by itself, from the existing direct perception of the natural world by a human being.

> "The notion of climate change is not a notion which belongs to our sphere of experience. It is a concept. Events with a broader existence, i.e. which deploy and reveal themselves only on broader geographical and temporal scales, are generally not perceivable. Our perception needs to be assisted by our intelligence to appropriate these phenomena. [...] Per se, climate and meteorology are planetary phenomena, but our experience of these phenomena is local, rarely planetary" (ibid., p. 252).

59 Cf. Kirchberg, Kagan and Behnke 2007.

60 Cf. Knebusch 2008.

Knebusch poses this issue as a phenomenological challenge for the arts.

The phenomenology of climate
Our insensibility to climate change stems, Knebusch argues, from the very phenomenology of the climate. Knebusch bases his analysis on recent publications of German phenomenologists Gernot Böhme and Hermann Schmitz. Like Abram, Böhme points at the experience of the weather and climate as an experience of a *landscape*.[61] The difference between weather and climate is that weather refers to a temporary state of the atmosphere and "ephemeral sensations" thereof, while climate "refers to a kind of large meteorological time such as seasons". The personal, direct experience of climate amounts to a "sensation of stability of meteorological time" and of the overall balance of this time across cyclical weather changes (ibid., pp. 246-248). This affects the experience of climate modifications: Unlike the "flowing of days [...] subjectively, a season is always experienced from its centre, at its point of balance [and] we experience seasons as atmospheres which are bathing us. I do not experience 'gradients' of Summer or Autumn which could be established from a scientific point of view" (ibid., p. 249). Consequently, the direct perception of climate as balance, affects the perception of changes in the climate (rather than climate change). "Sometimes, these fluctuations reveal us seasons as defective, affecting our trust, maybe distressing us even if seasons remain time structuring" (ibid., p. 250).

An art & technology-based sensibility to climate change

The first challenge to art is then "to integrate particular climate phenomena in a more global vision in order to consider them as revealing the global phenomenon [and] to invent new means of *approaching* these phenomena". Furthermore, the impression that the "weather is producing itself without us" should be overcome, so that "[r]eal perception of the phenomenon would also be perception of our own participation in climate change" (ibid., p. 253). The latter challenge (i.e. sensing our participation in climate change) already was met in Abram's plea for a participatory-animistic subjective perception. However, the first challenge (i.e. sensing the global) implies, Knebusch argues after American phenomenologist Don Ihde, a turn away from Husserl and Merleau-Ponty's relative defiance for modern-scientific experience. Here, Knebusch's analysis turns its back to Abram's, insofar as Knebusch follows Ihde in formulating "two conditions for a phenomenological perception of global warming, i.e. a vision of science as technoscience and a vision of Earth as planet Earth" (ibid., p. 260):

61 Cf. Knebusch 2008, p. 245.

- The former especially "means that science is technological and allows connecting oneself to the world" (ibid.). From this perspective, contemporary Western technologies do not only numb our direct perception as Abram argues, but also represent some opportunities for new perceptual insights. A pragmatic use of technologies would thus make sense, for aesthetics of sustainability.
- The latter implies e.g. that we not only consider our satellite images of Earth as representations, but also that we "look through them and [...] participate in the visualized reality" (ibid.).

Examples of the use of technologies to render climate change sensitively perceptible, can be found in some examples of contemporary art concerned with climate change. "We notice that since the end of the 1990s, a certain number of contemporary artists are trying to visualize, sonify or realize in another way, meteorological and climatological data in order to *translate* these realities in a different manner than with the well-known graphs and colored maps we know from school" (ibid., p. 257). For example, the American artist Andrea Polli is "using her computer as a music instrument which enables her to choose meteorological and climatological parameters which she is translating and *recomposing* into sound waves compositions [... s]he translates the variations of temperature in loudness, pitch, length, timbre, etc. as well as in a colored graphic (website-)interface" (ibid.). In an interview with Knebusch where he asked her whether "an art of climate change [can] be something else than an 'illustration' of climate change [...] and become a sort of 'presentification' of climate change", Polli acquiesced and replied: "[N]o one can stand in front of a piece for 10 years to listen to how, for example, climate in the arctic has changed, even though in climate terms that's a very short period of time [...] we have recently recorded a 40 minutes sound work that interprets the data from the last 5 years" (ibid., p. 258). The work of New-Zealand artist Janine Randerson further points at "the role of new technologies in our perception of weather and climate and the modifications they induce", and she wants "to provide other approaches for perceiving not necessarily visible processes like carbon emissions which normally are only accessible through abstract data" (ibid., p. 259). Randerson argued to Knebusch in an interview that "when one encounters a screen to walk under and becomes immersed in the image and sound, the mediated data becomes reinvented as a physical experience" (ibid.), and she "added that she would also like to activate other senses in the future – the haptic and sensations like hot and cold" (ibid.). Knebusch further describes one specific installation by Randerson: "At sunset, the Museum visitor of *Anemocinegraph* may walk under big screencups and admire satellite imagery of weather and images of microscopic air and water whirls and currents and listen at the same time to a sound composition. The sound is not a real time transmission but again compressed sounds of the actual data generated by CO^2" over a longer period of time (ibid., pp. 258-259).

Such artistic perspectives stand in contrast to other climate-change-related contemporary artistic work, which keep the perception merely at the individual and immediate level. Knebusch quotes David Buckland, the leader of the 'Cape Farewell' expeditions which brought several star-artists to the Arctic: "When unpacked, each artist has in some form responded to this cold, Arctic place and the way it is changing as the Earth warms. They have all told the story on a human, rather than a planetary scale" (Buckland quoted in Knebusch 2008, p. 257). But as Knebusch remarks, "putting things this way is implying that the planetary scale (the one of the climate models) is not a human scale" (Knebusch 2008, p. 257). On the contrary, Knebusch argues that humans should "appropriate such a complex phenomenon [as] climate change", at the very phenomenological level, if they are to develop a sensibility for the search-process of sustainability.

From a representation to a presentation of nature?

Another author recently pointed at the importance of a "green aesthetics" developing a sensibility to complex ecologies: Loïc Fel, in *L'Esthétique Verte* (2009). I will partly echo his discourse, which main interest lies in its focus on the benefits of science for the aesthetic perception of nature. The reading of Bateson's understanding of the "sensibility to the patterns which connects" as an intuitive knowledge beyond rationality and even shared with other life forms, of David Abram's plea for an animistic phenomenology, earlier in this chapter, and of chapter 1 with its warning against modern science's disjunctive thinking and linear causality, may leave to the reader the inaccurate impression that science can contribute no good to an aesthetics of sustainability. However, as shown throughout chapters 2 and 3, a science of systems and of complexity does bring foundational elements for cultures of sustainability. And as Loïc Fel clearly demonstrates, ecology as a science of complexity does bring insights that are eminently relevant to an aesthetic appreciation of a complex and dynamic nature, a "Dyonisian" rather than "Apollinian" nature (as Fel calls it, following young Friedrich Nietzsche's typology), i.e. a constantly evolving system of systems marked by phenomena of emergence, complex inter-relations and multiple spatial and temporal scales as well as contextuality.[62]

---

62 See especially Fel's developments in pp. 201-208, 253-257, 290-291. However, like many others, Fel confuses "holism" and "complexity", and therefore sometimes falls in the trap of simplistic holistic accounts of natures' "harmony." But he does so only rarely, and overall his work denotes an elaborate attention to the complexity of natural processes. The most concerning case is Fel's uncritical reporting of Stan Godlovitch's simplistic holistic "natural aesthetics" which claims to bring an "a-centric" aesthetic mode whereby only the wholly whole is experienced and all subtleties of perceptions have to be lost – because perceptions are all-too-human for Godlovitch (cf. ibid., pp. 276-277). At another point (p. 193),

These complex realities of nature, Fel argues, are beyond our normal capacities of perception and can only reach our experience thanks to the insights gained from the sciences of ecology.[63] And even with the insights from the science of ecology, the aesthetic experience of nature's complexity is not easy, as noted by Malcom Budd:

"In either case, the fact that an observer will perceive only a small time-slice of an ecosystem, and even then only a small part of what is contained within that time-slice, presents a problem for the appreciation of the aesthetic value of that ecosystem, a problem that cannot be avoided by emphasis on the transformation of perception by knowledge, the ecologically informed observer perceiving events and states within an ecosystem as stages in circular movements of energy through different forms of life. For, in addition to the difficulty presented to an observer of encompassing the totality of an ecosystem in its spatial extent, the temporal duration of an ecosystem is likely to exceed, often greatly so, the time one might give to observing it, precluding the realistic possibility of one's appreciating that value, no matter how much one's perception of things or events in it might be informed by relevant ecological knowledge or how vividly one might imaginatively realize the biological processes that underlie and are responsible for the visual or other appearance of the system." (Budd 2002, p. 105, and quoted in Fel 2009, pp. 215-216).

But Fel then turns this difficulty around and argues that such dimensions beyond human imagination can allow an aesthetic appreciation of the "sublime" in nature.[64]

Therefore, the main insight that I retain from Fel's discourse is: not to forget that contemporary aesthetics of complexity do benefit from, and integrate the insights of sciences of complexity (as I did myself integrate these insights in chapters 2 and 3).

On the other hand, Fel's discourse fails to explicitly recognize the intuitive level of the sensibility to the patterns that connect, as described by Bateson and Abram.[65] In Fel's view, 'green aesthetics' are a consequence of

---

Fel falls – but only shortly – into the older trap of physical reductionism (which could have been avoided, if his discourse had included Maturana and Varela's theory of autopoïesis).

63 In his analysis, Fel (2009) explores in more details the scientific and aesthetic insights brought by scientists such as Alexander von Humboldt (cf. especially pp. 108-109, 134-135), Aldo Leopold (cf. pp. 308-314), and more recently by researchers in the field of landscape ecology (cf. pp. 179-186).

64 Fel's notion of the sublime is not exactly comparable to Kant's sublime.

65 Fel only rarely and maybe almost accidentally, points at "an intuition" (p. 137) but without giving it any further consideration. At one other point, after a short discussion of Merleau-Ponty's phenomenological insights and of his warning against the impoverishment of experience by scientific knowledge, Fel notes that natural complexity can be beyond human intellect (p. 229); but even at that point

the knowledge of complexity brought by the science of ecology, and he does not consider its possible presence in earlier (and/or non-Western) cultures. For him, "without the adequate scientific concepts [...] the experiencing subject cannot come to a sensible comprehension of the complexity of nature" (Fel 2009, p. 313).

Although several parallels can be drawn between Fel's discourse (which insists, as I did in the previous pages, on systems thinking and on the perception of complexity allowed by the science of ecology) and my own development, I cannot follow his main line of argumentation, which is the claim that "green aesthetics" have shifted aesthetics "from the representation to the presentation of nature."

Fel's claim starts with a number of arguments which are convincing and legitimate: Just as the sciences of systems and complexity (as e.g. ecology) are no longer focusing on single objects but on processes, interrelations, i.e. organizations in the *in vivo* context of ecosystems (cf. Fel 2009, p. 158), "green aesthetics" no longer contemplates objects of aesthetic gaze but engages immersively, into inter-actions with natural systems, allowing the perception of their complex and spontaneous dynamism, "the experience of interdependent complex phenomena" (ibid., p. 219).[66] He further makes a helpful distinction between images of the *natura naturans* which merely depict formal appearances of natural objects, and images of the *natura naturata* which re-enact/simulate natural processes and their principles.[67] This brings Fel to his most convincing argument in favor of a "presentation" rather than a "representation" of nature: While a representation can only show objects (whether static or in time) to which it is heterogeneous, a presentation can consist in "perceptible systems" like for example computer simulations (ibid., pp. 176-177).

However, in his further argumentation, Fel considers the experiences of what he calls the "presentation of nature" as "a direct relation with natural objects" (p. 148), an unmediated contact with an "objective" nature, untainted by any cultural, social, or subjective filters. This astonishingly naive assertion is basing itself first of all on the outdated discourse of Vladimir Ivanovich Vernadsky, a natural scientist who in the 1920's argued in favor of "empirical generalization", an inductivist discourse which claimed naively to be void of scientific hypotheses.[68] Since the second half of the 20th century however, inductivism has been criticized as an illusory pursuit, in the works of e.g. Karl Popper, Theodor Adorno (i.e. the two main protagonists of the German "Positivismusstreit" who at least agreed on the rejection

is no mention to be found, of a human intuitive knowledge beyond-rationality à-la Bateson.

66 For example, Fel points at Allan Kaprow's *happenings* as aesthetics of relationships (cf. Fel 2009, pp. 222-223).

67 Cf. Fel 2009, p. 175.

68 Cf. Fel 2009, pp. 163-165.

of inductivism), Imre Lakatos, Paul Feyerabend (1993) and Paul Watzlawick (ed., 1996), among others.[69] The gravest mistake in Fel's discourse is the claim that "[t]his method [initiated by Vernadsky] does not necessitate anymore the prerequisite conception of nature, it saves [the subject] from the construction of a representation by starting from the facts and mere observation" (Fel 2009, p. 164). Such a claim ignores the overwhelming insights from cognitive sciences, epistemology and social sciences which have repeatedly established that:

- The perception of reality by the human mind is never a direct perception but always already a re-presentation (by an autopoïetic system as discussed by Maturana and Varela, or by a mind that re-constructs "the difference that makes a difference" in Bateson's formulation);[70]
- Any 'empirical observation' is therefore always pre-framed by assumptions, and the absence of explicit hypotheses can only mean the unaware and naive reliance on un-scrutinized a prioris/prejudices;
- Scientific observations are tainted by cultural and ideological a prioris held by the scientists, and more generally by the reigning paradigm in a given scientific discipline/field at a given point in time (Kuhn 1962);[71]
- And in general, reality as we perceive it and experience it is always at least for a part, the result of social constructions which filter and frame perceptions and cannot be entirely done away with (Berger and Luckmann 1966).

Therefore, Fel's following argumentation in favor of an aesthetics of presentation of nature which would directly access objective reality, is untenable and impossible to take seriously into consideration... as is untenable and irrelevant the equally epistemologically naive "aesthetics of the natural environment" from Allen Carlson as discussed by Fel, i.e. a positivist discourse on a so-called objective aesthetics, supposedly based on science (or rather, on an epistemologically invalid misrepresentation of scientific knowledge), which I will thus leave entirely outside of my attention.[72]

---

69 Besides, Fel (p. 163) mistakenly interprets Popper's falsifiability as the idea that hypotheses are true until falsified, and thus that hypotheses are bad and should be replaced by direct induction (i.e. he misunderstands Popper in order to bring support to Vernadsky's inductivism). However, what Popper (and his school of "critical rationalism") did argue is that "positive evidence" is no evidence at all and that only claims to knowledge which are potentially falsifiable, are scientifically valid.

70 See also e.g. Smith and Jenks 2006, pp. 96-97.

71 On the role and importance of a paradigm, from the perspective of Morin's complexity theory, see Morin 2008, pp. 1477-1478, 1808-1844.

72 The interested reader can ponder over Carlson's arguments in Fel 2009, for example pp. 123-126. To his credit, Fel does not exactly follow Carlson, but he

However, despite this epistemological weakness, Fel's insights on the importance of complexity in the science of ecology, and the influence thereof for ecological aesthetics, remain valid.

To be fair, the argument I just made against Fel's epistemological mistake, is also valid against David Abram's belief in an unmediated relationship to natural reality (echoing Smith and Jenks' rejection of Merleau-Ponty's belief in such a "correspondence").[73] Once again, neither the autopoïesis approach (the black box), nor a naïve phenomenology, are acceptable, but a concept of autoecopoïesis is required, that can incorporate the ecological insights from the animistic-phenomenological intuitions of David Abram into a theory of complexity.

CONCLUSION

As this chapter argued, aesthetics of sustainability constitute both an invitation and a challenge to contemporary arts practitioners. In order to foster cultures of sustainability (cf. chapters 2 and 3), the aesthetic experience should foster a sensibility that would acknowledge the shared process of creativity between natural phenomena and the artist, highlight the interpenetration of nature and culture, and more generally function as a "sensibility to the patterns that connect". However, such a sensibility should neither turn into a holistically simplified perception, nor into a merely individualized and localized perception, but should strive to become a sensibility to complexity. The challenge of aesthetics of sustainability, as a phenomenological challenge, implies both a revival of our basically animistic participation to the presence of the living natural world around us, locally, and the construction of a participative perception of planet Earth as the basis for a planetary citizenship.

Such aesthetics of sustainability would concur to the development of a "literacy of complexity": I introduced this expression at the 2009 Summit of the World Alliance for Arts Education, inviting the assembled international arts education associations to work toward an integration of artistic education and education for sustainable development (which is promoted at the UNESCO through a "UN Decade of Education for Sustainable Development" from 2005 to 2014). As I then argued:

"The domain of the trans-... constitutes an epistemological frame, which together with the methodology of action-research and the new literacy of ecology and complexity, is suggesting to orient arts education towards the education of citizens-

---

does comment extensively over Carlson's writings, as well as on other philosophers such as Malcom Budd and Denis Dumas who brought some corrections to the inept arguments from Carlson (cf. Fel 2009, pp. 267-273).

73 Cf. Smith and Jenks 2006, pp. 142-143.

artivists-artscientists, integrating human capabilities in their 'unity in diversity' in order to face the challenge posed by the global crisis of unsustainability" (Kagan 2009a).

In the next two chapters, a number of recent and contemporary artistic practices will be explored in the light of the notions of 'culture(s) of sustainability' and of 'aesthetics of sustainability', assessing how some discourses and practices in the arts (and more specifically in the art world of the so-called visual arts) relate to the themes, perspectives and challenges articulated so far.

5. Ecological Art

Introduction

How far does the aesthetics of sustainability, proposed in the previous chapter, relate to actual practices and discourses in the art worlds? Although a thorough empirical examination is not attempted at in this work, an exploratory overview will be conducted here, both for illustrative and exemplary purposes and to highlight the diversity of perspectives and issues in the field. The overview will focus especially on the interplay between art and ecology, with the emergence of the movement of 'environmental/ecological art' since the late 1960's and with (in the next chapter) the art world's relative interest for the notions of ecology and sustainability in the past decade.

Neither will this overview be exhaustive[1], nor will I do justice to the wider spectrum of artistic movements and perspectives in the 20th century, whose involvement with political, economic and social issues are relevant to the search process of sustainability, and may (or may not) be fostering elements of aesthetics of sustainability.[2]

I will also have to leave aside of my focus, many other movements in contemporary artistic practices, which relevance to the question of cultures and arts of sustainability would however deserve to be discussed. Among them are notably:

---

1 For example, this overview will be very much focused on North-American and West-European cases (and biased by the use of English-speaking literature mainly). For some hints on artistic practices in Central and Eastern Europe relating to the theme of sustainability since the 1970's see Fowkes and Fowkes 2009. Furthermore, an ongoing research process about art and ecology in Asia will be leading to reports by Asian researchers (expected for 2011), thanks to a commission by the program 'Connect2Culture' of the Asia-Europe Foundation.

2 For a brief overview of precedents in the visual arts in the late 19th and in the 20th century, see for example Laville and Leenhardt (1996), pp. 42-58. The same authors also discuss ecological art (pp. 59-77).

- "Relational Aesthetics", as theorized by Nicolas Bourriaud (1998), which I only shortly evoke in this volume, but also less superficial forms of "intervention art" in contemporary art, of which I will discuss only one example in chapter 7: the collective Wochenklausur;[3]
- "Institutional Critique", which I also discuss only shortly across this volume;[4]
- "Culture Jamming" and "Tactical Media", which are especially relevant to social, political and ecological issues (e.g. with the work of the collectives 'Critical Art Ensemble' and 'Bureau d'Etudes');[5]
- The "Theatre of the Oppressed" (with the forms of "Forum Theatre," "Invisible Theatre" and "Legislative Theatre") as developed by Augusto Boal in Brazil (who published several books on the subject) and further practiced worldwide in the past decades, which inaugurated especially relevant strategies for socially participative art-activism;
- "Community Arts", which are especially relevant as socially transformative cultural interventions;
- and the "documentary" form in art (e.g. in photography and in film-making).[6]

An overview of the emergence of ecological art over the past 4 decades is especially warranted, when one notices how the discourses in the art world of 'contemporary art' easily tend to ignore this movement, even when an author claims to promote today an "ecoaesthetics" as does the artist, curator and founding editor of *Third Text*, Rasheed Araeen in a recent article (Araeen 2009) in which no attention is given to ecological art, while land art and Joseph Beuys are subjected to a short critical mention. Ecological art deserves more attention than that...

This chapter will thus focus the attention on the past development of environmental and ecological art, from pioneers of the 1960's and 1970's (Hans Haacke, Joseph Beuys, Helen Mayer Harrison and Newton Harrison, Alan Sonfist and Mierle Laderman Ukeles) to some of its current practitioners especially in Europe and the USA. I will also further discuss accompa-

3 Other Intervention artists of great significance could have been further discussed, such as for example Jeanne van Heeswijk in Rotterdam.

4 Cf. ed. Welchman 2006

5 Cf. Lasn 1999 for Culture Jamming; concerning Tactical Media, cf. Comiotto et al. 2003; I also experimented with Tactical Media interventions with students and colleagues in 2007 (under the name 'Cultura21 Lüneburg'), with the project *Leuphana21* (cf. http://www.leuphana.de.vu), which was presented at a side-event of the Venice Biennale in 2007.

6 On the comparative advantage of the documentary form, against the superficiality of relational aesthetics in contemporary visual arts, see Baqué 2004. On the relevance of documentary film for sustainability, I wrote a short article based on my own experience as director of the film *Land of Thorns* (Kagan 2009b).

nying art-historical and curatorial discourses that contributed to the rise of ecological art in the past decades.

SECTION 1: LAND ART, ENVIRONMENTAL ART, ECOLOGICAL ART

About the terms in use: land art, earthworks, environmental/ecological art

In this chapter, I will mainly discuss the work of so-called environmental artists and ecological artists. To prevent any confusion, I will first disambiguate them from the categories of 'land art' and 'earthworks':

In trying to sort out the different expressions in use, i.e. 'land art', 'earthworks', 'environmental art' and 'ecological art', the curator Sue Spaid pointed out that:

- "Land art, the most general category, encompasses any work that activates the land, however temporarily. Earthworks, ecological art and environmental art are all examples of land art" (Spaid 2002, p. 10);
- "Earthworks, as an art historical category, was devised to describe works like [...] Michael Heizer's *Double Negative* (1969-1970), Robert Smithson's *Spiral Jetty* (1970) [and others who created] primarily permanent, large-scale, non-natural forms sited in 'wide open spaces' as opposed to particular natural environments [...] or in an urban setting" (ibid., pp. 10-11);
- "Environmental art [...] is generally less monumental and tends to employ nature as a medium, so as to enhance the viewer's awareness of nature's forces, processes and phenomena, or to demonstrate an indigenous culture's awareness of nature's sway" (ibid., p. 11);
- "Ecological artists consider issues of sustainability, adaptability, interdependence, renewable resources, and biodiversity, but", she adds, "they don't necessarily attempt to transform the local ecology" - the latter being characteristic of what she calls "ecovention" (ibid., p. 12). The specific expression "ecovention", which corresponds to an exhibition under the same name curated by Amy Lipton and Sue Spaid in 2002, will be discussed later in this chapter (in section 6).

As Spaid's definitions illustrate, the difference between 'environmental art' and 'ecological art' is not clearly established, and several artists seem to alternate between the two denominations.[7] Indeed, Spaid's view quoted above

7 For example, the Harrisons (who are introduced below in this chapter) use both expressions to characterize their art.

would tend to characterize the expression 'environmental art' as oriented towards working with nature as art's medium, and the expression 'ecological art' as oriented towards addressing ecological issues, but this distinction is hardly tenable as a distinction between two separate terms, as these two dimensions are depending on each other and including each other in themselves (as will be illustrated in the following pages). Rather, such a distinction helps to see two related angles on the same activity.

But other definitions have been offered. Several artists (such as e.g. Aviva Rahmani) use the term 'ecological art' (rather than 'ecovention') to point specifically at projects that are interdisciplinary and "result in direct intervention in environmental degradation" (Rahmani, on www.ghostnets.com). Amy Lipton and Patricia Watts, founders and curators of Ecoartspace (a USA-based organization promoting and curating ecological art), define "ecoart" as

> "not merely an art movement – it is a paradigm shift in how we define what art is. This shift in the viewer's perception is in challenging the predominant value system that determines what an artwork can be. These artworks cross the line from traditional art production and institutionalization, into the larger context of human and non-human natural communities. They are focusing on our relationship with the larger, living ecosystem, recognizing that our existence depends upon its survival" (Lipton and Watts in eds. Strelow, Prigann and David 2004, p. 94).

Sam Bower, the director of greenmuseum.org (a very resourceful online platform for ecological and environmental art, featuring mainly North-American and European artists), dedicated an article on the issue of the "profusion of terms". He notices that "many artists and critics use the words [...] interchangeably or at times use the same term to mean different things" (Bower 2009). He also notices the recent rise of other expressions such as 'green art' or 'sustainable art', which try to counter a feared isolation of 'environmental' issues from other issues relevant to the question of sustainability. He himself advocates for a relative standardization of the terminology and proposes to settle for:

- eco-art (ecological art) as a term characterizing contemporary work in the field, "a term which distinguishes contemporary activist approaches from earlier land art and earth art";
- environmental art as an "umbrella term to describe the whole movement".[8]

8 It is not clear in his text, whether Bower includes in "environmental art", the relatively 'un-ecological' instances in land art. Cf. below for a discussion of 'un-ecological' instances in the history of land art.

To support his view, Bower argues that the term "ecological" relates only to ecosystems and biological cycles and is less inclusive than the term "environmental" which more generally expresses a relation to surroundings. Although his argumentation here is plain wrong (i.e. 'ecology' is as inclusive a concept as 'environment': cf. Gregory Bateson's "ecology of the mind" or Edgar Morin's notion of an "ecology of ideas" as examples), my goal here is only to mention these different understandings of the expressions within the ecological/environmental art community, and not to denounce them.

Concerning land art, Sam Bower departs from Sue Spaid's wide definition and prefers to restrict the term to the precedents of environmental art, i.e. "it ought to sensibly be considered an early chapter in our evolving notion of the environment as it grew from meaning 'space' to a broader concept [beyond land art] which included ecosystems" (Bower 2009). This, Bower concedes, is a retrospecting redefinition and restriction of the term 'land art': Indeed, as the expression "ecological art" did not come into use before the 1990's, he notes, in the 1970's and 1980's the term "land art" also generally covered ecological reclamation project.

The organization SFEAP (South Florida Environmental Art Project) builds further upon Sam Bower's definitions and uses the expression "ecoart" as the preferred term for its practices. It sees ecoart as "the most engaged form of environmental art" and as "meld[ing] aspects of environmental art, activist art and community animation/mobilization art with engineering and science-originated processes of restoration of damaged ecosystems" (www.sfeap.org). SFEAP highlights especially the requirements of site-specificity, public commitment, intervention and collaboration (i.e. interdisciplinarily and with communities): These keywords will come back regularly along this chapter.

To come back to the two further expressions used by SFEAP in its characterization of ecoart, besides the already discussed 'environmental art':

- By "activist art", SFEAP refers back to the "Bread and Puppet Theatre" of the 1960's and its anti-Vietnam war actions, to artists who worked on AIDS-related political issues in the USA in the 1980's, to the Guerrilla Girls' feminist critique of museum institutions and to Suzanne Lacy's work.[9] SFEAP relates ecoart to activist art insofar as it allies with environmentalist movements, "engages the public, primarily outside traditional art venues (but also in them), and often in ways that attract media attention".
- By "community engagement or 'animation' art", SFEAP points at the movement of community arts (cf. e.g. the "Community Arts Network": www.communityarts.net) and its focus on social change, refers to artists

9 Suzanne Lacy's work will be further discussed at a later point in this chapter.

> such as John Malpede in the USA.[10] For SFEAP, ecoart is rooted in community art insofar as it engages into "community mobilization and animation", and not only "awareness raising", because "[b]uilding community to steward the environment will be crucial to the deep cultural changes necessary to enhance the chances of the planet's survival – and our own" (www.sfeap.org).

The more detailed discussion of the work of several ecological artists, in this chapter, will offer a more detailed and differentiated overview of the discourses and practices in this field.[11]

The critique of Land Art as a possible turning-point

A critical turn against certain un-ecological practices and discourses in land art is often considered to have contributed to the constitution of a movement of environmental/ecological art in the 1970's and 1980's.

The beginnings of ecological/environmental art would thus be rooted in a critique of the un-ecological, or even anti-ecological practices and discourses of certain prominent land artists in the 1960's and 1970's. Suzi Gablik gives two vivid examples of land artists whose work (or some part of their work) did not demonstrate ecological concerns: "In 1969, the same year that he poured a truckload of asphalt down the side of a hill, Robert Smithson was prevented from dropping broken glass on an island in Vancouver by environmentalists fearful that it would harm the birldlife of that area" (Gablik 1991, p. 140). Gablik further quotes Smithson's *Collected Writings*: "The ecology thing has a kind of religious ethical undertone to it [...] There's no need to refer to nature anymore. I'm totally concerned with making art" (Smithson, quoted in Gablik 1991, p. 140). She also quotes Michael Heizer, whose *Double Negative* sculpture consists in two gigantic cuts in the desert, displacing 249 000 tons of earth, as saying "I don't care about landscape. I'm a sculptor. Real estate is dirt, and dirt is material" (Heizer, quoted in Gablik 1991, p. 140).

However, it would be unfair to categorize these land artists unequivocally as un-ecological. Especially in the case of Robert Smithson, some of his work could be characterized as manifesting some degree of ecological awareness, most especially his *Spiral Hill/Broken Circle* (1971), a reclaimed open sand pit in Emmen, the Netherlands. To a lesser extent, his *Spiral Jetty* could be interpreted from an ecological perspective, although that work did contribute to environmental damages on an already degraded site. More

10 John Malpede's work will be mentioned again when discussing Suzi Gablik's analysis, at a later point in this chapter.

11 Relatively intelligible terms will be helpful in this chapter, but art historical labeling is not a major concern in my analysis.

generally, Smithson's view was that artists should not 'clean up the mess' left by industry, but he argued that art should be an arena to critically discuss ecological issues: "Art can become a resource, that mediates between the ecologist and the industrialist. Ecology and art are not one-way streets, rather they should be crossroads" (Smithson, quoted in Spaid 2002, p. 53). Smithson tended to work on degraded sites (and attempted repeatedly to work for mining companies), and aimed to make the damage visible rather than attempting at remediation. Given this strategic choice, Smithson however also decided to provoke and oppose artists making different choices, for example asking: "Could one say that art degenerates as it approaches gardening?" (Smithson, quoted in Spaid 2002, p. 54). This provocative question also reveals an elitist concern for the preservation of the "high" status of art in society and its presumable higher qualities. As remarked by Maja and Reuben Fowkes, Smithson's discourse tends to accept environmental degradations "as a necessity" and they conclude: "Although his vocabulary draws heavily on natural metaphors [...] these linguistic devices betray little concern for the increasingly precarious conditions of the environment under growth-oriented capitalism" (Fowkes and Fowkes 2009, p. 674).[12]

Similar concerns transpire in the case of the artist Robert Morris: At a panel discussion alongside the Seattle Art Museum's 1979 exhibition *Earthwork: Land Reclamation as Sculpture*, Morris positioned himself against certain ecological artists, complaining that "art was going to cost less than restoring the site to its 'natural condition'. What are the implications of that kind of thinking [...] that art should be cheaper than nature? Or that site-works can be supported and seen as relevant by a community only if they fulfill a kind of sanitation service" (Morris, quoted in Spaid 2002, p. 4). I will come back later to this typical critique of, and taboo against, usefulness and social responsibilities in contemporary art, when discussing Suzi Gablik (in section 4).

The critical review of land art continues to this day, as shows Rasheed Araeen's recent text promoting an "ecoaesthetics". About Robert Morris, Araeen notices that Morris

> "contemplated growing a crop on farmland and turning it into a work of art. But he did not execute the project. It seemed he faced the problem of his nar-ego. The conflict between the individualism of the intellectual and the collectivity of manual work prevented the artist from recognising the farm workers as his equal partners – in

12 Marga Bijvoet tries to excuse or justify Smithson's ambiguities with unconvincing arguments (e.g. she argues that Smithson died before he could read Lovelock's *Gaia*): See the section on "Smithson and 'Mother Earth'" in her online text "Greening of Art" available at: http://www.stichting-mai.de/hwg/amb/goa/greening_of_art_06.htm#07 [last accessed: 10.07.2010].

terms of both its production and its ownership. The only choice Morris left himself was to abandon the idea" (Araeen 2009, p. 682).

Araeen sees nevertheless one positive potential development thanks to land art (which has indeed often been recognized in the work of Robert Smithson – see also Kester in eds. Strelow, Prigann and David 2004): "a paradigm shift from the idea of *representation* to a process of continual *transformation* that would have allowed art to become part of living processes of productivity of the land itself as well as its inhabitants" (ibid.). However, Araeen fails in his text to realize that this development did start occurring with the emergence of ecological art, which he ignores while he rightly criticizes land art for its re-assimilation by art institutions: "despite this historical shift in the conception of art, most of the works ended up as photographs in museums, mere objects of gaze. What should have become the living process of productive land ended up in museums as frozen, reified art" (ibid.).

Four decades of art exhibitions

As usual in the visual arts (i.e. 'contemporary art'), curated exhibitions mark important steps in the development of certain art movements and certain themes of interest. The following very short overview aims to highlight a few insightful exhibitions, several of which will be discussed more specifically at later points in this chapter (but it does not claim to establish a selection of the 'best' or 'most significant' exhibitions on art and ecology or sustainability).

In the USA, the group exhibition *Fragile Ecologies* (1992, curated by Barbara Matilsky, at the Queens Museum of Art), was according to Sue Spaid "the first exhibition to focus exclusively on ecological art" in the sense of ecovention, i.e. projects concretely realized on sites. In the earlier group exhibition *Earthworks: Land Reclamation as Sculpture* (1979, at the Seattle Art Museum), several proposals for site reclamations were outlined, of which the proposals by Robert Morris and Herbert Bayer were later realized. Before this, especially important group exhibitions devoted to land art and earthworks were *Earth Art* (1969, at Cornell University) *and Elements of Art: Earth, Air and Fire* (1971, at the Museum of Fine Arts in Boston).[13] In Germany, Heike Strelow curated *Natural Reality*, an important exhibition of ecological art in 1999 at the Ludwig Forum für Internationale Kunst in Aachen. In 2002, the exhibition *Ecovention* (at the Contemporary Arts Center in Cincinnati) marked an important step in the recognition of ecological

13 A historical overview (for North-America and Europe) of ecological art related exhibitions, from 1951 to 2002, and a historical overview of ecoventions between 1955 and 2002, are included in the catalog of *Ecovention* (Spaid 2002, pp. 133-136).

art. In 2008, the exhibition *Greenwashing* (curated by Ilaria Bonacossa and by 'Latitudes') at the Fondazione Sandretto Re Rebaugendo (in Turin), aimed to critically address environmentalist discourses, and distanced itself from ecological art.[14] The same year, in my vicinity i.e. in a southern Hamburg district (Wilhelmsburg), was organized the exhibition and interventions project *Culture|Nature* (curated by Anke Haarmann and Harald Lemke) which invited ecological artists, 'Tactical Media' artists and local community groups to intervene together in public spaces.

SECTION 2: PRECURSORS AND PIONEERS

A relatively small number of prominent artists who were especially active from the 1960's or 1970's onwards, developed areas of artistic work that opened up the field of environmental/ecological art: Hans Haacke, Alan Sonfist, Joseph Beuys, Helen Mayer Harrison and Newton Harrison, and Mierle Laderman Ukeles will therefore be introduced in this section, before discussing the work of other artists and of curators who further explored the interplay of art and ecology.

Hans Haacke

"An artist is not an isolated system. In order to survive he has to continuously interact with the world around him... Theoretically there are no limits to his involvement" (Hans Haacke, quoted in Gablik 1991, p. 115). The work of German artist Hans Haacke was influenced by the ideas of writer, artist and curator Jack Burnham in the 1960's on "Systems Esthetics" (which were themselves influenced by Von Bertalanffy's 'General System Theory'), discussing the boundaries of art and positing that "when an art work crossed its boundaries, it became part of a context, or a system", and thus that "a systems esthetic was no longer solely limited to the specific concerns of the art world. In Bunrham's view, this meant that an analysis of these art works only made sense if one took into account "their assigned context." As context, he not only defined the immediate literal art space, but also the larger space of social, political and technological/scientific contacts and collaborations with which the artist, who had moved into the world, had to deal with" (Bijvoet 1994, p. 68). In the mid-1960's Burnham and Haacke had numerous exchanges, and Haacke took up the concept of systems analysis very seriously, as a cornerstone of his own artistic development which was already heading in a similar direction. Bijvoet comments that Haacke's work "shows a conscious application of its concepts from a relatively simple

14 That exhibition will be discussed in the next chapter, as it poses itself as clearly not-belonging to ecological art.

level toward an ever more complex layering of the issues researched" (Bijvoet 1994, p. 73). The relatively simple level was explored by Haacke in the first half of the 1960's, with water, wind, light, heat and cold, set under predetermined conditions: "These were the condensation boxes, simple plexiglass boxes in which the viewer could register a continual cycle of waterdroplets evaporating, condensating, falling, [i.e.] cyclical changes [which] were primarily dependent on the outside light and heat sources" i.e. a very specific/limited link with the environment. In the second half of the 1960's, Haacke introduced biological systems in his work, making grass grow on soil placed on a plexiglass cube (*Grass Cube*, 1967) and on an earth mound (*Grass Grows*, 1969, at the *Earth Art* exhibition). At the *Earth Art* symposium alongside the exhibition, Haacke stated: "I'm more interested in the growth of plants – growth as a phenomenon which is something that is outside the realm of forms, composition, etc., and has to do with interaction of forces and interaction of energies and information" (Haacke, quoted in Bijvoet 1994, p. 74). In 1971, Haacke further specified: "I use the word 'systems' exclusively for things that are not systems in terms of perception, but are physical, biological, or social entities, which I believe, are more real than perceptual titillation"; and he dissociated himself from Minimal Art: "they were interested in inertness, whereas I was concerned with change [as] the ideological basis of my work [...] there's absolutely nothing static [...] the status quo is an illusion, a dangerous illusion politically" (Haacke, quoted in Bijvoet 1994, p. 75 and p. 76).[15] In the same interview from 1971, Haacke detailed his general understanding of systems, which showed a clear holistic-systemic perspective, lacking the attention to complexity that Edgar Morin later developed (as discussed in previous chapters): "the functioning of either one [i.e. nature and technology] can be described by the same conceptual models, and they both obviously follow the same rules of operation. It also seems that the way social organizations behave is not much different. The world [...] is one supersystem with myriad subsystems". (It should be noted, however, that most cyberneticians and system theorists of the time were having the same holistic bias.)

From 1969 onwards (starting with *Gallery-Goers' Birthplace and Residence Profile*), Haacke's work "took a major tun toward the exploration of social or economic networks and their specific function as mediators of power structures" (Bijvoet 1994, p. 76). Haacke's critical investigation of the institutional structures of the art world played an influential role on the development of the movement of 'Institutional Critique' in the subsequent decades (i.e. an important reflexive turn allowing the self-analysis of art in-

15 Haacke turned his back on phenomenology. As I discussed in the previous chapter with David Abram, however, an 'animistic' phenomenology is crucial, for an embodied ecological sensibility, and aesthetics of sustainability.

stitutions in sociological terms), but this dimension lies outside the scope of the current chapter.[16]

Some of Haacke's projects since the 1970's do have a clear ecological dimension, especially concerning the soil and water:

His *Bowery Seeds* (1970) consisted in an circular area of earth, uncultivated and open to receive airborne seeds. His proposal for the buildings of the German Ministries of Education, Science and Justice in 1973-1974 *(Vorschlag "Niemandsland")* consisted in a circle with a diameter of 25 meters, to be carved in the pavement and on which earth would be randomly deposited. Sue Spaid pointed out that this proposal elaborated "a fluid alternative to the political system's rigid structure", as not only did it celebrate indeterminate, living, unpredictable, changing natural processes, but it also questioned the territorial sovereignty of the State: "He even requested the German government sign an internationally binding treaty that relinquished all rights in and to this territory, thus pledging to grant everyone access to this no-man's land" (Spaid 2002, p. 28). A later project of Haacke, which did get implemented (despite the protests and active obstruction of the CDU and CSU – German conservative political parties), also involves the German soil. *Der Bevölkerung* (from 2000 on) consists in an installation at the Bundestag (in the northern Courtyard), which expresses nationality as based on the German soil, rather than the German blood, with a collection of soils from all German Länder (federated states), which he asked Members of Parliament to contribute (New MPs are asked to bring new soil, and some soil is removed when a MP's term expires). Spaid commented that the "new ecosystem mirrors the population's inherent diversity, while nature's tug-of-war symbolizes the democratic process" (Spaid 2002, p. 30). Haacke himself observed: "In an extremely controlled building, the ecosystem of imported seeds in the Parliament's courtyard constitutes an enclave of unpredictable and free development. It is an unregulated place, exempt from the demands of planning everything. It is dedicated TO THE POPULATION" (Haacke, quoted in Spaid 2002, p. 30). The site is covered by dense vegetation, without any gardening, and a fauna of snails, spiders and insects. Over seven years, Haacke has been taking pictures of this flora and fauna, at various seasons of the year. This led to the installation *Bundesgartenschau* (2008), exhibited at the Galerie Sfeir-Semler in Hamburg. The gallery commented on the title of this work: "While the biennial Federal Horticulture Show [i.e. the actual "Bundesgartenschau"] is a competitive exhibition of gardeners and landscape architects, the ecosystem of DER BEVÖLKERUNG does without horticultural interventions."[17]

16 Institutional Critique, however, brings very valuable inputs, which are contributing necessary insights and critical tools, for art worlds to play a role in wider social transformations.

17 Quote retrieved at: http://www.re-title.com/artists/HANS-HAACKE.asp [last accessed: 01.03.2010].

Haacke also addressed the topic of water pollution in his *Rhinewater Purification Plant* (1972, at Museum Haus Lange, Krefeld): Some Rhine water (released from the Krefeld sewage plant) was "pumped into a container where it was filtered and purified before entering a large rectangular basin housing goldfish [...] The presence of a large fish bowl and the picture-window view into the wooded landscape served as a point of contrast between a life-supporting ecosystem and one on the verge of collapse" (Barbara Matilsky, quoted in Spaid 2002, p. 30). Spaid adds that "[b]y pumping the water through an additional filtration system and using the surplus water to water the museum's garden, he introduced gray-water reclamation" (Spaid 2002, p. 31).

Alan Sonfist

The US American visual artist Alan Sonfist started, from the 1960's onwards, to experiment growing micro-organisms and re-staging natural conditions in art galleries. As Margaretha Bijvoet argues, Sonfist "intended to make the spectator aware of 1. nature's natural processes, patterns and structures, recurring time-cycles, 2. of the environmental changes that took place throughout the earth's history, both with and without the human hand" (Bijvoet 1994, p. 114). For example, in *Colony of Army Ants* (1972), the ants gradually redrew patterns first traced by Sonfist in the arrangement of food. In *Crystalline Enclosure* (1969), he experimented with the re-crystallization of gazes in a glass sphere, influenced by the changing heat, light and air currents, upon which the exhibition-spectator too had an influence. Thereby, Sonfist revealed some aspects of complexity: "The growth of the work is characterized by patterns of indeterminacy, since a multiplicity of natural variables creates unforeseen occurrence" (Sonfist quoted in Bijvoet 1994, p. 116). "Most of my pieces deal with this idea in an ecological sense in which I present the change in form caused to bring change in another related form" (ibid., p. 118).

Bijvoet further argues that Sonfist's work became increasingly concerned with time as not-only linear, but also cyclical. Sonfist himself argued in 1972 that "any one visual image must be seen as only one segment in a total process. Any one form must be seen as an essence in a continuum that is shaped by other, perhaps visually absent, factors. Landscapes present the fragmentation of a continuum" (ibid., p. 117).

In 1969, Sonfist monitored air quality at several intersections of New York City and posted results for passers-by. In 1973, he drew the concept of *Seed Catcher* ("pool of virgin earth to collect the seeds of nearby forests through wind and animal migration")[18], i.e. a concept similar to Hans Haacke's *Bowery Seeds* (1970), and realized it concretely in 1975 as *Pool of Earth* (at Artpark in Lewinston, New York).

---

18 Sonfist quoted in Spaid 2002, p. 90.

Sonfist increasingly worked outside of art galleries, in public space, redefining "public monuments" in order to include the natural world in western culture's self-celebratory tradition, with e.g. *Rock Monument* (Buffalo, 1978) and *Time Landscape* (1977-78). "Public monuments traditionally have celebrated events in human history – acts or humans of importance to the whole community. Now, as we perceive our dependence on nature, the concept of community expands to include non-human elements, and civic monuments should honor and celebrate the life and acts of another part of the community: natural phenomena" (ibid., p. 118).

The *Time Landscape*, Sonfist's most famous work, re-creating elements of a pre-colonial forest in New York City, compresses several time-dimensions, with a cohabitation of different stages in forest-cycles with the contemporary urban environment. Furthemore, according to Sue Spaid, Sonfist's *Time Landscape* had a long-term influence on New York City's Parks Department, which later developed the project of "Greenstreets" across the city, i.e. transforming parts of streets into mini-landscapes.[19]

Sonfist also advocates for a pragmatic use of technologies, from an ecological perspective: "Technology can even visualize aspects of nature outside the range of the human eye, such as public outdoor projections of telescopic observations" (ibid., p. 120). But more generally, Sonfist highlights the higher value of nature when compared to man-made artefacts, even if these are 'art' objects. In 1978, he created *Natural and Bronze Time Enclosures*, i.e. a bronze branch valued at $4 and a real tree branch valued at $4000, requiring them to be purchased together. Sue Spaid sees in this work a demonstration of Sonfist's difference with the discourse of a Robert Morris (as quoted above) who does not work on valuing nature very highly.[20]

Labelled by 'Greenmuseum' as "one of the early pioneers of eco-art", Sonfist was characterized by Jack Burnham as "believ[ing] that man-nature stability will come only when we have become acutely sensitive to the natural changes around us" (Burnham quoted in Bijvoet 1994, p. 120). Indeed, as Sonfist argued, his goal is not just to repair the landscape but to "repair the hole in the psyche which is left when all traces of our biological are ecological roots are obliterated" (Sonfist, quoted in Spaid 2002, p. 8).

Joseph Beuys

In the wide-ranging work of Joseph Beuys, who became one of the most prominent German artists of the 1970's and 1980's, the ecology played a major role, and his influence on ecological and social/activist artists is still very strong especially through his concept of "social sculpture". The very

19 Cf. Spaid 2002, p. 9.

20 Cf. ibid., p. 5.

short overview attempted here will only superficially introduce Beuys' significance, in the context of ecological art.[21]

In his performances, Beuys explored the connection between humans and nature: In *Eine Aktion in Moor* (1971), he ran and swam through swamps, celebrating the value of endangered marshes. In *Coyote. I like America and America likes me* (1974), Beuys spent several weeks locked up with a coyote in a New-York gallery and cohabited/communicated with the animal, which is also symbolic of Native American culture.

Beuys also stressed the political and socially transformative dimension of ecology and art. He was an active member of the German Green party, running for office in 1976, and engaged in the anti-nuclear movement. In his 1981 *Appeal for an Alternative*, reprinted in the Documenta 7 catalog, Beuys eloquently denounced the "modern industrial civilization" and its "expansive growth". At Documenta 7 (1982), Beuys started the *7000 Oaks* project for the city of Kassel (i.e. the planting of 7000 trees of different species, paired with basalt columns, and inviting individuals to sponsor the trees), denouncing deforestation and calling forward a development of urban ecology policies. In 1983, upon invitation by Hamburg city's cultural department, he proposed *Gesamtkunstwerk Freie und Hansestadt Hamburg* ("Total Work of Art Free and Hanseatic City Hamburg"), a plan "to plant more than 800.000 seedlings on the Altenwerder spoil fields with the goal of renaturalizing the area and to establish an Office for Environmental Matters downtown" (eds. Haarmann and Lemke 2008, n.p.). The plan was rejected by the Senate of the City-State of Hamburg in 1984 (and the location became instead a container terminal for the harbor), but the art community in Hamburg still remembers the project.

Joseph Beuy's concept of "social sculpture" inspired many artists interested in contributing to social and ecological change, in the following decades (see e.g. Shelley Sacks, discussed below in section 3). In the words of Richard Demarco (nd): "Joseph Beuys regarded 'social sculpture' as a way to reshape and restructure society, according to the understanding that 'every human being is an artist' and by utilizing the healing and transformative powers of art in the lives of individuals and in the lives of communities, large and small." Joseph Beuys explored the definition of 'sculpture' and

"reflecting on sculptural form (plastisches Gestalten) always led him to question what the essence of sculpture might be. He was not satisfied with the usual answer that sculpture was a three-dimensional entity in space. He therefore continued to experiment with and reflect on the whole notion and nature of sculpture. In doing so he realised that every three-dimensional entity not only has a spatial dimension, but also a temporal and a thermal dimension. In addition to this he also maintained that sculpture could not be identified with material objects alone, but rather that the sculptural

21 For a transcribed account of Beuys' own discourse in one long interview, See e.g. Harlan 1986.

element encompasses other, invisible entities as well. And so it became clear to Beuys that, using the language of art, social processes could equally be described as aesthetic processes. However, in order to be able to perceive the sculptural qualities of these processes, new organs of perception needed to be developed" (Zumdick, 2008).

Senses, and organs of perception, were also to be linked to thinking (and imagination) as a plastic, forming process. "It is in my thinking that the process begins which, through my bodily organs and other instruments, then comes into the world as an impression and a form that is able to inform" (Beuys, quoted in Zumdick 2008).

Although the influence of Joseph Beuys' actions and of his concept of "social sculpture" are very often credited by ecological artists, some critical voices also have sobered what would otherwise sound as a Beuysian Gospel, or Apology of Beuys. For example, the Austrian collective Wochenklausur explicitly departs from Beuys' depiction of the artist as a shaman, and from his grand-social revolutionary discourses, preferring to focus on 'real' small-scale concrete interventions in social reality, fostering clearly identifiable social transformations.[22] For his part, *Third Text* founder Rasheed Araeen, while recognizing that Beuys "opened up a space for future developments", however (also with his tendency to develop his own personality-cult) "failed to resolve the problem of art trapped within both the artist's nar-ego and the institution that does not allow art to become part of the collectivity of people's everyday life" (Araeen 2009, p. 682).

Helen Mayer Harrison and Newton Harrison

Another 'pioneer' of the eco-art movement was the US American Harrisons couple (Helen and Newton), who shifted their art practice in the early 1970's to environmental concerns. In 1970, Newton made high-quality organic compost, in *Making Earth*. Helen and Newton's series called the *Survival Pieces* (1971-1973), were experiments in the artificial reconstruction of eco-systems within art contexts (i.e. at exhibitions). They grew shrimps, catfish and plants under different conditions. Although this early work was criticized, from an environmental perspective (e.g. by Jonathan Benthall for its dependence on fossil fuel – cf. Bijvoet 1994, p. 124), it did visualize ecological processes and the dependence of life on ecosystems. The Harrisons themselves also recognized the limitations of this work, as "they realized that farming under lights was too energy-expensive" (Spaid 2002, p. 33).

The Harrisons' subsequent work, the *Lagoon Cycle* (1972-1984) included a much more thorough scientific work (for which they received a science grant from the Scripps Institute of Oceanography) on the complexity of ecosystemic conditions necessary for sustaining the breeding cycle of a

22 Cf. ed. Zinggl 2001.

specific species of crab (a question then unresolved by scientists). As an increasingly careful attention to complexity became part of their work, its ecological relevance also expanded. As Bijvoet observes: "The small-scale portable farming pieces extended into planning whole ecosystems. What were at first personal explorations in search of a new art form, developed into large-scale research projects into global survival plans, with proposals and plans, collaborative actions and political and social discourse" (Bijvoet 1994, p. 125). European consecration to the US-American couple came later than in the USA, most especially with their invitation at the European Mecca of contemporary art, i.e. the Documenta in Kassel: The Harrisons presented at "Documenta 8" (1987) an urban renewal project aiming to connect the city-center of Kassel to its river, the Fulda. From 1996 to 1998 was realized *The Endangered Meadows of Europe*, a roof-garden/meadow on top of Bonn's Kunst- und Austellungshalle des Bundesrepublik, which consisted in a transplantation of an endangered meadow from the region of Eifel. Its success in terms of biodiversity preservation led to the extension of the meadow into the Rheinauen Park (also in Bonn) in 1997.

About the Lagoon Cycle

The *Lagoon Cycle* worked as a contradictory narrative, with an exchange between a "Lagoon-Maker" proposing technological solutions for ecosystemic restoration, and a "Witness" critically assessing and questioning these proposals. Other ideal-typical characters are also involved, characterizing the working of market economy. The *Lagoon Cycle*, a story-telling process involving drawings, maps and a wide-ranging research-background, aims to

"create a 'world' that reaches out into many different 'regions' (territories, disciplines, space and time, etc.) both real and imaginative [... which] can be perceived in relationship to one another, including the artists themselves. These relationships, however, are subject to processes, and to change [... and] relative positions come forth in the dialogues/discourses between the two protagonists in their views of structure and content vs. process and context" (Bijvoet 1994, p. 126).

Furthermore, the *Lagoon Cycle* works towards a way of thinking that is less atomistic, and more relational, as Bijvoet suggests (ibid., p. 127). And the attention to these relations is what characterizes the art of the Harrisons, as Michel de Certeau argued in the exhibition catalog to the *Lagoon Cycle*:

"Simply paying attention guarantees the transformation from a nature supposedly asleep to the work that displays nature's strange vitality. Art is what attention makes with nature. The work of Helen and Newton Harrison is based on that premise. [… M]aps, metaphors, narratives, dialogues, politics. Those elements call out to each other. Together they form a cycle. They intertwine. To distinguish them is to bring out the movement that puts them in play together" (de Certeau, in Mayer Harrison and Harrison 1985).

Or, as de Certeau writes later in his text: "*Between* is the keyword". The relationships set in motion are not only at the level of observations. Indeed, as put by de Certeau, "[t]he art of observing is combined with the art of inventing. The vision that analyzes is joined with the vision that prophesies" (ibid.). Furthermore:

> "Everywhere the Harrisons introduce into their representation of things another way of seeing them: there are photographs, but they're painted; there are scientific maps, but they're revised (re-envisaged) by the future that is drawn on them. The technical representation of an area is a tool the artist uses to see as it will be, or as it could be. To see things in that way, in the movement that opens up another space for them, is to see them metaphorically" (ibid.).

Another critique addressed to the Harrisons in the late 1970's (e.g. by critique Kristine Stiles – cf. Bijvoet 1994, p. 129) was their lack of genuine involvement with the local communities which they were addressing in their work. Such critique was less relevant in their later work, the *Baltimore Promenade* (1981), where the Harrisons proposed several walkways through disregarded areas of Baltimore. They organized "a number of 'physical walks' (with city officials, among others), as well as meetings to instigate public discussion" (Bijvoet 1994, p. 129), thereby engaging with the community and confronting the people's mental maps of the area with the physical features of the area as explored by the artist-couple. As Bijvoet comments, this social-political level of involvement "has added another dimension to their work as well as to their function as artists".

In 2004, the Harrisons themselves characterized their work as

> "address[ing] the co-evolution of biodiversity and cultural diversity most often, though not always, at watershed scale. [...] We believe that in a well-functioning system, cultural diversity and biodiversity exist in a state of mutual interaction – the former self-conscious and able to intend and transform, and the latter the pattern of self-organization from which we all spring and to which we all return, and which ultimately determines the possible" (Mayer Harrison and Harrison 2004).

Several of their projects exemplify this mutual relationship between biological and cultural diversities, e.g. their above-mentioned project with meadows in Bonn: Indeed, meadows are a biologically diverse co-production by hundreds of years of human activities (e.g. forest clearing, livestock grazing, hay cutting) and the co-adaptation of other species.[23] From their perspective, valuing these diversities, the Harrisons take a critical stance towards globalization insofar as it reduces diversity and thus "the possibilities to improvise".

The Harrisons explicitly pay tribute to the influence of systems thinking and complexity theory on their work: "Aside from the influences of our basic education in literature and mythology, the visual arts and ecology, we consider ourselves the beneficiaries of systems theory and systems thinking" (Mayer Harrison and Harrison 2004). They mention Norbert Wiener and Von Neumann (i.e. early cyberneticians), as well as Maturana and Varela (the later generation of system theory based on 'autopoïesis') and the cognitive scientists Lakoff and Johnson (who base their work on complexity theory).

The Harrisons typically determine the boundaries of the area under focus in a project, with the question "How Big is Here?", which determines which kinds of patterns will emerge at anthropic and ecological levels. They also ask "How long is our Now?", addressing one or several time dimensions of phenomena.

The Harrisons insist on being able to pose those initial questions themselves, following an open invitation: "always, the overarching concept is envisioned by ourselves". They benefit from artistic autonomy in this sense, i.e. not entering into a project where "the space, time, context and agenda are determined by others" from the outset. This is an art-related convention that the Austrian art group Wochenklausur also stresses.[24]

The Harrisons' approach is problem-solving oriented, concerning the cases they are working on, but is also metaphorical (as de Certeau argued), insofar as they consider their artistic work as transcending the single ecosystem they work on in a given project: "By understanding a simple system, the Harrisons sought to understand, and create understanding about the larger more complex systems of life" (Bijvoet 1994, p. 130).

As exemplified in the *Lagoon Cycle* (cf. boxed text above), the work of the Harrisons is based on the "belie[f] that every place is telling the story of its own becoming" and more generally on a belief in storytelling, with "a sense of the mythic and iconic" (Mayer Harrison and Harrison 2004). But it does not take the form of a mere monologue-discourse.

"By using the conversational form, with diverse proposals embedded in it, we have been able to, in the main, avoid much of the sense of the authoritarian or coercive

23 The importance of livestock grazing for biodiversity, especially for plant species, and especially in times of climate change, is also a topic on which I myself worked recently with colleagues and students, in the framework of the "Karamoja Campaign" and with the art installation *Caution Border*: Cf. Eds. Knaute and Kagan 2009 and the website www.karamoja.eu.

24 Cf. ed. Zingg 2001.

that can be present in public projects or planning operations. [Furthermore,] the conversations we begin, and often evolve in elaborate ways, can then be more easily adopted by others. Thus it can drift away from us and develop a life of its own. [This allows] diverse outcomes for any work [and] is a way of being non-possessive and sharing authorship quickly and easily [and it also allows] to surrender the idea of closure, of finishedness" (ibid.) usually associated to an identifiable 'work of art'.

Non-possessiveness is necessary for effectiveness in influencing policy-making, as the Harrisons explain about the project *Green Heart Vision* in the Netherlands: "The work continued to be ours in the catalog or exhibition or brochure, but, in order to be adopted, it has to reappear as the idea of others who take ownership" (ibid.). The core concept of the Harrisons' *Green Heart Vision* was to set up a biodiversity ring / biodiversity corridor all around the famous 'groene hart' surrounding the Randstad, balancing economic (e.g. housing needs) and ecological concerns.

The work of the Harrisons has become increasingly political with the years, as they aim to influence and modify political decision-making from the local level, as with *Baltimore Promenade*, and regional/state level, as with *Green Heart Vision* and with *Barrier Islands Drama: The Mangrove and the Pine* (at the Ringling Museum in Sarasota, Florida in 1982, which "was partly responsible for the banning of the so-called Australian pine from South Florida", according to Spaid 2002, p. 34), to international levels, as in *Peninsula Europe: The Force Majeure* (from 2008 on), which aims no less than to reform the ecological-political governance of the whole of Europe in the context of climate change and other inter-linked ecological crises.[25]

"Always the strategy has been to impact the planning process in a region, as the cultural landscape is so often formed by them. [...] Our methodology [...] fundamentally it is designed to address, infect, invade, transform and expand planning processes. [... T]he work is often necessarily text heavy. Typically what we do must convince planners, politicians, ecologists, and diverse bureaucrats, as well as the art community and everyday people, that what we propose is worth doing" (Mayer Harrison and Harrison 2004).

Such a process can take a long time, as in the case of *Peninsula Europe: The Force Majeure*, which went through several phases and is still under way, as the artists' website documents:[26]

- *Peninsula Europe I* (2000-2003), whereby the challenges of climate change for the high grounds of Europe and their watersheds, were stud-

25 Cf. personal communication with the Harrisons about the "Force Majeure" series of work, October 2009.

26 Cf. http://www.theharrisonstudio.net [accessed January 21st 2010].

ied (in collaboration with European scientists) and led to the proposal of a Europe-wide reforestation, with a "3-4000 square foot exhibition" together with the Schweisfurth Stiftung (a German association supporting sustainable agriculture). One of the elements was a video, realized in 2001 by the Harrisons, and based on research by Prof Dr Georg Grabherr (Director of the department of vegetation, ecology and conservation biology at the university of Vienna) and his students. The video tells the story of plants migrating up the slope of a mountain (the Schrankogel), with rising temperatures, eventually reaching the top and disappearing altogether. This video, entitled *The Mountain in the Greenhouse*, was also shown in 2007 as part of "Weather Report", an exhibition on climate change curated by Lucy Lippard for the Boulder museum of contemporary art, in collaboration with EcoArts.[27]
- *Peninsula Europe II* (2007), with further research-based work commissioned by the UNEP, on possible climate change-induced droughts in Europe and on mitigation opportunities related to European river systems. The work was exhibited at the Nobel Peace Museum in Oslo.
- *Peninsula Europe III* (initiated in 2008), a research-proposal for a sort of Master-Plan based on European high grounds rivers ability to act as a buffer against global warming, thanks to a trans-European forest yet to be conceived.
- *The Force Majeure* (2008-2009), an exhibition with installations combining the Harrisons' research on "Peninsula Europe" and two other climate change-related projects of theirs (*Tibet is the High Ground* and *Greenhouse Britain*).

The Harrisons use some technologies in their work, but do not thereby allow any techno-optimism. "The idea that technology is able to buy us out of problems is an illusion. We are going to have to make vast changes in our consciousness and behavioral patterns" (Newton Harrison, quoted in Bijvoet 1994, p. 126).

Mierle Laderman Ukeles

The work of the US American artist Mierle Laderman Ukeles is often cited as exemplary of social-ecological art intervention.[28] In 1969, Ukeles wrote a manifesto entitled "Maintenance Art—Proposal for an Exhibition" addressing gender stereotypes about 'housewife' roles and pointing at ecological issues (with the notion of "Earth maintenance"). In her manifesto, Ukeles quoted a Balinese Saying: "We have no art, we try to do everything well". She opposed the "death instinct" of modern art to the alternative of a "life instinct": "The Death Instinct: separation; individuality; Avant-Garde par

27 Cf. eds. Lippard and Gerdes 2007, pp. 56-57.

28 Cf. e.g. Gablik 1991, pp. 69-75; Dieleman 2008, pp. 129-130.

excellence; to follow one's own path to death—do your own thing. [...] The Life Instinct: unification; the eternal return; the perpetuation and MAINTENANCE of the species; survival systems and operations; equilibrium" (Ukeles 1969). To the death instinct of modern art, Ukeles further associated, in her manifesto, the notions of "Development: pure individual creation; the new; change; progress; advance; excitement; flight or fleeing". To the life instinct, she associated the processes: "preserve the new; sustain the change; protect progress; defend and prolong the advance; renew the excitement; repeat the flight", already formulating the theme of dynamic balance, rather than linear movement, which is also an element of an ecological literacy as formulated by Capra (cf. chapter 2, section 3). She also proposed to link the "personal" (e.g. her personal maintenance work in everyday life), the "general" (i.e. maintenance in social life, from sanitation work to education and health work) and the "Earth" (i.e. the general ecosystemic support of human life, which pollution urges maintenance work).

From then on, she called herself a "maintenance artist".

"By accepting the reality of her situation as a necessary role in maintaining the household, she discovered the reality of maintenance as a means to the survival of personal freedom, art and all other social institutions. In other words, maintenance art was a necessary part of the human condition. Through this approach to the problem, Ukeles began to extend the references in her work outside of a purely feminist content in order to reveal the conditions of work, and the stereotypes handed to maintenance workers on all levels, whether in public, private, or corporate enterprises. [...] As this position extended from the household system to that of public sanitation, the artist performed in museums and office buildings and eventually in the streets of New York City" (Morgan 1998).

In 1976, she realized "I Make Maintenance Art One Hour Every Day", working for two months with sanitation workers, i.e. assisting 'cleaning women' and janitors in cleaning floors and elevators. From 1978 onwards and for over 30 years, Ukeles has been working as nonsalaried artist-in-residence at the New York City Department of Sanitation. In 1979 and 1980, over eleven months, eight hours per day, she did the intervention/performance "Touch Sanitation", whereby she shook hands with all the workers of the Department (i.e. so-called 'garbage men'), i.e. 8500 persons across five boroughs of NYC. She did not just shake their hands, but performed each time a sort of "ritual in which I faced each person and shook their hand; and I said, 'Thank you for keeping New York City alive'" (Ukeles, quoted in Gablik 1991, p. 70). The handshake ritual connected the complex social system of a megacity's sanitation efforts, with the personal human level embodied by each worker's hands. The workers union thanked her, making her a honorary teamster. During that work, she also heard from the workers, about the bad names they were being called. This led to another installation-performance work, in 1984 at Ronald Feldman Gallery, in New

York ("Cleansing the Bad Names"): One element of the installation was that the gallery's windows were covered with these 'bad names', and Ukeles handed sponges to everyone present at the exhibition's opening, inviting them to help her clean those windows right away. She also realized in 1983 "The Social Mirror", covering a fully-operational garbage collection truck with mirrors, thereby reflecting back to the passers-by, in the streets of NYC, their own part in the process of garbage production.

Since then, Ukeles continued her work, for example working on the rehabilitation of diverse landfills, being commissioned to work further on sanitation issues by the city of Givors in France ("Re-spect", 1993) and approaching other topics such as peace building ("Unburning Freedom Hall", 1997). Ukeles is interested in waste landfills as "the basis of our whole culture – buy, buy, buy, and waste, waste, waste" (Ukeles quoted in Spaid 2002, p. 93). She considers landfills as social sculptures, to which entire societies contribute.

"The design of garbage should become the great public design of our age. I am talking about the whole picture: recycling facilities, transfer stations, trucks, landfills, receptacles, water treatment plants, and rivers. They will be giant clocks and thermometers of our age that will tell the time and health of the air, the earth and the water. They will be utterly ambitious – our public cathedrals. For if we are to survive, they will be our symbols of survival" (Ukeles, quoted in Spaid 2002, p. 94).

SECTION 3: EXEMPLARY DIRECTIONS IN ENVIRONMENTAL/ECOLOGICAL ART

Across the past four decades, the interplay of art and ecology was explored along different, but complementary directions: To introduce this creative diversity, I will focus on the cases of eight artists whose work (from the 1970's onwards) exemplify these different ecological dimensions in contemporary art: Patricia Johanson (who was a pioneer at the crossroads of environmental art and urban planning), Herman Prigann (who coined the notion of 'metamorphic artworks' and realized numerous site reclamation projects), Aviva Rahmani (who developed a methodology of ecological restoration), Lynne Hull (who introduced 'Trans-Species Art'), David Haley (who developed an understanding of 'ecopoiesis'), Shelley Sacks (who further developed Joseph Beuys' concept and practice of 'social sculpture'), James Turrell (as an example of the interest of Earthworks) and Andy Goldsworthy (as an example of the interest of 'art in nature').

Patricia Johanson

In a monography by feminist and ecological artist and writer Caffyn Kelley (based in Vancouver) about the work of US American artist Patricia Johanson, curator Lucy Lippard introduces Johanson as "a seldom-acknowledged pioneer [of eco-art]. Her writings of the late 1960's, when she was still in her twenties, are a cornucopia of possibilities for environmental art and planning that are still being 'discovered' today" (Lippard, in Kelley 2006, p. vii).

Along the 1970's Johanson complemented her artistic expertise with studies and work in the fields of engineering and architecture, and was developing ideas and designs for public infrastructures, parks and walkways. Until the late 1970's, none of these ideas were turned into concrete projects, but they eventually "paved the way for an entirely new approach to integrating nature and the urban infrastructure" (Spaid 2002, p. 65). Johanson started realizing concrete public projects from the 1980's onwards, thanks to public art commissions, while most of her ideas and designs were still initially shown in art galleries and discussed there as "visionary fantasies". Kelley notes that "[e]very facet of a Johanson project is designed to perform multiple functions: cultural, social, infrastructural and environmental", and observes that Johanson's designs therefore generally aim to "cleanse water and create habitat for many species, while providing multiple economic and social benefits for human inhabitants" (Kelley 2006, p. 3 and p. 10). Lippard further claims that Johanson's work offers multiple levels of experience: "The shifting patterns and kinesthetic macro-energy unfold into a series of experiences rather than a single picture. She talks about the very multiplicity of every place, which allows everyone (all living things) an entrance into it, when the way is shown" (Lippard in Kelley 2006, p. viii).

Lippard also praises Johanson's ability to "communicate [her] sense of the local [...] to the people who will live with her work", as well as her engineering ingeniousness (ibid., p. xi). This sense of the local is based on careful research on each site. "I never design until I have discovered the meaning of the place. Each place has a unique set of conditions, and we need an intimate understanding of what it has been, is now, and will become in the future, in order to create a design that is more than a willful act" (Johanson, quoted in Kelley 2006, p. 19).

Furthermore, Johanson is attentive to natural processes and does not try to design as-in control, the whole process. "The most important aspect of my art is in the parts I do not design" (ibid.). In this, she had to go against her art education.

"We are so unwilling, as artists, to let anything interfere with our grand vision. But anything that's alive has its own set of patterns and needs, and it's important not to expunge that in the process of creating our ideal works. [...] Motherhood taught me that the role of the creator, while important, is limited. You are heading into a long process. This is very different from what I was taught in art school – that you aim for perfection; your genius is recognized; and your work will be preserved by a grateful world forever. For me, incorporating unplanned experiences makes the world richer" (ibid., p. 53).

This also implies, as Kelley notes, that "Johanson relinquishes control of her work not just to nature, but also to budgets and community processes" (Kelley 2006, p. 53).

Johanson claims to also focus on the linkages between the local and the global:

> "Beyond accomplishing any particular project, I always have goals that reach out to the whole world and its future. Small projects ask large questions, and point out that the world is of a piece. You can't do something in one place, whether it's positive or negative, without repercussions throughout the whole system. Each of my projects is a model for an inclusive, life-supporting, self-sustaining world" (Johanson, quoted in Kelley 2006, p. 14).

Among the concretely realized public projects discussed at length by Kelley (2006), is *Fair Park Lagoon* (Dallas, Texas, 1981-1986), whereby Johanson rehabilitated a large body of water in the middle of the city – which was murky, polluted and even dangerous to passers-by – literally bringing it back to life with the (re)introduction of local species and design of attractive walk-paths on the water, which "sculptural forms simultaneously control bank erosion, serve as paths and bridges over water and create microhabitats for a wide variety of" plant and animal species (Johanson quoted in Kelley 2006, p. 25). These species also improve the water quality, and the "lagoon also serves as a municipal flood basin, thus familiar forms and paths of travel are frequently altered by fluctuating water levels" (ibid.).

Johanson's research (together with the Dallas Museum of Natural History) prior to intervening on this site led her to understand that the "lagoon had died because its food web was out of balance [...] middle food-web species were not present, largely due to the absence of a littoral zone, which is composed of vegetation and supports 75% of a pond's life [...] Her 1981 design, which seems like common practice today, was totally experimental back then" (Spaid 2002, pp. 66-67). Her plan included not only the introduction of selected native species, but also the removal or reduction of other species from the site (moving Asiatic ducks to another location, preventing the proliferation of floating algae and encouraging the fishing of sunfish), and halting certain practices such as the fertilizing of strips of grass around the lagoon.

This project was also a success in terms of visitors satisfaction, and the place has become popular for families. Furthermore, according to an empirical study of the lagoon's visitors by Jacqueline Zanoni de los Santos carried out in 1996 and 1997, the observed visitors not only enjoyed the site but also discovered the species and biological and ecological processes on the site.[29] More generally, Kelley characterizes the public spaces created by Johanson, as "democratic meeting grounds" where human and non-human

29 Cf. Zanoni de los Santos 1997, cited in Kelley 2006, p. 25.

communities are catered for, and encountering each other (Kelley 2006, p. 40).

However, Johanson not only aims to contribute to ecological and social restoration and to visitors well-being, but also to confront them to ecological imbalances, and thus some of her projects aim to visualize problematic ecological issues (e.g. *Toxic Rainbow*, *Garden of Organized Killing / Soil Fertility*).[30] In 2000, she stated: "I have become increasingly interested in landscapes that confront us with the world as it exists, rather than those that think only in anthropocentric and aesthetic terms, which is ultimately not to our benefit" (Johanson, in Kelley 2006, p. 26). In the San Francisco bay, she concretely realized the project *Endangered Garden* (*Sunnydale Facilities –* 1987-1997), a sewer with a bay-walk and life-supporting habitat for endangered species. About this project, Kelley argues that it "shows that art can support functioning ecosystems rather than perpetuate an ideal view of nature" (Kelley 2006, p. 29). More generally, she argues that Johanson's projects, "[i]nstead of conceiving of the wild as a space preserved outside ordinary experience", rather integrate the "wild" nature into "the physical structures of our contemporary human society [...] becoming part of the everyday reality of our lives" (ibid., p. 38).

About *Endangered Garden*, Kelley also describes how Johanson had to fight her way through the engineers and administration, to be accepted as co-designer of the entire sewer-system and not just "decorate" some part of it.[31] Johanson also had difficulties with environmentalists who opposed her conception of a mutually beneficial cohabitation of humans and other species, in this project with the specific case of the shellfish. Kelley recalls how these environmentalists actually destroyed the shellfish population with policies forbidding local Native Americans from performing their traditional annual harvesting (with the scientists realizing the mistake only too late). Johanson comments: "The preservationists always want such complete control that they diminish what they have. I have as much trouble with people who are narrow preservationists as I do with people who don't understand anything about ecological communities" (Johanson, quoted in ibid., p. 35). Wise words indeed, echoing the culture of sustainability outlined earlier (in chapters 2 and 3).

Another quote of Johanson illustrates her sensibility to complexity as expressed in her aesthetic work:

"Creating overlapping patterns within the same composition is an idea I use again and again in different ways. Because these patterns unfold in time, you may need to reach the end of the composition before you see how the beginning is connected. With accumulating knowledge, the patterns keep shifting, so you are putting it to-

---

30 However, these two projects have, to my knowledge, not been concretely realized.

31 Cf. Kelley 2006, pp. 26-29.

gether in your mind as you go - very much as you experience music" (ibid., pp. 45-46).

Johanson (who studied first music at college) refers to music here in a way similar to Edgar Morin in his metaphor of the "musical ear" (cf. chapter 4, section 3).

Herman Prigann

Hermann Prigann was a prominent German (and Spain-based) environmental artist from the 1980's onwards, with both large-scale landscape restoration projects (especially on former industrial sites) and smaller, more "minimalist solutions" work as he used to call it, in other forms of landscape (i.e. closer to the "art in nature" kind of work as by e.g. Goldsworthy). His best-known large-scale work is the *Skulpturenwald Rheinelbe* (1996-2000; part of the *Terra Nova Project*), in Rheinelbe, Ückendorf in a former coal mine very close to the city of Gelsenkirchen that had been turned into a garbage dump. A land reclamation project, *Terra Nova* turned the site "into an archeological field replete with traces of former buildings, stone sculptures, and a major landmark called the Skystair" (Margolin in ed. Smith 2005, p. 23). The 2004 publication edited by Heike Strelow et al., *Ecological Aesthetics*, to which I already referred in the previous chapter, is largely dedicated to, and focusing on Prigann's work. Before his environmental work of the 1980's, Strelow explains, Prigann was already active in the 1960's and 1970's, being for example at the origin of "Release", a self-help organization for drug addicts (with artists and social scientists involved) which opened in 1969 in Hamburg and spread throughout West-Germany, but was forced by the authorities to dissolve itself in 1974. That group was experimenting with alternative lifestyles, based on a critique of "consumer culture and throwaway society and its resulting harm to the environment as well as the authoritarian behavior patterns of the state".[32] Strelow compares that social experiment to Joseph Beuys' concept of 'social sculpture'. From the 1980's onwards, Prigann developed ideas on "Metamorphic Objects / Sculptural Places" both implemented on derelict industrial land such as former mining areas, and in urban public space (as with *Duftender Meiler*, 1985, a fuming charcoal pile left for five weeks in front of Vienna's city hall). For the smaller works in nature and accompanying workshops/symposia (e.g. *Parler de la nature – c'est parler de l'amour*, 1996, Luberon, France; *Perception and the Experience of Landscape*, 1997, Ankara, Turkey; *Perception and Pattern*, 1999, Salamanca, Spain), Prigann worked in collaboration with the artist Vera David, who in eds. Strelow, Prigann and David 2004

32 Quote extracted from Herman Prigann's website Terra Nova, available online at: http://www.terranova.ws/releas1.htm [last accessed: 06.01.2010]. Cf. also eds. Strelow, Prigann and David 2004, p. 10.

(pp. 50-55), develops a discourse on the experience of nature and poetic dialogue with nature developed in these workshops and installations: a discourse which comes very close in spirit to the understanding of "aesthetics of sustainability" as I proposed it in the previous chapter: Vera David too refers to phenomenology (with an interest in "sensitizing attentiveness, wakefulness") as well as to Bateson's *Mind and Nature,* and she argues that "the concerns of ecological art are the perception of nature's beauty as a landscape experience, its atmospheres, the great variety of interaction in natural processes and processes between humans and nature, transformation and metamorphosis" (David in eds. Strelow, Prigann and David 2004, p. 50). To come back to Prigann: In a text from 1998 about 'art in nature', Prigann formulated the nature/culture relationship especially suggestively: "Nature is the Dance of Evolution and Entropy. Culture is One of the Dancers" (Prigann 1998).

Prigann's focus on metamorphosis stems from an ecological philosophical discourse: He argues that "even though we suppress it [with the Platonic Ideas, reason and concept of the stable human soul], we humans still are and remain "ephemeroi", subjects of change and metamorphosis. The aesthetic of the ephemeral, the integration of metamorphosis into working artistically in dialogue with nature can help us to once again see this truth of our existence in the world" (Prigann in ibid., p. 74). Prigann proposes an aesthetics of metamorphosis as the basis for

> "a new capacity for aesthetic judgement [which] is being demanded [in the context of the ecological crises]. The integration of metamorphosis as a deciding principle of creation ('Gestaltungsprinzip') in artistic work alters our interpretation of art and expands our perception of the environment. The environment is then no longer merely panorama for the artwork. Instead both pervade one another. Thus nature and art become one entity in dialogue" (ibid., p. 75).

Prigann aims to make this metamorphic process visible, to make "presence palpable" and have "the artwork communicate its connection to the living world" with "the traces of its history, the amorphous, the changeability, that lends things their presence. Thus it is precisely the metamorphic artwork that evokes memories of the transience and the change in and of everything" (ibid.). In the case of the Rheinelbe former mining site, Jörg Dettmar describes Prigann's "way of working [as] using materials typical of the space, installing new objects based on relics from the site's former use and demolished material from old industrial buildings, occupying spaces and then leaving them to change 'naturally'" (Dettmar in eds. Strelow, Prigann and David 2004, p. 130). Prigann worked on several other comparable projects in the late 1990's, such *as Die gelbe Rampe* (1993-1995, on an open cast mine near Cottbus, Germany), *Türme der Wandlung* (1995, on industrial wastelands near Athus, Belgium), *Das große Steinfeld* (1999, on Mont Cenis Mine, Herne, Ruhr area of Germany), and *Wasserstände* (at the Marl chemical complex, Ruhr area) and more... Some designs/proposals were not concretely implemented, such as *Die Mottbruchhalde* (1999, on the Mottbruch Coal Stocks, Gladbeck, Ruhr area), and most especially the 'reclamation art' project *Museum der verlorene Wünsche* (1996-1999): the project of a gigantic mountain/volcano and a pyramid made of landfill (i.e. waste), integrating a museum, a research lab and a café, on 170 hectares for the city of Berlin.

Prigann's work neither 'camouflages' the traces of industrial history, nor 'imitates nature', but is "culturally conscious and therefore artificial" (Prigann in eds. Strelow, Prigann and David 2004, p. 214). Discussing his overall concept for the "creative reshaping" of "destroyed landscapes" (under the title *Terra Nova*), Prigann proposed to develop an interdisciplinary working methodology with the integration of an on-site established "campus", landscape art, new job opportunities in the environmental field, experimental "large-scale wetlands", "watergardens" (for water management and water purification), and reed cultivation, aiming at an "aesthetic-ecological transformation" of depleted landscapes.[33]

Aviva Rahmani

Aviva Rahmani's artistic work started in the 1970's, as a performance artist working in collaborations (e.g. with Allan Kaprow, Judy Chicago, Suzanne Lacy) and interested in feminist issues. I will however focus here on Rahmani's work since 1990, when she moved in to live on the Island of Vinalhaven (on the coast of Maine), and initiated a cycle of research and work that now amounts to two decades of ecological restoration.

In her ecological art work, Aviva Rahmani links the micro, the meso and the macro geographical levels, and her ongoing work is shaping a theory of "trigger points":

- at the physical-geographical level, specific highly-degraded areas which ecological restoration can have exponential effects at a wider level: local restoration on Vinalhaven Island, hopefully affecting further coastal wetlands (*Ghost Nets*, 1991-2000); multi-regional restoration hopefully affecting the entire Atlantic Ocean and the planet (*Cities and Oceans of If*, from 2001 on);
- at the social level, specific social processes in which intervention may have global impact, such as the International Climate Conference (the so-called 'COP'), in which Aviva Rahmani is directly involved since 2009 (in this respect, Aviva's trigger points also bear a relationship to

33 Cf. Prigann in eds. Strelow, Prigann and David 2004, pp. 214-215.

> the general notion of 'tipping points' which was already explored in the social sciences as well as in climate science).[34]

The trigger points theory is rooted in an intuition of the 'pattern that connects', and in Rahmani's affirmative proposal that, if the environment has been lost by increments, it can also be restored by increments. It builds on 20 years of work so far, and is still a work-in-progress (as of 2010), as Rahmani embarked on a PhD thesis at the 'Z-Node' network to further formulate her theory, as a transdisciplinary art&science-based research process.[35]

With *Ghost Nets*, Rahmani worked over ten years on a specific area of the American Eastern coastal wetlands, hoping that the reconstruction of wetlands at specific locations in the Gulf of Maine would start a wider recovery process, with not only ecological benefits, but also sustainable economic benefits for a fishing industry in decline.[36] *Ghost Nets* consisted in the transformation of a town dump on Vinlahaven island, into a salt marsh. Rahmani comments:

> "It was a connecting corridor for wildlife in a larger regional hotspot. I calculated that the restoration would reinvigorate the larger landscape for the Guld of Maine, that connected the North and South American continents via sea animal migration patterns. I also speculated that when we were done, we might calculate how many square feet of restored wetlands would generate how many saleable fish to promote such restoration amongst landowners within the fishing industry" (Rahmani, in eds. Kagan and Steinbrügge 2007, p. 21).

---

34 On tipping points in climatology, see Lenton et al. 2008; the term was introduced in sociology in Grodzins 1958. This term, which was already used in physics earlier, was also related to the concept of "phase transition" in thermodynamics (i.e. the transformation of a thermodynamic system from one state of matter to another).

35 A description of Aviva Rahmani's work of the past two decades, at a much more detailed level than possible here, was carried out by Rahmani herself in a publication I edited in 2008: Cf. Rahmani in eds. Kagan and Kirchberg 2008. My exchanges with Aviva Rahmani started in 2007 when she participated (virtually, by videoconferencing, to avoid flying) in the conference that we organized at the University of Lüneburg for the European Sociological Association's Research Network for the Sociology of the Arts, and (also virtually) in the day of talks and workshops that we (i.e. the newly created "Cultura21 International") organized at the Arsenale Novissimo in September 2007 by invitation of curator Lucrezia de Domizio Durini, as part of a side-event at the Venice Biennale.

36 The word "ghost nets" stands for these lost or abandoned fishing nets, left drifting in the gulf of Maine (and worldwide), which are not biodegradable, and slowly kill ever more fish, as an absurd sort of "ghost fishing" phenomenon.

Starting in 1991, Rahmani conducted preparatory research with natural scientists for six years, and immersed herself in the local fishing community: "She cultivated relationships with her neighbors by going out on fishing boats, conducting extensive interviews, and singing with the local church choir" (Spaid 2002, p. 115). In 1997, the concrete bio-engineering work was implemented by a team of ten people over three days, and then improved the following year. It consisted first of all in cleaning up, "by removing garbage and other barriers to natural water flow [...] sixteen truckloads of granite rip rap were taken away [which] had been dumped on the site" (Rahmani 2009); and then in re-vegetation (with e.g. salt marsh grasses) and several wave-attenuation barriers. After a decade of work, the ecological results were very positive, as Rahmani observes: "In 2000, Dr. Michele Dionne, Director of Research at the Wells National Estuarine Reserve, Wells, Maine [...] monitored the site and found all 18 indicator species of healthy estuarine function" (Rahmani 2009). This allowed to replicate the method for another, ten times bigger site (10,5 hectares) with a half-million dollar budget from the US Department of Agriculture *(Blue Rocks*, 2002, presented as part of *Site Specific* curated by Pat Nick on Vinalhaven Island*)*.

With *Cities and Oceans of If*, Rahmani is "imagining the watersheds on either side of an ocean as a linked system, rather than as separate waterways, and then understanding the waterways as chakras with vital trigger points" (Spaid 2002, p. 117). Rahmani argues: "Just as the human body has meridians of electrical energy that can be activated with acupuncture, ecological studies tell us there are 'hot spots' of biological diversity in the global body upon which larger systems depend" (Rahmani, in eds Kagan and Steinbrügge 2007, p. 21). Rahmani's work for *Cities and Oceans of If* was conducted over several years mainly virtually, over the Internet (with "virtual residencies" in Korea, Italy and India in 2006), in order to reduce flying. Since hurricane Katrina, and until her decision to fly again for the COP15 Copenhagen conference in late 2009, she committed strictly to her no-flying constraint, despite the constant (and typical) pressure-to-fly characteristic of the working and "life" style of the art world of contemporary art.[37] The research, carried out in collaboration with natural scientists, involves the use of GIS (Geographical Information Systems), satellite and other mapping technologies, and evaluations on the ground, to identify the most degraded areas and prepare their restoration, in different regions around the world.

---

37 More generally, the absurdly extreme mobility of art curators and performing artists is a known issue for the art world, and has become an increasingly problematic issue even for the art organizations dedicated to promote artistic mobility, in the context of climate change (personal communication with Mary-Ann de Vlieg, Director of the IETM network, and Guy Gypens, Director of the KAAI Theater in Brussels).

From the experience of *Ghost Nets* and *Cities and Oceans of If*, Rahmani (2009) describes the phases in her working process as:

- Conceiving and choosing the site (and analyzing the site together with scientists);
- Cleaning the site (and bioengineering);
- Watching the site (i.e. monitoring its ecological evolution) and "contemplating the site" (aesthetically, sensually and slowly);
- Applying lessons from the site (for other sites);
- Connecting the sites (i.e. identifying their linkages, e.g. with water ways);
- Proposing a new site (which sometimes can take the form of an open invitation to exhibition visitors, as in *Pervious, Porous, Integrated*, 2006, at Exit Art Gallery, NYC where the installation included "indigenous plants for sale, to be planted on Tenth Avenue [...] to hold the soil when flooding might occur");
- Comparing sites (as in *Trigger Points / Tipping Points*, 2007, a comparison of the Nile, Ganges and Mississippi as conflict zones related to major rivers, together with Dr. Jim White of the University of Colorado at Boulder);[38] and
- Networking the site (discussing across sites, with virtual Internet-based technologies – video conferencing and desktop sharing, with stakeholders, "to find new insights and affect policy", as in *Gulf to Gulf*, 2009, again a collaborative project of Rahmani and White, comparing the Gulfs of Maine and of Mexico and other gulf sites around the world).

In December 2009, Rahmani was an accredited participant (with the University of Boulder, Colorado) at the COP 15 in Copenhagen, i.e. the "Conference of Parties" which tries to establish a world policy in response to the global challenge of climate change. This recent extension of Rahmani's work is ongoing as I write these lines, and Rahmani is currently preparing for COP16 which will be in Cancun (Mexico) in November 2010.[39] From the Copenhagen experience, one of the social dimensions which retained her attention is the importance of legal language in the elaboration of international treaties, and its connections to ethical dimensions of language and to

38 That comparative work was featured in several forms, at different locations: A DVD presentation was premiered at a one-day event organized by Cultura21 (Francesca Cozzolino, Davide Brocchi and myself) as part of a side-event (curated by Lucrecia de Domizio Durini) at the 2007 Venice Biennale, and shown again at *Feeling the Heat* (Deutsche Bank, NYC, 2008). An installation was made for the exhibition *Weather Report* (2007) curated by Lucy Lippard at the Boulder Museum of Contemporary Art, and shown again at *In Transition, Russia* (National Center for Contemporary Art, Moscow, 2008).

39 In the end, Rahmani had to renounce going to Cancun for COP16.

artistic work on the polysemy and potentials of language. She is now exploring the opportunities and implications of introducing certain words (such as "aspirations") into the treaty language.[40]

Lynne Hull

The artistic work of formerly Wyoming-based (and nowadays Colorado-based) environmental artist Lynne Hull is not merely intended for human enjoyment, but also for enjoyment by other species: Since 1983, she practices a new category of art: "trans-species art". As her argument goes, human civilization has done a lot to benefit itself, and must now work to benefit other species. "And if art is a leading edge of civilization as we hope it is, was it possible that this idea of trans-species action could be done as art?" (Hull quoted in Gablik 1991, p. 88). Lynne Hull indeed belongs to the movement of ecological and environmental artists who critically reflected on the precedent of Land Art. "I've had a long-standing in earth or site-specific art, but too often it seemed so egocentric -on a grand scale, to go out and abuse the land in the name of art [...] I felt a growing need to make a positive gesture to the earth. Couldn't there be a small-scale, nurturing, perhaps even 'feminine' land art?" (Hull quoted in Gablik 1991, p. 89). Like the Harrisons who argue that, whoever invites them, their real client is the land itself, Hull states on her website that "her client list includes hawks, eagles, pine martin, osprey, owls, spider monkeys, salmon, butterflies, bees, frogs, toads, newts, bats, beaver, songbirds, otter, rock hyrax, small desert species, waterfowl and occasional humans".[41]

Hull's work involves the creation of etchings on rock surfaces in deserts, that are deep enough to retain water or snow-melt for animals to drink (she calls these "hydroglyphs"); "raptor roosts", i.e. painted wooden sculptures on which birds of prey can perch without being electrocuted; small nesting islands for water birds; and other "habitat" sculptures which restore wildlife habitats damaged by human interventions (Hull calls this process "eco-atonement").[42] She also conceived interpretive trails and wildlife observation areas. The effectiveness of her work for the intended species was studied by wildlife biologists and zoologists. "Five summers in a row, a specialist observed young Ferruginous hawks, a threatened species, nesting in

40 Personal communication with Aviva Rahmani (December 2009, in Copenhagen alongside COP15), and email discussions on Cultura21's "COP16" mailing-list, which I created in January 2010. This lively online discussion, suggested by Aviva Rahmani, involves several ecological artists, social scientists and cultural activists from Mexico, the USA and Europe, around the question "which strategy for art at COP16?".

41 Artist's website: http://www.eco-art.org/ [last accessed:07.02.2010].

42 See Hull's artist statement at *Green Museum*, available online at http://greenmuseum.org/content/artist_index/artist_id-7.html [last accessed 07/02/2010].

Hull's *Lightning Raptor Roost* (1990) [...] This led him to deem the project a success on scientific terms" (Spaid 2002, p. 76). To mention a project that failed despite its ingenuousness: Hull worked together with Mayans living in a Yucatan forest in 1998 (in Punta Laguna, Quintana Roo, Mexico), on *Puente Monos (treetop monkey bridge)*, attempting to help spider monkeys cross safely above a road, thereby supporting both biodiversity and eco-tourism. However, despite the technical solutions found by the Mayans (combining light tree branches and nylon cord), the monkeys did not trust the construction, i.e. didn't make use of the bridge (and Hull regretted the early departure of a biologist from the project, who might have been able to help solve this issue).[43] Not content with supporting one specific species, Lynne Hull also aims to put "biofeedback loops" in place, i.e. to also support other species indirectly through her intervention, and thus help rather than hinder biodiversity.

The ecologically-minded art critic Suzi Gablik commented on Hull's work, in 2000: "Where early site-specific sculpture was in a sense a parasite of nature, absorbing surrounding beauty but contributing little to its continuity, Hull's work functions within that environment, exchanging beauty with that of the surroundings" (Gablik quoted in Spaid 2002, p. 76).

In her work, Lynne Hull tries to harmonize the needs of human and non-human local communities. Laura Millim, Director of the Missoula Museum of the Arts (in Montana), who commissioned a project, observed that "Lynne is quite remarkable in her ability to communicate with people in all levels and sectors concerning her environmental ideas. She is able to convince public officials that her projects are practical and do-able. She takes simple problems and orchestrates public involvement to solve them. And she's non-confrontational: gentle but determined" (Millim, quoted in Thornton 1993). In that specific project, Hull strategized a 'Modus Vivendi' between the local community and an invasive group of beavers. However, it seems that her strategy was not fully implemented, with the opposition of local environmentalists who, according to an account by Sue Thornton, were offended by having an artist involved on their 'terrain'. Nevertheless, Thornton argues that, whereas Hull's services to animals cannot directly save any ecosystem, the "work's greatest strength may lie in how it can educate and unite people [...] Hull's collaborators range from telephone company employees and wildlife biologists to fellow artists and the residents of a youth crisis center" (Thornton 1993). Thornton further argues that "Hull's projects nearly always require delicate choreography between government officials, private citizens, the natural world and the arts. The people who are involved must share each other's needs and knowledge, and this collaboration provides an opportunity for all parties to grow" (ibid.). This insight

43 However, Hull realized further projects with the Mayans that were more successful in concrete terms, also in terms of eco-tourism: Cf. Spaid 2002, pp. 78-79.

begs further analysis: The question of effective social change in art projects involving communities, will be addressed in chapter 7.

David Haley

The Manchester University based, British ecological artist David Haley, bears a peculiar position in this analysis, as his thinking contributed an especially important turn in my own thinking for the current work: It is his use of the notion of "ecopoiesis" that freed me from the intellectual trap of Niklas Luhmann's concept of "autopoïesis" (as a descriptor of modern social systems and especially of the art system), together with Edgar Morin's concept of eco-auto-organization (which became the grounding for my own theoretical development). I first used Haley's "ecopoiesis" in 2007-2008 (e.g. in eds. Kagan and Kirchberg 2008) before developing my own notion of "autoecopoïesis" (introduced earlier, cf. chapter 3, conclusion).[44]

In an article written for the catalog of the exhibition *Everything will be fine / Alles wird gut* (which I curated in 2007 together with Bettina Steinbrügge), Haley imagines a future characterized by an "eco-culture" and "ecopoiesis" as "the new way of life" and no longer the new paradigmatic-prophetic task of ecologists and eco-artists.[45] He then discusses how art and science may contribute to the move towards such a future. Global crises alone, such as climate change, may not be enough as drivers of change, because "eco-centricity is not just a matter of political, ideological, corporate and civic preferences, it's a way of understanding and working with our 'embodied ecology'. So, restoration of the habitats of others becomes the realisation of our complete connectedness" (Haley in eds Kagan and Steinbrügge 2007, p. 19). Such a change requires art, and not only science, be-

---

44 My first encounter with David Haley (and with the Harrisons, as well as a few others) occurred in Budapest in 2006, at a symposium on Sustainability and contemporary art organized by curators Maja and Reuben Fowkes at the Central European University. Listening to Haley's interventions and arguments in the discussions unfolding there, I was inspired and convinced most by his approach, in contrast to those of the other contributors. Since then, we have been exchanging and collaborating on a regular basis: I have invited Haley in March 2007 both as keynote speaker at the conference of the European Sociological Association's Research Network for the Sociology of the Arts in Lüneburg and as artist in the exhibition *Everything will be fine* which I curated alongside the conference; in October 2008 in Beijing as speaker-participant at the Asia Europe Dialogue on Art, Culture and Climate Change (on which I was advising the organizer, Katelijn Verstraete of the Asia Europe Foundation); and in August 2010 as workshop organizer in the program of the first International Summer School of Arts and Sciences for Sustainability in Social Transformation, which I am directing.

45 Haley's interest in 'Futures Studies' is directed especially at Richard Slaughter, the author of *Futures Beyond Dystopia: Creating Social Foresight* (2004).

cause "ecology, like Gaia, is not a science, but an art – 'rt'. Science could only ever gain data about Gaia and ecology. Art is the understanding." This claim is based on David Haley's definition of art as "rt" which bases itself on "an Indo-Aryan noun/adjective of the *Rg Veda*, meaning the dynamic process by which the whole cosmos continues to be created" (Haley 2008, p. 201). But he is aware that such a definition has little to do with the mainstream definitions of art, which are encapsulated in a culture of unsustainability (cf. chapter 1): "It must be said, however, that seldom did art in Western Culture come anywhere near this understanding. The problem was that linear forms like progress and development were used as moral values by societies unwilling to recognise the complex, non-linear world [...] Much art, even eco-art became trapped in the paradox of radical gestures performed in reactionary contexts" (Haley in eds Kagan and Steinbrügge 2007, p. 19).

Haley further points at the introduction of uncertainty in our world views, as a normal condition in much art, as an already dated novelty in science (which came in force throughout the 20th century, starting with the fundamental uncertainty in quantum physics, and later with what have become the sciences of complexity), and as a new challenge in society.[46] "Uncertainty has never been a problem in daily life. Uncertainty became a problem when people couldn't let go of their 'classic' mindsets and binary opposites." Haley quotes a question from F. David Peat: "In moving from certainty to uncertainty, how will we begin to represent and envision our new world?" (Peat quoted in Haley 2008, p. 197).

Discussing ecological crises as systemic social risks, Haley addresses the social dimensions of (un-)sustainability. He wonders "why there was no revolt or civil unrest in the United States following the blatant exclusion from provision of the poor and in particular the African American poor" following hurricane Katrina (Haley 2008, p. 197).

Haley further engages his fellow humans to envision "graceful withdrawal" in front of the rising waters (the expression is borrowed from the Harrisons) and at times, his discourse grows very critical of any civilizational prospects for sustainability, and entertains the possibility of learning through "collapse" as a preliminary to "reorganisation and recovery – or evolution": In the context of the sixth great extinction of species in the history of life on Earth, which our species is causing (as Haley points out after E. O. Wilson), Haley affirms: "Sustainability ceases to have any meaning in this context" (ibid., p. 199).

David Haley is both a poet and an eco-artist, as well as a researcher and educator (director of the art&ecology research unit at Manchester Metropolitan University and leading the MA program 'Art as Environment'). He conducted art projects in the UK, China, Taiwan, Germany and the USA, such as *A Walk on The Wild* Side (from 2004 on; eco-urban art-walks and

46 On uncertainty as a normal condition in art, cf. also Haley and Iyer 2009, p. 10.

films that consider Manchester as a living organism) and *Rivers from the Future* (a critical exploration of suburbia, with a focus on freshwater). He also collaborated with the Harrisons on the project *Greenhouse Britain:Losing Ground, Gaining Wisdom* (2008).

The limit I see in Haley's approach, and which led me to differentiate my concept of autoecopoïesis from his ecopoiesis, is that his discourse is sometimes too holistic, insisting too much on unity and not sufficiently on diversity, and thus insufficiently addressing complexity (even though he does insist on the importance of complexity, repeatedly). This can be seen for example in his critique of Felix Guattari's "three ecologies" (a book which has indeed become a fad in the British art world in recent years, and which I will discuss in chapter 6, section 3), where Haley argues that "there is only one ecology – the ecosystem of the whole and this embraces all diversity", and in the process, ignores the complex tensions between unity and diversity, and the different levels of emergence in reality that allow to speak of different, though inter-dependent, ecologies (an ecology of life, an ecology of the mind as initiated by Bateson, and an ecology of ideas and communication in human societies). Haley misses this point and reinforces his position with a quote from David Bohm which sounds especially holistic (and thus simplistic): "If [man] thinks of the totality as constituted of independent fragments, then that is how his mind will tend to operate, but if he can include everything coherently and harmoniously in an overall whole that is undivided, unbroken, and without a border then his mind will tend to move in a similar way, and from this will flow an orderly action within the whole" (ibid., pp. 200-201). Although this argument is not wrong (and even has some value as a counterweight to classical reductionism), it only tells a neat, "orderly" story indeed, and thus falls short of understanding complexity.

I wrote above that Haley's discourse is sometimes too holistic, but, to be fair, this is not systematically the case. In a recent text co-written with the Indian artist Jaya Iyer for the Asia Europe Foundation's program "Connect2Culture", he insists on "see[ing] the rich complexity of relationships between all the challenges as a joined-up entity requiring many diverse ways of thinking and inventing. Here the arts make an essential contribution, to 'keep the discourse plastic'. Art has the ability to exceed expectations, to maintain a flexible mind, and make new questions" (Haley and Iyer 2009, p. 8).[47]

47 More recently even, in the Spring 2010, Haley expressed his interest in Edgar Morin's work after I introduced him to it (David Haley, personal communication).

Shelley Sacks

The South African, UK-based artist Shelley Sacks, who studied at the Akademie für Bildende Kunst in Hamburg and at the Free International University in the 1970's as a student and collaborator of Joseph Beuys, works on the basis of Beuys' concept of "social sculpture". She founded the Social Sculpture Research Unit (SSRU) at Oxford Brookes University. "The unit defines 'social sculpture' as an interdisciplinary art practice that involves the audience in shaping social processes. Since thought and discussion are its core materials, it recognizes that all humans are artists capable of shaping a democratic, sustainable world" (Weintraub 2006, p. 75). The website of the SSRU further claims:

"Social sculpture is a field of transformation. It is a place of questions. Social sculpture is a territory where inner and outer work co-incide. It emphasises the role of imagination in transformative social process and the centrality of alternative modes of thought. Social sculpture is concerned with exploring strategies, methods and principles that enable us to see, understand and act imaginatively to shape a humane and ecologically viable world. It is about:

- perceiving interrelationships
- joined up thinking and practice
- enabling connective understanding."[48]

Sacks is best known for her work *Exchange Values: Images of Invisible Lives* (from 1996 on), an installation (typically exhibited in art spaces) based on an extensive research: Sacks purchased many bananas in the UK where she lives, dried them and stitched them together, turning them into 20 dark sheets.[49] She labeled each sheet with the "grower identification number" of the crate from which they came. She then traced back the origin of the bananas, and found back the specific farmers who cultivated the bananas, in the Carribean Windward Islands. Sacks interviewed the farmers about their working and living conditions, and co-organized meetings with them. The audio recordings of the interview were associated to the sheets, for the installation. In the center of the installation, 10,000 dried banana skins are laid on the floor. Linda Weintraub comments:

48 http://www.social-sculpture.org [last accessed: 02.03.2010].

49 As Sacks describes, her project resulted from a very long gestation: She started drying banana skins and collecting them already in the early 1970's in South Africa, reflecting on their producers and on economic networks. For more details on the detailed process of Sack's work, see: Shelley Sacks. "Banana is not an Easy Thing." Available online at www.exchange-values.org [last accessed: 02.03. 2010].

"While the sweet smell of bananas calls to mind the familiar pleasures of a comfort food, the taut appearance of these sheets evokes anguish. This aesthetic contrast dramatizes the economic disparity between consumers and producers participating in the banana market. Memories of eating pleasures are thereby linked to the farmers' indignation and despair over their victimization by the worldwide banana market [...] Individual voices reporting the unsettling details of their lives shatter the anonymity of global economic distribution" (Weintraub 2006, pp. 71-72).

The installation reveals "complex networks of human relationships" (ibid.) and the different interests and viewpoints of producers, traders and consumers. Cultural geographer Ian Cook (Cook et al. 2000) characterized *Exchange Values* as "connective aesthetics" - an expression that comes very close to the notion of "aesthetics of patterns that connect" that I proposed in the previous chapter after Gregory Bateson.

The theme of Sacks' work is connected to the "banana wars" (whereby the EU was condemned by the WTO because of the preferential access to EU markets accorded to Carribean producers, which displeased Chiquita - the monopolistic Banana-multinational and infamous historical supporter of several Latin-American dictators). As Shelley Sacks commented, so-called 'free trade' is a joke when "the enormous multinationals can afford to flood the market for weeks at a time with underpriced bananas" and achieve high yields by imposing extremely polluting agricultural practices and inhuman working conditions (Sacks quoted in Kelley 2008, p. 24). From its beginning in 1996, the *Exchange Values* project was clearly associated to a promotion of the ideas of "fair trade", which in the meanwhile have greatly gained in popularity and in market-share across Europe. In 2008, Caffyn Kelley notes that "Oxfam has initiated an international campaign for a fair and sustainable banana industry. In March 2008 the Windward Islands farmers association announced a contract to sell bananas through their national fair trade organizations" (Kelley 2008, pp. 25-26).

The installation by Shelley Sacks, confined to the typical art space of the exhibition (and exhibition visitors), is not comparable to the ecological art, embedded in communities, of e.g. Patricia Johanson, nor to the policy-influencing work of the Harrisons. But Sacks claims that it is transformative too, because

"although the consumer standing listening to the voice of the invisible producer is not, in that moment, involved in changing the status quo in any concrete way, responses suggest that the experience of absence is so tangible - of a producer whose 'skin' is stretched before us, whose voice is inside us - that it stirs one imaginatively, provoking an inward movement that we carry outwards into the world. People describe how the experience has given them a sense of their power to see things differently, and to explore ways of getting involved in shaping a better world."[50]

As her opening speech for *Exchange Value*s at the South African National Gallery, Cape Town, South Africa, on April 21st 2001, the writer Karen Press commented:

"Is this social sculpture project doing anything that a pamphlet, a documentary film, or any other educational campaign by fair trade activists is not doing? What does it offer me? What does it expect from me? The first response to these questions must be that this is, whatever its other dimensions, an aesthetic experience: I am asked to look, to listen, I will want to touch (though of course I won't, because of the rules of art galleries), I will stand still and let my senses be stimulated, my perceptions rearranged. Something will happen that I can't predict - maybe the smallest heightening of attention, maybe memories triggered, or associations awoken. Connections will happen [... Shelley Sacks] reminds us of the deep connections we have with lives and labour and economic forces and organic processes in ever-widening circles around us, just by participating in the ordinary business of our days. This sense of connection is the first step in doing what the banana farmers' story tells us has to be done: taking hold of the world, and changing it."[51]

Furthermore, the exhibited installation itself is not the only aspect of *Exchange Values:*

"A number of 'social sculpture' forums have taken place during the course of the project: whilst collecting the skins from members of the public over two weekends in the centre of Nottingham, where free bananas were given out in exchange for their skins; in the Windward Islands with growers and sustainable development activists, and in the social sculpture forums that take place alongside the physical installation in each venue."[52]

Overall, the project presents itself as an example of "social sculpture", according to its promoters:

"The whole project is an excellent example of social sculpture - combining transformative social process (her re-envisioning work with the producers in the Windward Islands), connective aesthetics (reconnecting us with the world around us through the aesthetic form of the installation), and a 'permanent conference' (discussions about

50 Shelley Sacks. "Exchange Values Six Years On". 2002. Available online at www.exchange-values.org [last accessed: 02.03.2010].

51 Karen Press. Opening Talk. 2001. Available online at www.exchange-values.org [last accessed: 02.03.2010].

52 Excerpt from the website www.exchange-values.org [last accessed: 02.03.2010].

our role as creative beings and positive forms of globalization, through the social sculpture forums that take place alongside the installation)" (Demarco nd).

Among the latest projects initiated by Shelley Sacks is *Ort des Treffens* (April to September 2009, as part of Hannover's *Garten Region Projekt* program), a series of events built around interviews of inhabitants of Hannover, who were "invited to reflect [...] about what they think they are doing in the world".[53] The project involved social sculpture artists Anja Steckling and Nicholas Stronczyk (Hannover), philospher Wolfgang Zumdick (Oxford/Aachen), philosopher and composer Alex Arteaga (Berlin), and Lukas Oertel (Bonn).

James Turrell

A good example of "Earthworks art" is Californian artist James Turrell's work in progress "Roden Crater". The Roden Crater is an extinct volcano in the Arizona desert which the artist has been transforming into a sky observatory since 1974. Turrell aims "to make visible the manifold phenomena occurring in the sky without us being aware of them under normal conditions [and] to enhance awareness of the perceptual qualities of the light [...] Turrell relies on whole systems of interdependent phenomena in the sky and on the earth" according to daily and seasonal cycles, weather conditions, and celestial phenomena (Bijvoet 1994, p. 98). A practicing Quaker, Turrell gives an explicit religious connotation to his work on the perception of light. About the experience of "Roden Crater", Suzi Gablik writes: "This is a setting in which to gather up and assimilate the grand harmonies of the cosmos – a place where you can experience geologic, rather than man-made, time. You have a powerful sense of standing on the surface of the planet. The feeling of being part of the physical world is very strong" (Gablik 1991, pp. 82-83).

Turrell also realizes museum installations also based on the perception of light. With all his light installations, and not only with "Roden Crater", Turrell tries to offer an immersive experience, a long moment of contemplation, to the viewer. But he is aware that such expectations are in most cases not realistic in the setting of art museums, because spectators tend to spend very little time exploring each exhibited artwork: "I saw the Mona Lisa when it was in L.A., saw it for 13 seconds and had to move on. But, you know, there's this slow-food movement right now. Maybe we could also have a slow-art movement, and take an hour" (Turrell quotedin Douglas 2005). In the earthwork "Roden Crater", this is more likely to occur, although, or rather because, the earthwork is not easily accessible, and thus

53 Excerpt from the SSRU website at http://www.social-sculpture.org/developments/projects-processes-and-transactions1/ort-des-treffens.htm; see also http://www.ortdestreffens.de/ [last accessed 02.03.2010].

demands efforts from the visitor coming up to the crater. That such efforts are required, was also formulated by Abram in his plea for a phenomenological-animistic experiencing of the landscape (cf. chapter 4, section 5).[54]

Andy Goldsworthy

The British sculptor and photographer Andy Goldsworthy is one of the most widely known environmental artists (also associated to land art), in terms of media coverage, including in the mainstream cultural media. In 2000, he was appointed officer of the Order of the British Empire (cf. Adams 2007).

In his work, Goldsworthy creates sculptures using, exclusively, materials found on the site. "I can't edit the materials I work with. My remit is to work with nature as a whole" (Goldsworthy quoted in Sooke 2007). Goldsworthy both realized ephemeral works, with his bare hands, and permanent sculptures with the use of machines and stones. Most of his works, though, are ephemeral, and go through a process of decay. In such cases, he takes pictures of the work at a specific moment in the process (and his work became popular in the media mainly through these photographs). According to Suzi Gablik: "His approach, too, is premised on respect rather than on domination; his gestures are delicate and unobstrusive. [...] The challenge is in tuning in and adapting to different landscapes and seasons, establishing a dialogue with the place, cooperating with its subtle web of interrelated processes" (Gablik 1991, p. 91). Goldsworthy's own discourse confirms Gablik's interpretation: "I cannot stop the rain falling or a stream running. When I work with a leaf, rock, stick, it is not just the material in itself, it is an opening into the processes of life in and around it. When I leave it, these processes continue... These things are all part of a transient process that I cannot understand unless my touch is also transient – only so is the cycle unbroken, the process complete" (Goldsworthy quoted in Gablik 1991, p. 92). Goldsworthy also learns techniques from traditional/indigenous sources, as for example in "Touching North" (1989), a work at the North Pole based on snow-cutting & packing techniques he learned from Looty Pitjamini (an Inuit based in the Arctic island of Ellesmere).

Artistic work such as Goldsworthy's (and e.g. Nils Udo's) is more often refered to as "art in nature", rather than as environmental or ecological art. As Sam Bower notices, "Art in Nature seems to find more inspiration in a type of Romantic Minimalism, revealing the abstract beauty and decorative potential of ephemeral natural forms. As such, it usually lacks overt feminist, ecological or political content" (Bower 2009). But it would be unfair to characterize Goldsworthy's work as only 'decorative': As his own discourse

54 However, Turrell's work does not address ecological issues or social issues directly, and therefore may not be labeled exactly as 'ecological art' (but art historical labeling is not a major concern in my analysis).

quoted above suggests, his work is indeed also dealing with the transience of inter-related natural processes, and thus can contribute to eco-literacy.

SECTION 4: THE "REENCHANTMENT OF ART" ACCORDING TO SUZI GABLIK

A new movement in art: connective and responsible

In her book *The Reenchantment of Art* (1991), art historian and critic Suzi Gablik announces the emergence of a new movement in art, away from modernism and from postmodern nihilism.[55]

> "If modern aesthetics was inherently isolationist, aimed at disengagement and purity, my sense is that what we will be seeing over the next few decades is art that is essentially social and purposeful, art that rejects the myths of neutrality and autonomy. The subtext of social responsibility is missing in our aesthetic models, and the challenge of the future will be to transcend the disconnectedness and separation of the aesthetics from the social that existed within modernism" (Gablik 1991, pp. 4-5).

Gablik's criticism of modernism is not my main focus here (cf. chapter 1). Rather, especially insightful here is Gablik's stressing of a conception comparable to the 'aesthetics of the patterns that connect' that I introduced in chapter 4 after Bateson. "There is a need for new forms emphasizing our essential interconnectedness rather than our separateness, forms evoking the feeling of belonging to a larger whole rather than expressing the isolated, alienated self" (ibid., pp. 5-6). This implies a different aesthetic relationship than the traditional observer-image relationship: "The vision we need to develop is not one that observes and reports, that objectifies and enframes, but one released from these reifying tendencies and rooted instead in a responsiveness that ultimately expresses itself in action [and] is not purely cognitive, or purely aesthetic, but is opened up to the body as a whole" (ibid., pp. 99-100). And it also implies a different understanding for the 'roles' of art in society: "In the past, we have had much of the idea of art as a mirror (reflecting the times); we have had art as a hammer (social protest); we have had art as furniture (something to hang on the walls); and we have had art as a search for the self. There is another kind of art, which speaks to the power of connectedness and establishes bonds, art that calls us into relationships" (ibid., p. 114). "As artists learn to integrate their own needs and talents with the needs of others, the environment and the community, a new foundation for a non-self-conscious individualism may emerge" (ibid., p. 144).

---

55 Gablik's critique of Modernism and of postmodern nihilism in art, was presented in chapter 1, section 2.

Gablik understands the self in a way that echoes the notion of autoecopoïesis. She argues that "[b]y redefining the self as relational [...] we could actually bring about a new stage in our cultural and social evolution. The restructuring of the Cartesian self, and its rebirth as an ecological self-plus-other or self-plus-environment" (ibid., p. 177) is an autoecopoïetic movement. Gablik also sees this as taking part in a larger cultural shift: "The emerging new paradigm reflects a will to *participate* socially [and] involves a significant shift from *objects* to *relationships*" (ibid., p. 7). She also acknowledges the origins of such a shift, when mentioning that "the essence of the new paradigm emerging in physics, general systems theory and ecology changes our whole idea of reality with the notion of interconnectedness" (ibid., p. 22). She explicitly cites Fritjof Capra on this paradigm shift and stresses the importance of elaborating a rich ecological discourse.[56] "The new ecological, non-Cartesian consciousness must be articulated very clearly, for it is both the model for personal transformation and the analytical framework for social criticism and change in the ecological age" (ibid., p. 165).

Gablik argues that the shift already started (when she was writing her text, i.e. already in the 1980's), more in artistic practices than in theoretical discourses. According to her, "the fact is that many artists now conceive their roles with a different sense of purpose than current aesthetic models sanction, even though there is yet no comprehensive theory or framework to encompass what they are doing [and she thus calls forward] the emergence of a more participatory, socially interactive framework for art" (ibid., p. 7). Gablik further characterizes such a framework as

> "a framework for reconstructive postmodern practice, which, although less visible in the mainstream than deconstructive art, implicates art in the operative reframing of our entire world view [...] Reconstructivists are trying to make the transition from Eurocentric, patriarchal thinking and the 'dominator' model of culture toward an aesthetics of interconnectedness, social responsibility and ecological attunement" (ibid., p. 22).[57]

56 Cf. Gablik 1991, pp. 163-164.

57 Gablik is here pointing at a 'feminine principle', not at gender issues. "It is crucial to understand that there is a serious difference between "feminist" issues and the cultural recovery of the feminine principle" (Gablik 1991, p. 130). Gablik's aim is to achieve "an integration of masculine and feminine energies into a more creative partnership as the very ground of our whole culture" (ibid., p. 176).

The feminine principle after Riane Eisler, 'integralism' and the risk of holistic simplification

In her discussion of a "feminine principle" to balance with a "masculine principle", Gablik (1991) refers to Riane Eisler's view of 'dominator' vs. 'partnership' models in human history, in the evolution of civilizations. Eisler (1987)'s gender-holistic "Cultural Transformation Theory" opposes two basic models of civilization across the history of Europe and the Mediterranean world: A "dominator model" (whether patriarchal or sometimes matriarchal) ranking and hierarchizing gender, and a "partnership model" where gender is no basis for superiority or inferiority and which results in a less authoritarian, more peaceful and egalitarian organization of society. Eisler's two models are not about women vs. men per se, but about two systems of values. In her best-selling book *The Chalice and the Blade,* Eisler presents historical examples of the partnership model (e.g. in Neolithic times, the Cretan civilization, in the actions and discourses of the Christ -according to her interpretation, also based on lately rediscovered 'heretic' gospels), and discusses how "gylanic resurgence" was repeatedly repressed (e.g. in classical Athens, among early Christians, in medieval Europe).

Eisler's work offers interesting, valuable insights at a conceptual level, but the supporting body of evidence is weak, and she often fails to offer a nuanced, less dichotomous account of European history. It raises the general issue of a risk of over-simplification by holism, which is high in the works of so-called 'integralists'.[58] Thirteen years after *The Reenchantment of Art,* Suzi Gablik (2004) refers to Ken Wilber, another famous 'integralist' author, overall less simplistic than Laszlo, though probably still over-holistic.[59] Gablik borrows from him the notion that "the world is one undivided whole".

However, in the same text Gablik also refers to Basarab Nicolescu's *Manifesto of Transdisciplinarity* which, while very open to spiritualist and religious discourses (as is Wilber), clearly rules out any holistic thinking model, while talking of an "open unity" of reality and referring to Edgar Morin's work. From Nicolescu, Gablik retains the difference between multi-, inter- and transdisciplinarity, as well as the notion of multiple "levels or reality", which, she notes, "has a unique goal: to propel us beyond either/or thinking into a co-existence of nested truths" (Gablik 2004). Gablik sees that Nicolescu and Wilber both focus on multidimensionality in reality, but she seems however to ignore the differences between the 'complexity' approach (Morin and Nicolescu, as introduced in chapter 3 and 4) and the integralist discourse.

58 Cf. my criticism of Laszlo in chapter 2, section 4.

59 Cf. Wilber 2007.

Reconstructive postmodernism

Gablik argues in favour of a reconstructive, rather than merely deconstructive postmodernism, because "we cannot reject this old view [i.e. modernism] until we have a new view that seems more convincing. Change is most likely to occur [...] through people who are as far removed from cynicism as they are from utopianism" (Gablik 1991, p. 23). She adds that "labeling as idealistic, utopian or naive those who believe change is possible can be seen as the most effective way to make sure that things are left exactly as they are" (ibid., p. 25). Such a comment was still relevant ten years after the publication of Gablik's text, especially against analyses that continued to defend ironic deconstruction as the only possible form of artistic discourse about sustainability: Art sociologist Ulf Wuggenig offered a symptomatic discourse going in that direction, arguing that the reactions of "advanced contemporary art" to the dominance of economic and techno-centric thinking in the discourse of sustainable development, "are today only conceivable in critical, deconstructive or ironic form" (Wuggenig 2001, p. 257).

But Gablik does not advocate for an end to deconstruction; rather, she argues for reframing deconstruction as a moment in a reflexive process informing possible choices rather than negating all choices.[60] She further characterizes *"responding* to what is going on", on the artist's part (as opposed to merely observing), as taking responsibility: "Responsibility implies that one is carrying out intentions, shaping the environment, influencing others. So the question is, how much responsibility are we willing to take for exercising intentionality in the world?" (Gablik 1991, p. 33). Gablik clearly opposes Baselitz's irresponsibility claim (cf. chapter 1, section 2).

Gablik also opposes the relationships-based art she promotes, to the autonomy-based art of modernism:

"[I]f the principle of linking, or partnership, is to become the basis of a new consciousness, then the notion that art and society are at odds with each other – the old adversarial relationship – will need to be revised. And if all levels of experience and the world are now perceived in terms of *relationship*, it represents the paradigmatic defeat of radical autonomy and the old avant-garde mandate for oppositional practices" (ibid., p. 68).

Shamanic art

Gablik longs for "the ability to shift mindsets and thus to perceive other realities – to move between the worlds, as ancient shamans did" (ibid., p. 42), echoing David Abram's phenomenological/animistic discourse. From this perspective, Gablik celebrates the work of Chicago-based artist Fern

60 Cf. Gablik 1991, p. 27.

Schaffer who enacts rituals, e.g. at the winter solstice in the waters of lake Michigan, and of Californian artist and dancer Rachel Dutton, who makes sorts of totemic sculptures and claims: "I was able to enter the [...] dream world directly through my art" (ibid., p. 51).

Gablik argues that shamanistic rituals are relevant to modern individuals because they "provide a possible basis for reharmonizing our out-of-balance relationship with nature" (ibid., p. 45), and because they break the "isolation of the ego-mind from the archetypal unconscious" (ibid., p. 52 referring to Karl Jung). Gablik's discourse should not be mistaken for a so-called 'luddite' rejection of all modern technology, or for superficial exoticism. "It is not a matter of trying to imitate an archaic cultural style so much as fostering psychic mobility – opening oneself up to a range of visionary experience in a culture whose mind-set has made the very idea of other worlds unthinkable" (ibid., p. 47). Gablik's discourse should also not be mistaken for 'New Age', esoteric spiritualism. "The sacredness of both life and art does not have to mean something cosmic or otherworldly – it emerges quite naturally when we cultivate compassionate, responsive modes of relating to the world and to each other" (ibid., p. 181). Gablik also characterizes such "visionary" experiences as

> "ego-deconstruction – a practice [...] in which the self experiences directly the deep connection between the human world, the plant world and the animal world [...] illusions of duality dissolve, and with them, old assumptions about a distinct and separate ego-self codified by our culture [... I]n the nondualistic view, everything in the universe is understood as dancing energy patterns interweaving a single continuum [of which] we are [...] cocreators" (ibid., pp. 54-55).

This bears consequences for the notions of self and of freedom: "We need to think about 'practices of the self' that do *not* separate the self from society and withdraw it from social responsibility [and] that *understand* the essential intertwining of self and other, self and society, that are aware of the subtle complexities in this intertwining" (David Michael Levin, quoted in ibid., pp. 64-65).

Gablik (1991) further refers to the paintings of Gilah Yelin Hirsch (from Venice, California) and sculptures of Richard Rosenblum (based in Newton, Massachusetts) – especially the latter's *Manscape* (a walking figure made of roots of dead trees -and then cast in bronze): For Gablik, "these figures seem to see through the earth's eyes. Uprootedness is part of modern alienation; Rosenblum creates a metaphor of belongingness, of rerooting ourselves in the universal bond of biological existence [...] *Manscape*, in particular, reconnects us with this *prima materia* of the earth [...] From it we learn that matter *can* give birth" (Gablik 1991, p. 56). Here, Gablik's discourse joins Morin's cosmological discussion on the poïesis of matter (and Prigogine's dissipative structures), although Gablik ignores Morin's work.

Gablik further discusses James Turrell's work "Roden Crater" as allowing a transcending experience of connectedness: "The viewer who comes to Roden Crater, in the tradition of the vision quest or pilgrimage, since it is not easy to get there – makes the transition from spectator to participant [...] we get inside the landscape and can develop our own affective ties with it" (ibid., p. 83). Here, Gablik's experience of "Roden Crater" can be further informed by Abram's discussion of a phenomenological-animistic quality of a landscape-experience.

In the same line, Gablik also discusses the work of Ontario-based composer R. Murray Schafer, for whom "modern theory teaches that revolutions in art occur when styles are challenged. But they ignore the bigger revolution of context" (Schafer quoted in Gablik 1991, p. 86), i.e. the physical environment in which an art-experience is offered.[61] Gablik discusses Schafer's open-air Opera "The Princess of the Stars", which takes place on a lake at late night / early morning (until the very exact moment of sunrise). The audience has to arrive to the lake-side around 5am, and is immersed in the performance of a "mythopoetic story" involving musicians, singers, actors and dancers on the lake, in canoes, in the surrounding trees, also eventually singing bird-calls to the local real birds who then begin to wake up and sing themselves. At dawn, the rising sun itself is integrated as a character, the Sun god rescuing the princess. "It will be an effort to get up in the dark, drive thirty miles or more to arrive on a damp and chilly embankment, sit and wait for the ceremony to begin. And what ceremony? Dawn itself, the most neglected masterpiece of the modern world" (Schafer quoted in Gablik 1991, p. 87). Gablik further discusses Schafer's book *The Tuning of the World*, where he addresses issues of sound-pollution, and also explains his interest for outdoor sceneries where he cannot control all the sounds but must work "on nature's terms" as Gablik phrases it. Schafer considers it a duty of musicians, who are "most attuned to sounds", to "take responsibility for orchestrating and controlling the whole acoustic environment [...] help[ing] alleviate noise pollution through refashioning our soundscape so it is more in tune with natural models" (Gablik 1991, p. 88).

61 At the level of art-sociological analysis, a more detailed focus on the importance of the physical and social context is present in Hans Abbing's comparison of the 'high art', introspective atmosphere of the typical classical music concert hall, vs. the 'new art', more informal atmosphere of a pop music venue (cf. Abbing 2006, 2009). But to Schafer, both contexts are equally uninteresting, as they offer, in his view, little more than two different forms of comfortable, non-demanding "entertainment". While Abbing acknowledges the entertainment value of art, Schafer demands of his audience an even greater immersion in the art-experience than the traditional classical music artists do.

An over-harmonious, holistic view

Like Edgar Morin, Suzi Gablik does not aim to abandon disjunctive rationality altogether, but rather to replace it in a new context: "We have yet to discover what new "wirings" are possible when the visionary enters into active collaboration with the rational" (ibid., p. 57).

Unlike Morin, however, she tends to over-emphasize harmony, up to the point of following a Habermassian tendency to negate differences and impose a notion of consensus, in her apology for the archetype of the 'shaman' vs. the modernist archetype of the 'hero'. For example, she writes:

"Whereas male myths, and the myths of modernism, typically have focused upon tasks of separation and mastery of self over environment, within the model of partnership it is a question of trying to realize a context in which social purposes may be served [...] 'by finding beautiful ways of harmonizing interests rather than sublime ways of detaching oneself from others' interests'" (ibid., p. 67, with a quote from Habermas).

Gablik, in her essay, repeatedly goes very far into an apology of consensus, celebrating "an open continuum for interaction, for a process of relating and weaving together – a flow in which there is no critical distance, no theoretical violence, no antagonistic imperative, but rather the reciprocity we find at play in an ecosystem, that is essential to skillful functioning" (ibid., p. 73). The discussion of complexity already demonstrated that a solely harmony-based understanding of the ecosystem is far too simplistic (cf. chapter 3, section 1). I will come back, later, to the fallacy of the notion of 'consensus', arguing myself instead for a non-Habermassian notion of 'compromise', whereby harmony and differentiation can be complementary, without the one being dissolved by the other (cf. chapter 7, section 2).

The connective self

Suzi Gablik sees an example of the "connective self" in the work of Mierle Laderman Ukeles (already introduced above). She quotes Ukeles:

"If art's function is to articulate a notion of human freedom [...] then the problem is how to make the notion of freedom relevant to everybody, not just an élite group. It has to be connected to the world, but then you immediately have restrictions. [...] Action painters were engaged in a notion of pure freedom. I love that notion of freedom, but you can only talk about freedom when you can deal with the air and the earth and the water [and the] people who *are* dealing with this" (Ukeles, quoted in Gablik 1991, p. 69).

Gablik vests very high expectations, or hopes, in such work. She argues, after David Michael Levin, that

> "our technological world is a reflection of gestures motivated by the masculine 'will to power' [...] If we could reverse these tendencies and develop instead gestures of 'reciprocal touching', gestures that bring together, receive and welcome, modest gestures of solicitude and tact, which belong to the maintenance of being, the possibilities are so profound as to imply a new social and cultural order" (Gablik 1991, p. 71).

"World healing, in this sense, begins with the individual who welcomes in the other [with a] lowering of the personal wall and expressing spontaneous empathy" (ibid., p. 75). The work of Ukeles, according to Gablik, moves in this direction, in starch contrast to Richard Serra's *Titled Arc* (cf. Gablik's critique of Serra, described in chapter 1, section 2).

> "Ukeles's extraordinary ability to empathically knit herself into the community of sanitation workers, and to transform the alien audience into the empathic audience, communicates, at least to me, the pleasures of creative attunement and interaction over those of autonomy and, as in the case of Serra, radical opposition of the self imposing itself on the other [... S]he merges her consciousness with the workers, converses with them, learns from them and becomes one with them" (ibid., p. 73).

Gablik celebrates artists who aim to connect with what Abram termed the "more-than-human", and especially with other species. She points at the work of Lynne Hull with her trans-species sculptures, but also at Andy Goldsworthy (whose work develops a "mystique of the land", according to her), and at the flower-paintings of New-York based Robert Janz (who redraws everyday, in a gallery, the life-cycle stages of a flower). The latter gives her the opportunity to argue in favor of "a revision of our time-horizon" away from short-term thinking and linear time, and into an appreciation of the time of ecological and biological cycles.[62] Gablik also points at the work of performance artist Rachel Rosenthal (based in Los Angeles), who portrayed the persona of the Earth in her piece *L.O.W.*.

But Gablik also discusses artists whose work focuses on the human social relationships, such as e.g. Krzysztof Wodiczko with his *Homeless Vehicle Project* (cf. Gablik 1991, pp. 100-102), and more especially Bervely Naidus whose work enacts social relationships and "succeeds in actually *realizing* this interrelationship; artist and audience form a relational dyad, so that the two parts can no longer be defined independently" according to Gablik (ibid., p. 158). The work of Beverly Naidus engages its audiences in discussing specific social and/or ecological issues, in the context of her exhibited installations. Naidus aims to involve the audience/participants emotionally and evaluates her success insofar as she can tap into the "audience's ability to identify and connect with what is being said" (Naidus, quoted in Gablik 1991, p. 161). The transformative value of Naidus' work, Gablik comments, lies in the value of genuine dialogue. Gablik further refers to

62 Cf. Gablik 1991, p. 90.

quantum physicist David Bohm, also known for his writings on the virtues of genuine dialogue vs. confrontational 'discussions' (Bohm 2006).

Coming back to the issue of homelessness, Gablik contrasts Wodiczko's art objects the work of another artist working with homeless people: Bradley Mc Callum. Mc Callum observed that the homeless people in his city (Richmond) were using grocery carts which were not very practical for them. Unlike Wodiczko, Mc Callum wanted to help the homeless as "a living artistic practice, an actual *modus operandi* that would go beyond a merely symbolic potential" (Gablik 1991, p. 171).[63] Mc Callum developed a cart and experimented it together with homeless people who were giving him feedback on how to improve the design. He developed the association "Collaborative Urban Sculpture", whereby individual carts and other temporary shelters can be crafted together with homeless people. Gablik especially describes one lengthy collaboration with a homeless woman named Jackie, who was uninterested with his carts and wanted instead to adapt some public benches to her needs. Gablik describes how they worked together until both were satisfied with the results (and not only Mc Callum). Gablik comments: "The healing power of what [Mc Callum] does rests not so much in the object exchanged, but in the path of mutuality and understanding that is created" (ibid.).

Gablik also celebrates the collective paintings by the teenagers group KOS ("Kids of Survival") around New-York artist Tim Rollins (cf. Gablik 1991, pp. 106-109), and the work of Los Angeles-based performance artist John Malpede, who founded and directs a theatre group composed of street people, basing their improvisational work on stories collected from the local community. Malpede's group is called LAPD (the Los Angeles Poverty Department), and gives, according to Gablik, "a direct experience of what it means to be a homeless person" with characters who are not necessarily nice or pitiful, and without political overtones. She comments:

> "No matter how accurately art may mirror back to society its own negative features, the perception that alienation subsides when we become aware of our connectedness with others leads inevitably to a different sort of artistic practice – oriented toward the achievement of shared understandings and [...] 'moral-transformatory processes'. [...] You cannot exactly define it as self-expression – it is more like relational dynamics. Once relationship is given greater priority, art embodies more aliveness and collaboration, a dimension excluded from the solitary, essentially logocentric discourses of modernity" (ibid., pp. 105-106).

63 Wodiczko's homeless carts made only for the art space, and which are considered impractical for real use by its critics, are also criticized by Wolfgang Zinggl of Wochenklausur who echoes "the suspicion that he is utilizing social destitution for the purpose of creating 'valuable exhibition pieces'" (ed. Zinggl 2001, p. 17).

This implies listening as much as telling. In what she calls "art in the empathic mode", e.g. in the work of performance artist Suzanne Lacy, Gablik sees an example of the "listening self", setting up "an experience with reciprocal listening" which is required in order to transform "the experience of exclusion and alienation from society into one of creative empowerment in the community" (ibid., p. 110). Gablik refers especially to Lacy's works *Whisper, the Waves, the Winds* (1984, a procession by 150 women followed by conversations between them and with the audience) and *The Cristal Quilt* (1986, a procession by 600 elder women, also followed by conversations and by a choreography of hands-movements suggesting quilt-making, in a pre-arranged installation with a larger surrounding audience).[64] As Gablik notes, "[F]or Lacy, the 'success' of the work is measured by whether or not the process of networking among the women continues once the performance itself is finished" (ibid., p. 111). Gablik further discusses how some of the women involved in the performance founded an organization dedicated to challenging gender stereotypes.

A critique of Critical Theory

Gablik also confronts how the art world of contemporary art, and especially the "critical theory" circles, have tended to reject work such as Dominique Mazeaud's, which Gablik characterizes as not-based on the transgression of prior aesthetic codes, but based on "another integrating myth entirely: compassion" (ibid., p. 121). Gablik tells how Dominique Mazeaud began in 1987, "The Great Cleansing of the Rio Grande River", ritually (but also concretely) removing trash and pollutants from the river once a month, and keeping a diary of the process. That project lasted until 1994, and was thus ongoing when Gablik was writing about it. After that project, Mazeaud later realized e.g. the performance "Point of Tears", which was performed internationally, and was invited by several academic and art institutions.[65]

Gablik claims that (at the time of writing her essay, in the late 1980's), Mazeaud's work is not recognized as art, by the "critical theory" circles, and she asks: "What makes using a bed for canvas, or walking a line in the desert, or exhibiting a manufactured urinal, more acceptable as art than hauling a withered sofa out of a dying river?" (ibid., p. 134). Gablik points at the resistance of the artworld to the idea of usefulness (which is to be connected to her overall critique of modernism and of postmodernism in art – cf. chapter 1, section 2). Chris Burden ("Art doesn't have a purpose"), David Smith (in his affirmation of his "arrogant independence to create") and Walter de

64 While Gablik talks of 600 participants in *The Cristal Quilt*, other sources talk of 430 participants (e.g. at http://en.wikipedia.org/wiki/Suzanne_Lacy).

65 Cf. artist biography at Green Museum: http://greenmuseum.org/content/artist_content/ct_id-202__artist_id-111.html; and artist website at http://www.earthheartist.com [last accessed: 10.02.2010].

Maria's essay "Meaningless Work", are stressed by Gablik as illustrations that "[o]ur culture's most cherished idea remains the aggressive insistence on freedom for its own sake, freedom without praxis – the kind of freedom that makes picking up the garbage valid as art only if you want to "romance" the trash (that is, use it for an aesthetic effect), but not if you step beyond the value vacuum to try to clean up the river" (ibid., p. 135). Gablik further echoes Allan Kaprow's critique of the predominance of "artlike art" over "lifelike art". Artlike art "is the art that is separate from life [and] engages in dialogue with other art, and is supported by galleries, museums and professional art journals, *all of whom* [...] *need artists whose art is artlike*" (ibid., p. 137). Gablik quotes Kaprow's account of the failure of "lifelike art" from the 1960's (e.g. Kaprow's own "happenings"): "We couldn't bypass the framing devices, perceptual clichés, and values of traditional Modern art. ... We were always obliged to put on a *show*. All the traditional esthetic habits of detached spectatorship [...] were brought to the new situation intact" (Kaprow, quoted in ibid., p. 138). Gablik criticizes the art world's refusal to give itself a "moral imperative", and she adds:

> "What we clearly do not have, at this point, is any working framework for a socially or ecologically grounded art – an art that is accountable to the larger whole, in the sense of being contextually rooted in a living connection with a containing organic field. And [...] we *can't* have such a concept as long as we remain hooked on the myth of pure creativity and the inherent purposelessness of art for art's sake [and the] ideal of an autonomous aesthetic culture" (Gablik 1991, p. 139).

But the situation is not blocked, according to Gablik.

> "When everything is perceived as dynamically interconnected [as in the nondualistic view advocated by Gablik], art needs to collaborate with the environment and a new sense of relationship causes the old polarity between art and audience to disappear. [...] Then meaning is no longer in the observer, nor in the observed, but in the relationship between the two [...] recognizing that when observer and observed merge, the vision of static autonomy is undermined" (ibid., pp. 150-151).

Usefulness or Uselessness:
A line of fracture in the art world of contemporary art?

The resistance to the very world 'useful', often accompanied by (legitimate) warnings against the threat of an 'instrumentalization' of art, was indeed a characteristic reaction which I also encountered repeatedly in the past few years, the last example being the director of the "Haus der Kulturen der Welt", during the Culture|Futures conference in Copenhagen in December 2009, alongside the Climate Conference: He repeatedly expressed anger against other Asian and European artists and theatre directors when they would come close to a wish to be "useful".[66] This strong emotional reaction points at a possible taboo. However, in the context of that international conference on the role of the arts in "the transition to an ecological age", his position was, it seemed, not shared by many of the participants.

A shift, and lines of fracture, may thus be appearing, which would point at a possible shift in the art worlds beyond the mere eco-art circles (but this would require further exploration). Gablik made such a claim in 2004, in her 'position paper' for the first Monongahela Conference, arguing that

> "the art world of today appears to have bifurcated into two completely different aesthetic paradigms, each one differing sharply in their view about the meaning and purpose of art. In the first instance [...] autonomy implying social separateness as the basic premise of art-making. In the second instance are artists who want art to have some worthy agenda outside of itself, and a socially redeeming purpose" (Gablik 2004).

Gablik observes that the strength of the latter lies in their "decentralized network structures", which she characterizes as "communities of practice" (pointing especially at the Internet), by contrast to the institutional networks of the more established art world organizations, which she describes as "hegemonic, competitive, institutional structures of modernism" (ibid.).

A critique of uselessness

Gablik recognizes that the notion of art's uselessness did historically have some limited value insofar as it constitutes a "slap in the face of utilitarian logic" and, quoting Haacke, acknowledges that "[i]n its time [the doctrine of

66 This observation stems from my direct participation at the ASEF experts committee at the "Culture|Futures" conference in December 2009. I was directly involved in such discussions, departing from the views expressed by the director of the "Haus der Kulturen der Welt". The reader should however know that the same "Haus der Kulturen der Welt", a major art institution in Berlin, is organizing from 2010 to 2012, an initiative entitled "ÜBER LEBENSKUNST: Intiative für Kunst und Nachhaltigkeit", with a budget of 3,5 Million euros from the German Cultural Foundation (Kulturstiftung des Bundes), and therefore seems to show a strong commitment to the question of art and ecology: Cf. http://www.hkw.de/de/programm/2010/ueber_lebenskunst/projekt_detail.php [last accessed 10.02.2010].

art for art's sake] did indeed perform a liberating role" (Haacke quoted in Gablik 1991, p. 141). But in contemporary society, and as I myself argued earlier (in my discussion of the Romantic Order in the Technological System, or RO-TS, in chapter 1), such art is alienating, and "in the name of radical autonomy, it is the pure and disinterested art work that can be most readily harnessed into the social process, and lends itself to easy cooptation by the economic apparatus" (Gablik 1991, p. 142).

By contrast, she celebrates Joseph Beuys' concept of the "social sculpture" for its focus on the moral and social dimension of artistic practice, and for going beyond earlier efforts at closing the gap between art and life, e.g. by Allan Kaprow. Gablik advocates for such an approach as Beuys' or Mazeaud's, and celebrates them as artistic work, because they combine a utilitarian with a spiritual function, thereby also connecting different levels of existence.[67] Gablik does not argue for a narrowly utilitarian art, but for an art that retains strong symbolic value and has the ambition to "heal a culture" (ibid., p. 143). In this sense, she quotes Kaprow as stating: "Lifelike art in which nothing is separate is a training in letting go of the separate self. [...] It is even possible that some lifelike art could become a discipline of healing [shifting the meaning of art] from being an end to being a means" (Kaprow, quoted in ibid., p. 145).

In her 2004 "position paper" for the Monongahela Conference, Gablik hints very shortly at the institutional dimension of the change process she heralds (i.e. a dimension which she failed to discuss in her 1991 book, offering herself to a possible critique from the 'Institutional Critique' movement, for her lack of attention to the inevitability and necessity of social institutions, also in the art worlds). She writes: "New paradigms demand new systems to support them. Just as Modernism brought us the white box gallery, I hope the environmental art movement encourages people to see the entire world as deserving of aesthetic attention" (Gablik 2004). But the "entire world" is too wide for being a social institution. As already noted, in the same text she points at the networks of the eco-art movement as "communities of practice" and at the example of the exhibition *Ecovention* in 2002 for its inter-disciplinary association of professionals from many disciplines beyond the art world. Such networks may indeed offer an adequately flexible and open institutional support structure for interdisciplinary, intercultural practices allowing 'entrepreneurship in conventions' as an implementation

67 Such a view is not necessarily shared by all. For example, the Austrian art group "Wochenklausur" rejects the shamanic figure of Beuys, and adheres to a more strictly utilitarian set of goals (personal communication from Wochenklausur members).

of the autoecopoïetic sensibility (cf. my model of 'double entrepreneurship in conventions' in chapter 7).[68]

Gablik's discourse is also echoed in the work of other artists whom Gablik (1991) did not mention, for example Patricia Johanson, who was introduced above, and who insists to see her art as "useful", by opposition to the self-referential discourses of contemporary art where "art feeds upon art, so you get an ever-diminishing dialogue" (Johanson, quoted in Kelley 2006, p. 104), i.e. as I argued in chapter 1, an increasingly autistic autopoïetic process. Johanson criticizes much of what is done under the name of "public art" nowadays, and advocates for a breakdown of the barriers "between aesthetic, functional and ecological – between what is art, architecture and landscape" in order to achieve ecologically and socially functional integrated projects (ibid.). About the risks of 'instrumentalization' of art, Johanson takes a pragmatic approach, deciding on a case by case basis, whether a project will be sufficiently contributing "toward a more benevolent world", rather than aiming for an abstract, detached critical-moral purity.

SECTION 5: THE MONONGAHELA CONFERENCE ON POST-INDUSTRIAL COMMUNITY DEVELOPMENT

In 2004, a conference in Pittsburgh gathered several (predominantly US-American) eco-artists and associated curators and critics/writers, who reflected together on their work as public art potentially bringing transformations to urban places. The documents from that conference illustrate both the issues discussed and the shared spaces of understandings, across different practitioners and discussants of contemporary ecological art.

Among the conference participants were the Harrisons and Suzi Gablik, who were already introduced earlier in this chapter, but also Tim Collins and Reiko Goto (the conference hosts), Ann Rosenthal, the collective Temporary Services, Grant Kester, the London-based collective PLATFORM, A. Laurie Palmer, Stacy Levy, Stephanie Flom (for 'The Persephone Project') Tom and Connie Merriman, Jackie Brookner and Erica Fielder.[69] I will discuss the "position papers" written by the above-listed persons and collectives as inputs for the 2004 Monongahela conference. But before this,

68 The international Cultura21 network, which I have contributed to build since 2006, is also working in this direction (i.e. flexible networks). See www.cultura21.net.

69 Besides, also participating at the conference were: the Pittsburgh collective "GroundZero Action Network", the author Malcom Miles, the sociologist Nicola Kirkham, the environmental reclamation artist Angelo Ciotti and the ecological artist and puppeteer Mark Dannenhauer.

I will discuss Tim Collins' follow-up paper which was written afterward as an attempted synthesis of the insights from the conference (Collins 2004b).

Collins' synthesis

In his post-conference paper, Tim Collins asks which specific contributions an eco-artist can offer, beyond what a social scientist, an architect, an engineer, a landscape architect or an urban designer can contribute already. He notes in passing, that the conference participants had difficulties to define their commonalities, due to their modernist-art education which trained them to over-emphasize their "unique creative differences".

From the conference discussions, Collins concludes that eco-artists consider art as an "interdisciplinary nexus in relationship to the areas of society and ecology" which is situated in, and/or shapes a "Shared Space that is occupied by both people and other species". At the conference, the expression "shared spaces" (pointing at social "relationships, interaction and discourse") was preferred to "public space" because of its better focus on common care and "stewardship through inter-relationship".[70] Eco-artists set as their goals, "to initiate collective care and advocacy" and "to incorporate aesthetic practice in all areas of life [and] the intellectual stimulation of new visions". Collins lists 20 goals from the Monongahela conference talks, among which are also the goals to:

- "infuse sites, places, regions with understanding";
- "revitalize ecosystems in relation to social systems", placing "art at the union between body, mind and watershed" and "address[ing] the co-evolution of biodiversity and cultural diversity";
- develop "focused critical engagement with questions of political efficacy as it refers to engaged art practice" (cf. e.g. the work of the Harrisons, as discussed above in section 2).

Collins further describes the methods of eco-artists as:[71]

- "framed in terms of critical thinking; as investigate-ers" and story-tellers of "alter-tales", "seek[ing] to identify conflicting and conflicted belief systems" (thereby dispelling the concern that ecological art would be lacking a critical edge);

70 The focus on shared space rather than public space, can also be related to the arguments of cultural economist Arjo Klamer (2004) in favor of "common goods" (such as a shared conversation) rather than "public goods" which abstraction and anonymity does not invite to caring relationships and community-engagements.

71 In total, Collins lists 53 methods items, categorized in three ensembles: primary, critical and applied (cf. Collins 2004b).

- "promot[ing] integralism" (thereby lending themselves maybe to a possible critique for the potential of a holistic lack of complexity in their approach, at least at the theoretical level);
- based on "systems knowledge": "we ask nature first, we seek networks, we try to understand the questions of scale, and the relationships between pattern and connection" (thereby echoing several aspects of the ecological literacy and literacy of complexity);
- introducing into projects an "unorthodox approach" which, while it can be an "instrumental method", also allows "to open doors and minds" (thereby paraphrasing Pascale Jeannée of Wochenklausur and pointing at the "inter-conventional" qualities of art, which will be further discussed in chapter 7).

Collins also lists a number of "tools" that eco-artists use in order to:

- frame a "discursive relationship": "trust, caring, respect, discourse, play and risk";
- "externalize creativity": "[h]umility, listening and visioning";
- and work practically "for material manipulation": "Time based, 2D/3D, and scale based tools".

Collins further appeals to the "theoretical ideals of social sculpture" and affirms that the difference between the artist and the "traditional planner" lies in the artist's "role [which] is to create an excitement about change and an investment and passion for perceptions, values and visions". The artist's role should also be to "intervene as advocate for shared spaces and ecosystems", "transcend[ing] primary authorship" i.e. fostering creativity, ownership and empowerment in the community rather than appearing as sole author of creative impetus. Collins is aware that the artist, however, cannot claim to be seen as a mere regular community insider (and thus should not be naively expecting this to happen by itself):

> "We cannot change the fact that we arrive as both outsiders and experts. It is up to us to take the various theoretical issues of agency, representation and dialogic equity to heart and mind and to work with care and consciousness in any resulting dialogue. We must act with full awareness of the fundamental need not to harm. We must act with the clear intent of dialogue, and we must act with a clear understanding of the relationships of power and our role in that context" (Collins 2004b).

These two aspects (at the level of the social interactions, and the political interactions within the art project) will be discussed in chapter 7.

Collins sums up his synthesis in the following "concept map" (figure 6). He also discusses the difference between initiating "shallow dialogue" and "deep dialogue" (from the experience of Reiko Goto and Tim Collins' own project "3 Rivers 2nd Nature", in Pittsburgh):

- Deep dialogue requires a long-term engagement with communities and ecosystems, to engage with all parties, identify research questions, formulate visions, and constantly rework on them.
- Shallow dialogue occurs over shorter time periods (some weeks) and engages the stakeholders together to identify issues and hopefully "instigate deep dialogues".

Figure 6: Collins' "concept map"

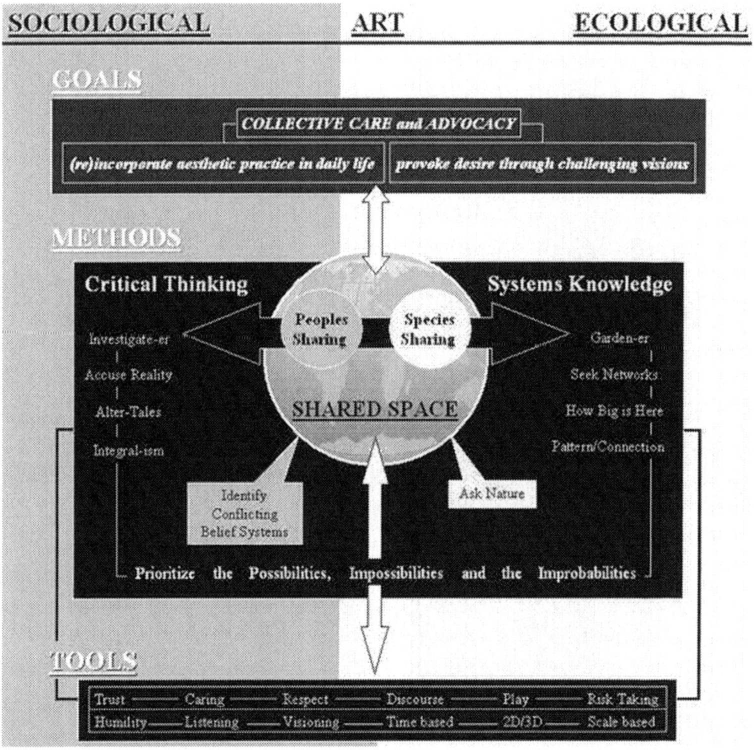

Source: Collins 2004b

The work of many ecological artists (e.g. Patricia Johanson, described above) aims to engage directly in deep dialogues, while some work more often amounts to shallow dialogues which aim to initiate deeper dialogues occurring beyond the artist's direct involvement (e.g. Wochenklausur, whose work will be discussed in chapter 7). Far from being clearly delineated or opposed, these two strategies are more to be understood as poles on a continuum and as (often) complementary.

The position papers

Monongahela conference hosts Tim Collins and Reiko Goto also wrote "position papers" prior to the conference, highlighting the insights that they retained from their own practice of ecological art:

Collins identifies ecological art as "a creative process that recognizes the historic dichotomy between man and nature, and works towards healing the human relationship to the natural world and its ecosystems" and as a practice which has to be "fundamentally interdisciplinary (relational)", i.e. "moving outside [artists'] discipline and institutionalized relationship to society" (Collins 2004a). Ecological art requires "for artists to educate themselves in multiple disciplines". In the project "3 Rivers 2nd Nature", Collins and Goto's "project teams develop strategic knowledge, and broad community discussions about rivers, streams and their attendant banks. The strategic knowledge helps us understand what is happening in and along our rivers and streams, and how we can either help or hinder a public realm recovery". The work is based on "rigorous field methods [...] gathering scientific data", studying "historical materials" and "conduct[ing] a map analysis of each subject area to understand the forces" at play. Collins speaks of "systems-based inquiry". But the knowledge development is not the end-goal of Collins, whose objective is to "[c]reate conceptually informed aesthetics experiences of complex systems" (ibid.).

Reiko Goto further explains the meaning of "2nd nature", i.e. "a cultural landscape, where humanity affects everything", as the basic idea of nature for their work on the nature/culture relationship. Adding to what Tim Collins already argued, Goto refers to the movement of "Bioregionalism" (and to the author Carolyn Merchant), and argues that the relationship to a place is something that needs to be worked on: "I am not an Eskimo rooted in my culture's endemic knowledge of place. I seek to understand nature, I must actively seek to become native to this place that I am occupying. I seek relationships with people who have been or are becoming native to this place. I seek a community of care" (Goto 2004). She reaffirms the importance of interdisciplinarity and "community participation", and further advocates for an openness to the unpredictable and a flexibility in the artist's working approach: "If an artist chooses a life of inquiry and social or ecological transformation, he/she must be flexible to work with many different kinds of people, life-forces, place, materials, and subject matter. Depending on what kind of opportunity, funding, and place, the artist must seek different ways to" work (ibid.).

In their "position paper", the London-based collective PLATFORM describes its own approach in "promot[ing] creative processes of democratic engagement to advance social and ecological justice" (PLATFORM 2004). PLATFORM aims to shape "communities of interest [...] to discuss complex issues", with interdisciplinary work as a small team engaging into "long-term initiatives that embody a deep commitment to London's ecology and peoples while at the same time exploring the city's impacts on the wider world" (i.e. working 'glocally'). The collective insists both on the long time needed "to work a subject out" and on the long-term thinking they aim at, linking "the communities of the dead and the unborn". They also insist that art for them is a "catalyst" i.e. clearly not an end in itself, even when they "create an imagined reality that is different from the present reality", i.e. activate a subversive imagination.[72] PLATFORM also states that for them "the end never justifies the means. The integrity of the process has always been more important than anything that may or may not result from it". One might however wonder, given that real-world actions necessarily imply compromises , how far this is possible, and PLATFORM's claim here might reveal either an extreme rigidity and naivety preventing action, or an exaggerate claim which does not reflect the actual practices of the collective.

PLATFORM identifies and describes seven dimensions in its work:

- *Dreaming*, i.e. having "visions" beyond what is usually considered possible;
- *Researching*, inter-disciplinarily and with communities to "[d]evelop in-depth understanding";
- *Selecting*, i.e. being "pragmatic [in c]hoos[ing] whatever strategy and medium is most appropriate to the aim of the work";
- *Forming*, i.e. setting in motion "a process like sculpture – molding, changing, experimenting";
- *Feeling*, i.e. to "[e]ngage with audiences [...] in the most intense and moving way possible [... m]ove beyond the rational alone [and e]ngage the soul as well as the mind";
- *Connecting* the local and the global, and "enabl[ing] individuals to understand their own power and ethical responsibilities"; and
- *Looking Long*, i.e. "a commitment to place and people over time", with certain works that extended over as much as 15 years.

In her "position paper", Jackie Brookner describes how she works to raise awareness about ecological cycles, with "biosculptures – sculpted ecosystems that demonstrate natural systems solutions to urban and rural water quality and water quantity problems" (Brookner 2004). She works since 1996 with moss, which covers her sculptures and acts, together with the bacteria living in the moss, as biochemical and physical filters for the water which is made to flow over the sculptures (*Prima Lingua*, 1996; *The Gift of Water,* 2001, was the first such sculpture installed outdoors).[73] Brookner also stresses that she works in interdiciplinary teams and with all "stakeholders" on projects, and that her work associates "ecological functionality"

72 On the notion of "subversive imagination" as understood after Marcuse, see ed. Becker 1994.

73 Cf. Spaid 2002, pp. 113-114.

with "sculpture's capacity for metaphor, imagery, form and concept". She describes how she aims to affect people both at conscious and subconscious levels and "through their bodies as well as their minds", and how she observes herself in her "own personal transformative process [...] consciously examining and experiencing my own fears, angers, vulnerability, feelings of insignificance, and negative projections". Brookner further defines her methodological approach as contextual and oriented on "look[ing] for the nodal points – places where multiple social and ecological factors intersect, and where solutions can be multifunctional" (ibid.).

In its "position paper", the collective Temporary Services (Brett Bloom, Marc Fischer and Salem Collo-Julin, formed as a collective since 1998 and based in Chicago) describes its work as "developing non-commercial methods of inserting ideas into publicly trafficked spaces", and its public projects as "temporary, ephemeral, or [...] operat[ing] outside of conventional or officially sanctioned categories of public expression [...] outdoor projects that are encountered by surprise rather than sought out with deliberation like exhibitions and special events [...] projects that do not have permission and challenge expected usages" (Temporary Services 2004). For example, they designed "an array of wearable inflatable devices [...] that could be used to create spontaneous spectacles or disturbances".[74] The collective organizes "spaces for dialogue" (e.g. "an autonomous radio station that runs off of a solar charged battery") and insists on the importance of intense communication and group collaboration rather than individual work. They oppose the "myth of the rugged individual. This myth deceitfully places emphasis on individuals rather than the complex web of people that makes individual accomplishments possible [and it] obfuscates the extreme exploitation of others that is often necessary to achieve this kind of personal gain" (ibid.). The collective aims to work "in relationships that are not competitive and are mutually beneficial". As a counter-narrative in an over-competitive context, this discourse can be very beneficial to social relationships, but, if seen as standing on its own, it severely lacks complexity if it ideologically excludes competitive relationships altogether (notwithstanding the many practical merits it may have in concrete realizations).

In his "position paper", art critic Grant Kester discusses his interest in "collaborative, process-based projects by groups such as Wochenklausur Platform, Suzanne Lacy, Stephen Willats, Littoral Arts, Helen and Newton Harrison, Temporary Services, Park Fiktion and others", because they not only "involve a deprivileging of conventional artistic identity (defined by a wholly autonomous capacity for critical reflection, unproblematized claims

74 This kind of work relates to a whole range of artistic-activist work: Tactical Media, Culture Jamming, Interventionist Art, Identity Correction, communication guerrilla, etc. with in the background also, the theoretical insights from the artistic movement of Institutional Critique, from the international situationist, from Hakim Bey's 'temporary autonomous zones' (TAZ), etc.

to cultural or social exteriority, and traditional models of agency and expressivity)", but also "worked to reconstruct artistic identity through collaborative exchange" (Kester 2004).[75] Kester further mentions his theoretical interest in the issue of the generativity of intersubjective opening (i.e. a concern which comes close to the notion of autoecopoïesis). Kester is critical of Habermas' "conspicuous lack of attention to the specific aesthetic conditions that facilitate discursive exchange across boundaries of difference. Instead we encounter an arid proceduralism that exiles the physical, somatic and nonverbal components of collaboration" (ibid.).

Grant Kester also dismisses critical discussions of 'engaged' art that "limit critical engagement to a straightforward calculation of political efficacy", and/or "fail to differentiate the work as such from the issue it addresses". Thereby, Kester aims to search for specific qualities of ecological & social/'engaged' art.

In her "position paper", Ann Rosenthal points at the search for "more sustainable solutions" and laments the lack of complex questions in contemporary societies. She characterizes her work as "an attempt to transform the way in which we relate to one another and the planet, including:

- shifting from instrumentalist to systems/relational thinking;
- perceiving the environment as 'self' rather than 'Other';
- extending 'community' to include non-human nature [Rosenthal refers to deep ecology author Aldo Leopold];
- rekindling our social desire for sustainable cultures and environments" (Rosenthal 2004).

In her discussion, Rosenthal points at the silence of postcolonial theory when it comes to analyzing the relationships to the "non-human nature", and praises the "significant contributions" from ecofeminist theorists such as Karen Warren. Rosenthal argues that eco-art in general, as well as herself in particular, are striving toward the values of:

- a "Land Ethic" based on a recognition of a community of interdependence beyond the human species' own interest (referring to Aldo Leopold);
- "Systems Thinking" (referring to Fritjof Capra);
- "Sustainability" (pointing at the WCED definition and its mention of future generations);
- "Social and Biological Diversity", acknowledging the value of diversity as "a prerequisite for systems health and resilience;
- "Social and Environmental Justice" (thereby claiming that "all species have a right to a clean environment");

75 Cf. also Kelley 2008, pp. 73-76 (see also the conclusion of chapter 6 below).

- "Collaboration – bridging the boundaries between disciplines, communities, cultures, classes, genders, and species"; and
- "Integrity – closing the gap between what we value and how we act".

In describing her own work, Rosenthal shows an openness to emergence, describing her work "process [as] more organic than linear [...] I've found that attempting to control everything in advance can produce lifeless results". Like most other ecological artists at the Monongahela conference, she works 'collaboratively' and combines interdisciplinary research, community collaboration and public interaction. However, besides this work, Ann Rosenthal states that she also exhibits "in a gallery/museum context". She is aware of the limitations of the "white box separation from daily life" but also sees benefits in this same character, as "offering a respite from the ferocious pace of postmodernity and inviting reflection" (ibid.). Rosenthal's pragmatic statement points at the interest of traditional art exhibitions if considered as a mere complement to the work in situ.

In her "position paper", Stepanie Flom, founding director of "The Persephone Project" (an initiative in Pittsburgh that promotes "gardening as a contemporary art practice [...] recognizing gardeners as artists"), describes her engagement with the community through the realization of "magic penny gardens [...] commissioned community art gardens created by artists using plants contributed from the perennial gardens of neighborhood residents", whereby the participants share not only their plants but also their personal stories (Flom 2004). Stephanie Flom advocates for an egalitarian conception of art and creativity, based on "honor[ing] the existing practices [of ordinary citizens], not imposing new ones". She argues that "all people should have the opportunity and encouragement to tap into their imaginations and into their personal creative practice", and sees her own role as

> "medium, catalyst and facilitator. [...] The artist can share a vision (be the medium) and if there is resonance, others will get excited (catalyst). [...] Now the idea or the project no longer belongs to the artist but to the community. Continuing to reach out to larger and larger concentric circles of community, the artist role switches to facilitator – facilitating the community will to make the project happen" (ibid.).

In their "position paper", Tom and Connie Merriman discuss the evolution of their own practice of "socio-political art" over twenty years. Over these years, they maintained a concern for ecological and social issues as inseparable from each other, as in the cases of "disenfranchised people" and of "the consequences of globalization". But they criticize their own early practice, which consisted in observing/documenting issues with "journalistic narrative art, which stood witness to global events" and was exhibited in art galleries. "In retrospect, this tactic was sincere but naive and not productive for making real change" (Merriman and Merriman 2004). In the 1990's, their work shifted and they became "more involved with the community".

The Merrimans work on issues with "an element of environmental or social conflict", carrying out research and interviewing the different "stakeholders" to understand both their perspectives and their agendas/interests. In some projects, they also conduct "environmental impact studies". The Merrimans are especially interested in the question: "How do you reconcile conflicting belief systems?" However, they state that their artistic work (in diverse (audio-)visual forms) on these conflicts "has not remedied any conflict". In their own judgment, a reason for this lack of effectiveness lies in their attachment to art galleries so far. In their position paper, they thus argue that their own practice must evolve, "go[ing] to places where it can more effectively influence change", and that the "audience must change" too. They further wonder whether it is effective for them to "translate" the views of stakeholders as they do, or whether they should rather "have the stakeholders communicate directly with lawmakers" (ibid.). These reflections are very valuable, as they illustrate a moment when a specific couple of artists comes to a reflexive turning point and aims to explore methodological changes in order to achieve effectiveness as ecological and social artists.

In her "position paper", A. Laurie Palmer highlights another aspect of complexity, which is not stressed so much in the other papers. She criticizes the reductionist bias in the search for "efficiency" and "usefulness", and therefore "aim[s] to maximize complexity instead, and highlight, rather than attempt to resolve, contradictions" (Palmer 2004). Palmer's discourse also works as a healthy skeptical warning against the reductionism of a simplistic holistic perspective: "Attempts to comprehend and visualize wholeness deny what is or what wants to be obscure, risky, dangerous, opaque – and severs a relation to the unknown. We are neither autonomous 'selves' (fully in control of our actions), nor are our intentions ever perfectly mirrored in their effects" (ibid.).

Palmer points out that "the impossible is a link to what we don't know (yet)", and tends to agree with the arguments of the late Marcuse in favor of art's subversive imagination "as a kind of parallel world". Palmer describes her own project (which are also involving communities with public spaces) as "open-ended structures – a certain conceptual autonomy accompanied by indeterminacy as to how, and to some extent if, each project would develop" (ibid.). She describes specific projects where the project space was later reused by participants in ways unforeseen by the artists. She insists on opposing "rationalized efficiency", and uses terms such as "indirection", "changing understandings", "not [...] a predetermined narrative" and "evolution", showing a sensibility to emergence.

Stacy Levy's "position paper" describes her interest in "natural processes" and how she works "to create visual metaphors for complicated systems" (Levy 2004): "Much of my work diagrams what is too slow or too fast, or too microscopic, or too vast to see or comprehend." Levy works in interdisciplinary teams on specific sites, where she is "working on making the layers and the connections of the site become more apparent". For example, she celebrates storm water runoff and aims to "celebrate the process rather than sweeping it under the rug".

However, Stacy Levy also confesses her difficulties in facing "the bottomless pit of stakeholders" in concrete projects, and her lack of patience with the "myriads of meetings" that she must attend. She therefore sees a challenge in "[l]earning new ways to work more directly with the community while retaining the time and space to create new ways of seeing the site" (ibid.).

In her "position paper", Erica Fielder also refers to ecofeminist philosophy (as Ann Rosenthal does), and stresses "the ecofeminist notion of care and relationship with all life" (Fielder 2004). She characterizes her own work as "intended to help dissolve a form of mind-body dualism maintained by our western social system [with a separation between] the dominant [...] 'rational' mind [and] nature as an abstract, female-like entity with resources to be manipulated and extracted". Fielder hopes to contribute to environmental healing by restoring the many senses and sensivities that have gone numb in western modern societies: "For example, watersheds, like neighborhoods, can be contemplated in tiny increments. Each one of us lives near a small wash, creek or stream, although it may be hidden within a culvert. With some attention, we can learn to feel the slope of the land and course of the water flow, sense the direction of the breezes and become acquainted with the plants and animals that live there" (ibid.). To this purpose, Erica Fielder developed *Walking In Deeper*, a set of handmade nature guides that aim to "open awareness to the sensual". She also made a *Bird feeder Hat*: "A participant wears a wide-brimmed, brushy hat covered with seeds. She or he must sit silent and still in order to feel the movement of birds on the hat" (ibid.). Such experiences, Fielder argues, can further awaken people to "a sense of kinship with the non-human world" (echoing the insights from David Abram discussed in chapter 4, section 5). With such an awareness, her hope is that "people will become conscious of the destructive impact that dualistic thinking has on their place, will begin to take action to reverse this condition and will fall in love with their home watershed" (ibid.).

SECTION 6: THE EXHIBITION ECOVENTION

Curated by Amy Lipton and Sue Spaid in 2002 at The Contemporary Arts Center (Cincinnati, Ohio) in collaboration with ecoartspace and greenmuseum.org, the exhibition *Ecovention* and its accompanying catalog *Ecovention: current art to transform ecologies*, is considered as a landmark in the self-definition and the recognition of ecological art (or 'ecoart') within the art world of contemporary art.[76]

Ecoventions

As expressed by Charles Desmarais, the Director of the Contemporary Arts Center in 2002, the exhibition demonstrated "the value of ecological intervention – 'ecovention' - as a contemporary art tactic" (Desmarais in Spaid 2002, p. i). Sue Spaid states that the word "ecovention (ecology + invention)" was coined in 1999, and defines the term as "an artist-initiated project that employs an inventive strategy to physically transform a local ecology" (Spaid 2002, p. 1), i.e. "designed with some intended ecological function" (ibid., p. 10). She points out that the exhibition *Ecovention* "focused on realized ecoventions, because artists' proposals, or 'visionary fantasies', rarely change public attitudes the way novel experiences do". This focus is of importance, as it differs from the common practices in art galleries and museums, which generally tend to prefer "visionary fantasies" to concrete interventions (as was illustrated above in the case of the galleries reactions to Patricia Johanson's work). Spaid also points out right away that such concrete ecoventions also imply that work is carried out in interdisciplinary teams and with the local communities (as was also repeatedly illustrated in the previous pages).

The artistic merit of ecoventions, Spaid argues, is that it "brings forth not beautiful images, but new realities, which are presumably original or inventive" (ibid., p. 3). Therefore, the "standard of inventiveness" has to be thought differently, i.e. not only in art-historical terms, but also in ecological, social, technical and other terms depending on the kinds of experiments being conducted on the sites. Spaid also argues that ecoventions, as art-based experiments in the real world, have comparative advantages vs. scientific experiments: She argues that "ecoventions are able to withstand a higher level of risk" (ibid., p. 4) and indeed are less constrained by a high pressure for scientific results and efficiency, thereby allowing broader and unusual experimentation. She quotes Lynne Hull, according to whom "ecological art will often differ from ecological restoration science in its process rather than in its intent [because, whereas] the scientist has to go through this scientific method, which can narrow perspective, and therefore he or she can lose track of the larger picture [the artist can be] open to all possibilities [and] can question and redefine anything at any step", allowing a more iterative process. Spaid adds that "[s]uch experiments usually cost less as works of art and garner broad support as community-building public pro-

76 Out of numerous informal exchanges with ecological artists since 2006, I noticed that many of them (and especially US Americans) were mentioning this exhibition. Several websites on ecological art and on art and sustainability, as well as artists' writings (e.g. Kelley 2008) also refer to *Ecovention.*

jects".[77] She gives as an example the work by Mel Chin, *Revival Field* (1990-1993), who in collaboration with a scientist from the US Department of Agriculture, made one of the first experiments in 'phytoremediation' (i.e. 'hyperaccumulator' plant-species absorbing toxic metals from soils). Spaid describes how the scientist earlier failed to obtain funding, and how Mel Chin managed, with great difficulties (and, interestingly, having to eventually develop an art-historical argumentation in terms of conceptual art), to convince the NEA to grant her $10,000. Spaid further tells of the success of the project and of the boom of the phytoremediation industry since then (which literally became a multimillion dollars industry in the meanwhile).

Ecoventions have remained relatively invisible in the art worlds, although a good number of concrete projects have been realized in the three last decades of the 20th century, Spaid observes, complaining that "it is truly amazing that so many built projects remain so invisible". She further points at some of the art world related reasons why such works tended to remain invisible: "Unlike a typical work of art that can move from one community to another, or is part of a body of work that can be discussed as a whole, most of these projects have impacted local communities in rather peculiar ways and therefore have remained local" (ibid., pp. 12-13). Furthermore, such work is often integrated "into public works (sewage and waste-water treatment plant, public gardens, public landfills)", is not easily fit for indoor exhibiting, nor for collecting. These are several hard characteristics that make ecoventions/ecological art indeed especially difficult to incorporate in the mainstream art world institutions as described e.g. by Howard Becker (1984). As a consequence, "the majority of [these] projects are still little known among the art world cognoscenti" (ibid.).

Five dimensions of ecovention

Lipton and Spaid's exhibition and catalog discuss a great number of realized ecoventions, and twelve artists produced site-specific ecoventions in and around Cincinnati alongside the exhibition. For the exhibition, Lipton and Spaid divided ecoventions in five categories, or rather five dimensions (as the concrete projects often involve several of these dimensions):

- "activism to publicize ecological issues / monitoring ecological problems";
- "valuing anew / living with brownfields";
- "biodiversity / accommodating species / studying species depletion";
- "urban infrastructure / environmental justice";
- "reclamation and restoration aesthetics".

77 I will not discuss here, whether this claim is confirmed in practice, as this would require extensive empirical research to check the comparative levels of costs and of community support in different projects.

With the first of the five dimensions, Spaid points at "artists disclos[ing] problems and often suggest[ing] solutions [and at] works [that] stand as a means to an end, unlike poetic practices that offer ritualistic acts of healing or consciousness raising and are conceived as ends in themselves", or in other words, Spaid points at "works that often weave their way into the fabric of society to intervene in situations in unexpected ways" (Spaid 2002, p. 21). As typical historical examples of such interventionism, Spaid points at Joseph Beuys, Hans Haacke and the Harrisons, but also at two groups with different approaches: "Ocean Earth" and "the Center for Land Use Interpretation", and at the work of artists George Steinmann, Basia Irland and Kathryn Miller in the 1990's.

An artist-run, incorporated research and development think tank creating commercial (i.e. sale-able) technologies, Ocean Earth Development Corporation was founded in 1980. They produce things as diverse as satellite imagery, water-based energy production or bodywear. Numerous artists worked in this group (e.g. Peter Fend). At their first exhibition (*Art of the State*, 1982), they proposed a world tax for pollution that would be based on satellite-image monitoring. They also used their satellite imagery, which they started selling to television channels (e.g. CNN, NBC, CBS and "several European channels") in the 1980's, to point at political issues (The Falklands War in 1982, Russian submarine bases, Pakistan's nuclear facilities), but also at ecological issues (e.g. Chernobyl in 1986, and the situation of the Amazon basin back in 1983). However, Spaid also notes that the group had difficulties to sell its more ecological satellite imagery to the mainstream media, and describes how the satellite-imagery based knowledge analyzed by Peter Fend was also misused, for example by Iran in the Irak-Iran war.[78]

Another artist-run group, but non-profit (unlike Ocean Earth), the Center for Land Use Interpretation (CLUI) also included numerous artists in its ranks since its inception in 1994. The CLUI carries out research on human uses of the land (e.g. at industrial and military sites), producing data at a "wildly prolific" level according to Sue Spaid. The CLUI also sponsors the "Land Use Museum Complex" (in Wendover, Utah), produces itinerant exhibitions and organizes guided tours. Their goal is to publicize "unusual and exemplary" industrial sites, without explicitly denouncing them, and taking a 'neutral' attitude (and Spaid draws a connection between Robert Smithson and the CLUI). Their argument for this choice is that "it may be more effective to put up a Suggested Photo Spot sign in front of a landfill than it would be to write an essay about why it (and other places like it) shouldn't be there in the first place – or why their presence is unavoidable" (Brendan Bernhard, quoted in Spaid 2002, p. 45).

More clearly supporting ecological transformations, the "Forum for Sustainability" created by the artist George Steinmann in 1996, also caught the

78 Cf. Spaid 2002, pp. 39-40.

attention of Sue Spaid.[79] A research center in the Priluzsky region of the Komi Republic (Russia), i.e. in the midst of one of the oldest forests of Europe, but also Russia's second largest energy reserves, Steinmann's "Forum for Sustainability" collaborates with communities and scientists in conserving the forest and promoting sustainable forestry. Steinmann also brought together Komi healers and a pharmacist for the development of phytotherapeutic products based on the Komi's medicinal knowledge. The initiative is supported by the Komi branch of the WWF.

Basia Irland's work *A Gathering of Waters: Rio Grande, Source to Sea* (1995-2000), consisted in a participatory performance along the endangered river, conducted by hundreds of stakeholders who collected small samples of water in a canteen and wrote down contributions in a field book, which were voluntarily passed from community to community. Irland also designed several rainwater collection systems, and founded the 'International Water Institute', a NGO "dedicated to archiving the use and abuse of water sources".[80]

The biologist and artist Kathryn Miller conducted several performances in the 1990's, seeing barren land in the tradition of Hans Haacke, Yoko Ono and Alan Sonfist, and distributed *Seed Bombs* (1999-2001). With Michael Honer, she realized in 1994 the *Desert Lawn* actions, dressing up as Emergency Room hospital doctors running around in the desert with a plot of grass on a gurney attached to an intravenous drip: As Spaid notes, in this action and later installations such as *Lawns in the desert* (1995, displaying the action's props together with the quantity of water required annually by this plot of lawn), Miller and Honer stress "the emergency caused by the lawns' pathological state of dehydration and its dependence on deadly chemicals" and question the absurdity of widespread suburban lawns (Spaid 2002, p. 52).

With the second of the five dimensions discussed in the exhibition and book ("valuing anew / living with brownfields"), Spaid discusses "artists who have altered views of particular sites and objects that most people consider degraded, dirty, contaminated, or useless. Artists have done so not by salvaging sites, per se, but by challenging the value judgments underlying words such as irreparable, impure or trash. They thus provoke us to see anew, as our perceptions, which are tied to our values, change" (ibid., p. 53). From this perspective, Spaid refers back to the legacy of Robert Smithson, and discusses the "Nine Mile Run Greenaway Project", and the works of Suzanne Leibovitz Steinman.

The "Nine Mile Run Greenaway Project" (1997-2000), in Pittsburgh, was conducted by three artists (Tim Collins, Reiko Goto and Bob Bingham), two of whom later hosted the Monongahela conference, and an attorney (John Stephen). They dealt with a flood plain that had become an illegal

79 Cf. ibid., pp. 46-48.

80 Cf. ibid., pp. 49-51.

dump site for millions of tons of 'slag' (a by-product of the steel industry) by the Duquesne Steel Company over much of the 20th century. The team aimed to find sustainable solutions for the site while strengthening community ownership. In the first year of the project, they provided training to volunteers and organized a community dialogue. In the second year, they involved experts into an interdisciplinary study, in close association with the community. In the third year, they presented the results to stakeholders with an exhibition in a Pittsburgh gallery (*Conversations in the Rust Belt: Brownfields into Greenways*, 1999) and with communication tools, and developed guidelines with the community participants. They came to realize that moving the slag to another space would not be a real solution, and even found out that "the slag heaps offer more biodiversity than typical lawns or city parks" (ibid., p. 62). In 1999, the team sprayed a mixture of nutrients and grass seeds on the slag, and they successfully advocated for safeguarding the site and promoting it for riverside bicycle commuting and other activities, while recognizing its role in carrying storm water.

A former student of Suzanne Lacy, Suzanne Leibovitz Steinman "is best known for facilitating massive public projects, which have involved her collaborating with" many different groups of stakeholders (ibid., p. 56). Spaid celebrates Leibovitz Steinman for her ability to "regularly change the function of even an ordinary site [with i]magination, persistence and team building" (ibid., p. 57). She reveals rather than conceals degraded sites (as e.g. in *Mandela Artscape*, 1998-1999, where she integrated the remains of a freeway collapse into a memorial garden, in West Oakland, California), and often provides or arranges for free trainings of community volunteers. Leibovitz Steinman also facilitated local urban agriculture, with *Food for Thought: Urban Apple Orchard* (1994), an experimental plantation of 'antique' apple species, and with *Gardens to go* (2001, also in West Oakland), an organic garden growing in recuperated bathtubs, beds, old wooden doors, on top of soils which themselves couldn't support the vegetables (because of lead pollution).[81]

Among her more recent work, Leibovitz Steinman was invited in 2008 in Hamburg, in Wilhelmsburg (a southern city district of Hamburg, home to a low-income immigrant population and cut from the rest of the city by the industrial harbor) by Anke Haarmann and Harald Lemke, curators of *Culture|Nature*: For the first 4 weeks of her stay, she installed a temporary garden (in containers) on a busy square near a 'S-Bahn' train station, inviting residents to pick the fruits, vegetable and flowers from there. In a second phase, and in collaboration with a local group called the "intercultural garden", she organized a planting action nearby, on a raised flower bed that is part of the city greens. As the catalogue of *Culture|Nature* comments: "Steinman wanted to demonstrate that gardening is possible almost anywhere, even without owning any land" (eds. Haarmann and Lemke 2009,

81 Cf. ibid., pp. 56-60.

n.p.). The third part of Leibovitz Steinman's contribution to *Culture|Nature* was the installation *Kitchen Table Diplomacy*, i.e. a simple table with chairs located outside the main exhibition location, where different gardening associations donated products, and where informal get-togethers were organized over several weeks, whereby gardeners and exhibition visitors exchanged seeds, harvested fruits and stories.

Leibovitz Steinman argued: "I love to do art that people might not even recognize as art, but they find it when they're going to the grocery store or to work" (Leibovitz Steinman, quoted in Spaid 2002, p. 60). This position can be found also in the work of other artists involved in comparable projects, some of them deeply embedded in neighborhoods, such as Rini Biemans and his group 'Creatief Beheer' in the Oude Noorden neighborhood of Rotterdam (whose work was empirically analyzed both by Hans Dieleman and by myself).[82]

With the third of the five dimensions discussed in the exhibition and book ("biodiversity / accommodating species / studying species depletion"), Spaid discusses artists whose works address biodiversity, from different angles. She stresses how Patricia Johanson's and Betty Beaumont's projects concretely enhance/restore biodiversity, discusses Lynne Hull's trans-species art, and how artists like Brandon Ballengée and Tera Galanti "reclaim species" (like others reclaim polluted lands). She puts all these artists in contrast to another artist who is much more appreciated by the art world gallery circles, as he spies and thrives on biodiversity, most of the time without contributing to it: Henrik Hakansson.

The interest of Hakansson's work lies in his video monitoring of species behavior, allowing museum visitors to peep in the lives of insects, birds, snakes and plants (and sometimes shortly showing visitors their own live-feed video, i.e. as exhibition visitors, allowing a reflexive turn, as in *Tomorrow and Tonight* at the Kunsthalle Basel in 1999). Spaid summarizes Hakansson's main point as follows: "Real feelings for nature require the right equipment to discern its intricacy. By fanning the flame with supposedly dispassionate instruments, viewers spy on nature's behalf" (Spaid 2002, p. 81).

Betty Beaumont studied environmental design at Berkeley and documented the steam cleaning in the aftermath of the Santa Barbara oil spill, in 1969. In 1977, her environmental art work *Cable Piece* consisted of iron cable from military surplus which, looped twelve times, was formed into a ring installed on a farm land (in Macomb, Illimois) and left to bury slowly into the ground. Infra-red aerial photography later confirmed that the installation hastens grass growth. The following year, she initiated the project *Ocean Landmark* (1978-1980), which aim was both to stabilize 'fly ash' (a by-product of coal) in water, and to sustain marine biodiversity by creating an artificial coral reef. 17 000 blocks of fly ash were dropped off the coast

82 Cf. Dieleman 2006, Kagan 2004.

of Fire Island, in the USA. The project involved a wide-scale collaboration with scientists, engineers and technicians from several universities. Since then, every five years the zone is being studied to monitor the evolution of its ecosystem.

Beaumont also later designed a series of flash cards to inform the general public about the mutations of fish exposed to nuclear waste, based on scientific studies *(Fish Tales*, 1991*)*.[83]

Brandon Ballengée's work is dedicated to amphibians, whose decline and genetic malformations he studies since 1995, in association with several biologists.[84] In conjunction with this scientifically-relevant research, Ballengée also works since 1999 at what he described as "species reclamation", experimentally breeding specimens to recreate a (nearly) extinct species of frogs from the Congo (Hymenochirus curtipes): "By controlled pairing of related species, he literally breeds the species backwards" (ibid., p. 84). Spaid also discusses another artist who, like Ballengée, is breeding a species 'backwards': Since 1997, Tera Galanti is breeding flightless silk moths to recreate flying silk moths. In doing so, she turns back countless generations of domestication by which the moths' wings had been purposefully atrophied, in order to facilitate silk cultivation. Galanti's breeding selects the strong-winged moths. Spaid discusses her work as a critique of the "need to control nature underl[ying] most of society's civilizing solutions [...] By forcing order to perfect their world, humans have thwarted biodiversity" (ibid., p. 88).

With the fourth of the five dimensions discussed in the exhibition and book, Spaid discusses artists who are "predominantly involved with urban infrastructures" and "exploring issues of environmental justice" (ibid., p. 89), i.e. she discusses the works of Alan Sonfist, of Mierle Laderman Ukeles, of Buster Simpson, of artist and bio-engineer Viet Ngo (who patented in 1983 'Lemna', a floating aquatic plant used to treat waste water, and since then made a very successful business out of it while designing site-specific water-treatment installations), of Laurie Lundquist (who participated in the development of urban infrastructures, like Patricia Johanson, and who organizes a summer residency for eco-artists), of the collective Superflex. Spaid also discusses in this context, some ideas of Mel Chin about inventive uses for burned houses in Detroit (Michigan).

Buster Simpson's work deals mainly with water in urban contexts, i.e. waste, sewer, grey water, and rainwater runoff. In 1978, he placed concrete casts of picnic plates at sewage outfalls into the Niagara river, and then exhibited the stains on them, created by the river's contaminants *(When the tide is out the table is set*). The same year, he created *Downspout – Plant Life Monitoring System*, grewing ferns in plumbing pipes on the side of a

83 Cf. Spaid 2002, pp. 72-74.

84 Since 2007, Ballengée also works on insects (e.g. *Love Motels*, Gunpowder Park, London).

building in Seattle, and placing limestone in the pipe's elbow to correct the water's pH level, thereby creating a cost-effective stormwater cleaning system before the water enters the public sewer (the efficiency of which was monitored for several years). He applied the same technique again in 1999 on two other buildings in Seattle. Buster also did a number of performances in rivers to sweeten overly acid rivers near Seattle and New York City, in 1983, 1984 and 1990, using hand-carved, soft-chalk limestone tablets. (Spaid notes that such a technique is nowadays common practice in environmental agencies.) In 1983, he also performed *David taking on global Goliath*, throwing a limestone, with a sling, in the direction of the World Trade Center in New York City.[85]

The Danish collective Superflex (whose work does not always directly address ecological issues, but more often focuses on social and cultural themes) has been developing since 1996 a biogas unit converting organic waste into gas and fertilizers, in collaboration with Danish and African engineers, in order to come up with a portable and simple solution. The units were tested in different contexts, in rural zones in Africa and Asia. Superflex is organized as a for-profit company, and developed its own development foundation, independent of public Danish development aid agencies.

With the fifth dimension, Spaid focuses on "artists [who] have developed and implemented novel strategies for reclaiming water and soil [and] contributed workable models for the fields of ecological restoration and landscape architecture": Betsy Damon, Jackie Brookner, Aviva Rahmani (*Ghost Nets*), Shai Zakai, Agnes Denes, Georg Dietzler and the group AMD&ART.

Betsy Damon won several environmental design awards and received commissions from numerous cities in the USA and China, thanks to her project *The Living Water Garden* (1995-1998) for the city of Chengdu in the Sichuan province of China: The water garden both cleans some of the water from the Fu and Nan rivers (only 250 cubic meters a day – but this already did positively affect local biodiversity) and educates visitors about water. An interdisciplinary collaboration with Chinese experts (including hydrologists and a microbiologist) and an American landscape designer, the park includes constructed wetlands with different water-purifying plants and flow forms designed by Betsy Damon, for the aeration of the water.[86] What Spaid (2002) does not mention about this work is how the first phase of Betsy Damon's work, before the realization of this park, consisted in a "pilgrimage to the headwaters of the river, in Tibet" and, back in Chengdu, in a performance (Washing Silk, whereby she rinsed white silk in the Fu-nan river, and the silk went brown-grey).[87] If I would interpret the work also in

85 Cf. Spaid 2002, pp. 97-101.

86 Cf. ibid., pp. 109-113.

87 Cf. Kelley 2008, p. 79.

Beuysian terms, I would argue that a social sculpture preceded and facilitated the ecovention (which itself also has a 'social sculpture' value).

Sue Spaid also points at the work of Israeli artist Shai Zakai, who addressed the issue of illegal dumping of concrete by concrete-mixer drivers into the stream of the Etziona. Instead of engaging in a legal proceeding against the quarry, Zakai managed to collaborate with the workers in stopping the damage, and she further worked on *Concrete Creek* as an on-site installation (1999 – 2001) and an exhibition and a festival in Tel Aviv (2001), questioning Israelis about their relationships with water streams.[88]

Shifting the focus from water to soil, Spaid discusses the work of Agnes Denes. In 1982, Denes planted, with the help of volunteers, *Wheatfield – A Confrontation* at the Battery Park Landfill (near the Twin Towers and the Statue of Liberty), and harvested it three months later. Denes wanted to confront the value of this agricultural field to the NYC context of multinational corporations, "an island of achievement-craze, culture and decadence" (Denes quoted in Spaid 2002, p. 121). Based upon an original design by Denes following a spiraling mathematical pattern (*Tree Mountain – Proposal for a forest*, 1982-1983), an artificial mountain was created in Finland (near Ylöjärvi) and on it were planted 11,000 trees (of an endangered species of pine) by 11,00 people who received inheritable 400 years certificates as custodians of their trees (*Tree Mountain – A Living Time Capsule*, 1992-1996). The expected benefits (besides preserving an endangered species of pine) included cleansing airborne pollutants and absorbing rainwater runoff.[89] That project was presented by Finland as the country's contribution for the 1992 Rio UN Summit on the Environment. Spaid however, fails to critically assess the issues at stake in Denes' mathematical design for *Tree Mountain*, which carries with it a typical modernist thought-pattern subsuming the natural elements to rational, simple mathematical modeling, instead of exploring the real complexity of nature-culture relationships. In doing this, the work runs the risk of hindering rather than facilitating an aesthetic experience of complexity.[90]

The German artist Georg Dietzler developed in 1989 a system to conserve acorn seeds by freezing them, and used acorn seeds for reforestation purposes (*7000 Acorns*, 1989-1990). From 1993 onwards, he experimented using oyster mushrooms to decontaminate soils polluted by PCBs (polychlorinated biphenyl). The mushrooms can indeed break down the chemical structure of the PCBs, and in the process, they generate compost as by-product. On the basis of these experiments, Dietzler created the installation

88 Cf. Spaid 2002, pp. 118-119.

89 Cf. ibid., pp. 120-124.

90 Such a sound critique is to be found elsewhere, in an article by Larissa Buchholz and Ulf Wuggenig (2002, pp. 157-158).

Self-Composing Laboratory (1999) in Aachen, for the exhibition *Natural Reality* curated by Heike Strelow.[91]

Finally, Spaid looks at the work of the interdisciplinary group AMD&ART, founded by the historian T. Allan Comp in the mid-1990's and conceived as "a community-driven design team" (AMD&ART, in Spaid 2002, p. 137). AMD stands for acid-mine drainage, a result of coal-mining which affects the groundwater, the local ecosystem and deprives the communities of potable water. In 2001, AMD&ART completed a land-reclamation project on a former mine in Vintondale, Pennsylvania, with an innovative design of ponds neutralizing the acidity.[92]

CONCLUSION

The historical emergence and development of environmental and ecological art marks an example in the way art worlds can play a constructive role in the construction of culture(s) of sustainability. With its diversity in approaches (from social sculpture to land reclamation) and with its common threads and principles (as discussed at the Monongahela conference), it seems that this movement fosters an advanced, elaborate form of aesthetics of 'the pattern which connects'. But the remaining problematic issue, in some instances, is whether or not it fosters an aesthetic experience of 'patterns that connect', and sufficiently develops aesthetics of complexity. Most especially, a simplistic form of holistic discourse is looming behind some of the artistic discourses introduced in this chapter.

However, this question cannot be settled here. Assessing the degree of sensibility to complexity, would require elaborate qualitative empirical assessments on a project basis (i.e. case by case). All that can be said, as a general statement, is that, although a sensibility to complexity, and aesthetic experiences of complexity, cannot be assumed 'a priori' to be present in all ecological art projects, they are likely to occur when and where the projects explore complex ecological and social processes and value emergence, instead of superimposing upon them some pre-designed models.

An overview of the 'contemporary art' art world's treatment of the relationships between art and ecology, and of its relationship to the search process of sustainability, as well as to the cultures and aesthetics of sustainability as I proposed to understand them, is however not completed, with a mere focus on 'environmental' and 'ecological' art, however widely this movement is defined. The following chapter will therefore introduce other approaches which, in the past decade, have dealt with the interplay of ecology, sustainability and art.

91 Cf. Spaid 2002, pp. 125-127.

92 Cf. ibid., pp. 128-129.

6. Sustainability and Ecology as keywords: a decade of contemporary art

INTRODUCTION

The overview of discourses and practices in contemporary art at the interplay between art and ecology, started in the previous chapter with a focus on the emergence of the movement of 'environmental/ecological art' since the late 1960's. This chapter will widen the perspective, to explore the 'contemporary art' art world's relative interest for the notions of ecology and sustainability in the past decade.

About the terms in use: art and sustainability

The terminology of 'sustainability' has been introduced slowly into artistic discourses and practices over the past decade. In parallel to certain researchers from the field of sustainability (such as the sociologist Hans Dieleman in the Netherlands)[1] who were attempting to introduce the theme in their exchanges with the art worlds, from the late 1990's onwards, curators and writers from the world of contemporary art started discussing the relevance and different dimensions of sustainability for art (e.g. Hildegard Kurt in Germany, Shelley Sacks in the UK, Maja and Reuben Fowkes in Hungary, Stephanie Smith in the USA).

Hildegard Kurt organized in 2002 a conference at the Akademie der Künste in Berlin, and edited (with Bernd Wagner) a collective publication on the relationships between sustainability, culture and art (eds. Kurt and Wagner 2002), as part of her ongoing research for a "future-able modernity" ("Suche nach einer zukunftfähigen Moderne"). The year before, in 2001, she was involved in the organization of a conference entitled "Ästhetik der Nachhaltigkeit" ("aesthetics of sustainability"), at the Evangelical Academy of Tutzing. The participants in Tutzing signed a "Tutzinger Manifest" which

1 Hans Dieleman started an "Art & Sustainability" program at the University of Rotterdam from 2000 onwards (in which I was enrolled in 2001, as student).

is now recognized as one of the early calls for the recognition of culture as a fundamental basis for sustainability.[2]

Kurt was among the first to explicitly discuss the notion of "aesthetics of sustainability" (Kurt in eds. Strelow, Prigann and David 2004, pp. 238-241). Kurt refuses to "see an aesthetic of sustainability as a completely new aesthetic", because it would both be "illusory" in the face of the "plurality of contemporary life-styles" and set up "a new edition of 20th-century constructivism, which inclined to totalitarianism" in her opinion. However, Kurt understands "aesthetics of sustainability" as "a new sensitivity to questions of aesthetics on the part of those involved in promoting sustainability". She further gives a varied number of criteria for aesthetics of sustainability:

- "to search for forms of the less, but also for forms of nature-friendly opulence" (i.e. a sensitivity to both frugality and an eco-friendly form of hedonism?);
- "to create cultural diversity";
- "to sensitize people to a new time culture and forms of spatial planning that are respectful of the environment";
- "to grant a productive constructive force to "sensual awareness" (Rudolf zur Lippe) and aesthetic competence, and use this force for designing life-sustaining futures";
- "to shed light on "forgotten nature" (Günter Alter)[[3]] in our culture" (i.e. to highlight that culture is part of nature, as stated in chapter 4 about ecological aesthetics);
- "to include the social dimension on this basis, especially with respect to justice, participation and community; for an aesthetics of sustainability will always be an aesthetics of participation as well".

Kurt also understands aesthetics of sustainability as "a new sensitivity to the fact that there is effective creative knowledge, beyond technical-instrumental reason", thereby contributing to a movement beyond the 'culture of unsustainability'.

Therefore, Hildegard Kurt's notion of "aesthetics of sustainability" already contained some of the elements of what I propose to view as 'aesthetics of sustainability' (in chapter 4), but I would argue that Kurt did not sufficiently articulate these different elements, and the insights she points at, into a coherent theoretical discourse.

Hildegard Kurt also developed the analysis of a "double deficit" ("zweifache Defizit") between the development of the concept of sustainability, and the art worlds, with:

---

2 The Manifest is available online at: http://www.kupoge.de/ifk/tutzinger-manifest/ [last accessed: 10.03.2010].

3 I believe Kurt meant actually Günter Altner, and that a typographic mistake slipped into her text.

- on the one hand an absence of cultural dimensions in the consecrated 'three-pillars model' of sustainable development in the 1990's and in the theoretical elaborations of the concept of sustainability which were based mainly on the natural sciences and on social sciences (including economics, political economy and public administration) but lacking any aesthetic elaboration;
- and on the other hand the absence of the notion of sustainability in the art worlds and in cultural policies, in the same time period (apart from the smaller circles of e.g. ecological art), and their lack of interest, or even suspicion against, the concept of sustainability.[4]

In Kurt's analysis, while the actors in the sustainability debate have a very limited understanding of art (as mere decoration or communication), the members of the art worlds tend to restrict the theme of sustainability to narrow environmental topics.[5] Furthermore, the common understanding of culture grounding cultural policy is still "based on a polarized opposition of nature and culture" (Kurt and Wehrspaun 2001, p. 82).

In parallel to Kurt's discussion on aesthetics of sustainability, the Hungarian curators Maja and Reuben Fowkes introduced the term "sustainable art" to qualify certain recent artistic practices, and have proposed some "principles of Sustainability in Contemporary Art" (Fowkes and Fowkes 2006), where they focus on notions highlighting the normative, ethical dimensions linked to aesthetics of sustainability :

- ecological citizenship, with "a shift away from [the] anthropocentric model" of e.g. land art, and a human responsibility in the biosphere "to include the non-human world in our moral universe";
- "a renewed sense of social responsibility" and a commitment to "grass roots democracy", with collaborative practices and attention to quality of life;
- and a gender perspective informed by "feminist critique of land and environmental art of the 1970's" and more generally of patriarchy's environmental consequences.

These notions especially echo Fritjof Capra's discourse on communities and civil society (cf. chapter 2, section 3) and his critique of patriarchal

4 Cf. eds. Kurt and Wagner 2002, pp. 15-17. Cf. also Kurt 2000, pp. 112-118.

5 Cf. Kurt 2002. In 2007, I co-supervised a "Diplomarbeit" in environmental communication at the University of Lüneburg, where the student empirically tested Kurt's claims of a "double deficit", in interviews with artists and members of the German environmental administrations. She partly confirmed Kurt's claims: Cf. Lang 2007.

cultures (cf. chapter 1 section 1 and chapter 2 section 3).[6] The notion of ecological citizenship echoes Morin's understanding of humans and nature "co-piloting" the Earth. The Fowkes further insist that sustainable art "draws on radical critique of art and society and the dematerialized practices of conceptual art to offer sustainable alternatives in art and life", pointing at the reflexive and critical qualities at play in the search process of sustainability (and which will be further discussed with Dieleman 2008: cf. chapter 7).

The Fowkes point further at artists "embodying both fantasies and possibilities of technological revolution" and echoing the discourse of ecological modernization which claims that "clean technologies and efficiency savings have the potential to solve environmental problems without the need for radical social change" (Fowkes and Fowkes 2006). Here, notably absent is a critique of the Technological System (cf. chapter 1), although the authors do explicitly warn against "the hand of authoritarian technocracy behind the optimism of ecological modernism".

However, a much more radical critique of technology is to be found in a later text of the Fowkes, where they are referring back to critiques from the 1970's (in the context of the the Club of Rome's *Limits to Growth* and of the 1972 Stockholm conference) on the part of the Croatian sociologist Rudi Supek (who denounced the "limitations of technological solutions" and advocated for "social solutions involving radical change to modes of production and human relations to nature") and of the artist Gustav Metzger who critically analyzed "the problematic relationship of artists to technology in modern times" from the Bauhaus to new media art (Fowkes and Fowkes 2009).

The Fowkes also do repeatedly use the term "interconnectedness" (or "experience of connectedness"), which I also discussed in chapter 4 and which numerous other authors and artists discussed in chapter 5, also used. They also re-use Suzi Gablik's argument in favor of 'reconstructive postmodernism', as in the text accompanying their 2002 exhibition *Human/Nature* (at Traffo Gallery, Budapest): "We witness a growing tendency to seek out new forms of spirituality and a rediscovery of ethics in art. Reconstructive postmodernism regards interconnectedness, social responsibility and ecological attunement as the crucial issues for human creativity" (Fowkes and Fowkes 2002).

Kurt's or Fowkes' arguments and terminologies, though often cited, are not taken for granted or generally recognized by the art world of contemporary art. Other terminologies are also made up 'ad hoc' for specific exhibitions, as e.g. for *Timeout* (2007, at Kunstmuseum Liechtenstein), where the definition of sustainability was associated specifically to the discussion of time and the societal goal to "slow down" human activities (such as in "slow food" and "slow cities"). Other understandings are developed in artists writ-

6 A critique of patriarchy in contemporary art was also conducted by Gablik (1991), which I discussed above, in chapter 5.

ings, such as in Caffyn Kelley's *Culturing Sustainability* (2008), where sustainability is understood as "the possibility or memory of cultural systems that integrate with – and enhance – the living and dying of planetary systems", pointing at ideals of non-violence, peace, decentralization, diversity, complexity, situated-ness, and "imply[ing] holding, suffering, nourishment, and interrelationship" (Kelley 2008, p. 12).[7]

A decade of art exhibitions

The notions of sustainability and sustainable development also gave rise to a wave of exhibitions in the past decade: For example, in 1999-2000, the Kunstraum at the University of Lüneburg was organizing Dan Peterman's *Nomadic Greenhouse* project, which also involved the exhibition of a few other artistic proposals dealing with the notion of sustainable development. In 2005, the Smart Museum at the University of Chicago organized the exhibition *Beyond Green: Toward a sustainable art* (curated by Stephanie Smith, who had already curated the exhibition *Ecologies* for the Smart Museum in 2000). In 2007 was organized at the Art Museum of Liechtenstein, the exhibition *Timeout. Art and Sustainability*, while in the same year I organized and curated, with Bettina Steinbrügge, the exhibition *Everything will be fine / Alles wird gut* (at the Halle für Kunst Lüneburg and Kunstraum of the University of Lüneburg) which was addressing the intersection of sustainability as a search process and art as social inquiry. In 2008, the exhibition *Greenwashing* (curated by Ilaria Bonacossa and by 'Latitudes') at the Fondazione Sandretto Re Rebaugendo (in Turin), aimed to critically address environmentalist discourses, and distanced itself from ecological art.

Several exhibitions in recent years were dedicated especially to the theme of global warming / climate change, such as most especially *Weather Report: Art and Climate Change* (2007, Boulder Museum of contemporary art), curated by Lucy Lippard (cf. ed. Gerdes 2007), as well as several exhibitions stemming from David Buckland's *Cape Farewell* project – a series of expeditions to the Arctic involving artists and scientists, and more recently expeditions to other climate change affected regions.[8] Alongside the

---

7 I should note, not just to credit myself but to point at the inter-relationships between discourses of artists and my own discourse as a researcher, that in her text Caffyn Kelley also commented: "I have previously described the word 'sustainable' as ruined by its association with 'development'. But I am encouraged to retrieve the concept from the compost by Sacha Kagan, who with Volker Kirchberg poses sustainability as "a new frontier for the arts and cultures" in a 2008 publication" (Kelley 2008, p. 12). This should serve as a reminder that the discourses of the social scientists can play a role within the art worlds, and therefore that researchers should deal responsibly with the "self-fulfilling prophecies" that they may foster, and not claim any neutrality.

8 Cf. Buckland et al. 2006.

UN climate conference in Copenhagen in December 2009 ('COP 15'), an unprecedented wave of exhibitions simultaneously took hold of the city of Copenhagen with numerous art institutions setting up exhibitions in parallel (e.g. *Rethink* at the National Gallery of Denmark; *(Re-) Cycles of Paradise* at DGI-Byen i.e. the site of the 'Climate Forum': the NGOs' alternative climate summit; *New Life Copenhagen:* a festival across several locations, organized by the collective 'Wooloo').[9] An increasing number of art institutions worldwide have been recently engaging into similar directions.

SECTION 1: "ECO-CENTRIC TOPICS" IN CONTEMPORARY ART

The exploration of relationships between art and ecology is not limited to the movement of environmental/ecological art, as discussed so far, but also involves artists whose work usually is not associated to ecological or environmental art. This is is especially clear in the discourse of curator and writer Linda Weintraub (2006), who discusses artists usually (more or less closely or loosely) related to ecological art (Shelley Sacks, the CLUI, Superflex), but also artists related to Culture Jamming (Reverend Billy Talen), and to the wider art world of contemporary art, some of whom are especially problematic from an ecological perspective (Eduardo Kac, Damien Hirst).

Ecocentrism in art

In 2006, the curator Linda Weintraub started a publication series on art and ecology ("Avant-Guardians: Textlets on Art and Ecology"). I will focus on her *Ecocentric Topics: Pioneering themes for eco-art* (2006) which presents a peculiar mix of artists working on ecological and biological themes, i.e. a selection of artists including work which is not usually associated with eco-art. Most especially, Weintraub's selection of artists includes a large proportion of art works that are especially suitable to indoor gallery exhibitions (and in several cases, constitute saleable items for the art market).

Weintraub bases her editorial selection on the notion of "ecocentrism", for which she claims authorship and which "describes humans relating to

9 For more details, see the website http://www.arts4cop15.org [last accessed 11.03.2010]. Among the many other forms of art interventions in the city, the Yes Men were present with *Good COP15* (i.e. yet another alternative climate conference: see http://www.good-cop15.org/), as well as many other collectives including Cultura21 (for which I was also personally involved, with the installation *Caution Border* (created by students at the CCC Program of the Geneva University of Art and Design, and students at Leuphana University Lüneburg) alongside Cultura21 Nordic's *Embassy of Sustensistan*).

the nonhuman environment in a harmonious, respectful, and pragmatic manner" (Weintraub 2006, p. 12).[10] Weintraub opposes the ecocentric to the anthropocentric, i.e. "the practice of interpreting reality exclusively in terms of human values and human experience. It omits ecological inclusiveness." She also opposes "the prefix eco-, which means home or habitat" and is "focused on ecological relationships", to the prefix ego-, i.e the "self-absorbed, self-centered, self-involved, and self-serving". Weintraub argues that "[s]witching prefixes from 'ego' to 'eco' directs awareness away from selfish interests and toward ecologically inclusive considerations." She further denounces the "fallacy of human-centeredness" and questions the value of honoring "humanists who utilize reasoned and critical thinking to promote the self-interest of our species." And she encourages artists to engage into "ecocentric themes" and thereby "introduce such fundamental considerations and their challenging implications into the public discourse" (ibid., p. 13).

This goal is very noble and relevant to the ecological crisis, but I have, once more, to point out the risk of holistic reductionism, if the egocentric dimensions of existence are negated. And I mean, not only the human egocentrism, but the dimension of ego-centrism which is present in each and every living organism, and inseparable from the ecological processes, leading Edgar Morin to talk, not only of an eco-auto-organization of life on Earth (cf. chapter 3, section 1), but, at a more detailed level, of a paradigm of "auto(geno-pheno-ego)-eco-re-organization", in which the prefix "ego" belongs to the complex reality of life.[11] Weintraub too comes to realize this, at a later point in her discourse, i.e. when she discusses the theme of power (see below).

Weintraub proceeds further with an insightful comparison on ecological science and ecological art. She starts with characterizing the terms traditionally, i.e. "ecology [as] the science of the relationships between organisms and their environments", and "art [as] a product of free creative expression" (ibid., p. 14). But if understood only as the "free creative expression" of an individual, such an understanding of art would follow the modernist definition of art and have little relevance for ecological art (see Gablik's arguments already discussed), producing in its collision with ecology, "a formula for dysfunction" as Weintraub mentions. Weintraub further proceeds to prove this assumption wrong (i.e. there is no dysfunction, but a very rich field of eco-art), but unfortunately without revising the definition of art. Nevertheless, her discussion of eco-art is insightful. Like every eco-artist discussed here, Weintraub too points out that eco-art is interdisciplinary and "acknowledges the multifarious capacities of the human mind." She sees

10 David Haley also uses the term "ecocentric" for a number of years. But the question of "primary authorship" is, as Tim Collins argued, of decreasing importance from the perspective of eco-art.

11 Cf. Morin 1980, pp. 351-354.

eco-art's complexity, arguing that "ecocentric art is an art of merger, subtlety, and richness. It has little in common with simplifying distillations and isolating separations. It is comparable to a symphony, an alloy, a conglomerate" (ibid., p. 15). She identifies five dimensions in which ecological art can bring something more than just ecological science:

- Merit: To scientific "accuracy", eco-art adds the extra "criteria of merit" of "imagination, expressiveness, relevance, aesthetics, skill, persuasiveness, and metaphoric significance" (ibid., p. 14).
- Outreach: As opposed to the peers-oriented reach of scientific ecology, Weintraub argues that "outreach is an essential component of art" and that "art provides the vehicle by which ecological insights can infiltrate culture and become a force of social change": This claim can be interpreted at several levels: (1) At the level of the typical contemporary art (with its circuit of exhibitions, galleries, museums, etc. - as in the case of several artists discussed by Weintraub) and of the traditional performing arts institutions, one may only expect to 'infiltrate' elite culture (i.e. those social groups with high levels of what Bourdieu defined as legitimate cultural capital), but it would be naive (i.e. ignoring several decades of art-sociological research) or dishonest to claim reaching out beyond that level. (2) At the level of most ecological art projects which are strongly embedded in specific communities however, a bottom-up process of outreach, initiated by an ecological artist, is imaginable (cf. also the next chapter) – but one should not forget that NGOs, ecological advocacy groups and the media play also a role at the general public level of course, and the artists are only one amongst many stakeholders capable of (community and media) outreach.
- Method: Weintraub points out that eco-art complements the "essentially impersonal methods" of scientific ecology, with "personalizing" and with all sorts of human emotions, because "[a]rtists have a license to employ all human communicative capacities."[12]
- Inquiry: While science focuses on the "how" questions, art can also address the "why" questions, and together they can engage into a rich philosophical framework.
- Role: While the social roles of scientists are most often limited, artists generally benefit from a wider repertoire of social roles: Weintraub mentions that they "can be scientists, but they also act as prophets, visionaries, moralists, pragmatists, tricksters, futurists, and revolutionaries"(cf. also my own categorization of roles of artists with regard to social change, in chapter 7, section 1).

12 Weintraub, as most other authors discussed in this chapter, ignores (or forgets to mention) that, in the social sciences, personalized methods are possible, and are practiced in some areas of anthropology and qualitative sociology, as well as in the humanities.

With her selection of artists in *Ecocentric Topics*, Weintraub explores "ecocentric interpretations" in art across ten themes: desire, newness, power, nature, globalism, diversity, mercy, death, decay and dirt, mixing well-accepted ecological arguments with some provocative statements. Weintraub brings several highly relevant insights for a complex aesthetics of sustainability, but also proposes a number of questionable, simplistic arguments, as is discussed below.

Desire

With the theme of desire, Weintraub compares and contrasts the "positive and negative urges", known as "tropisms" in biology, and which "are inherent to animals and plants", to the largely cultural development of human desires, most alarmingly hyper-trophied in a contemporary consumer culture inundated by advertisement and commerce, and I would add, by a general commodification of all human experience, as the Marxist critique and especially Guy Debord's *Société du Spectacle* exposed – but Weintraub does not discuss Debord an the Situationists. Instead, she highlights the pernicious cycle of crave-purchase-credit-stress-frustration-anxiety-crave... which characterizes the consumer experience. Under this perspective, she discusses the performance art of Reverend Billy Talen (creator and reverend of the "Church of Stop Shopping"), who "is summoning the persuasive strategies of evangelical preachers to instill negative tropisms toward excessive purchases, and positive tropisms toward empty closets" (ibid., p. 19). Talen preaches in diverse locations, including art spaces, churches and universities, but his favorite target locations are shops, especially "Starbucks, the Disney Store, GAP, Wal-Mart, Barnes and Noble, and Nike" (ibid., p. 21). The Church organizes 'Spat theater' (a staged drama in the shops "in which members of the choir pose as customers and pretend to argue about the ethics of consumption")[13] and 'flash mobs' (performances with dozens-to-hundreds of volunteers, coordinated by mobile phone), sings its "Stop Shopping" hymn, and celebrates the Buy Nothing Day (founded by Vancouver artist Ted Dave and spread worldwide by Adbusters and other groups). Talen "also conducts exorcism upon the iconic representations of sinful consumption, such as cash registers and credit cards." Weintraub notes that "[t]he Reverend has been arrested on numerous occasions. Evidently it is illegal not to shop in America" (ibid., p. 22). She further describes the different techniques employed by Talen to release people from material addictions (rightfully treating consumerism as a serious addictive phenomenon) and explore the "unclaimed space" opened by non-consumption, an exploration into "the unknown" for ex-consumers. This

13 This technique echoes the earlier development of "invisible theater" by Augusto Boal in the 1960's, whose 'Theatre of the Oppressed" inspired a worldwide movement.

space "opens up a sensual and peopled life, and it also has in it an acceptance of the unknown" (Talen, quoted in ibid., p. 24).

In her ecocentric exploration of the theme of desire, Weintraub discusses two further artistic strategies: the CLUI's guided tours to freight container hubs (cf. also Spaid's discussion of CLUI, in chapter 5, section 6), which she expects to have an educative impact, and Song Dong's performance art which is "honoring nonmaterialistic values", for example with *Writing diary with water* (from 1995 on): "a daily practice in which Dong sits with a particular block of stone, lifts a calligraphy brush, dips the brush in water, and writes a diary entry on the surface of the stone. In seconds, the text vanishes without a trace" (Weintraub 2006., p. 25). Despite its powerful poetic value, the later type of performance limits its 'outreach' to the art world and thus fits its elitist ethos better than most ecological art described in chapter 5.

Newness

Weintraub's ecocentric discussion of "newness" is focusing on the new as "the replacement of serviceable items and ideas" (ibid., p. 29). She doesn't argue for any simplistic rejection of the new, but engages the reader to be reflexive and engage in "an ecological cost/benefit analysis" on a case-by-case basis (to which, I would add, should be associated a social cost/benefit analysis if one is concerned with sustainability and not only with 'ecology' in a strict sense). In the case of art, Weintraub points at modern and contemporary art's unreflected preference for the new, whereas ecological artists, on the contrary, tend to reflect upon issues of fashions, quick turnovers and built-in obsolescence. She points at the work of Rob Fischer, who recycles his own unsold artworks into new artworks, echoing the ecosystemic cyclical principle according to which "the destiny of all material substances is ultimately to be absorbed and reused" (ibid., p. 31). Fischer also reveals the former uses of the elements, conferring an aesthetic value to the process by manifesting "temporary states of being" and connecting the histories and prospects of his sculptures. Furthermore, Fischer "rejects carefully engineered designs that guarantee predictable outcomes" (and he thereby adds an element of complexity to his eco-literate work). That aspect of his work is most developed in his studio's backyard, from 2001 onwards, in which he let rain accumulate, weed settle and an organic life invade his sculptures. Using this biological emergence as a part of his work, he transposed parts of his backyard ecosystem into gallery settings (e.g. with *Summary (Goodyear Ecology)*, 2004-2005). Weintraub notes that "Fischer weaves entropy and regeneration into his art system" (ibid., p. 34). She adds that Fischer's backyard, which gradually became a swamp, "demonstrated the power of biotic recycling and the resilience of life" (ibid., p. 35). She sees this work as redefining newness: "Fischer's swamp introduces the never ending newness of ecosystems that contrasts with the finite newness that characterizes most human manufacturing and art making. Responsiveness to evolving opportunities replaces a rigid adherence to original intentions."

However, the same observation about the limited outreach and community value can be made about Rob Fischer's work, as about Song Dong's.

Besides Rob Fischer, Weintraub's critical exploration of newness also involves a discussion of Jim Shaw, who collects and exhibits "junk art works" bought for very cheap, and of Mike Bidlo, who copies works from previous artists.[14] Strangely, she does not mention that Bildo's work builds on the precedent of Sherrie Levine, who did a relatively similar work two decades earlier. Also, she seems content with this critique of originality in art, while Suzi Gablik already discussed the limits of such deconstructive postmodern art, by comparison to reconstructive postmodern art (cf. above, chapter 1, section 2 and chapter 5, section 4).

Power

In her discussion on the theme of power, Linda Weintraub acknowledges that self-interest is not only a human attribute, but "a biological imperative" (ibid., p. 41). In this, as I noted above, the 'ego' comes back, as a biological reality. But, as Weintraub describes, this reality takes on a new level and significance with the human species and its increasing technological capacity, which gradually turned humanity's self-interest into a "capacity to dictate when and how nonhuman life forms, the nonliving environment, and other humans should be used for our well-being" (ibid. p. 43). Under this perspective, she does not introduce artists who critically address this issue, but rather points at the work of Superflex as "proactive in its effort to empower weak humans and weak ecosystems." This perspective on the work of Superflex is more insightful than Sue Spaid's discussion of Superflex for *Ecovention* (cf. chapter 5, section 6), which limited itself to the purely environmental aspects of the group's work. Weintraub characterizes "their entire practice [as] an offering of tools", and she describes the empowerment of users/beneficiaries of those tools, as in *Guarana Power* (from 2003 on; a fair trade energy drink company in the Brazilian Amazon), *Superchannel* (1999; an Internet TV project), and *Supergas* (from 1996 on; the biogaz project already discussed by Sue Spaid).[15]

---

14 Cf. Weintraub 2006, p. 38.

15 Cf. Weintraub 2006, pp. 45-50. Weintraub also echoes the arguments from Superflex, aiming to explain why their approach would differ from those of non-artistic development NGOs. However, having worked myself with the French NGO ACTED for the Karamoja campaign, I am not at all convinced by these arguments: For example, Superflex applies bottom-up organizational processes, but ACTED does also to a relatively large extent.

Weintraub also echoes Superflex's critique of power relations in the art worlds, which "support superstar status and elitism" (ibid., p. 45).[16]

Nature

Weintraub points out how the word "nature" is "suspiciously absent from the discourse conducted by professional ecologists" (ibid., p. 55), who are actually scared off by the historical, cultural (and Romantic) heritage of the term because it doesn't fit the "rigorousness" of their disciplines, and, I would add, because it introduces more complexity than they can accept to handle: In Morin's perspective, this is a sign of reductionism typical of the preference for "clear", i.e. non-complex concepts. Weintraub further points at two problematic biases toward nature:

- the human denial of nature and will of power against nature, ignoring that "life is more authentic in the absence of human will and manipulation", and
- the "infatuation with images of virginal nature [that] may prove to be as fatal as blatant control" (ibid., p. 57). Weintraub is indeed fully conscious that "before life began, nature consisted of chemistry and physics [...] Even the first single-celled organisms changed the complexion of nature [...] natural conditions perpetually change, even in the absence of human influence" (ibid., p. 63).

In this context, she discusses the work of Dave Burns and Matias Viegener who work with corn, i.e. a species with a long and difficult history of human domestication. In *Corn Study* (2004), they created an "academy for corn", literally sending plants of corn to a school, to supposedly help corn "resist its manipulation by humans" and especially the dubious GMOs imposed by Monsanto worldwide. The contents of the 'curriculum' focused on the history of corn's domestication. Weintraub speaks of "a humorous allegory", or a "parody of the manipulated state of 'nature'" (ibid., p. 60). But the "[c]orny parody was adopted to accomplish human instruction", she adds. She quotes Viegener, who points out that with this stratagem, they "want to pose questions in a way that doesn't automatically turn the public off", and whose discourse echoes her own careful position on the subject:

"We recognize that all species are always being modified, either artificially or naturally [...] we're critical of the strict naturalists who see all manipulation as evil [...] our view of nature is very inclusive – we include ourselves as well [...] The distinction we do seek to make is between wisdom and folly, which is a cultural and ethical category. We want to catalyze a movement in all species, in all of nature, toward

16 I will come back to the political dimension of art worlds in chapter 7.

wisdom and insight. The key to this understanding is to be attentive to the nonhuman world" (Viegener, quoted in ibid., p. 61).

And, in another text, Viegener and Burns add a critical note: "At this stage of global development, humans can no longer be entrusted with full stewardship of the environment. Perhaps, if other species can intervene, they will do a better job" (Viegener and Burns, quoted in Weintraub 2006, p. 62).

Globalism

With the theme of "globalism", Linda Weintraub addresses not only the contemporary phenomenon of globalization, i.e. free trade and the accelerated international flows of communication and people, but also the global ecological reality, along the "Gaia hypothesis", and the relationship between globalism and localism. In this perspective, she discusses Shelley Sacks' *Exchange Values: Images of Invisible Lives* (already discussed above: cf. chapter 5, section 3). Weintraub notes Sacks' skepticism regarding globalization, "based on her observation that global peace and prosperity are enjoyed by few at the expense of many. Global exchange does not guarantee equitable distribution [but] actually provides an opportunity for businesses to circumvent local interests in order to leverage the labor and resources of underdeveloped regions" (Weintraub 2006, p. 69). Weintraub also discusses Marko Pogacnik's work, who applies the "Gaia approach to globalism by tracing the biological, geological, chemical and hydroogical processes that account for the Earth's resemblance to a single self-regulating organism" (ibid., p. 76). As the Gaia hypothesis views the Earth as a living organism, Pogacnik literally implements a supposedly therapeutic procedure, i.e. an acupuncture for the planet in the form of "lithopuncture", placing pillar-shaped sculptures on sites "where he perceives blocked meridians of energy". However, besides the symbolic value of this work, the direct ecological relevance of this entirely spiritually-based approach is much less convincing than e.g. Aviva Rahmani's research on trigger points. Finally, Weintraub also evokes Gu Wenda's installation *United Nations* (from 1993 on; a monumental banner woven with hair coming from barbershops all over the world) as representing a "utopian vision of global unity [...] literally weav[ing] kinship among all peoples" (ibid., pp. 76-77).

Diversity

Addressing the theme of diversity, ecocentrically, Weintraub notes that, while the "number [of species] has been in constant flux since the origin of life on Earth [...] the current decline is both precipitous and preventable", and that "human brains [and] functional ecosystems thrive on diversity" (ibid., p. 79). She is aware that "diversified systems seem more productive, stable, and resistant to perturbations than simpler systems," and worries about the long-term consequences of "mono-agricultural practices". She however notes also that "deposits of new species occur too. Like extinctions, augmentations in diversity can occur" (ibid., p. 81). She also points out that the process is unpredictable. For example: "Transferring members of a species from one locale to another will decrease diversity if the newcomers out-compete indigenous life-forms, and increase diversity if they co-exist with indigenous life forms" (ibid., p. 89). She reminds her reader that, through breeding techniques, humans have also augmented biodiversity, not only reduced it. She even comes to qualify contemporary, "impatient" genetic engineering as "super evolutionary powers" (ibid., p. 82 – as I will argue below, the last claim is however utterly mistaken).

On that basis, she discusses the work of Eduardo Kac, a Brazilian artist who developed so-called "transgenic art". Kac describes his work as such: "Transgenic art, I propose, is a new art form based on the use of genetic engineering techniques to transfer synthetic genes ot an organism or to transfer natural genetic material from one species to another to create living beings. Molecular genetics allows the artist to engineer plant and animal genomes and create new life forms" (Kac, quoted in ibid., p. 83). Weintraub seems to follow Kac's claim, commenting: "He suggests that by creating new life forms, humans might offset the extinctions caused by their mismanagement of ecosystems" (Weintraub 2006, p. 82). Among the concrete works of Kac that Weintraub discusses are *GFP Bunny* (2002), "a live albino rabbit that glows bright green when it is illuminated with blue light"; and *Genesis* (1998-1999), an art work for which Kac introduced a gene of his own creation in bacteria (a translation of the sentence from the Bible "Let man have dominion over the fish of the sea, and over the fowl of the air, and over every living thing that moves upon the earth"). Kac subsequently subjected the multiplying bacteria to ultra-violet light, causing genetic mutations. Kac translated back the resulting text (from the mutated "art" gene) into English. Weintraub comments: "Even the Bible is subject to unexplored evolutionary trajectories" and cites Kac in pointing at "the dubious notion of (divinely sanctioned) humanity's supremacy over nature" (ibid., pp. 84, 85). Although the last step of *Genesis* and Kac and Weintraub's comments do evoke a clear sensibility to the theme of uncertainty and unpredictability in evolutionary processes, Kac's work is however extremely problematic, to an extent that Weintraub fails to express: The ambition to "engineer" biological processes, by (playfully or seriously) adding or removing specific single genes, reveals a dangerous confusion between the relatively simple rules of man-made cybernetic models (and its complication), and the real complexity of natural evolutionary (emergent) processes, i.e. biological-ecological phenomena beyond the level of modeling by molecular biologists. As I already mentioned in chapter 2, section 3, Fritjof Capra warns against genetic engineering as a misconceived and short-viewed practice that intentionally ignores the recent progress in molecular biology towards an understanding of complexity, and introduces GMOs without having any clue about the consequences thereof. In this light, Kac's discourse as reflected by Weintraub, cruelly lacks complexity, and it demonstrates grave irresponsibility, both in its perpetuation of human techno-scientific hubris and in its absence of any 'Precautionary Principle'.[17]

Much more promising and insightful are other artists discussed by Weintraub in the context of the theme of diversity, such as Patricia Johanson (already discussed in chapter 5, section 3), Natalie Jeremienko and Fritz Haeg: With *One Tree(s)* (1998), Jeremienko cloned one walnut tree a thousand times, planted it in different sites around San Francisco, and observed the growth of the clones: She found out that "as the trees grew, they were not perfect replicas of each other. In fact, they were remarkably distinctive. Each was shaped by its unique microclimate. The seemingly infinite variation of branching patterns, foliage, and growth rates of trees created from a single common ancestor proved that environment plays a significant role in generating biological diversity" (ibid., p. 86). Besides, what Weintraub does not mention is that genetic "codes" do not have linear consequences and that genetic "information" interacts in complex ways with cellular networks and with the environment (cf. chapter 2, section 3). The work of Fritz Haeg consists in replacing suburban lawns with more diversified plantings (e.g. edible species for the *Edible Estates* series), and criticizing the "puritanical homogeneity and mindless conformity" represented by the culture of the suburban lawn which simplifies both ecosystems and human experiences: "The lawn divides and isolates us. It is the buffer of antisocial no-mand's-land that we wrap ourselves with, reinforcing the suburban alienation of our sprawling communities" (Haeg, quoted in ibid., p. 88).

Mercy

Linda Weintraub's ecocentric treatment of the notion of "mercy" salutarily complexifies the term. She contrasts mercy, i.e. "compassionate treatment; an act of kindness designed to alleviate distress [...] a matter of sentiment", with justice, "a matter of mind". Mercy sounds ethically good and is often advocated by animal-lovers, but as Weintraub points out, "mercy may actually interfere with the resilience, adaptability and vitality of organisms" (Weintraub 2006, p. 92). However, Weintraub depicts a rather simplistic neo-Darwinist view of evolution by arguing that "only fit organisms survive to reproduce", while the reality is more complex and involves the survival of not-so-fit but inter-dependent networks of organisms (cf. chapters 2 and

17 Arguments of a similar order have been developed in more details, against the ecological insanity of Eduardo Kac's discourse on his *GFP Bunny*, by Loïc Fel: Cf. Fel 2009, pp. 291-293. However, I reject the last argument developed by Fel (pp. 293-294) against Kac, whereby Fel describes nature as exogenous to the social. However, if nature is not exogenous to the social, its complexity (including the complexity of the social) does exceed the embrace of purposive rationality.

3). Therefore, the implications and questions she draws from her "fitness" argument are relatively dubious. Weintraub also contrasts mercy with trust, i.e. charity with self-help: Can a specific stressed ecosystem heal itself, or does it require intervention (and which level thereof)? The same question applies to human communities. Under this perspective, Weintraub discusses the work of Jean Grant, who "questions intense human interventions that produce dependencies, even those that are motivated by compassion" (ibid., p. 93). In *Gardens of Desire* (1999), she worked with young women 'on probation' on planting seeds of wild flowers under two different conditions:

- in hydroponic gardens, "dependent upon human acts of mercy" and on controlled conditions (e.g. specifically selected soil nutrients, temperature, etc.); and
- scattered as outdoor thickets, left on their own.

The first results were that the "wilder" plants, in the thickets, were giving much better results.[18] Furthermore, after transferring the two groups of plants to the other environment, "very few of the ones reared inside survived" in outdoor conditions (Grant, quoted in ibid., p. 97). The experiment also brought some metaphorical insights to the young volunteers.

"The contrast demonstrated the disadvantages of being subjected to conditions that are imposed and enforced, even when those actions are motivated by goodwill. Like the plants, the girls too were being provided for within rigid institutional structures. They began to realize that their support system taught them how to conform to rules like showing up on time, but not how to gain their independence. [...] They began to understand how the wrong kind of care can makes things die" (Weintraub 2006, p. 97).

Furthermore, as Grant points out, the work on *Gardens of Desire* illustrated the deleterious effects of controlled environments for life (i.e. also for living humans): "We don't want to accept the industrial model anymore. These institutional situations reduce resilience. They deny the sensuous needs for growing. They deprive plants of the sensations they need. It is also demoralizing for humans to be placed in such controlled settings" (Grant, quoted in ibid., p. 98). As Weintraub notes, Grant also traveled to Cuba and points at the ordinary Cubans as positive examples of community creativity (in the organization of urban permaculture) under adverse conditions.

18 Part of the difference was due, not only to the better 'plant-schooling' conditions of the outdoor environment, but also to the fact that Grant and the volunteer girls experienced difficulties with the school sports-hall in which the hydroponics plants were cultivated, being not allowed to adapt the temperature and luminosity as would have been preferable (cf. Weintraub 2006, p. 96).

Death, Decay, Dirt

From an ecocentric perspective, death is not just the end of all things (the "terminus" for the dying system), but also "a passage" and a necessary moment in a cycle for the wider ecosystem in which it is embedded (echoing Morin's bio-thanatic principle – cf. chapter 3, section 1). This situation is further "complicated by civilization [and] culturally laden baggage", as Weintraub (2006, p. 105) points out. She introduces the work of Catherine Chalmers, who "exposes the inconsistencies that commonly prevail in attitudes surrounding death [by] utilizing nonhuman surrogates to expose human paradoxes regarding death." With close-up photographic prints of "murders" committed by insects (*Food Chain*, 1994-1996), she reminds the viewer of the naturalness of predation and killing for survival. With *Executions* (2003), she stages cockroach executions: While we tend to gas them massively and ignore the details thereof, Chalmers places single cockroaches on small electric chairs, burning at the stake and other humanoid execution forms. (To be noted, Chalmers does not actually kill these cockroaches.) This leads her to further reflections about the human-cockroach relationship: "Our hatred of roaches has, perhaps, grown in proportion to the solidity of the boundaries we have erected between ourselves and the natural world. These animals are one of the few now that can, at will, cross over and challenge those barriers. I think at a fundamental level their trespass upsets our confidence in our ability to successfully control and transform nature to suit our needs and desires" (Chalmers, quoted in ibid., pp. 109-110). Coming back to human modern reactions to the death of animals, Weintraub further quotes the writer Michael Pollan: "The disappearance of animals from our lives has opened a space in which there's no reality check, either on the sentiment or the brutality" (Pollan, quoted in Weintraub 2006, p. 111). From an ecocentric perspective however, it is less the death of single organisms that should be cause for concern (as they "assure the health of an ecosystem by providing molecules to recycle into new life forms"), but the death of entire species ("because they reduce diversity" as already discussed). Weintraub points at an artist who works on commemorating now extinct species: Rachel Berwick, who also linked them to extinct cultures in *may-por-e* (1996), an installation in which two parrots speak an extinct South-American language.

Furthering her discussion of death, Weintraub comes to the process of "decay", which consists in ecosystemic recycling of organic matter and thus "belongs to the cycling process of the life force [...] Still, few of us welcome decay as an opportunity for renewal". Weintraub however highlights how beneficial and co-creative decay is, by opposition to entropy: "Whereas decay means matter is being dismantled and readied for reconstitution, entropy disperses energy and makes it less useful. Entropy charts change in the direction of less organization and less complexity, while life supports organization and complexity. Decay is integral to life processes" (Weintraub 2006,

p. 117). In this perspective, Weintraub focuses her attention on the work of the star British artist Damien Hirst, who literally "exults over decay": In Hirst's own words: "You don't have to buy a microscope to see how fabulous it is. The real gear, the stuff that we're living in, rots. And things that rot are so fucking colorful" (Hirst, quoted in ibid., p. 119). Weintraub believes that, with the confrontational approach of works such as *A Thousand Years* (1990 – an installation including the rotting remains of a cow head), Hirst aims to "purge viewers of their prejudice about decay" (while other critics would argue that Hirst merely seeks provocation). "Whether or not the sensationalism of the art work is a sign of the artist's lunatic lust for publicity," she argues, "it addresses society's lunatic denial that matter decomposes" (Weintraub 2006, p. 119). Hirst also addresses such denials in other works, such as *Pharmacy* (an installation of various medical drug pills), about which Hirst commented: "We all die, so this kind of big, happy, smiling, minimal, colorful, confident facade that medicine and drug companies put up is not flawless – your body lets you down, but people want to believe in some kind of immortality" (Hirst, quoted in ibid., p. 121).

However, to Damien Hirst and his ultra-rich colllectors, the art market is more important than the process of decay: when another one of Hirst's works started decaying (a shark conserved in a tank of formaldehyde, purchased for $12 million), its collector engaged into great efforts to prevent further deterioration.[19] Other contemporary artists also approached the theme of decay. Weintraub discusses especially the collective Gelitin, for their *Hase* (2005 – 2025, Colletto Fava), a 55 meter tall pink rabbit on a mountain in the Italian Alps: Made of straw-stuffed fabric, the sculpture is expected to decay over twenty years, as a theatrical spectacle. Weintraub quotes Gelitin member Tobias Urban: "The rabbit is feeding a lot of other kinds of life. There are already mushrooms and grass growing out of it [...] I saw birds nesting. Cows want to lie down on the straw. The straw is radiating warmth as it decays. [...] It makes all these animals feel like home" (Urban quoted in Weintraub 2006, p. 124).

Finally, Weintraub discusses how "dirt" (she actually speaks of organic humus), the product of much organic decay, plays a "wondrous" role in healthy ecosystems. She points out also that the 'health' or 'sickness' of such dirt is relative, and that humans (i.e. not only "farmers and gardeners") play a role in enhancing or degrading "soil health". By contrast, modern societies value "the sanitized products of technology, engineering and industry" (Weintraub 2006, p. 128). She would prefer humans to "create soil health by attending to its needs as one might attend to a beloved pet. Dirt is living matter. It thrives and multiplies when it is fed and tended" (ibid., p. 129). Some artists do just that: Weintraub points at Joe Scanlan's *Pay Dirt* (2003): Scanlan researched for five years on the best way to use common consumer by-products (retrieved locally) to produce high-quality earth. In

19 Cf. Weintraub 2006, p. 123.

2003, he turned an art gallery into "a mini-processing plant that actually produced dirt. The dirt was packaged and sold to visitors as part of the exhibition" (ibid., p. 131). Scanlan obtained a Patent for the high fertility of his earth (labeled "IKON EARTH"), and he encourages the collectors of his art works (i.e. IKON EARTH bags) to not just let the earth sit in its bag, but to use it for their gardens. (I shall remind the reader of the already-mentioned work of the Harrisons *Making Earth* which also consisted in the making of high quality humus, although Weintraub does not mention it.) More generally, Weintraub celebrates the artists who "advocate on behalf of a substance that sustains life but receives little recognition or care" and who, "instead of relying on representations and symbolic expressions [...] allow the medium to convey their message" (ibid., p. 135).

Redrawing the art worlds maps

With these ten themes and the artists she highlights, Weintraub aims to demonstrate how "little by little, art maps and cultural maps [i.e. attention-framing priorities] are being redrawn to acknowledge ecological considerations [...] address[ing] responsibilities for the short- and long-term well-being of ecosystems" (ibid., p. 138). She identifies "three categories of approaches" in these ecocentric works:

- "Shared ecological information", which she also labels as "shared truths" such as: "all life depends upon sunlight," "ecosystems benefit from diversity," "change is inevitable," "in a functioning ecosystem, there is no waste," "organisms are interdependent," "the Earth's material resources are limited," "biological matter inevitably degrades", "all organisms participate in the food chain," and "all organisms cause changes in their habitat through their life processes" (ibid., p. 139). She also, in the process, mentions some "truths" which lack complexity and are thus questionable, such as: "nonhuman life forms tend to acquire only as much energy and materials as they require for survival" (even the most conservative Darwinian scholar would criticize such an assumption, if only on the grounds that "sexual selection" requires energy waste; and complexity approaches in evolution show much more instances of 'waste' of resources), and "form optimizes function" (cf. also complexity approaches in evolutionary theories, e.g. Wesson 1991).
- "Theories" or "personal ecological opinions" and preferences of the artists, for example about the rather chaotic or rather organized character of ecosystems, the different space-scale of attention (whether it is a single cell, a watershed or the solar system), or the focus on different specific natural phenomena.[20] While Weintraub argues that such choices "can't be proven", I would strongly disagree, and insist, as I did in the previous

20 Cf. Weintraub 2006, p. 140.

chapters, that an aesthetics of sustainability imposes *not* to choose a single spatial scale as better, but to link several scales and levels; *not* to claim that natural reality is either chaotic of organized, but is both and is, beyond both, complex! Weintraub argues that "no work of art can be all-inclusive" (ibid., p. 141), but this shall be no excuse for remaining only within what Morin labeled as "disjunctive thinking."
- "Blueprints" or "new cultural imperatives [...] outlining future prospects for civilization: systems of production and exchange, technologies of production, methods of education, forms of literacy, concepts of designs, definitions of wealth, measures of success, concepts of time, forms of celebration," etc. (ibid.).

Weintraub's analysis, as is clear by now, focuses for a great part on "art space" artists, i.e. artists who mainly produce work at art galleries, unlike the "ecovention" artists heralded by Sue Spaid and Amy Lipton. Even though such gallery-artists have less credibility in terms of outreach and as "collaborative art" with communities, and are unlikely to bring about much structural change in the very problematic institutional structures of art worlds, Weintraub nevertheless demonstrates that they do bear some limited relevance, at least as contributors to a relative potential for change in perspectives among the international cultural elites that do pay some attention to contemporary art.

SECTION 2: EXHIBITIONS ON ART AND SUSTAINABILITY

For the past decade, a number of curators in North-America and Europe have set up exhibitions of contemporary art around notions of "sustainable art" or "art and sustainability".[21] I already discussed Hildegard Kurt's and the Fowkes' discourses earlier in this chapter. I will now give a closer look at the exhibition *Beyond Green* (2005, curated by Stephanie Smith) which has become a reference on the subject.

Beyond Green: The curatorial discourse

The exhibition *Beyond Green: toward a sustainable art* (2005, at the Smart Museum of Art, University of Chicago), with its accompanying catalog, was the first internationally acknowledged contemporary art exhibition to place itself primarily under the label of 'sustainability'.[22] The exhibition makes

21 In the current book, I am limiting my scope to North-America and Europe. However, important exhibitions occurred e.g. in Asia, such as *48°C* in Delhi in 2008 (see www.48c.org).

22 Since 2006, the exhibition also traveled to other art museums across the USA.

the argument that "sustainability has spread from the field of contemporary design into the arena of contemporary art" (ed. Smith 2005, back cover page). The exhibition also aims to focus on

> "artists [who] offer counterpoints to established forms of environmentally conscious art: rather than large-scale interventions, they explore sustainability at a more modest, portable level. Some adopt proven principles of 'green' design. Others propose small scale, alternative modes of living. Still other incisively highlight the problems and contradictions of the very discourse of sustainability" (Anthony Hirschel and Judith Olch Richards, in ed. Smith 2005, p. 9).

About the later (and also in the case of the later exhibition *Greenwashing*, discussed below – cf. section 3), I will try to assess whether the critical pretensions of some artists (which are a "must-do" in the arrogant ethos of contemporary art) do reveal potentially relevant insights for cultures of sustainability. Although the exhibition presents artists who "all [...] in fact, worked site-specifically, with particular communities, ephemerally, or outside the boundaries of art museums" (i.e. the curator wants to highlight she didn't go for purely white box art, i.e. art made for the gallery exclusively), it does focus on gallery-compatible productions, with "a specific curatorial choice to feature another side of these practices: structures, objects and processes that can be used and reused in a range of contexts and can be experienced directly by visitors at each exhibition tour venue," implicitly marking a contrast with the exhibition *Ecovention*. Smith herself (the curator), argues that the difference of the chosen artworks, as compared to site-specificity, is that they convey "a more nomadic sensibility exemplified by the mobile structures, objects, and processes/networks featured in this exhibition. Such works [...] can mutate and adapt over time and in new places" (Smith in ed. Smith 2005, p. 15).[23]

Another quote reveals a clear difference of view on art, as compared to *Ecovention*: When arguing that the works on display are "artwork, not [...] design, architecture, or activism," the authors argue: "They are for the most part provisional rather than practical, polemical and opportunistic rather than reasonable" (Anthony Hirschel and Judith Olch Richards, in ed. Smith 2005, p. 11). The insistence on not-being-practical reveals the kind of taboos that Gablik denounced and tried to deconstruct. Such an attitude, fueled by the fear of 'non-artness', also reminds me of Robert Frost's frenetic neighbor in the classical poem *Mending Wall,* whose disciplinary attitude repairs old walls and holds on to the unreflected claim that "good fences

23 Contemporary art is fascinated with nomadism and developed an unsustainable habit of hyper-mobility. Indeed, nomadism both breeds potentialities for sustainable practices (as exemplified by Spanish traditional sheep-herders – cf. Beckmann and Garzon Heydt 2009) and for unsustainable ones such as the current forms of hyper-mobility.

make good neighbors".[24] However, the curator herself (Stephanie Smith) seems to be less constrained by this taboo, as she qualifies her selection of artworks as "metaphoric, practical, speculative, ironic and playful," thereby blending the practical and the non-practical in her discourse. Guest author Victor Margolin is even more directly addressing, without the above-taboo, the "never-ending debates on the difference between art and design," acknowledging that some artists "want to achieve social results without identifying themselves as designers", while asking, then: "how should the critical community respond, and why is the artist's work given special status in a museum or gallery if its aims are predominantly practical?" (Margolin in ed. Smith 2005, p. 25).

In her opening essay, "Beyond Green," curator Stephanie Smith celebrates "sustainable design" as "an approach that balances environmental, social, economic, and aesthetic concerns [...] emphasiz[ing] the responsible and equitable use of resources and link[ing] environmental and social justice" (Smith in ed. Smith 2005, p. 13). She opposes this approach to "a prior generation of more narrowly eco-centered or 'green' approaches." Smith also warns against the wave of "greenwashing" that carries the trend for "green products" and "green lifestyles" across the commercialized 'developed' world in recent years, and against its superficiality as "a fleeting surface treatment rather than a deep and demonstrable good." Smith's problem with 'green' is made even clearer when she argues: "Green tactics only address one strand of a complex problem. In these globalized times, a more holistic approach seems a sensible and necessary response to the deep interconnection among human activities and other 'natural' systems" (ibid., p. 14). She insists that "a purely green approach [...] considers environmental questions in isolation from other factors." (However, if this critique may be relevant for some environmental groups, such an accusation is unfounded and counter-factual, when it comes to the work of several of the ecological artists discussed in chapter 5.) Smith refers most especially to Michael Shellenberger and Ted Nordhaus, advocates of so-called "post environmentalism" who accuse the environmentalist movement of having a limited, inadequate view of nature and ecology (I will come back to this discussion when introducing the ideas and curatorial work of Max Andrews, who is also heavily influenced by the discourse of "post-environmentalism").

Smith warns that the selected artists in *Beyond Green* "have no desire to be labeled as 'eco' or 'green' or even 'sustainable' artists" (ibid., p. 13). She nevertheless links them, in art-historical terms, to three streams of the previous three decades:

24 Robert Frost's poem is available online at: http://writing.upenn.edu/~afilreis/88/frost-mending.html [last accessed: 04.03.2010].

- the development of environmental/ecological, site-specific art;
- the "rise of critical practice" in the 1990's: pointing especially at artists collectives in the 1990's who were "contesting or sidestepping traditional notions of authorship", at Grant Kester's notion of "dialogical art" (i.e. "art that takes form not through objects but rather through platforms or processes meant to foster dialogue") and at Nicolas Bourriaud's *esthétique relationelle* engaging audience participation within the art-space context;[25]
- "and the fertile crossover between art and design" critically engaging with design in the 1990's (aware of the unsustainability of design's mainstream popularity, accelerating hyper-consumption with ever more marketing niches).[26]

Applying the spirit of "critical practice" she mentioned, Smith partly engages in a reflexive critique on her role as curator in an art museum. She does pragmatically assert that the "museum exhibitions can provide material resources and recognition that may be useful to the artists as they pursue their own independent projects" and hopes that her work will be "securing a place for [them] within official records of art history" (ibid., p. 17). She even makes the promising claim that "museums can learn" from such an experience and thereby strive to become "more sustainable places" (a critical call further articulated in a later article by Smith)[27], and acknowledges that "some of the projects sit a bit more comfortably within the white cube than others, and there's always a risk that the museum setting could overdetermine the ways that visitors respond to these works." And she acknowledges the risks associated to "ironic detachment" (as opposed to "earnest engagement") and to merely "enact[ing] change" (as opposed to "finding the creativity, courage and resources needed to sustain it over time"). But Smith also emits the sociologically naive – but so typically curatorial, claim that such an art museum exhibition is "introducing this work to wider audiences."

---

25 However, while failing to criticize the superficiality of Bourriaud's 'relational aesthetics', Smith (in ed. Smith 2005, pp. 15-16) also fails to mention the movement of Institutional Critique which is far more relevant as a "strong channel for critical practice" (to use her own expression). To her credit, Smith did mention Institutional Critique in other publications (e.g. in Smith 2001, and in ed. Andrews 2006).

26 In this respect, Smith points at Hal Foster's critique of contemporary design, and at the utopian heritage of the Bauhaus and constructivist (cf. Smith in ed. Smith 2005, p. 16).

27 Cf. Smith in ed. Andrews 2006, pp. 191-195.

Margolin's discussion of art and sustainability

In his contributed essay for the exhibition catalog, the design historian Victor Margolin marks out the term 'sustainability' as a product of international politics (reminding its genesis at the WCED and across diverse UN meetings), and thus as a more pragmatic notion than those stemming from the ecological movements (i.e. Arne Naess' deep ecology, "spiritual ecology, which puts a particular emphasis on the capacity to experience oneness with the planet; James Lovelock's Gaia movement; and social ecology, which [after Murray Bookchin] emphasizes social organization and collaboration with nature" - Margolin in ed. Smith 2005, p. 21). Margolin proposes, after the WCED, to stress as the goal of sustainability, "to promote harmony among human beings and between humanity and nature", i.e. "social justice" and "environmental justice". He echoes Hildegard Kurt in denouncing the double "cultural deficit" in the development of the notion of "sustainable development" in the 1990's (mentioning the economics/development-dominated discourse of UNESCO's *Our Creative Diversity* report in 1995), and in art worlds' paralleling lack of interest for the notion of sustainability.

However, I have to point out that, by the time *Beyond Green* was produced, the situation had already started to change, and not only in the art world as Smith's exhibition exemplified: In May 2004, at the 4th Forum of Local Authorities for Social Inclusion in Porto Alegre (as part of the first Universal Forum of Cultures), the United Cities and Local Governments (UCLG) adopted the "Agenda 21 for Culture" and further developed it since then.[28] But this process was only in its initial phase then and was probably unknown to Margolin and Smith.

About the understanding of a qualification such as "sustainable art," Margolin comments that such art covers three "categories: art that engages with the land or landscape; art that incorporates sustainable practices such as recycling; and art that responds to social issues through the production of objects or discourse" (ibid., p. 22). In his first category, he discusses land art and ecological/environmental art. In his second category, Margolin points at a scattered heritage of recycled-material artworks, from Kurt Schwitters' collages to the street artists of several African countries, and mentions the work of Ukeles. In his third category, he focuses especially on the political work of Joseph Beuys (from his *Organization for Direct Democracy*, in 1971, to his participation in the German Green Party).

Margolin discusses Hildegard Kurt's coining of the term "aesthetics of sustainability", wondering what art can "provide as a basis for aesthetic judgment" in terms of sustainability. He comments: "Kurt's view of art in a modernist context leads her to characterize it as a 'form of knowledge'". This definition enables art to bring "aesthetic competence into the cognitive process – which makes it different from science and at the same time its

28 Cf. the website www.agenda21culture.net [last accessed 04.03.2010].

equal". It is not clear what antecedents in modernism's past Kurt is referring to when she characterizes art as knowledge, but one might imagine conceptual art, the Situationists, and some Fluxus activities."[29] Margolin further complains that "though Kurt's emphasis on art as bearer of cognition brings it into relation with a discourse on sustainability, it does not clarify sufficiently what the boundaries of this discourse are, nor does it explain the contribution that art might make to it" (ibid., p. 24).

Margolin further raises questions as to the interpretation and aesthetic evaluation of artistic practices based on action and participation, such as Beuys' *7000 oaks* and Mel Chin's *Revival Field.* He asks: "Where is the aesthetic dimension?" (ibid., p. 26). Discussing Beuys' notion of social sculpture and its further development by Shelley Sacks, Margolin finds no answer to his question. He however observes: "The central focus of the projects appears to be the creation of an experience for the participants." His hint comforts me in arguing, as I did in chapter 4, that an "aesthetics of sustainability" actually deals with an inter-subjective experience of complexity. Margolin, however, argues against a notion of 'aesthetics of sustainability':

> "The prevailing division between art and design practice is one of the biggest obstacles to holistically envisioning a new sustainable culture [...] By separating art too rigidly from complementary practices that engage the same issues and situations, one runs the risk of maintaining a misleading cultural hierarchy in which art projects are understood to carry a heavier discursive load than more pragmatic designs" (ibid., p. 27).

Margolin may be right to point at this risk of elitism in Kurt's discourse. But, if we take Dewey as the basis for the definition of the aesthetic experience (as I did in chapter 4), we also follow Dewey's argument that aesthetic experiences may also abound in everyday experiences and are not to be found in the sole confines of artistic practices.

Margolin also points out that the higher legitimate-cultural-capital value of art can be pragmatically used, in the way the artist Dan Peterman does (e.g. at *Beyond Green*): "By presenting his kiosks in an art exhibition, Peterman performed a service in that he called the need for such structures to public attention, and one could well argue that he used the cultural capital of art's discursive power to call attention to a social need" (ibid.). Wochenklausur (also presented at *Beyond Green*, and whose work I will discuss at more length in chapter 7) also makes a similar argument, for the pragmatic use of art's cultural capital toward socio-political aims. But as Margolin remarks, such strategies open difficult issues, as they maintain in place, paradoxically, cultural hierarchies which are incompatible with a "culture of sus-

29 In her conferences and talks, Kurt actually often refers to the social sculpture of Joseph Beuys, and further back in time, to ideas emanating from the Bauhaus as 'antecedents' (though she indeed is not doing so in her 2004 article).

tainability".[30] Margolin is thus critical of Joseph Beuys, who "was strategically brilliant in trading on his recognition as a gallery artist to gain attention for his action projects [which thereby] have been drawn into an art discourse." Margolin advocates for a very radical movement, i.e. "old categories need to collapse before we can begin to create a different dialogue on aesthetics in a sustainable culture".[31] And, to start with, he makes a convincing call for ecological designers to be discussed, shown and recognized together with 'sustainable artists' (i.e. at eye level with 'art'), which is not often enough the case in the art world of contemporary art. He for example points shortly at "Nancy Jack Todd's and John Todd's New Alchemy Institute, particularly their 'living machines' that have successfully been used for water treatment and other purposes but also their ecological designs for urban spaces."

Social and Environmental justice

One of the themes addressed by Smith's exhibition is social-environmental justice (which is indeed a crucial theme for sustainability): This theme is most strikingly present in the work of Allora and Calzadilla, and to a lesser extent, in the exhibited works of Michael Rakowitz, Wochenklausur and Marjetica Potrc (i.e. in a relatively small proportion of the works exhibited, surprisingly).

Among the works on display in the *Beyond Green* exhibition is a video by Jennifer Allora and Guillermo Calzadilla, which was also selected for the Venice Biennale (in the Arsenale), entitled *Under Discussion* (2005): The video features a table turned upside down and used as a 'fishing boat'. On the video, an inhabitant of the island of Vieques uses this boat to navigate along coastal areas where the land status is contested. Vieques is a small island off Puerto Rico that the American army and the NATO used over 60 years for bombing exercises, causing chemical pollution and health problems to the locals. In 2003, after years of protests by the local population (and by the two artists), the bombings stopped, but the US Department of Interior, Fish and Wildlife services took over, imposing a top-down policy involving wildlife refuges and plans for tourism. Allora and Calzadilla stand with the local people, as expressed in their different art works exhibited internationally. In the case of *Under Discussion*, they argue that turning the "negotiation table" upside down, also "takes the debate into new, unexpected directions" (Allora and Calzadilla, in ed. Smith 2005, p. 36). They hope that the video, shown at the Biennale, can thereby "reach out to a

30 In his text, Margolin uses the term "culture of sustainability".

31 Although he does not refer to it, Margolin's call for a collapse of cultural hierarchies echoes the critical discourse of John Lane (1996), against the very existence of professional arts, and calling for the collapse of art as a distinct social category.

global network of supporters." They also stress that their work aims to contribute to a more complex understanding of environmental issues, revealing issues of North-South relations, colonial history and neo-colonialism, political power, and social justice alongside environmental issues.[32] About *Under Discussion*, Yates McKee commented:

> "In liberal thought, 'sitting down at the table' suggests an ideal-space of conflict-resolution through rational dialogue [...] Yet this ideal fails to account for the inequalities that underwrite the space of the table to begin with, such as the hierarchical division between scientific expertise and local ecological knowledge [...] *Under Discussion* is an experimental device for publicizing such counter-knowledge" (McKee, in ed. Smith 2005, p. 32).

Unrealized fantasies and social commentary

The exhibition also presents works which are not connected to any realization, i.e. which present themselves as "fantasies". Such is the case of Nils Norman, who realized a mural for the exhibition. Norman is interested in issues of urban regeneration, urban interventions and utopian experiments. But, as it seems, only as an artist observer. Although the mural illustrates Norman's interest in "utopian communities" and suggests the exact design of a "research vehicle" with which researchers and artists could "study and observe public spaces across America" and especially "utopian community sites [...] urban agriculture, permaculture centers and squats" (Norman in ed. Smith 2005, p. 83), Norman had no intention to realize this project, claiming "I'm more interested in the ideas and research than in the vehicle itself. The vehicle is just a framing device through which to view the content" (ibid., p. 91). Norman seems to be ignoring the interaction between the concrete research process (i.e. the "framing device"), and the "contents" of one's research. Or maybe, if we follow Smith's interpretation, he shows "a certain skepticism – or maybe just a dash of black humor – about the possibility of actually implementing any of these progressive structures" that he imagines on his murals (ed Smith 2005, p. 84).[33] By comparison to the works of ecological artists like Patricia Johanson, and to the "reconstructive postmodernism" advocated by Suzi Gablik (cf. chapter 5, section 4), Norman's approach as presented in *Beyond Green,* shows the face of conservatism and irresponsibility.

32 Cf. Allora and Calzadilla, in ed. Smith 2005, p. 39.

33 However, other artists do try to implement relatively similar projects, as has done e.g. AVL Ville in Rotterdam for a few years (Joep van Lieshout), and as Cultura21 Nordic (Oleg Koefoed, Kajsa Paludan and Daniel Tjäder) started to experiment from December 2009 onwards with the *Embassy of Sustensistan* (based in a low-cost vehicle, i.e. a refurbished caravan).

Another relatively detached attitude is to be found in the work and "statement" of Andrea Zittel for *Beyond Green*. The artist expresses her 'difference' as an artist by claiming: "I am not a designer – designers have a social responsibility to provide solutions. Art is more about asking questions" (Zittel in ibid., p. 146). As in the case of David Haley, "questions-based learning" can be very insightful indeed, and can be as engaging as "solutions". Unfortunately, I looked in vain for insightful questions in Zittel's statement, apart from uninspiring personal ramblings about technologies and materials for her installation.[34]

The installation by Frances Whitehead, *Primary Plus* already gives more food for thought: An installation made of a number of commercially produced inflatable objects that are in use in cases of disasters (borrowed or rented temporarily from the companies, for the duration of the exhibition), the work gives at first sight the spectacle of a neat, almost minimalist, art installation (i.e. without much interest from the perspective of sustainability). But at second sight, if we follow Stephanie Smith's interpretation and Whitehead's text, the work points at "the many-layered ambiguities of these containers", which on the one hand do work as "ingenious devices" for disaster remediation, but on the other hand, "they are also emblematic of a culture that offers surface solutions rather than seeking to address root causes" (ed. Smith 2005, p. 128) or in Whitehead's own words, we should rather "design proactive systems on the front end, moving out of a reactive mode" (Whitehead in ibid., p. 131).

Another exhibited artwork at *Beyond Green* which consists also mainly in a social commentary based on the exhibition of a commercially ready-made object, is Marjetica Potrc's *Power Tools*. Potrc is a Slovenian artist, quite successful in the art world of contemporary art, who does installations that mainly accompany her research on informal social practices in urban contexts such as e.g. the Balkans and the barrios of Caracas. With *Power Tools*, Potrc focuses her attention on some commercial devices that "apply sustainable design strategies such as durability and self-power (through body movement or solar power, for example) to real social needs such as lack of easy access to electricity or running water" (ed. Smith 2005, p. 108). Potrc's exhibited work is informative, and well-documented. But it hardly is transformative in any way.

Recycle, Upcycle, Re-use

Several works exhibited as part of *Beyond Green* focus on the theme of recycling, upcycling or re-use, elaborating more concrete practices than the artists mentioned in the previous paragraph: The collective Learning Group (constituted of Brett Bloom who co-founded Temporary Services, Julio Castro who co-founded the Mexican group Tercerunquinto, and Rikke Luther

34 Cf. Zittel in ed. Smith 2005, pp. 150-151.

and Cecilia Wendt who cofounded N55) presents an installation based on their *collecting systems* (e.g. urban dwelling systems composed especially of discarded plastic bottles, paper and/or cardboard) which they set up in 2005 on the outskirts of Monterrey (and in 2004 in Japan).[35]

Under the brand name "People Powered", the artist Kevin Kaempf mimics corporate marketing to introduce small sustainability innovations in daily life, supporting real grassroots communities rather than the artificially created, brand-based 'communities' of major consumer brands. At *Beyond Green,* several works of his were presented, among which *Soil Starter* (from 2002 on), a small-scale composting network based on the packaging of compost in sort of tea-bags by Kaempf; and *Loop* (from 2003 on), whereby Kaempf collects half-empty cans of paints, remixes and repackages them into cans which are redistributed and/or which are to be used in order to realize murals.[36]

The Austrian collective Wochenklausur, known for its socio-political interventions (cf. also chapter 7), created for *Beyond Green* (with the help of students of Chicago) an upcycling system for waste materials of art organizations in Chicago, turned into new uses, for example as furniture (the first, test cycle resulted in a line of furniture delivered to a Chicago women's shelter). The system was turned into a dedicated organization, "Material Exchange", which the students took ownership of after Wochenklausur's residency.[37]

A different perspective, about the issue of re-use and of the lives of objects, was cast by Dan Peterman's installation for *Beyond Green: Excerpts from the Universal Lab*, a "recycling" of an earlier exhibition by Peterman in 2000, already for an exhibition curated by Stephanie Smith at the Smart Museum (cf. boxed text below). Peterman's installation does not consist of any recycling per se, but rather in a collection of objects with a long history, i.e. objects from a scientific laboratory, which were then scavenged by amateur scientists, and finally recuperated by Peterman. He discusses how the University of Chicago, and the art spaces he approached, had difficulties properly disposing of especially the hazardous materials from this collection of objects. He also discusses how, once re-used for art installations, the objects know "a continual shift of polarities [...] between waste and resource value and valuelessness", thereby also pointing at the art world's issues with its ever growing collections of objects.[38]

35 Cf. ed. Smith 2005, pp. 62-71.

36 Cf. ibid., pp. 94-99.

37 Cf. ibid., pp. 136-143.

38 Cf. ibid., pp. 102-105.

The exhibition *Ecologies*

Before *Beyond Green*, Smith curated in 2000, also at the Smart Museum, the exhibition *Ecologies: Mark Dion, Peter Fend, Dan Peterman* (for which the catalog was published in 2001). Dion, Fend and Peterman were invited, as artists-researchers, or interdisciplinary researchers, to create works that address three ecologies, i.e. "interrelationships between particular organisms and specific sites: a museum building, a river landscape, a university campus" (Smith 2001, p. 24):

- Dan Peterman worked on the waste/remains of a former research lab (as already mentioned);
- Mark Dion literally conducted a 'bug hunt' in the museum building *(Roundup: An Entomological Endeavor for the Smart Museum of Art)* to demonstrate "that life on earth can be found where one least expects it" (and thus that even the most disconnected, whitest white cube is irremediably embedded in an ecology);
- Peter Fend (already mentioned in chapter 5 when introducing Ocean Earth in the context of *Ecovention*) plotted a grand scheme for the Yangtze River in China (a work which bears a relative comparison to the work of the Harrisons at first sight, but which unlike their work, contains no "conversational drift", no cautionary tale warning against human technological hubris, but rather exposes the gigantic geo-engineering schemes of Fend without an ounce of self-doubting – Fend's self-confidence in his "artist statement" for *Ecologies* is especially concerning and scary).[39]

With this exhibition, Stephanie Smith highlighted the research value of the work of these three artists, as instances of very interdisciplinary or even (although she does not use the term) transdisciplinary research, "setting emotional, intellectual, and experiential associations into play through combinations of actions, words, and objects" (Smith 2001, p. 26), and blurring disciplinary lines. In the process, Smith also makes an insightful further argument: "Maybe instead of becoming a new kind of interdisciplinary expert, the artists are becoming an old type of amateur" (Smith 2001, p. 27), i.e. amateur in the nobler, 16th-17th century sense of the term, when early modernity still allowed researchers to be collectors of cabinets of curiosities and learners of all dimensions of reality, keeping a sensibility to complexity and fuzziness that declined with the later developments of "Enlightenment rationality and its rigorous scientific methodology, scholarly discipline, and museological classifications." Smith points, with this hypothesis of a Renaissance of the noble Ama-

39 Cf. Fend in Smith 2001, pp. 69-77.

teur, at an avenue of reflection for a practice of transdisciplinarity rooting itself in arts-based research practices.

Peterman's Greenhouse

Although Smith does not discuss Peterman's other works, I shall point out that, in 1999-2000, Dan Peterman also conducted a project at the University of Lüneburg, with the university's "Kunstraum" team: A *Nomadic Greenhouse* which changed location on the university campus, offering a space for exchanges about the theme of climate change (and alongside which the Kunstraum organized a symposium on "art, ecology and sustainable development" in 2000). Buchholz and Wuggenig (2002) discussed that project, pointing at Michel Foucault's apology of nomadism: "Develop actions, thought and desires by proliferation, juxtaposition and disjunction to prefer what is positive and multiple, difference over uniformity, flows over unities, mobile arrangements over systems. Believe what is productive is not sedentary but nomadic" (Foucault, quoted in Buchholz and Wuggenig 2002, pp. 159-160). In contrast to this inspiring line of discussion, Buchholz and Wuggenig also undertake a defense of Peterman's irresponsible-provocative-artist discourse and practice, quoting Peterman (about tree trunks from Chicago, used as part of the Greenhouse installation): "It is, of course, crazy to ship these trunks around the world when even more energy has to be consumed and more carbon released into the atmosphere, but this is meant as provocation to further discussion" (Peterman, quoted in ibid., p. 167). Buchholz and Wuggenig expect this "provocation" to elicit a reflexive change at the level of "institutional structures of contemporary art production." However, the art world is highly accustomed to such kinds of artful provocation, which repeatedly fail to foster social transformation in the practices of art institutions: Here the postmodernist, merely 'deconstructive' and ironic discourses show their limits and their untenability.[40]

Ecological design as social intervention

Finally, *Beyond Green* also includes several works exemplifying practices of eco-design and design as social intervention: In terms of eco-design, an installation by JAM (Jane Palmer and Marianne Fairbanks) is presented, which consists of bags (often constituted of waste materials and) equipped

40 And so does the formal site/non-site dialectic developed by the authors after Robert Smithson. (Buchholz and Wuggenig uncritically follow the site/non-site discourse, failing to perceive the problematic issues linked to Smithson's discourses and practices.)

with solar panels, associated to a video screening of a flash animation (by Arthur Jones) which "highlights JAM's concerns about the interconnections between energy consumption and military action" (ed. Smith 2005, p. 54). Contrasted to the video, the self-powered bags gain further significance as an alternative to centralized power grids.

Another type of empowerment, not only energetic, but also informational, is proposed with the *Audio Relay* (from 2002 on) by Brennan McGaffey in collaboration with the collective Temporary Services: The exhibited object is a portable unit that allows live radio broadcast and also contains an archive of earlier radio audio works (which can thus also be re-broadcasted with the device). The unit can be powered, either with an electric plug, or with a car-battery and solar panels. This object reflects the goals of both collaborators, i.e.:

- McGaffey, whose objects typically "allow the individuals who use them to temporarily disrupt much larger systems such as the electrical power grid and radio waves"; and
- Temporary Services, who "through short-term projects [...] seek to create dialogue and to emphasize the relationship between aesthetics and ethics" (ed. Smith 2005, p. 74).

As pointed out by Brett Bloom (of Temporary Services), such a portable radio emitter "is very hard to control and can be moved rapidly [...] it is one of many means toward building our own culture that can't be shut down by dominant powers" (Bloom in ibid., p. 81).

Another work at the crossroads of design and social intervention, is Michael Rakowitz's *paraSITEs* (from 1998 on), inflatable structures made from cheap available materials (and tailored-designed for homeless persons who enter in contact with Rakowitz) and then inflated with the waste heat vented from buildings. So as not to institutionalize unsustainable situations of homelessness, Rakowitz however insists that his structures are only "temporary, imperfect solutions" and bring attention to their users' "untenable situations".[41]

Finally, *Beyond Green* also featured a work that both works as an art installation, as an intervention-kit for usage in (super)markets and as an online information campaign: *F.R.U.I.T.* (for "Fruit Route User Interface", 2005), by the collective Free Soil (which includes Amy Franceschini, the founder of Futurefarmers), focuses on the paths taken by oranges, from the production sites to the fruit stands. In the gallery space, the work takes the form of an installation: a fake fruit stand, with fake oranges wrapped in a paper wrapping which contains graphics and information about the campaign and promoting urban farming. The wrappers can be taken by museum visitors,

41 Cf. ed. Smith 2005, pp. 120-125.

and can also be downloaded from the website. The artists encourage the further use of the wrappers, to be wrapped around oranges in stores.[42]

The exhibition Timeout

The keyword 'sustainability' was also the focus of an exhibition at the Kunstmuseum Liechtenstein in 2007, organized and curated by the Museum Director Friedemann Malsch, with curator Christiane Meyer-Stoll: *Auszeit. Kunst und Nachhaltigkeit (Timeout. Art and Sustainability).*

In the introduction of the exhibition catalogue, Friedemann Malsch describes the recent popularity of the adjective "sustainable" simultaneously in a great variety of social fields, "in the vocabulary of conservative as well as progressive, right-wing as well as left-wing politicians [...] in the world of business" as well as among ecologists (ed. Malsch 2007, p. 18). Malsch characterizes this as a "gradually mounting number of initiatives and voices over the past twenty years that speak out against forced acceleration in all areas of life yet but do not reject our social structures as such," and he refers specifically to the "slow food" and "slow city" movements (ibid., p. 6). "The present popularity of the term 'sustainability' thus appears to be the symptom of a growing need as well as a growing awareness of the fact that the demand for / development of faster and faster organization and production have sped themselves into a crisis" (ibid., p. 18). Malsch points at the historical acceleration process linked to industrialization, and later post-industrial technological developments. There is no question of any radical paradigm change, thus, in his understanding of sustainability as developed in this exhibition, but his chosen perspective does explore a critical theme which constitutes a dimension for a potential paradigm change: time. "What all these movements have in common," Malsch claims, "is that they reject constant acceleration as an ideological foundation for modern life. They are primarily concerned with the deceleration of production, communication and other life processes" (ibid., p. 6).[43] Unfortunately, Malsch fails to concede that this theme of deceleration is only one dimension in the search process for sustainability. In this, his discourse offers only a very limited understanding of sustainability, although the developed insights are valuable if reintegrated in a wider context.

Furthermore, some critics argue that the calls for "slowing down" and for non-paradigmatic change are harmful to the search process of sustainability: As do the promoters of 'Cradle to Cradle' (McDonough and Braungart), economist James Greyson too warns against such a perspective:

42 Cf. ibid., pp. 42-51.

43 Malsch's focus on deceleration for the exhibition *Auszeit*, was shared with a parallel exhibition at the Kunstmuseum St.Gallen (in Switzerland), entitled *In the Eye of the Storm.* The two exhibitions shared the slogan "Vaduz – St. Gallen Doubly decelerated".

"Conventional 'solutions' try to slow down the conveyor belt by constraining its emissions (or other unwanted effects) when the underlying problem is that the entire model is obsolete" (Greyson quoted in Kelley 2008, p. 91).

Malsch's discussion of the theme of deceleration brings up as a theoretical background, the sociological work of Norbert Elias (in an essay from 1974, *Über die Zeit*) who describes different models of time colliding interculturally and the worldwide imposition of the Western, "linear time model" through colonization.[44] On this basis, Malsch hints at the possibility of a systemic conception of time, where "the various components of a specific constellation are related to one another with regard to meaning, and [where] any change in one element bears clear consequences for the overall system" (ed. Malsch 2007, p. 22).

The exhibition and its catalogue engage into a typical art-historical discourse about modern art's relationships to time, from Romanticism to the Italian Futurists' mythology of speed and acceleration (i.e. their idealization of industrialism), and more recently (and included in the exhibition): from Arte Povera starting in the 1960's (exhibiting works of Giovanni Anselmo who "addressed himself to the long-term aspect of natural processes"[45] and of Piero Gilardi) to the works of some younger contemporary artists who engage in a "critical revision" of the phenomenon of acceleration (e.g. Clemens von Wedemeyer and Maya Schneider, or the Afghan artist Lida Abdul). The exhibition also includes works by historical conceptual artists, such as On Kawara who "focused on the connection between temporal structure and individual life span" (ibid., p. 28).

However, I will not focus more closely on this exhibition, which constitutes a relatively traditional contemporary art, 'white box' exhibition for the most part, with a focus on a very specific theme (and not on the overall search process of sustainability). The only exception to this general character of the exhibition *Auszeit*, is the latest project by the German artist Jochen Gerz, which is also presented in the exhibition: *2-3 Strassen (2-3 Streets)* has been in the meanwhile realized, as a large-scale social experiment over several months, conceived by the artist (whereby the daily life of a few streets are the subject of artistic attention), and which is being implemented in 2010 in Essen as part of the program of the European Cultural Capital 2010. But the implementation of this project, in a non art space, was not what interested Malsch, who argued that, "to a certain extent it is of minor importance whether the project is ever realized or not," and claimed that the interest of the work lies in its conceptual critical inquiry and its "awareness-raising" capacity as part of the exhibition (ibid., pp. 28-30).

44 Cf. ed. Malsch 2007, pp. 20-22. However, Malsch makes no mention of Paul Virilio, whose concept of "dromology" is a reference work on the phenomenon of acceleration in what he defined as "hypermodern" societies.

45 Ed. Malsch 2007, p. 26.

SECTION 3: POST-ENVIRONMENTAL ART: THE EXHIBITION *GREENWASHING*

Of special interest is an exhibition produced in 2008 (and its accompanying catalogue), which under the title *Greenwashing*, aimed to elaborate an artistic critique of environmentalist discourses, founded on a recent movement called "post-environmentalism".

In 2008, the Fondazione Sandretto Re Rebaugendo (in Turin) organized the exhibition *Greenwashing. Environment: Perils, Promises and Perplexities*, curated by Ilaria Bonacossa and Latitudes (Max Andrews and Mariana Canepa Luna), and published an accompanying catalogue, with the aim to highlight critical artistic perspectives on environmental issues. Two years earlier in 2006, Max Andrews had been already the guest editor of a noted publication initiated by the Arts & Ecology programme of the Royal Society for the encouragement of Arts, Manufactures and Commerce (RSA), and entitled *LAND, ART: A Cultural Ecology Handbook.* In that earlier publication (which contains an insightful diversity of approaches and discourses on art and ecology), can already be found the first formulations of what turned into a full-force 'post-environmentalist' discourse elaborated in *Greenwashing.* In the following pages, I will heavily criticize this discourse, which is flawed in so many ways that several pages will be required to expose it.

The exhibition catalogue of *Greenwashing* gives the following definitions of its title word:

- "a neologism indicating the activities of companies, industries, political bodies or organizations aimed at creating a positive image in oder to conceal or take attention away from their own responsibilities in terms of negative environmental impact" (Patrizia Sandretto Re Rebaudengo in eds. Bonacossa and Latitudes 2008, p. 11);
- "a disparaging term for the dishonest representation of ecological merits" (Ilaria Bonacossa in eds. Bonacossa and Latitudes 2008, p. 19).

The critical analysis of 'greenwashing' phenomena thus constitutes the starting point of the curatorial discourse.

The Institution's discourse

In his opening text for the catalogue, Francesco Bonami, the artistic director of the foundation organizing the exhibition, sets a discourse in place which strikingly lacks subtlety in its critique of environmentalism: He compares CO2 emission trading to the sale of indulgences by the Catholic church in the 13th century, and further files complaints against "catastrophism and disaster predictions" as "a very lucrative business" (cf. Bonani in eds. Bonacossa and Latitudes 2008, pp. 13-15). The lumping together of Bonami's arguments show a certain degree of confusion and incoherence, e.g. blaming the ecological 'catastrophists' en bloc (and Al Gore in particular) for not "working on immediate solutions" and for not focusing on current "tragedies of everyday political crises, from Darfur to Chad, to Kenya". His polemic logic, posing short-term and long-term issues in terms of alternatives, shows a clear lack of comprehension of the "patterns that connect". Moreover, the appeal to Darfur and Kenya furthers shows a focus on mass-media hyped, immediate crises, which, to say the least, does not reveal a deep reflection on the global crises of our civilization. Bonami further instills some doubts on the ecological crisis, speaking of "a lot of speculation" and of "ecology, homeopathy and organic food" (note the lumping-together, once again) as "a huge unregulated scam". He further suggests that the "fundamentalist environmentalist" are at least as much interested in "mobiliz[ing] an enormous amount of money" as in "chang[ing] the world".

However, Bonami moderates his attacks, 'in extremis', by stating that his text is "playing devil's advocate of course". His main goal, he then claims, is indeed to highlight that "fear sells more than sorrow" and that an "ecological business" is thriving on this basis and contributing to a "consumerists' fever". He points at the incoherence of many consumers of 'green' products ("hundreds of SUVs have filled the parking lots of environmentally sound hotels and restaurants"). Bonami here points astutely at the *contraditio in terminis* of green business, i.e. the perpetuation of consumerism through ecological arguments and images. But instead of explicitly stressing the need for more coherent ecological lifestyles, he argues that "the alarm about the environment will eventually cause more environmental problems" and places his faith in Mohammed Yunus' microcredit, also pointing at "one hundred dollar laptops and affordable cell phones in isolated areas [which] can make the difference for hundreds of thousands of people".[46] In short, his hopes are placed in technology and the markets. The absence of a thorough systemic critique appears again. Besides, the juxtaposition of a critique of the "ecological business" and of a praise for the business as usual of "cellular phones" business and of microcredit markets, seems not to bother the author (who also conveniently fails to mention the linkage of the cellular phone industry to the Koltan-related conflicts in Congo).

Bonami ends his text by arguing: "to over-elevate human responsibility or focus exclusively on macro-issues could undermine attention to short-time responsibilities and deny the resolution of immediate damage". This might seem at first sight to be a reasonable warning, pointing at real con-

46 I happened to conduct a research project on unsustainable development in such an 'isolated area' i.e. Karamoja: Cf. eds. Knaute and Kagan 2009. A thorough analysis led our research team to conclude that such 'band-aid' measures such as microcredits and cell phones will bring a region like Karamoja, nowhere near a genuinely sustainable development, and our results confirmed the need for a systemic change. For a radical critique of microfinance, see e.g. Bateman 2010.

cerns about environmentalist discourses. But is long-term thinking really the hegemonic way of thinking in current policies and economic strategies worldwide, crowding-out short-term thinking, as Bonani suggests? No: Such a claim is extremely 'exotic' (to use a polite word). The rhetorical logic of Bonani's text also demonstrates climate-negationism, a rejection of the whole systems perspective of ecology and a narrow focus on a short-term perspectivism, i.e. a straightforward defense of the autism of Modernity.

The Curatorial discourses

The exhibition catalogue offers "a conversation between the curators", i.e. between Ilaria Bonacossa on the one hand, and Latitudes (Max Andrews and Mariana Canepa Lupa) on the other hand, which allows a look into the curatorial discourse of the exhibition.[47] The curators share a critical perspective on environmental discourses, which brought them to work together on the exhibition "Greenwashing", but they do not stress exactly the same values and interests in their respective replies:

Ilaria Bonacossa stresses the qualities of artists' critical reflexivity in addressing the idea of the environment: "I guess we can say that art has the power to create situations that rewire our normal understanding of reality. Art can oblige us to reconsider common assumptions about many issues that are constantly discussed in the public arena only in a superficial way".[48]

She mentions her disinterest for "just presenting 'green' or 'eco-conscious' artists" and her preference for "a provocative and ironic approach". Bonacossa is not just looking for an ironic attitude, but for "an active articulation of some of the contradictions and assumptions that we come across individually and as a society [and for] critical inflexions that could be said to interrogate human behavior". Latitudes shares this perspective, arguing that the exhibition is "not a call to arms, a denouncement or a manifesto" but an exercise in "self-reflexivity".

Bonacossa apparently shows a sensibility for complexity and for the 'patterns that connect', pointing out "that old-fashioned concepts of 'nature' cannot properly comprehend the complex ecological debate of the present" and that "cultural, political, social and economic factors are all implicated" in "ecosystemic collapse". She also points at the complexity of balancing personal and collective, local and global, short-term and long-term responses, and warns that "our understanding of what is at stake is very reductive". In a long quote of Slavoj Zizek, Bonacossa points at the gap between the scale of the ecological crisis and the limitations of our common sense:

47 Cf. eds. Bonacossa and Latitudes 2008, pp. 19-31.

48 However, the question remains, whether contemporary art does not too often address these issues in even more superficial ways (cf. chapter 1).

"So, back to the prospect of ecological catastrophe, why do we not act? It is too short to attribute our disbelief in the catastrophe to the impregnation of our mind by scientific ideology, which leads us to dismiss the sane concerns of our common reason, i.e. the gut sense which tells us that something is fundamentally wrong with the scientific-technological attitude. The problem is much deeper, it resides in the unreliability of our common sense itself which, habituated as it is to our ordinary life-world, finds it difficult really to accept that the flow of everyday reality can be perturbed. Our attitude here is that of the fetishist split: 'I know very well (that the global warming is a threat to the entire humanity), but nonetheless... (I cannot really believe it). It is enough to look at my environs to which my mind is wired: the green grass and trees, the whistle of the wind, the rising of the sun... can one really imagine that all this will be disturbed? You talk about the ozone hole - but no matter how much I look into the sky, I don't see it - all I see is the same sky, blue or grey'" (Zizek quoted in ed. Bonacossa and Latitudes 2008, p. 30).[49]

Here, Zizek describes the limitations of the human individual experience, in front of phenomena that encompass multiple scales of time-space and may even escape our senses (in the following lines of his text, not mentioned by Bonacossa, Zizek cites the terrifying example of the invisibility of radioactivity around Chernobyl); and, in passing, he also shortly but explicitly mentions the issue of modernity's techno-scientific paradigm. Bonacossa, with this quote, may have also meant that art's reflexivity may also allow to address the issue of common sense's limitations as discussed by Zizek. If this interpretation is correct, then Bonacossa's implicit message here joins the insights of Knebusch (2008) about art's role in the transformation of the perception of climate (cf. chapter 4, section 5). But such a message would then contradict Latitude's discourse: The curatorial team rather tends to argue that only a rational, scientifically based opinion makes sense, but that such opinion is definitely out of the reach of citizens: "How can anyone make anything more than an emotional decision about the viability of various alternative energies without being an independent scientist?" (eds. Bonacossa and Latitudes 2008, p. 22).

Bonacossa focuses her critique on "business and political agendas that try to capitalize on" ecological guilt. But she also criticizes some radical green arguments: Bonacossa expresses this critique only in half-tones, e.g. arguing that "the familiar refrain to limit growth, particularly in the developing world, seems in some way unrealistic [and that] ecological concern is much greater in rich societies". This last argument is pushed further by Latitudes, who claims that environmentalism is "the luxury of affluent societies" and "simply does not exist as a perceivable category" in the so-called Third World. If the concern for global social justice and for welfare in the southern hemisphere is laudable among these curators, their apparently blunt ig-

49 Zizek's text is integrally available online at: http://www.lacan.com/zizecology2.htm [last accessed: 05.03.2010].

norance of green movements in these countries is misleading both themselves and their readers.[50] The ignorance of Latitude's discourse is also exemplified in the counter-factual concrete example they give to criticize the ecologically conscious purchase of locally-grown food: They argue, as if it were a matter of fact, that "an individual in Germany, for example, feels better about buying locally-grown strawberries which have in fact required considerably more artificial heating and light than those flown in from Spain" (ibid., p. 22). The problem is that their provocative assertion, here too, is counter-factual: heated strawberry-production is an exception in Germany, not the rule.[51]

Latitudes stresses repeatedly that the selected artists are "not eco-artists" and "not necessarily interested in what has been termed 'eco-art'", and rejects the heritage of Joseph Beuys. The curators-team associates itself explicitly with "post-environmentalism", a movement rejecting the "romantic representation of nature". Latitudes refers specifically and heavily to the writings of Nordhaus and Shellenberger (whom Stephanie Smith already referred to in her 2005 catalogue) and criticizes "how the discourse of environmentalism is based on the understanding that humankind and culture are somehow seen as being separate from this thing 'out there' called the environment" (see also the boxed text below on Shellenberger and Nordhaus).

As seen in chapters 2, 3 and 4, the systems and complexity based discourses about ecology, sustainability, and the "ecological aesthetics" of ecological artists, do not deserve such a criticism, although they do indeed recognize some qualities in Romanticism, i.e. in opening up to the mystery of nature, not as a separate entity from culture, but as a dimension of human

50 In the case of Mexico, see for example Enrique Leff's "Saber Ambiental" which is based partly on native American inspirations (cf. Leff 1998). Besides, the global movement of biological farmers 'Via Campesina' is for a great part stemming from the so-called 'developing countries', and so are leading figures of the global environmentalist movement such as Vandana Shiva. Not to mention the "World Social Forum" where activists from these countries are very much present, and which was organized in these countries (i.e. in Brazil, India, etc.). In the light of such strong movements, the claim by Latitude is untenable.

51 I checked this both with local farmers and in national statistics: I happen to know farmers in the county of Winsen (near Hamburg) who grow strawberries, and as I could check for myself, they do not need any "artificial heating" to grow their strawberries, as the mere tunnel-growing technique is sufficient. And non-heated strawberry production is the standard practice in the region-state of Lower Saxony where I live, as I could check in the Lower-Saxonian State's statistics: Cf. http://www.nls.niedersachsen.de/Tabellen/Landwirtschaft/ernte03/texte/Info1009.pdf [last accessed: 06.03.2010]. Besides, the same is true for the whole of Germany, as the national statistics show: Cf. https://www-ec.destatis.de/csp/shop/sfg/bpm.html.cms.cBroker.cls?cmspath=struktur,vollanzeige.csp&ID=1024674 [last accessed: 06.03.2010].

nature that modernist simplification cannot comprehend. As seen in chapter 1, the progressive separation of human culture from nature is an outcome of Modernity rather than a creation of the Romantic or Environmentalist imagination. Maybe, the environmentalists are being accused of acting as the carriers of bad news, tending to Modern Westerners a mirror image that some of the so-called post-environmentalists refuse to acknowledge.

Latitude accuses environmentalism of considering "all human activity [...] as a disruption and a transgression" of Nature. This accusation confines to a gross caricature: Most environmentalist discourses develop a critique of the industrial age (and in Ellul's and Illich's cases, of the 'Technological System') but most often not of "all human activity".

Latitude also criticizes the romantic claim about a "harmony" of Nature, and wants to highlight the conflictual realities of the natural realm. While being valid, and even salutary at a certain level of analysis (against holistic simplification, as I argued repeatedly in the previous chapters), this critique, in the way it is expressed and framed, reveals an incapacity to think the complexity of ecosystems where harmony, conflicts and chaos are not just contradictory but are interrelated and allowing an overall self-regulation of the biosphere (cf. chapter 3).

Not so surprisingly, Latitude's account of Nordhaus and Shellenberger sounds very Modernistic (besides bending toward irresponsibility): "the politics and mindset of restrictions, fears, limits and prevention should be deposed by a new vision of innovation and growth where, for instance, climate change is not seen as the consequence of too much development, but not enough of the right kind of development" (eds. Bonacossa and Latitudes 2008, p. 27).

Much more surprisingly, the curators seem not to perceive the contradiction between their condemnation of 'green business' and their apology of growth, innovation and development! This may be a symptom of a lacking reflexivity on the question of Modernity as a Western, historical contingency. Symptomatically, Latitudes follows Nordhaus and Shellenberger in perceiving the landing on the moon as one "greatest achievement of mankind", revealing a naively technological vision of the 'progress' of humanity, and in resting their hopes in "the rapid development of clean affordable energy", ignoring the pressing issue of a way out of consumerism (probably because they do not want to think of 'restrictions' or 'limits' to human hubris!).

Bonacossa, who in the conversation invites Latitude to expand on these arguments, seems not to express any critical doubts about the discourse of 'post-environmentalism' that Latitude is uncritically following.

Post-environmentalism

In his earlier publication (ed. Andrews 2006), Max Andrews published an interview with Michael Shellenberger and Ted Nordhaus, to give a basic introduction to their ideas and analysis. Shellenberg and Nordhaus's critique of 'environmentalism' is rooted mainly in their critique of a form of environmental discourse that is "defining humans as outside of environment [...] But if humans are part of the environment then the concept of environment is meaningless" (ed. Andrews 2006, p. 198): The binary logic of this argument illustrates a rather simple binary thinking, in terms of 'either/or', while a complex understanding of the "environment" requires attention to the multiple forms and scales of otherness and to the properties of boundaries marking self from other. But the authors' critique of a rigid definition of 'environment', excluding humans, remains a valid critique (which they elaborate much better in their 2004 essay "The Death of Environmentalism").[52] They define "post-environmentalism" as "the thought movement of former environmentalists who view the so-called 'ecological crisis' as conceptually and politically inseparable from the human crisis, and who believe that environmentalism is incapable of turning this crisis into an opportunity" (ibid.). They criticize environmentalism further for its lack of attention to social/human sciences (a criticism which does have some ground), and for its "soft misanthropy" as well as "its outmoded liberalism, which believe humans are and should be rational, materialist and self-interested individuals; and its gloomy, anti-aspirational focus on 'limits' and 'restrictions'." Of the last two points of criticism, the first sounds perfectly well-grounded (and would apply as well if not better, as a point of critique against most of the science of economics); the second is a critical point shared by Braungart, the co-founder of the 'cradle to cradle' approach).[53] Nordhaus and Shellengerger want to encourage "creating, manufacturing and unleashing a totally different kind of economic growth – one that is not merely 'less destructive' but rather entirely good for both human and non-human worlds. The technology exists for this kind of economic growth; the problem is that our politics haven't caught up" (ibid., p. 199). I already discussed this discourse in chapter 2. I will only repeat now my warning against over-optimistic technological discourses (see chapter 2, section 3). However, it should be noted that, in their 2004 essay "The Death of Environmentalism", the authors were

52 Cf. Shellenberger and Nordhaus 2004. One should note that much of the criticism in the 2004 essay is valid for the USA, and not necessarily for the European ecological movement (although Bruno Latour 2008 argues otherwise). But it includes a number of insightful points of critique and ideas, which are not mentioned in their 2006 interview with Max Andrews, such as their criticism of US environmentalist groups' fighting one issue at a time, instead of advocating for core values (whereas the neo-conservatives excel at values-based discourses).

53 Personal communication with Braungart.

criticizing the ineffectiveness and small-mindedness of 'techno-fixes'. In that text, was not yet clearly appearing on the foreground a belief in a big technological revolution, but rather, a more open belief in the "need to tap into the creative worlds of myth-making, even religion, not to better sell narrow and technical policy proposals but rather to figure out who we are and who we need to be" (Shellenberger and Nordhaus 2004, p. 34).

However, not all is questionable in the 'post-environmentalist' discourses as echoed by Latitude and Bonacossa: The argument that "biodiversity cannot be separated from economics or pollution from unemployment" points at a serious and complex issue, but Latitude's tendency to then set them as alternatives rather than interconnected dimensions, is itself disappointedly simplistic. Here Bonacossa offers a more valuable perspective, stressing about one of the works exhibited (Sergio Vega's *Paradise on Fire*): "The impulse to conserve and protect the forest and a concern for the social conditions of its inhabitants are seen to be intricately related" (eds. Bonacossa and Latitudes 2008, p. 28).

Latitude also critically points out that "possibilities for action are being increasingly depoliticised under the guise of 'ethical living' and lifestyle trends". The team argues that a serious shift towards a more ecological society will require massive governmental involvement. Indeed, individual 'green' consumers alone cannot be expected to bring about the level of change that is required (however important their reflexive 'agency' may be, in Giddens' words).

These valuable aspects in Latitude's discourse, actually do not sound especially original or provocative: They simply correspond to the basic assumption of the concept of 'sustainability', i.e. that "everything is connected" and that a sustainable development requires 'squaring the circle' i.e. reconciling ecological, social, economic, political and cultural imperatives of justices and well-being.

Some of the arguments of the curators-team are especially contradictory: On the one hand, Latitude scorns some practical, everyday efforts towards green practices, as being turned towards "sacrifice and self-denial" (ibid., p. 26). But on the other hand, the team celebrates (and selected for its exhibition) the initiative *RAF* (Reduce Art Flights) initiated by Gustav Metzger – which asks participating artists to take the train instead of the plane – as "refreshing and uncluttered honesty" (ibid., pp. 25, 125-127).

Besides, the team resorts to some old 'tricks of the trade' in the art world, such as scorning artists who would be simple enough as to "attempt to be commonsensical, or to merely use art to reach a wider audience." Indeed, the elitist art world caters well to its own needs, mending its own fences against the 'popular' and the 'vulgar'.

Latitude repeatedly describes the exhibition and the curatorial work with ecological metaphors, as an "organic process", talking of "the nutrient-rich substrate of the exhibition-habitat" and claiming that their "process-based thinking has clear affinities with the way energy circulates in living systems" (ibid., p. 21). How far this aspect reflects an ecological literacy, a show-off (of a relative awareness of ecological terminology) or an ironic play, is open to interpretations. For her part, Bonacossa echoes Latitude's 'ecological literacy' in the description of the exhibition's qualities: "the process-based and speculative approaches of some of the artists in the exhibition can articulate energy and its material transformation, which is a fundamental principle of ecology" (ibid., p. 25).

Finally, Latitude refers to some arguments set by Félix Guattari in *Les trois écologies*. As they are by far not the only ones to refer to him, but quoting Guattari has become a fashion in the British contemporary art circles interested in questions of ecology and sustainability (mostly in positive terms, but also sometimes in negative terms such as in the case of David Haley who argues that one should speak of only "one ecology of the whole"), I will conduct a separate discussion of Guattari's essay at the end of this section. Coming back to Latitude: the curatorial duo points most especially at Guattari's call for transversal thinking (which at first sight, echoes the notion of transdisciplinarity which I explored in chapter 3): "Now more than ever, nature cannot be separated from culture; in order to comprehend the interactions between ecosystems, the mecanosphere and the social and individual Universes of references, we must learn to think 'transversally'" (Guattari quoted in ibid., p. 29). Max Andrews already made a similar reference to Guattari in his introduction to *Land, Art*, talking of "different registers of ecological thought and action", while also acknowledging the interdisciplinary tradition in ecological thinking (referring explicitly to Naess, Lovelock and Bateson), which is not "separating out the human species" from the rest of life (ed. Andrews 2006, p. 20).

The exhibition

Do the selected works in the exhibition *Greenwashing* reflect the same weaknesses and problematic arguments as the curatorial discourse that bundled them together?[54]

Several artists in the exhibition are presenting works questioning the supposed "harmony" of nature, and focusing critically on cases where certain areas have been turned 'back to Nature/wilderness' and where such decisions entailed questionable consequences:

54 Two of the artists in the exhibition (Ettore Favini and Jorge Peris) are not discussed here, but their inclusion in the discussion would not have changed the analysis significantly.

- Jennifer Allora and Guillermo Calzadilla's works in the Puerto Rican island of Vieques explores how locals fought ecological destruction by the US Navy but also how the creation of a wildlife refuge, while aiming to restore the site, also destroys "the memory of those who lived and worked there" (eds. Bonacossa and Latitudes 2008, p. 37). In his commentary on their work, Max Andrews pulls Allora and Calzadilla into his own discourse, proclaiming that their "artworks witness the death of environmentalism" (cf. section 2 above for a different interpretation).
- Lara Almarcegui's work *Wasteland* questions why the Dutch government leaves some sites "unplanned" and back to wilderness, and more generally why they consider Nature as "that which has not been sullied by humankind" i.e. a supposedly naively romantic view of nature (ibid., p. 43).
- Cyprien Gaillard's videos and paintings aim to highlight chaos, entropy, violence and disharmony in nature, Mariana Canepa Luna argues, noting that Gaillard shares with Smithson a "consideration" for entropy (but failing to see that living systems are not merely about entropy, but about the very different processes of neg-entropy, decay and ecological cycles, as Capra, Morin, Prigogine and Weintraub clearly expressed).

Some of the selected artists are typical provocateurs:

- The Mexico-based Spanish artist Santiago Sierra is already a star of contemporary art, for his extreme provocations which notoriously push the limits of human dignity in the most troubling of ways. Following-up on a previous work in 2006 on the same subject (*245 m3,* in Pulheim, Germany, an exhibition which had to be closed down because of public protest), Sierra's work on show at *Greenwashing* was *Two black vehicles with the engine running inside an art gallery.* The cars did run continuously for the first 3 days of the exhibition. To potential protesters, the curators oppose (citing Gustav Metzger) that the same people would not protest against the "normal" use of the car. (However, given that the art gallery is by no means a "normal" space, especially not for its curators, such a defensive argument sounds especially hollow.)
- The Cuban artist Wilfredo Prieto, who is interested in doing "purposeless art" (and who in 2003 parked a tanker truck and a mobile electrical generator outside a gallery for *Mucho Ruido y Pocas Nueces*, i.e. merely to water and lit a small potted plant inside), realized for *Greenwashing* the installation *Estanque*, "in which a congregation of crude oil barrels have seemingly been transformed into an idyllic, 'eco-friendly' lily pond habitat with the addition of water puddles and a live frog" (ibid., p. 122).

Other artists in the exhibition are presenting cases where technological development literally brought happiness to the people, or where techno-futurist dreams are made up:

- Maria Thereza Alves, in her film *Il Sole*, tells the story of the village of Viganella (in the Italian Alps), which thanks to a computer-controlled mirror, can now receive sun rays during the 3 months of winter (whereas in the past, the lack of sun caused depressions among the population); this case, according to curator Mariana Canepa Luna of Latitude, demonstrates "a unique place where technology has changed the climate for the benefit of its inhabitants" (ibid., p. 47). However, the lack of critical reflexivity of the curator is problematic. How far the case of Viganella matters in assessing the role of new technologies in climate change, or in making sense of issues of sustainability, is questionable. It rather seems likely that the nice story only serves a rather superficial praise of technological short-term fixes.
- The exhibition includes the techno-futurist dreams of flying cities *(Air-Port-City*, imagined as floating on airborne structures, and extra-national) of Frankfurt-based Argentinian artist Tomas Saraceno – which Bonacossa claims "is not a utopian [project] but a kinetic reality" (ibid., p. 135).
- Another technological dream featured at the exhibition is Nikola Uzunovski's *My Sunshine* (from 2007 on): the artist would like to create a 'copy' of the sun to illuminate the winter in Lapland.
- The exhibition also includes an installation by Simon Starling, whose exploration of technological-biological cycles is inspired especially by McDonough and Braungart's *Cradle to Cradle.*
- Two of the artists exhibited use technologies to actually criticize the growing importance of digital realities in human experience (James Yamada) and the subsequent alienation from our own bodies (Norma Jeane). The curators acknowledge these concerns, but then conveniently focus on the "paradox", according to them, that these two artists do use technologies for their critical work.

Some of the artists in the exhibition offer ambiguous accounts and perspectives:

- Ibon Aranberri's work focuses on the abandoned project of a nuclear plant in Lemoniz by the government of Franco; about this work, curator Mariana Canepa Luna comments on the "conflicted future of nuclear's claims to carbon-free energy despite the 'not in my backyard' responses that it invariably provokes" (ibid., p. 53). That nuclear energy poses more problematic issues than the single climate change question, is not addressed by the curators, who limit their analysis to the NIMBY phenomenon: How much does this contribute to an understanding of complexity?
- Indonesian artist Fiona Tan's video *News from the Near Future* (2003) was also shown, a video montage evoking sea level rise and flooded cit-

ies, but with the use of film archives from the Amsterdam Filmmuseum, eliciting an ambiguous telescoping of past and future.

Many of the artists in the exhibition however present works which more clearly advance several aspects of a culture of sustainability while critically addressing the theme of corporate greenwashing:

- Two works of Amy Balkin are exhibited: *This is the Public Domain*, whereby she purchased a portion of land in California and legally proceeded to return it to the public domain, promoting the values of open source and public space; and *Public Smog*, whereby she bought carbon emissions credits (normally purchased by polluting industries) and put them "out of use", symbolically declaring the clean air purchased as "clean air parks" in the atmosphere. Balkin writes: "it's important to consider the air as a political construct, legally and financially manipulated by polluting entities and nation-states, who continue to sabotage attempts to ameliorate climate change" (Balkin quoted in ibid., p. 57). Balkin also applied to have the entire atmosphere declared as UNESCO heritage site. Here, the denunciation of carbon trading as greenwashing echoes a clearly critical reflection which is however not lured by technological sirens.
- The installation *Beyond Pastoral (Shroud of Turin)* by art collective The Bruce High Quality Foundation (BHQF), addresses BP's re-branding as an 'ecologically correct' company (Former 'British Petroleum' renamed itself 'Beyond Petroleum' and adopted a new green logo, to cover up its bad image and over-emphasize its renewable energies activities). Max Andrew gives a detailed account of this fascinating art installation:

"The installation grows out of a project it initiated for an exhibition in New York in 2007. It consisted of a 1/5 scale model of the BP petrol station located opposite the gallery, underneath which thousands of lemons and limes were arranged on the floor in the form of the BP logo. Each fruit was wired with electrodes and together they generated enough electrical current to illuminate the model. What appeared at first to be an earnest demonstration of a viable and alternative energy source was soon undermined as the citruses inevitably rotted, posing a health hazard. It also became apparent that transporting the fruit had used hundreds of litres of fuel. Seemingly embroiling itself with a message that was as falsely virtuous as the clumsy eco-marketing which it ostensibly critiqued, the project now necessitated its own form of re-branding. [...] The BHQF exploits the power of faith, images and advertising in the new found religion of 'green' to further test the sustainability, credibility and authenticity of both corporate critique and supposedly miraculous technological promises. It seems to ask: how can both art and business go beyond the facile or false communication of green credentials? In the words of Bruce High Quality, 'These are good and tasteless times. Good and tasteless times call for provocative reflections and murderous predictions'" (eds. Bonacossa and Latitudes 2008, pp. 62-63).

- The installation *Constellation* by Chu Yun, a collection of second-hand electronic appliances in a dark room, "is a testament to the sheer pace and inbuilt obsolescence of the consumer goods market", argues curator Max Andrews of Latitude, i.e. a simple but very relevant issue. However, the other work exhibited, *Green Water*, "a monumental human-sized glass tank of virescent liquid – meanwhile seems to demand an explanation of the presence of what looks like alarmingly polluted water, yet no such justification is forthcoming" (ibid., p. 67): If we are to follow the curator's intepretation of this *Green Water* and search no further, not much insight emerges from the work apart from the Minimalist aesthetics and the formal qualities of the green colour; such a work then seems typical of the superficiality typical of many minimalist artworks.
- The *Cone of Silence* workshops organized by the collective A Constructed World (Jacqueline Riva and Geoff Lowe), aim to "provide moments of dialogue, interaction and discussion about complex and urgent issues." The title of the workshop is meant to denounce the inaction of decision-makers, despite ample evidence of the global ecological crisis, as the artists clearly say: "We all know our habitat is under threat, the Club of Rome first published numerous articles on the subject, which left no room for doubt [...] Pretending that something is uncertain when it isn't has led to the difficult situation in which we find ourselves today" (Riva and Lowe quoted in ibid., p. 72). The artists also placed in the exhibition space, modified versions of 'green issues' of *Vanity Fair* and *Time Magazine* ("bring[ing] a critical edge to the pages").
- Several works of Mexican artist Minerva Cuevas are included, which target brand names with clear social critique in defense of public space, but show no clue of the curators' 'post-environmentalist' discourse (e.g. *Del Montte*, 2003: from the Guatemalan dictator Montt and the fruit brand Del Monte, a mural in the form of a Del Monte advertisement, with accompanying information detailing the history of colonialism and land expropriations; *Egalité*, 2003: re-branding small Evian water bottles as 'equality' bottles, implicitly questioning the social-environmental injustices of the privatization of the commons of water).
- Tue Greenfort is also present in *Greenwashing*. The Berlin-based Danish artist realized concrete measures in previous exhibitions (e.g. *Exceeding 2 degrees* in 2007 at the Sharjah Biennial, raising the temperature by turning down the air conditioner in a building), and denounced greenwashing tactics by local authorities (e.g. with his work for Münster's 2007 Sculpture Project: *Diffuse Entries*, a manure spreading truck, used to spray onto a lake polluted by manure and fertilizers, with a chemical solution that only makes the sediments sink, and thus "makes the water only appear to be cleaner": a typical short-sighted political response). For *Greenwashing*, Greenfort redesigned the garbage-bins in the neighborhood around the museum, making them transparent, thereby

forcing more transparency about the inhabitants' effective waste disposal habits.

- *Chomskian Abstract*, a video-interview of Noam Chomsky by Cornelia Parker, about environmental crises is also in the exhibition (and the transcription of it is in the catalogue). It is especially amazing that the curators chose to show that piece, which clearly contradicts in several ways the discourse of Francesco Bonami (Chomski being especially critical of governments, companies, but also of technological faith and of consumerism).
- Sergio Vega's video *Paradise on Fire* (2007) shows formally beautiful images... of what reveals itself to be a part of the Amazonian forest burning (at the Brazil-Bolivia border). The video documents the extensive habitat degradation of a unique area. By contrast to this work, Cyprien Gaillard's videos which are playfully showing smoking landscape, appear even more futile.
- An installation by the Chinese artist Wang Jianwei, *Spectacle* (2005), consists in a room with hundreds of plastic bags whirling around (blown by fans placed at several locations): while the sight is "hypnotic and evocative," according to the curators, the work is a "reference to unbridled consumerism and to the pollution it has led to" especially in China where the careless use and throwing-away of plastic bags in urban areas has reached catastrophic proportions in recent years. The work further refers to the "human desire to have the artificial take the place of what is natural – an ambition [...] that is so dominant today in China" (eds. Bonacossa and Latitudes 2008, p. 171).

Overall, unlike the curators who propose an unconvincingly extremist discourse turning into a witch hunt against environmentalism, many of the exhibited artists do contribute valuable, and relatively subtle/complex critical insights, especially into the issue of 'greenwashing'.

Guattari's three ecologies

Guattari's "three ecologies" have been cited as a philosophical basis by several artists and curators in the past decade, including Latitude and Bonacossa for *Greenwashing*, but also Heike Strelow in her *Ecological Aesthetics* (2004). Some artists have emitted criticisms against this trend, accusing Guattari of compartmentalizing ecology, where it should bring a unified understanding of reality (e.g. Haley 2008). But what is Guattari's 'actual' discourse (as far as I do interpret it myself)?

Félix Guattari, French psychoanalyst and philosopher, is most famous for his books written in collaboration with the philosopher Gilles Deleuze in the 1970's (such as *L'Anti-Oedipe*, 1972, *Rhizome*, 1976, and *Mille Plateaux*, 1979). A relatively shorter essay of his from 1989, *Les Trois Ecologies*, gained much attention in the circles of contemporary art interested in ecology (especially in the writings of Bourriaud and in the UK) in recent years.

In this essay Guattari discusses the effects of major "techno-scientific transformations" causing both "ecological disequilibria" and a "progressive deterioration" of "human ways of life" (Guattari 1989, p. 11).[55] The "relationship of subjectivity with its externality," Guattari argues, "is compromised by a general movement of implosion and regressive infantilization". Guattari points at the impoverishment of local community relationships and family life, at media-induced relationships and at the standardization of experiences (e.g. in tourism). Whereas the political parties and governments are only pointing at the ecological crisis from a "technocratic perspective," Guattari argues that what we have to better understand nowadays is "the way to live, from now on, on this planet, in the context of techno-scientific mutations and demographic growth." What he proposes is to elaborate "an ethical-political articulation – which [he] call[s] *ecosophy* – between the three ecological registers, that of the environment, that of social relationships, and that of human subjectivity" (ibid., pp. 12-13).[56] This would then be the basis for "an authentic political, social and cultural revolution reorienting the goals of the production of material and non-material goods" at a planetary level, away from the simplifying values of global markets and from their allies, the heavily militarized States,[57] and more generally, away from "the 'unidimensionalizing' value-systems of the West" (ibid., p. 16). Guattari is also explicitly critical of "structuralism, and then postmodernism, which have gotten us used to a world view denying the relevance of human interventions in the form of concrete politics and micropolitics" (ibid., p. 32).

Guattari is calling for a "collective and individual re-singularization," i.e. a respect for, and promotion of complexity at the individual and social levels, by contrast to the simplifying reality of the mass-media (ibid., p. 21). And this, Guattari argues, should be done through "micro-social and wider institutional experimentation", against uniformity, following "the ways [of] the artist" (ibid., pp. 22-23). Guattari calls forward "ethical-aesthetically inspired paradigms" which would bring forward "intrinsically evolutionary, creative dimensions" which the social sciences have been missing, following the insights of "Goethe, Proust, Joyce, Artaud and Becket rather than Freud, Jung and Lacan" (ibid., p. 25). Guattari however warns that he does not aim to favor the "affect" by opposition to the "concept", but to reintegrate them in their complementarity.[58] Guattari also points out that the crea-

55 The quotes from Guattari (1989) are translations from the original French text by the author.

56 It should however be noted that he term "ecosophy" was already used earlier by the founder of 'deep ecology', the Norwegian philosopher Arne Naess.

57 Cf. Guattari 1989, pp. 14-15.

58 Cf. ibid., pp. 26-27.

tive, evolutionary potential of the human psyche requires an openness to the rich "fields of virtuality" of the subconscious (which is not just attached to an archaic past).[59] He cites as another example of the aesthetic paradigm, from the scientific field of physics, the arguments by Ilya Prigogine and Isabelle Stengers according to whom a "narrative element" must be introduced in theories of evolution in order to understand irreversibility.[60]

With his "three ecologies", Guattari does not ask for a disjunctive consideration of three different realities, apart from each other (as David Haley's critique would suggest). On the contrary, Guattari advocates for transversal thinking (as Max Andrews points out), denouncing the "fallacious character of a compartmentalization of a certain number of domains of reality. It is not right to separate action from psyche, socius from environment" (ibid., p. 32). "Less than ever, can nature be separated from culture, and we have to learn to think 'transversally'" (ibid., p. 34). This is not a thinking in terms of fixed patterns and structures, but in terms of dynamic processes involving "deterritorialization".[61] This is also not a thinking that seeks for consensus (unlike the communicative logic of Habermas) but a process that maintains complex tensions of diversity: "Far away from pursuing an infantilizing, dumbing-down consensus, the point [of a mental and social ecology] will be in the future to cultivate *dissensus* and the singular production of existence" (ibid., p. 44). "The individuals must become both solidarian and increasingly different" (ibid., p. 72).

Guattari's discourse, which calls forward a process of "hetero-genesis," comes very close to Edgar Morin's about complexity (although he does not mention Morin):

> "The eco-logic does not impose to 'solve' contradictions as Hegelian and Marxist dialectics required. [...] This new ecosophical logic [...] connects with that of the artist who can be led to review his work by the intrusion of an accidental detail, an event-incident that suddenly provokes a bifurcation from the original project, to make it drift afar from the most assured previous perspectives. A proverb says that 'the exception confirms the rule', but it can also reorient it or recreate it" (ibid., pp. 46-47).

Guattari also shortly mentions a "logic of the included third" (without explicitly referring to Lupasco), and refers to an 'ecology of ideas' where "the boundaries of the minds do not coincide anymore with the individuals partaking in them" (ibid., p. 50, citing Gregory Bateson).

At the social level, Guattari advocates for "the promotion of an affective and pragmatic investment in human groups of diverse sizes" (ibid., pp. 58-

59 Cf. ibid., pp. 27-28.

60 Cf. ibid., p. 31.

61 Cf. ibid., p. 36. Deterritorialization is a concept created by Deleuze and Guattari in *L'Anti-Oedipe* (1972). As an understanding of cultural change, it is often associated with the notion of "reterritorialisation".

59). He calls forward "a transition of capitalistic societies from the mass-media era to a post-media era, i.e. a re-appropriation of the media by a multitude of subjects-groups, capable of managing them toward re-singularization" (ibid., p. 61). Here, Guattari's insights on the "technological evolution of the media" (ibid., p. 62) foreshadows the role played by the Internet in the evolution of 'Tactical Media' practices and for the constitution of a global ecological-social movement (with the world social fora as some of its multiple expressions) and of an alternative economy (with free and open source software).[62]

However, more problematic is Guattari's unreflected optimism about "new techno-scientific means, potentially capable of solving the dominant ecological problems" (ibid., p. 17), inevitably leading in his view to "a logarithmic growth" of productivity gains in coming decades (ibid., p. 48), and his fascination for "the prodigious expansion of a computer-assisted subjectivity" (ibid., p. 28). But Guattari's fascination for technology is tempered by his awareness of "the limits of the techno-scientific powers of humankind" (ibid., p. 32), which itself is linked to an awareness of the irreversibility of the technological development of contemporary societies.

Overall, Guattari's *three ecologies* offers an insightful, though very dense, introduction to thinking the ecology in complex terms. But, probably because of the very short length of this essay and of its lack of articulation of the complexity of understanding complexity, Guattari's work was maybe not appreciated by all its readers from the world of contemporary art, to its full value.

The growing popularity of the prefix -eco

Apparently, the prefix "-eco" has risen in popularity in recent years, in art discourses in contemporary art. I will take, as the most recent example as I am finalizing the current chapter, *Third Text* editor Rasheed Araeen's short article "Ecoaesthetics: A Manifesto for the Twenty-First Century", which was published in September 2009 (expanding on an earlier version from October 2008). I already mentioned in the previous chapter, his critical remarks on land art and on Beuys, and that Araeen ignores the whole ecological art movement.[63] But I did not yet describe how Araeen conceives of an "ecoaesthetics": His proposal is to revive the goal of integrating art and "the collective struggle of life today" (making an implicit reference back to Allan Kaprow's goal to work toward life-like art), away from "the silly games of the so-called Neo-Dada" and calling upon artists to "enhance not only their own creative potential but also the collective life of earth's inhabitants" (Araeen 2009, p. 684). In his text, Araeen especially points at the social di-

62 See also my short discussion of Internet and the global civil society movement, after Capra and Himanen (cf. chapter 2, section 3).

63 Neither does Araeen make any reference to community arts in his article.

mensions of global ecological crises such as climate change, and at the risk of environment-based conflicts in the 21st century. He calls for a strategy of "empowerment" of the creativity of all people, especially in the global South, and not only of the artists themselves. He also shortly points at the example of solar energy for ecological-technological cycles, but only to embed it in an unconvincing and superficial discussion of "desalination plants as an artistic idea" (ibid., pp. 683-684). Nevertheless, however superficial Araeen's discourse is, it gives one more sign of a growing awareness, in the art world of 'contemporary art', of the insights that the prefix 'eco-' (for 'ecological') can bring to artists and curators concerned about sustainability and willing to critically reflect on their practices.

CONCLUSION

A transversal overview of themes addressed in contemporary art reveals the emergence of 'eco-centric' perspectives and insights, even coming from unlikely sources. Furthermore, the keyword 'sustainability' is gradually integrating the vocabulary of a number of art curators, and generating discourses and practices exploring the interplay of social and ecological critique and contemporary art. The growing debate is lively and diversified, even though some of its elaborations reveal especially questionable discourses (such as post-environmentalism, or un-reflexively pro-technological practices and discourses).

I already mentioned in the introduction to chapter 5, several movements and themes in art which I have not directly addressed, although they would be relevant to the discussion of sustainability as a search process, to the notion of culture(s) of sustainability and to the question of aesthetics of sustainability. Further areas which should be addressed are explored in *Culturing Sustainability*, an online publication by the Canadian artist and writer Caffyn Kelley (2008). Among these thematic areas are:

- A critical exploration of how subjectivity and objectivity are being, or can be rethought, in complex ecological terms and with special attention to the role of the subconscious;[64]
- A more in-depth discussion of "collaborative art" vs. "conventional art", following reflections from Grant Kester and others, and of participatory intelligence vs. individual knowledge;[65]

64 Cf. Kelley 2008, pp. 19-62. The role of the subconscious in art was only shortly evoked in chapter 4, section 2 after Bateson, when discussing the aesthetics of patterns that connect. See also appendix 1.

65 Cf. Kelley 2008, pp. 73-76. This theme was only shortly mentioned earlier in the previous chapter, when discussing Grant Kester's "position paper" for the 2004 Monongahela Conference.

- A discussion of gender and race (in relationship with nature) as "points of instability" with a potentially high leverage value in systemic terms;[66] and furthermore, an exploration of a "queer ecology" promoting an "ecology of strangeness", taking inspiration from Queer Studies to suggest that 'queer is nature', being more attuned to "complexity, fluidity and interconnections" than traditionally gendered culture;[67]
- The shaping of "spaces of possibility" through art, which could overcome 'psychic numbing' without falling back into 'positive', unreflexive practices.[68]

The relative absence of these themes in this and the previous chapter constitutes a caveat which, hopefully, I will be able to correct in future research work on art and (un)sustainability.

Apart from Weintraub's discussion on Eduardo Kac (and my doubts about this), I did not explore in depth either, the apparently especially technologically-oriented work on ecology and/or complexity and science by groups, artists and scientists such as Makrolab (Marko Peljhan), FoAM (http://fo.am), Roger Malina (and the LEONARDO network), Victoria Vesna (and her questionably uncritical work on nanotechnologies), or the MIT's Visual Art program (currently headed by Ute Meta Bauer).

The question of 'collaborative art' (which I mentioned after Kester and Kelley in the previous and this chapter), as well as the question of artistic intervention outside the confines of the art world's institutional spaces (and of art in/with/for/by communities), begs a further exploration of the artist's possible role within processes of social change. This question will be explored in the following chapter.

66 Cf. Kelley 2008., pp. 97-99.

67 Cf. ibid., pp. 111-112; indeed, sexuality in nature is much more complex, polymorphic and changing than previously conceived: Cf. also Daniel Lebrun cited in Fel 2009, pp. 253-254. But some authors are still unaware of this, e.g. Smith and Jenks (2006) who prefer to refer to an outdated publication on the "functional evolution of sexuality" in order to dismiss queer identities (p. 134 – to be fair, their attack is aiming more generally at the weaknesses of postmodernism, and makes a great number of valid points; it is not directed at ecological arguments for a queer ecology, which they simply ignore).

68 Cf. ibid., pp. 124-206.

7. Fostering Change: Art and Social Conventions

Introduction

Paraphrasing Merleau-Ponty, Suzi Gablik affirms "that it is not enough for philosophers – or [she] would add, artists – to create or express an idea; they must also awaken the experiences that will make their idea take root in the consciousness of others" (Gablik 1991, p. 108). Such a demand points at several questions, about learning, and about how certain experiences, arising in social interactions, may affect their beholders. Such questions interest the social sciences as well as psychology. Therefore, my attention will now focus on the concept of 'conventions' in the social sciences, and on specific insights from systemic psychology as developed by Gregory Bateson (with the concept of deutero-learning).

These insights are relevant to understand and analyze what may be happening at the points of interactions between artists and the rest of society, and at the border zones where art worlds meet the outside worlds.

The possibility to perform 'entrepreneurship in conventions' (i.e. to transform social conventions) will be discussed. This discussion will lead to the notion of 'double entrepreneurship in conventions', pointing at the challenges posed by the art worlds' own conventions. The possibility to perform 'double entrepreneurship in conventions' will be shown to depend also on the political setting of the art worlds, in the second section of this chapter which will introduce an analysis of 'polity conventions' in art worlds.

Considering art as a field of experimentation

If art practices related to the aesthetics and cultures of sustainability shall not be understood as restricting themselves, irresponsibly, to provocative attitudes, thoughtful statements and/or critical observations from within a mere inconsequential playground or sandbox, they shall however not be understood either as totally undifferentiated from everyday social interactions. When sustainability is understood as a continuous search process, art is then to be constructed as a field of experimentation, engaged in everyday life and culture but with certain liberties allowing a sufficient flexi-security (i.e. the flexibility of allowing experiments, but with 'safety nets' in place). But why would such a field of experimentation be required, for the search process of sustainability? Gregory Bateson answered this question with the metaphor of an acrobat walking on a wire:

"The healthy system, dreamed of above, may be compared to an acrobat on a high wire. To maintain the ongoing truth of his basic premise ('I am on the wire'), he must be free to move from one position of instability to another, i.e., certain variables such as the position of his arms and the rate of movement of his arms must have great flexibility, which he uses to maintain the stability of other more fundamental and general characteristics. If his arms are fixed or paralysed (isolated from communication), he must fall. [...] Note, in passing, that the analogy of the acrobat can be applied at a higher level. During the period when the acrobat is learning to move his arms in an appropriate way, it is necessary to have a safety net under him, i.e., precisely to give him the freedom to fall off the wire. Freedom and flexibility in regard to the most basic variables may be necessary during the process of learning and creating a new system by social change" (Bateson 1973, p. 474).

The question then is: How can such a field of experimentation be practiced in, in a connective way (rather than in isolation), i.e. allowing the experimentation to be expanding beyond the delimited confines of art worlds?

SECTION 1: DOUBLE ENTREPRENEURSHIP IN CONVENTIONS

How may social processes occur through artistic processes, that would potentially generate social change? To address that question, I developed and refined from 2004 onwards, a 'model of conventions', based on sociological research (and interdisciplinary social sciences / economics research) by the so-called 'School of Conventions' which operates across sociology and economics.[1] On this basis, I proposed an understanding of the 'artist as entrepreneur in conventions' (cf. Kagan 2004), which I later refined into an un-

1 The roots of these uses of the concept of conventions plunge much deeper and allow different streams of interpretations, pointing at earlier sources such as David Lewis' definition of conventions, Herbert Simon's work on cognition and rationality, René Girard's 'triangular desire', Karl Polanyi's 'tacit knowledge' Keynes' 'conventional judgement' and further back with Ludwig Wittgenstein's 'rules' (implicit and explicit rules), American pragmatism (especially John Dewey), Max Weber and all the way back to David Hume's understanding of convention as reciprocal referentiality.

derstanding of 'double entrepreneurship' taking more into account the internal challenges posed by the art worlds to social agents operating from and/or with the art worlds (cf. Kagan 2008a).

Conventions: a sociological and economic concept

The study of conventions calls forward a sociological and economic understanding of the social construction of reality departing both from the perspective of an autonomy of individual choice (as promoted by the methodological individualism of standard economics) and from a strict heteronomy of individual behavior (as would be inferred from a purely structuralist perspective).[2] The rationality and efficiency of individual behavior does not depend on an individual social agent's own logical computations alone, but also on their degree of coherence and compatibility with the social environment. Rationality is contextual: There are no inter-temporal universal rules for individual computation. And rationality is procedural: it is constituted through an interactive social process.[3] Interactive imitation or "Mimesis" in the definition of René Girard (1961), rather than computation, is the core human behavior. Mimesis works preliminary to interpretations and computations. Mimesis also structurally integrates the social context into the cognition of the individual. The individual's decision has meaning only relatively to his/her environment. This environment is constituted of conventions which give the individual points of reference to guide him/her.

I propose (as in Kagan 2008a) to define conventions as collectively constructed units of understanding of reality, organizing beliefs and habits around moving structures of interaction. They institute the coordination of action. They are constituted of both routine, non-reflexive relations and of more reflexive interactions through which participants may remodel their collaborations. Owing to an emergent process of rationalization in which individuals involved are inter-dependent, a convention supports discourses and devices constituting an information system (or information screen) that allows interpretation and evaluation of the social environment and of one's own behavior. Thereby, conventions overcome uncertainty.

Conventions allow relative stability. In a given convention, a number of beliefs (as well as a number of habits) are stabilized and considered as 'common sense': beliefs about the goals one is expected to aim at (in the social activities relating to the convention); beliefs about what kind of means

---

2 The conceptualization of conventions on which I am working, was launched through the 1980's and 1990's in France among an interdisciplinary team of economists and sociologists and is labeled as "economics of conventions" or "economic sociology of conventions": Cf. ed. Orlean 1994, and eds. Favereau and Lazega 2003. See also Biggard and Beamish 2003 for a short introductory review of this movement.

3 Cf. Simon 1976.

should be available (in order to attain one's goals); and beliefs concerning the perception of how things work and, subsequently, the causality models (or theories) one should use to translate means into ends in an appropriate way. It is the conviction that a certain convention is a reference for the others (the people one is interacting with in a given context) that makes behavior efficient for an individual acting in and/or upon the convention. It is the common trust in the convention that fosters this individual conviction. It is the coherence of the information system of the convention that fosters common trust in the convention. The information system is made of discourses and conversations (with a rhetoric highlighting values, principles, assigning roles to people and defining the boundaries of what is the normal realm of the convention) and of material settings that support those discourses and conversations.[4]

The actions by different individuals vis-à-vis the existing conventions, affect the coherence of the information system. Individuals can play strategically upon the information system to try and change the convention, to replace it by a 'suspicion' of that convention that becomes an alternative convention, or to move to (or create) a convention of another kind.

The concept of conventions in Howard Becker's *Art Worlds*

In the context of the sociology of the arts, the concept of 'conventions' was developed by Howard Becker (1982), who described the importance of conventions for the very functioning of 'art worlds'. In an art world, "the interaction of all the involved parties produces a shared sense of the worth of what they collectively produce. Their mutual appreciation of the conventions they share, and the support they mutually afford one another, convince them that what they are doing is worth doing" (Becker 1982, p. 39).[5]

Conventions as understood by Becker, provide a useful art-historical concept for explaining the artists' ability to contribute to the social construction of art works which bring emotional and cognitive responses in audiences (and expectations) because both sides share knowledge of an experience. The shared knowledge of bodies of conventions is part of a common culture. Members of the art world can therefore rely on earlier agreements and/or understandings that have now become customary, conventional. And different groups know different parts of the total body of conventions used by an art world. Conventions also regulate the relations between artists and (the well-socialized) audience, specifying the rights and obligations of both sides. Conventions undergo constant adjustments in response to changing conditions and changing strategies of individuals and groups, but they are overall relatively stable in the art worlds, and as Becker observes, revolutionary changes in conventions are relatively rare.

4 Cf. Gomez 1996.

5 Exact page numbers refer to the 1984 paperback reprint.

Knowledge of the relevant conventions and routine interactions following the terms of these conventions, define the outer perimeter of an art world. Traditionally in the arts (understood as an identifiable social activity in modern societies, whether 'high art' or more popular arts), the diffusion of new conventions coming from the creative core of an art world involves a process in which the first to understand (have the ability to decode) the worth and values of what is produced are the ones most involved in the production (creators, interprets), followed by 'serious audience members' who share some of the conventions (or the directly involved participants, in more participatory art forms), and through them (and their interactions with wider segments of society), only later do 'well-socialized members of the society' and the general public share an understanding and value the 'products' or other outcomes of the artistic practices.[6]

New conventions introduced by a successful artist work out as a new language.[7] Over time, they inter-relate and form a specific self-referential body of conventions, which audiences can learn 'by experiencing them'. However, more than mere experience and accustoming, the processes of construction of conventions involve rhetoric in the building of conviction and suspicion, in symbolic fights over conventional coherence (as will be explained below).

How do conventions work?

Individuals are driven to imitate one another and distinguish themselves from one another and thereby create a network of interdependent behaviors. The individual is able to perceive indications on how to take decisions, because he or she can find landmarks in the behavior of others; he or she can therefore position him- or herself towards these behavioral guides, and decide whether or not to imitate them. In Howard Becker's terms: "Even when you don't want to do what is conventional, what you do want to do can best be described in the language that comes from the conventions" (Becker 1982, p. 57).

The evolution of conventions depends both on:

- the intentional strategies of those individuals who decide to play on the rules rather than in the rules, to modify the terms of imitation rather than just imitate others,
- and on the unintentional behavior of individuals (or unintended consequences of intentional behavior) especially at systemic 'points of instability'.

6 Cf. also Bowness 1989.

7 By the way: Language itself constitutes a typical case of convention (cf. Lewis 1969).

I will focus in this chapter more especially on the first case, i.e. intentional strategies. Insofar as an individual is able to perceive (parts of) these rules of generally recognized behavior, he or she reaches a level of reflexive awareness. However, playing intentionally on the rules requires also an awareness of the availability of alternative sets of rules and a competence in handling alternative conventions (see below, the competences of the entrepreneur). Given these qualities (that would relate to the agent's "practical consciousness" in the structuration theory of Anthony Giddens, i.e. a capacity to reflect on one's routines, allowing to potentially act differently upon them)[8], the individual social agent is then relatively freely engaging in a social space that offers him or her the landmarks around which a relative freedom (or 'agency', in sociological terms) can unfold. However, some others have to be relatively open to the possibility of playing on the rules; this is comparable to what is described as the tolerance of peers for role-deviation, allowing role-making rather than role-taking, in George Herbert Mead's sociological theory.[9] A new or modified convention only survives when it is adopted widely enough to become a norm at a given level of social organization. A convention is not imposed upon people, it is not a law (although it can attain as much authority as a law). It is rather a self-fulfilling prophecy. It is not an exogenous constraint upon the social agents, but an endogenous constraint adopted to allow behavior.

A convention has the following characteristics:

- A situation of uncertainty disturbing or even overwhelming individuals unless they refer to 'normal' behavior;
- A 'normal' solution readily available to individuals (without them having to define it by themselves), being both stable and tacit (unlike a negotiated contract): The concept of convention involves a trend of organized social interactions "that develops over time a path-dependent way [with] conventions that maintain stable relations" (Biggart and Beamish 2003);
- A shared trust in the fact that the convention is being complied with by its adopters and that adopters will anticipate that others will comply too;
- The fact that another solution could exist for the problem the convention is solving.

These characteristics are partly taken from Lewis (1969). I am however not strictly following his widely cited definition, for the reason that Lewis as-

---

8 Cf. Giddens 1984.

9 Cf. Kirchberg 2007, pp. 116-117, 132. This interplay between self and peers, with the constraint of being accountable to peers, is to be found back not only in Mead's notion of the self as "generalized other", but also in Goffman's dramaturgical approach, Garfinkel's ethnomethodology, and in Dahrendorf's focus on peer group pressure (as noted by Kirchberg).

sumed that all social agents are fully aware of the convention (it is 'common knowledge'). My conception of conventions rather assumes that conventions (1) offer 'incomplete rules' that are open to interpretation and (2) conventions interplay with the individual social agent both at conscious and subconscious levels.[10] A convention allows to rationalize one's behavior. "This is an emergent process, a performance of rationality that is constructed in interaction with others and is rational in the sense that it appears rational to self and others within a social setting but not necessarily in some objective external sense" (Biggart and Beamish 2003). This emergent, performative, interpretive process is not only involving the conscious mind of the individual, but also involves subconscious levels. The concept of rationalization points at the procedures and techniques through which individuals confer reason to their decisions.[11]

How do practices become recognized as rational then? One rationalizes in front of others, one gives a reason. But then, how can different individuals use the same rules for computation and thus rationalize in the same way? From the point of view of an individual in a convention, how does one come to comply? The key concept is individual conviction: the individual belief that others act in a certain way in a certain circumstance. Individuals do not check continuously that others follow the rules, because they are convinced that they do so. This trust is a foundation of social life.

A convention has to be convincing, so as to become self-evident (so that even deviant behavior still refers to the convention). A belief is effective when it is shared widely enough to have effects. An effective convention is a convention that is adopted as true. The study of conventions looks at the justifications of choices made. At a wider (meso- to macro-social) scale, it looks at historical processes, at the evolution of systems of conventional rules. Some systems of inter-related conventions become more 'hard programmed' than others; they then become 'social institutions'. To come back to the micro- to meso-social level, the study of conventions also looks at individual strategies aiming at changing conventions by introducing suspicion about them. By definition, a convention can be challenged by an alternative convention (lingering as its antithesis). There is therefore a suspicion on the convention, that could take over and destroy the convention if conviction was to fail. A dynamic process takes place. Rationalization changes as the convictions change, and the evolution of conventions can explain the evolution of behaviors. Becker (1982) observed that one small change often requires many other changes. As this small change introduces suspicion in existing conventions, these 'many other changes' are more than a mere technical requirement. In Systems Thinking terms, the small changes with the

10 The first argument, about conventions offering 'incomplete rules', was already made in ed. Batifoulier 2001. The second argument, which was suggested to me by Tasos Zembylas, draws on the insights from Wittgenstein about 'rules'.

11 Herbert Simon (1976) spoke of "procedural rationality".

largest-scale consequences would be labeled as having a 'high leverage' on the system (i.e. on the institutions as systems of interrelated conventions), vs. the 'low leverage'of the small changes with lesser consequences. And according to the different versions of systems thinking and complexity theory (as in the concepts of "tipping point" and Aviva Rahmani's concept of "trigger point" discussed earlier – cf. chapter 5, section 3), there exist points of instability in systems, where small changes are more likely to yield high-leverage. The challenge (for agents of social change) is then to identify possible points of instability, and to take the risk of working on them.

As Gomez (1996) argued, "the social world being woven with conventions, individuals can refer to different conventions, and so justify behaviours that are different because rationalized differently." Apart from the case of a convention and its potential anti-thesis, there are also many cases of tensions between altogether different conventions solving different problems but being raised simultaneously, involving some of the same people and/or having some characteristics in common. Here again, the individual will choose the convention that he or she judges to be most convincing (and subconsciously feels attracted to). But these interference zones between different conventions will also have the effect of throwing suspicion on each competing convention. Therefore, there is an interaction between the internal dynamic of suspicion and the external competition from other conventions. However, a special case can occur, when different conventions are associated through a double morality, i.e. paradoxical combinations of apparently contradictory conventions under a discourse justifying double-standards and rhetorically solving its internal contradictions (as do for example the cultural policy advocates who claim to pursue 'elitism for all'). Altogether different conventions can refer to very different normative principles for making sense of reality, i.e. 'orders of worth'.[12] In such cases the competing conventions are based on entirely incompatible paradigms and their confrontations result in contextual compromises rather than in a general synthesis.

Acting as informational screens, conventions prevent individuals from having to manipulate a large amount of information continuously. The more complex a society, the larger the number of conventions that will be used, because they allow easier information management. Because of conventions, information is organization. And because of conventions, only new, dissonant, abnormal information gets the attention of individuals.

What grounds the conviction of individuals about the generalization of a given convention? To find it out, one has to observe the relationships between the convention as screen and the individual cognition and dispositions: to observe by which means behavior-defining common beliefs and habits are formed, transmitted and modified. This set of relationships is the "information system" of the convention. Discourses and conversations qual-

12 Cf. Boltanski and Thévenot 1991.

ify the convention, i.e. what is (in)compatible with the convention, where are its borders, how convincing it is. Discourse can be conscious, intentional and direct but also subconscious, unintentional and indirect. However, the information system is not based only on a rhetorical content, but also on the material characteristics of the social interactions it organizes. These characteristics affect how the discourse is conveyed and how conversations and other social interactions are organized materially. The power of conviction, and therefore the strength of the convention, depends on the coherence of the discourses, conversations and material settings of social interactions making up the convention as a screening system.

The dynamic of the relationship between the individual conviction and the convention tends to have an exponential character: It normally works into a self-reinforcing feedback loop both strengthening individual conviction and reinforcing the existing convention against change. Repeated interaction between adopters of the convention and the influence of some authority figures within the convention both contribute to keeping 'normal' behavior on its track. But as soon as noticeable elements of suspicion bring dissonance (especially in the early days of a convention when conviction is still fragile, and in environments involving heightened competition between conventions), incoherence may start gripping the mechanism and turning the dynamics inversely, in an exponential rhythm, so much so that a self-reinforcing feedback loop weakens both individual conviction and the status quo. Such a 'vicious' or 'virtuous' cycle can lead to an increased discrepancy between the expectations nourished by the new convention and the satisfactions provided by the existing (and increasingly suspicious) convention. The discrepancy can in turn provoke 'relative frustration' and a higher probability of social violence, protest, exit or other forms of opposition against the existing convention. But discrepancies can also lead to a diversity of conscious or subconscious strategies of reduction of 'cognitive dissonance'.[13]

A convention cannot be controlled by a single individual, given that each and every individual is participating to its social construction, to some relative degree. But collective strategies of change are possible: groups of individuals (or single individuals with groups of followers) can play on the discourse and/or on the material devices of a convention in order to increase or decrease its power of conviction. However, such strategies are very unlikely to have linear effects. Unintentional effects and subconscious processes are to be taken into account. Furthermore, the farther away one moves from the actions of specific individuals, groups or organizations, the more unpredictable the outcomes: As already seen in chapters 2 and 3, in a social

13 On relative frustration, cf. Gurr 1970. On 'protest, exit and voice', cf. Hirschman 1970. On the long and complex discussion of 'cognitive dissonance' and strategies for its reduction, since Leon Festinger's first formulation in 1956, see Harmon-Jones and Mills 1999 and Cooper 2007.

reality which is hyper-complex, any claim to predict, control or manipulate large-scale social movements into a specific direction (however high the leverage point one works on), should be considered as preposterous.

Conventions, complexity and communities of practice

The analysis of conventions brings now back to the foreground, the specific complexity of human social reality, with its duality of formal and informal structures, which was already evoked after Capra in chapter 2, section 3 as a duality of design and evolution.[14] By looking at the complex evolutionary dimension of social organization, Capra places the question of social change within the general context of living systems. Such an endeavor is not biological reductionism, as was feared about so-called 'sociobiology' three decades ago. As Morin argued, human culture is emergent while it is integrally part of life: Recognizing this does not amount to reducing the specific complexity of human society, but brings to sociological analysis, the complexity of life, at a transdisciplinary level as well as into the discipline itself (and thus it should not be confused with the rigid, lifeless structuralism and systems analysis of a Talcott Parsons). The understanding of life's evolution brings into full force, the vision of a society which cannot be entirely designed a priori according to rationalized schemes, and brings attention to the large degree of unpredictability in macro-social evolutions, which is in a complex balance/tension/opposition/complementarity with the relative predictability of specific (and less-complex) formal dimensions of social reality (i.e. involving design, hierarchy, centralization, dogmas and stability).[15] In the words of Edgar Morin: "Life carries in itself the idea, not of irrationality, but of irrationalization, and it is irrationalization which resists to instrumentalization." (Morin 1980, p. 432).

Therefore, the analysis of conventions should be understood within this framework of living human complexity, working at the interface of the formal and the informal, as well as at the interfaces of the explicit and the implicit (or 'tacit'), the conscious and the subconscious, the voluntary (or purposive) and the involuntary.

Consequently, the analysis of conventions, in its attention to informal social processes, also bears parallels with the concept of "communities of practice" by educational theorist Etienne Wenger: "Wenger defines a community of practice as characterized by three features: mutual engagement of its members, a joint enterprise, and, over time, a shared repertoire of routines, tacit rules of conduct, and knowledge" (Capra 2002, p. 108). As to the first feature ("mutual engagement"): For Wenger (1998), collective relationships arising from mutual participation in certain activities, bind individuals

14 I will not repeat here all that was already discussed in that section of chapter 2. I therefore invite the reader, to look again into those pages.

15 Cf. Morin 1980, pp. 414-438.

together as social communities. The second feature ("joint enterprise") relates to the shared understanding that binds together the members of the community (in terms of conventions: cf. below about the 'convention relating to purpose'). The third feature (the "shared repertoire") points at the communal resources produced by the community of practice and then used to achieve certain goals. These three dimensions are present in the analysis of conventions while the analysis of conventions offers probably a more thorough account of the dynamics of social interactions building communities of practice. The concept of communities of practice is especially useful to describe the forming of informal social groups, i.e. one of the dimensions of the working of social conventions. Wenger, McDermott and Snyder (2002) also developed the analysis further, considering how the concept of communities of practice may help formal social organizations, tap in to 'tacit knowledge' and thereby improve learning among their members (see also Capra on organizations as learning beings, as discussed earlier in chapter 2).

Purpose and Conviction

The understanding of the dynamics of conventions, in both formal and informal dimensions, depends partly on the concept of the 'entrepreneur in conventions' which I developed after Gomez (1996). This concept points at the 'agency' of social agents, i.e. when they play not only *in* certain conventions but also *on* certain conventions.[16]

Any form of social organization is characterized by at least one common convention: a convention relating to purpose. The existence of an organization is linked to a generalized conviction about its role. Every member, every participant needs not have the same goal, but each participant does think that the objectives (including personal objectives) can only shape out of collective action. How are social agents lured into a specific instance of collective action and effectively convinced of the fact that there exists a common purpose? But before I answer this question, I need to clarify why I am asking it: The relative coherence of the conviction will be a determinant factor for the relative strength of the organization. Indeed, as Capra expressed it, "force or energy are not the issue; the issue is meaning. Meaning-

16 The word 'entrepreneur' has strong connotations, linking it to market values and the figure of a business 'genius', as popularized by Schumpeter as "Unternehmergeist". The term is indeed used the most in the economic and management literature. However, a different use of the term is made here, linking it to sociological notions of social 'agency' beyond the traditional realm of the economic entrepreneur. This approach is not totally new, as illustrates e.g. the rise of the expression "social entrepreneur" in recent decades. For a short account of definitions of entrepreneurship, social entrepreneurship and cultural entrepreneurship, see Kagan 2004, pp. 39, 68-69.

ful disturbances will get the organization's attention and will trigger structural changes" (Capra 2002, p. 112). And meaningfulness starts with the conviction of a few social agents, or 'adopters' of a convention.

One answer to the question about the conviction of social agents, has to do with the concept of *Entrepreneurship in Conventions*, i.e. strategies of conviction-building affecting the new adopter (cf. the following pages).

Other, complementary elements of answer come from the history of the incoming adopter: i.e. in sociological terms, his/her habitus, his/her repertoire of dispositions (which is not to be assimilated entirely to Wenger's "shared repertoire" at the level of the community of practice: see boxed text below).

The social agent's repertoire of dispositions

Bernard Lahire (1998, 2003, 2004) elaborated the concept of "repertoire of dispositions" as a correction of Bourdieu's habitus, focusing on the plurality of socialization at the individual level, and thereby contributing to an understanding of the complexity of micro-social reality (through what he calls "a sociology at the level of the individual"). For Lahire, habitus is indeed internalized by social actors through socialization, but it is neither unified, nor necessarily sustained uniformly through an individual's existence, nor entirely activated under all everyday circumstances. It is formed by varied experiences and offers a diversity of alternative behaviors. The diversity of experiences (of roles) a social actor lives through, brings him/her a multiplicity of thought-habits and action-schemes, shaping a repertoire that becomes available according to diversified contexts.

Going away from the assumed notion of a coherent set of dispositions making up the habitus of social agents (after Bourdieu), Lahire argues:

- "that social agents have developed a broad array of dispositions, each of which owes its availability, composition and force to the socialization process in which it was acquired" (Lahire 2003, p. 329);
- that "dispositions to act" and "dispositions to believe" have to be analytically separated, in order to perceive their discrepancies and the subjective feelings of uneasiness or even guilt resulting from such discrepancies;
- that "the intensity with which dispositions affect behaviour depends also on the specific context" of social interactions (ibid.).

Lahire is interested in "individuals as being products of pluriform social processes", not in group categorizations. His insights allow to question and assess the relative coherence of dispositions, how they are acquired and how they can change. Dispositions have a 'life'.

Of great importance is the question: How do different dispositions (acquired through different experiences of socializations, at home, at school, with friends, etc.) make up a whole "repertoire" of dispositions and an individual's Habitus? Here the distinction between dispositions to act and dispositions to believe becomes important. "Agents may have internalized specific norms, values, or ideals without ever being able to develop the habits to act that would allow them to attain their ideals [...] In such cases, beliefs are powerless because there are no conditions under which dispositions may turn them into reality" (Lahire 2003, p. 337). For example, when one holds ecological values highly, but does not manage to act accordingly, this gap can lead to self-depreciation, frustration and guilt. This aspect points at one important warning for the discussion of a culture of sustainability: It is not enough for social agents to develop the 'dispositions to believe' in the moral sake of a certain way of life, as long as they cannot acquire "dispositions to act" that will have incorporated relevant habits, for such a way of life to be experimented, practiced, enacted.

It is important to be aware of the plurality of dispositions of each individual social actor, as analyzed by Lahire (2004). Conventions function as incomplete rules and are open to interpretation (as already mentioned with reference to ed. Batifoulier 2001). Therefore, the specific repertoire of dispositions of each individual social actor may lead to specific capabilities for entrepreneurship in conventions and/or to a specific receptivity (as 'adopter') to someone else's entrepreneurship.

Entrepreneurship in conventions

Entrepreneurship in conventions is the successful construction of a common purposeful convention. "An entrepreneur is a convincing individual [...] To undertake enterprise is to convince the actors whom are necessary to the realisation of a common objective" (Gomez 1996, pp. 224-227). The entrepreneur looks for ways to convince others. He therefore plays on the information system of the organization's convention, and tries to modify it to make his own agenda look more convincing. The entrepreneur is the promoter of a renewed convention, but if strategies of change are possible, they cannot be imposed by a single individual: they require convinced groups to act upon the discourse, conversations and material settings of the convention. Therefore, the entrepreneur is in no way a human 'Deus ex Machina', and his entrepreneurship starts with convincing selected individuals to shape a closer group or network of allies (which can be either a formal group, or an informal community of practice).

How can the entrepreneur in conventions be successful (i.e. be convincing)? And first, who can be an entrepreneur? Entrepreneurship in conventions is not a priori reserved to a given category of individual (i.e. the 'entrepreneur' as understood in business language). Rather, entrepreneurship in conventions can come from any individual with the means and the strategy to play on the rules rather than in the rules, to modify the terms of imitation rather than just imitate others. Insofar as the individual is able to perceive (parts of) these rules of generally recognized behavior, he/she is (relatively) free to try changing them.

An important part of the entrepreneurship in conventions lies in the process by which the individual 'entrepreneur' gains consciousness, making some conventional rules emerge from the subconscious to the conscious level. That emergence of consciousness is not necessarily a straightforward process: conventional rules may be played with while they are still lying mostly at the subconscious level: The entrepreneur may be playing with a conventional 'iceberg', and such a process will require competences beyond linear rationality, i.e. involving also forms of 'hermeneutic' and 'ontological' reflexivity: Hermeneutic reflexivity is explicitly linked to the concept of conventions by Hans Dieleman, as it "is situated in communities and [...] explores day-to-day life, routines and conventions. It is based on comparing existing particular practices, routines or conventions with potentially new ones" (Dieleman 2008, p. 119). The concept of hermeneutic reflexivity, which Dieleman uses after Scott Lash, deals with "looking into the looking glass self" (ibid., p.127). In systemic terms, it can be understood as a dimension of second-order observation, i.e. the observation of observation. In Bourdieu's terms, it would relate to gaining awareness of one's "feel for the game" i.e. unearthing one's habitus. Hermeneutic reflexivity requires a process of estrangement and detachment, comparable to that of the ethnologist. As Dieleman notes, it "is an ethno-methodological concept that [...] exists since long in symbolic interactionism, the phenomenology of Garfinkel, the praxeology of Bourdieu and the structuration theory of Giddens" (ibid., p. 127). As for ontological reflexivity, Dieleman uses the term to refer to 'lateral thinking', i.e. associative thinking rather than linearly rational thinking.[17] In Bateson's terms, it can be understood as 'abduction' (vs. deduction and induction), i.e. knowledge generated by metaphorical processes fostered by the aesthetic as understood by Bateson (i.e. a sensibility to patterns that connect: cf. chapter 4, section 2). These competences involve not only conscious, strategic capabilities but also subconscious, sensitive 'lateral thinking' capabilities such as can often be found among artists (and potentially – but not readily – available to anyone).

However, such a "reflexive capital" (Dieleman 2008, p. 117), is not enough: The entrepreneur in conventions is not only reflexive, but also driven by a commitment to and competence for social change. His/her success will then depend on his ability to convince others (which in the context

17 Cf. Dieleman 2008, pp. 131-136.

of the art worlds, typically starts with constructing a circle of recognition or gaining influence / seizing power in an already existing circle).

To have such means to perceive more distantly and play on the rules, is not readily open to anyone: An analysis of the specific cultural capital and social capital (in Bourdieu's sense) and history of socialization (and resulting repertoire of dispositions, in Lahire's sense) of the entrepreneur, for a given convention, would be insightful. It can be of great help for an individual to be well-socialized in a number of different conventions and therefore able to handle different information systems: this offers more opportunities to question a given convention. This quality bears comparison to that of the "marginal-sécant" in Crozier and Friedberg (1977): someone who is relatively at the margin of a social organization but who retains specific power resources, thanks to his/her access to outside information (or rather in the model of conventions: access to other screens of information). While Lahire's concept of repertoire of dispositions, can help identify such a capability, Lahire's work (2004) also points out that dissonant cultural profiles can be found among a wide variety of social categories (and not only among "cultural omnivores" from the upper social classes as the cultural sociology of Richard Peterson suggested in the 1990's). As Capra noted, "in our daily activities, most of us belong to several communities of practice" (Capra 2002, p. 109). Under favorable circumstances, cultural dissonance may serve as a seed for the emergence of hermeneutic reflexivity and may be met with a commitment to change.

An important dimension of entrepreneurship in conventions is to be able to bring suspicion upon the already existing conventions which the entrepreneur aims to modify or eliminate. Then, an important success factor is the ability to get to and weaken the second and third conditions of Lewis (1969), i.e. to instill a growing belief that compliance is already less than general and a growing belief that compliance should be less general indeed. To be effective in this enterprise, the would-be entrepreneur should look for opportunities on both internal and external dynamics of suspicion: I already discussed above the interaction between the internal dynamic of suspicion within the information system of a convention and the external competition from other conventions. An effective entrepreneur is someone who can effectively play with this interaction. He/she finds the relevant balance between internal and external suspicion and the appropriate extent, targeting and quality of suspicion he/she will employ in his/her attempt to promote certain changes.

Next to suspicion building, the effective entrepreneur holds the rhetorical qualities that will convey conviction. Especially, he/she is able to provide a (seemingly) coherent discourse and to interact with others (in terms of the material settings of his/her action) in ways that are also coherent with the discourse and the conviction aimed at. When the entrepreneur want to support an already existing convention, he/she will work to increase its coherence, fine-tuning the discourses, the style and contents of conversations, the settings of social interactions and the values they lift, etc.

My analysis after Gomez (1996) bears the comparison with a similar development in the field of the institutional theory of organizations. The rhetoric of conviction and suspicion by the entrepreneur in conventions echoes the "human agency" of "institutional entrepreneurs" playing upon legitimation/delegitimation of institutional "rationalized myths," as analyzed by Paul Colomy (1998, pp. 272-276). Colomy, and after him, Jaffee (2001, p. 237) call his model a "micro-corrective" to institutional theory, stressing the role of individual agency.

However, purposeful, conscious entrepreneurship in conventions can only achieve 'so much', i.e. as I already noted (but I need to insist on this point, to prevent any misinterpretation), it is very unlikely that one single individual social agent would manage to entirely purposefully manipulate the reorientation of a social convention on his or her own. Effective conventional change requires that a whole group of people (whether a formal organization or an informal community of practice) be engaged in the change process. At its point of emergence, such a process can as much be set in motion inadvertently as it can be pushed by a single entrepreneurial individual. And once in motion, such a process is likely to be non-linear, i.e. not following exactly the direction that any single social agent may have designed/desired for it by his or her intervention.

There can occur, especially at points of instability in social systems, an "emergence of new order", as a typically meso-social phenomenon, not controlled by a single entrepreneur in conventions directly, but set into motion by communities of practice: See the long quote from Capra in chapter 2, section 3 (p. 127).

Therefore, the concept of entrepreneurship in conventions, although it brings especially useful insights, should not be conceived as a linear strategic tool, but as one dimension within a complex understanding of social evolutions.

The artist as entrepreneur in conventions

Many of the conventions that the individual usually refers to in the context of his/her own social environment (i.e. in the different fields in which he/she operates), are less coherent when confronted to the behavior of other people with different backgrounds. The interaction becomes all the more difficult when these people are not pushed to interact intensively and for a long time, by an organization to which they belong (in which case common conventions will eventually emerge). In such situations, where the forming of communities of practice is not fostered within a given social-institutional setting, the difficulty is increased even more by the individual's relative drive to reduce cognitive dissonance. A socialized individual who is not used to being confronted to dissonant conventions, will be often likely to overlook them. However, some individuals have a relatively higher tolerance to cognitive dissonance and therefore are more open to experiences that may question their regular assumptions (i.e. their current conventions).

That difficulty would present itself extensively in the case of an art project involving an artist and a given social target group, which is not already related to the art world or used to the practice of community arts. The group will a priori be likely to overlook dissonant experiences from the artistic intervention that would risk bringing suspicion into their current conventions. However, that difficulty may turn into an essential opportunity for the artist to attempt at an entrepreneurship in conventions. This may be the case of a number of artists intervening in society and claiming to be agents of social change.

As already mentioned above, within a given convention, most information is not processed, and only abnormal information really manages to catch attention. But the first reaction of many individuals confronted to such abnormal information will in all likeliness be one of rebuttal, denial and/or neglect, so as to avoid the unpleasant experience of cognitive dissonance. Therefore, a powerful entrepreneurship in conventions would then be indispensable. But why would artists be especially good at that game? How can artists be entrepreneurs in conventions? More specifically: How can an artist successfully throw suspicion upon some of his/her audience's regular conventions and bring conviction about alternative conventions?

Rhetorical abilities

An artist as entrepreneur in conventions would need strong rhetorical abilities. These 'abilities' have both to do with the socialization of artists in European societies and with the expectations about artists' roles in our societies. The qualities I am discussing below are not inherent to being an artist at any time, anywhere. They are social constructs that may variate according to the local and regional context, and that may at some point in the future, evolve into something else.

I propose five components for rhetorical abilities that I claim to be relevant nowadays, in North-Western societies (i.e. Europe and North-America):

- Attention catcher: the artist is able to catch the attention of her interlocutors more effectively than the average social actor (one's attention is caught when one's "mind focuses on the phenomenon" calling for attention). Especially, the artist as entrepreneur is able to catch "high quality attention": attention that involves "active mental or intense emotional activity" (Klamer 2002).

- Sacralizer, eye opener and rule cracker: the artist is able to open a sacred arena around the 'aura' of the word art.[18] The artist is able to make the invisible visible, or rather to bring in the conversation things (or ways to look at things, perspectives and angles) that formerly were absent from it. And the artist is able to break some rules/habits that most people comply to.
- Itinerant broker: the artist does not fully belong to most groups (s)he interacts with, and is also able to insist on playing the outsider at times and the insider at other times, but only up to a certain extent (as being a complete outsider will work against conviction).[19]
- Exaptation facilitator: Because the artist often leaves open both the meanings, interpretations and the functions of his or her work, this can open up the "purpose" of any project carried out together by teams of artists and non-artists. This allows a process of 'exaptation' (see the definition further down) to occur.
- Structurator: The artist is granted the social role to be 'creative' i.e. to restructure reality in a different way. This attributed role can be manipulated by the artist in a process of entrepreneurship in conventions: The artist can portray himself as a (re)structurator, i.e. a 'creator' and thereby acquire a legitimacy to propose alternative conceptualizations and practices.

Some of these components (attention catcher, rule-cracker) are necessary requirements for any social actor to have the ability to act as 'entrepreneur' within a convention. The component 'itinerant broker' is allowing the artist to act, within the setting of the art project, as a "marginal-sécant", with one foot among the target group and one foot outside. The ability of the artist to act as a "marginal-sécant" is also being reinforced by some of the other components (sacralizer, eye-opener, structurator, exaptation facilitator), which are essential to the power of conviction, in a community, of elements coming from outside (including from conventions stemming from the art worlds).

Inter-conventionality and art as social process

The artist as entrepreneur in conventions may even engage himself in an intervention project implying interactions and conversations gathering and rallying a wide diversity of people, i.e. different people who are themselves socialized into different conventions. In some cases, those people would otherwise probably not interact with each other (they would even avoid re-

18 Cf. Abbing 2002, pp. 23-25, 28-30.

19 The concept of the 'itinerant broker' comes from social network analysis (cf. de Nooy et al. 2005). About the risks of being a 'complete outsider', as I identified it in an empirical study, see Kagan 2008a, pp. 184-187.

peated contact, here again to prevent cognitive dissonance). An artist successfully forming such a group, has rhetorical abilities that help both attract attention from, and interact with, these different people (and also, prevent the group from disbanding). Thanks to these abilities, the artist may then foster links between those different people, by creating arenas for conversations and other social interactions where these people, stepping out of their own spheres (or fields) of interaction (ruled by their own conventions), may experience an effective inter-conventional interaction. By 'effective' I mean: minimizing misunderstandings (substantially and sustainably) and moving towards the shared construction of a common social reality (including a 'joint enterprise' in Wenger's terms).

The networking abilities of the artist may give him the possibility to be an itinerant broker between very different people; he or she often has the possibility to be 'creative' in networking (i.e. to establish links that the average social actor would not make). The artist would then be not only an entrepreneur in conventions, but also an entrepreneur in inter-conventional interaction.

The rhetorical and networking qualities of the artist as entrepreneur in conventions are linked to the idea that art-related social interactions can provide some valuable tools for entrepreneurship in conventions. In what ways can art provide such tools?

After Dieleman (2001), we can look at art as a social process bringing an awareness of social reality "beyond-rationality", or rather: beyond linear rationality, and beyond "formal" types of rationality as understood by Max Weber.[20] The art group WochenKlausur claimed along a very similar line: "Through certain freedoms that art has been granted, an area is opening for art where the deficiencies of codified politics can be pointed out and their resolutions can be paradigmatically demonstrated. [WochenKlausur points at] art's opportunity to approach a problem unconventionally, naively and open-mindedly" (ed. Zinggl 2001, p. 132).

In the efforts to get attention and instil conviction and/or suspicion, the experience of art as social process by participants can play a role. In such a perspective, some specific experience processes may occur. Dieleman (2002) identified five processes:

- Enchantment: providing an immediate experience of a desired state of reality;
- Detachment: confronting the participant to persisting routines (that may grow undesirable when looked upon) and 'unfreezing frozen frames' (i.e. constructions of reality involving emotional attachments);
- Empowerment: changing one's self-image and perceived capacities to exercise influence and make a change, reducing inhibitions;

20 On the latter, see Kalberg 1980.

- Subversive imagination: unleashing imaginative powers that both envision the present state of the world as a critical reflection and remind the participant of desires that have been buried by his/her current conventions;
- Healing: reducing fear and stress induced by the social context.

Among those five processes, two are especially relevant for both introducing suspicion in a convention and pointing at other existing conventions: subversive imagination and detachment. Besides, the three other processes may contribute to building conviction about some new convention promoted by the entrepreneur.

Dieleman's process of subversive imagination also points at Herbert Marcuse's understanding of a subversive power of art in its *Aesthetic Dimension*. Marcuse (1978) pointed out that "the need for radical change must be rooted in the subjectivity of individuals themselves." As Carol Becker (1994) observed, belief in the regenerative abilities of imagination was fundamental to Marcuse's understanding of social change. Beyond mere sensibility, Marcuse saw in the aesthetic dimension of art a potential to transcend art's social determination and sublimate the given reality in its "overwhelming presence" in order to negate the established, 'objective' social order. According to the late Marcuse, only by an estranging form can art contradict the status quo and envision alternatives.[21] Therefore, an experience of the 'subversive imagination' of art requires that art retain a relative degree of autonomy from social order. However, I do not follow Marcuse's and Becker's claim that "the political potential of art lies only in its own aesthetic dimension. [...] The more immediately political the work of art, the more it reduces the power of estrangement and the radical, transcendent goals of change" (Becker 1994, p. 120). Such a claim, if it points at the fragility of processes of estrangement and detachment, responds to this risk with a fearful insulation from reality which perpetuates the culture of unsustainability of strictly autopoïetic, autonomous art worlds (as discussed in chapter 1, section 2). A line of complex tension here emerges: the experience of subversive imagination requires a relative, procedural autonomy, in accord with an autoecopoïetic process of engagement with social reality, rather than a fixed, substantive autonomy of art.

Another dimension of art as social process, not included in Dieleman's list, is *art as exaptation*: If art is not declared useless (as a Romantic Order of fully autonomous art would have it), art does not then always have a priori functions, i.e. functions determined at the outset of the process. The functions can come in the process itself. This process can be linked to the concept of 'exaptation' from evolutionary psychology: A property that appeared for some reason develops new functions for itself and fulfills unfore-

21 Cf. Marcuse 1978, p. 7.

seen goals (this is a process of extension of functionality).[22] In the case of conventions allowing 'common goods' to appear, the exaptation of art can be very helpful, because it can indeed bring different people together, allow those different people to agree to interact with each other where there is no clear purpose imposed on them, and subsequently help them shape a common function (common valuation, common utility) out of it.[23] If art as social process allows (more easily than other forms of, more functional, social activities) exaptation to take place, this can help understanding how common goods can be created and valued by different people who do not know beforehand that such common goods and values can be found through their interaction. It can therefore help conceiving why art can function relatively well as a social activity that offers opportunities for entrepreneurship in conventions at an inter-conventional level.

A number of material conditions (contributing to structure the material setting of conventions) are necessary for the artist to have the opportunity to expose a target-group to his/her entrepreneurship in conventions: The organizational setting of the art project should be coherent enough for the art project itself to become an efficient conventional framework (i.e. a convincing set of beliefs). It should offer enough stability to elaborate a common convention providing a common good. It requires that a formal group, or better, a community of practice be taking form.

Chances to effectively perform entrepreneurship in conventions will be much higher if the intervention by the artist takes an intensely participative form, i.e. is experienced *with* the community of practice, rather than produced *for them* and delivered *to them*. Participants in an art-project will have the opportunity to experience change in conventions more than mere audiences receiving and interpreting a product of those changes (and, if not prepared, likely to have few if any clues about how to decode the new conventions).

The problem of art world conventions

If an artist may have some opportunities to perform entrepreneurship in conventions, these available capabilities and opportunities can however be undermined by a number of difficulties, biases and traps coming from the art worlds themselves: the artist is indeed not a creator beyond conventions, but is embedded in the web of conventions of a specific art world. The dilemma for the artist aiming to bring about social change through direct intervention in society is the following: How can he or she behave as an entrepreneur in social conventions and remain an agent of the conventions of his own art world?

---

22 Cf. van de Braak 2002, pp. 42-43.

23 On the construction of 'common good' through art, see Klamer 2004.

Close attention should be drawn to conventions among the group in which an artist tends to belong. It is no news to the art sociologists, nor to critical theory in art (and especially artists practicing Institutional Critique) that the production of art does not exist in a conventions-free vacuum. The processes taking place within such groups constrain both the thinking and actions of the artists, whether we describe this social reality in terms of Pierre Bourdieu's fields, Howard Becker's art worlds, subcultures or smaller "circles of recognition" i.e. smaller specific artistic/cultural groups within an art world or across art worlds (as will be discussed in the next section). The production of art, which Becker (1982) showed to be a result of collective action (therefore involving organization), needs agreement among participants, about materials used, types of abstractions to convey ideas or trigger sensitive responses, the form in which these abstractions will be organized and the relations between the social actors involved (e.g. 'artists', 'participants', 'audiences'). Thanks to conventions, art works are able to evoke an emotional response in audiences. But because of such conventions, the ability of the artist to behave as an entrepreneur in conventions is relatively restrained, as long as he or she is tied to these conventions of his/her art world. How the conventions of art worlds, in general, impose restraints, has already been discussed earlier with the discussion of autopoïesis and of the 'Romantic Order' within a 'Technological System', framing the beliefs in autonomy and authenticity (cf. chapter 1, section 2).

One difficulty stemming from this context, with relationship to the goal of social involvement outside the art world, is the art world's distinctive distance from the general public. Because of the complex conventions of his/her art world, demanding a high level of specific cultural capital, the artist may experience great difficulties to be understood in his/her dealings with outsiders. Therefore, for those participants (who are themselves outsiders relatively to the art world's conventions), the artist is very much *marginal* but not *sécant* at all. The difficulty of conveying the elaborate own 'language' of a specific art world or circle of recognition, is actually also shared with academics, scientists, and many intellectuals, when confronting other social worlds. It is the inner circle of the audience (who themselves may have been on the artist's side once) who are the "most understanding and forgiving" (Becker 1982, p. 54). Hence the ever-present temptation to deal exclusively with one's own circle of recognition.

Another difficulty is that the artist him- or herself is likely to be trapped in the conventions of the own art world (or at least of his/her circle of recognition). The interdependency of the conventions in an art world makes it difficult to step aside. As Becker observed, "Conventions place strong constraints on the artist [...] because they do not exist in isolation, but come in complexly interdependent systems, so that making one small change often requires making changes in a variety of other activities" (Becker 1974, p. 772) This constrains the artist in his/her ability to bring about conventional entrepreneurial changes in his/her work.

As long as the artist remains unknowingly captive of the romantic/technological ghetto of the art worlds as described in chapter 1, his or her ability to effectively act as an entrepreneur in conventions will be severely limited.

At one extreme, a given artist will lose recognition and status as an artist. For example, in the French art worlds is often used a denigrating way to qualify some artists and projects as "Socio-Cu" (meaning 'sociocultural', art for and with the communities with a supposedly lesser artistic value). I do not imply myself that people involved in community art, art for the community and so-called sociocultural activities in France, necessarily perform better in terms of conventional entrepreneurship for social change. This is another question, and it may very well be that some other factors (including the conventions of social workers) cripple the artist's effectiveness as entrepreneur in social conventions and change agent, in the context of these practices. The risk of a loss in terms of 'subversive imagination' (in Marcuse's terms) and of other rhetorical qualities of the artist is also non-negligible.

At the other extreme, the artist would have so much incorporated the conventions of the art world, and would be so careful to maintain and improve his/her recognition as an artist among his/her circle of recognition and/or art world, that his/her conventional entrepreneurial abilities would be severely limited.

The entrepreneur's dilemma is thus now clear: it is about instigating change on two fronts, both inwards and outwards. This means *double entrepreneurship*. But to start with, this means finding some working tools proving to be relevant for both fronts:

Some working conventions of entrepreneurial artists

The artist may be able to use, as working conventions, elements allowing both some inward legitimacy in the relevant art world and some effectiveness in the outward-oriented entrepreneurship in conventions. The rhetorical qualities described above can indeed be appealing to the art worlds' Romantic Order. As a relative outsider (itinerant broker) the artist can claim to retain some autonomy. As a sacralizer, attention-catcher and rule-cracker, he or she can be portrayed as authentic. As an eye-opener, he or she can "make the invisible visible" as Paul Klee notoriously affirmed. As a rule-cracker, exaptation facilitator, structurator and relative outsider, he or she can tactically mimic a rhetoric of relative social uselessness when required (i.e. when survival in the art world is depending on it).

One relatively successful group of Intervention Art in the 1990's openly advocated using some working conventions as opportunities to work outside the current set of conventions of art worlds but with the legitimacy conferred by the art world of contemporary visual art: the Austrian art group Wochenklausur, which drew itself since its inception (in 1993) into direct interventions in society (see boxed text below).

Wochenklausur's working conventions

The Austrian art collective Wochenklausur became, within a decade, a much-discussed example of "intervention art". They realized many concrete interventions in society, such as *Homelessness*: a mobile clinic that provides healthcare free of charge to more than 700 patients monthly (Vienna Secession, 1993); *Drug Problems*: a shelter created for drug-addicted women who earn their money through prostitution (Shedhalle Zurich, 1994); *Language schools*: in the wake of the Balkan wars, 8 language schools for refugees in Macedonia and Kosovo (Venice Biennale 1999), etc. My focus here will be not directly on the outcomes of their concrete interventions, but on the working conventions which they developed, as an art group involved in social-political interventions.

Indeed, Wochenklausur saves its artistic status through a limited number of very rigid conventions:

- First at a practical level: accepting invitations only when they come from recognized art institutions: This convention is explicitly part of a strategy:

> "Just as traditional artworks, material objects, whether paintings or bottle drying racks, cannot initially be art per se, but rather are awarded this appellation through special sanctioning, perfectly normal actions or socio-political interventions can be given this appellation. Following their presentation within the context of art and after the acceptance of their petition to be recognized as art, these actions mutate and suddenly are art. […] Powerful institutions like museums, schools and media are decisive for what becomes art" (Ed. Zinggl 2001, p. 130).

- Then at a conceptual level: keeping faith in the artist's mission and creative potential for social change – inline with Joseph Beuys' faith in Social Sculpture and Paul Klee's invisible made visible. Wochenklausur celebrates an ode to "the psychology of the artist and to the abilities that differentiate him or her from others. [They claim that] artists' sensibilities lead them to notice where trends are heading before others do, that they have the ability to draw attention to problems that are not seen by others, that they make finer distinctions in certain areas, originate issues that attract attention and the like"(Ed. Zinggl 2001, p. 133).
- Besides, the group argues that it "points to new ways of acting not dominated by market-orientated mechanisms". Such defiance towards the villainy of commerce (independently of its potential relevance for a critique of unsustainability) is typical of the dominant

Romantic Order in art worlds (and reminds of the typical "denial of the economy" as described by Bourdieu and by Abbing)[24].

However, because WochenKlausur did reach a high level of awareness of the conventions of the art world of contemporary visual art, these working conventions were considered as a means rather than an end in itself.

"Because there are no universally valid norms for art, there can also be no absolute 'artistic quality' […] According to these notions [that are dominant in the art world], art is something of genius. It goes beyond conventional standards and thus cannot be measured by these standards. WochenKlausur's activist art has little in common with such conceptions of artistic quality […] Whether art has quality or not is merely a question of whether it conforms to certain predefined criteria" (Ed. Zinggl 2001, p. 130-131).
"In view of the American philosopher Richard Rorty's claim that concepts are continually being implemented as means of achieving certain purposes, all that remains in the end is the question: What is the word art used for? Who achieves what with it?" (ibid., p. 129).

Working conventions are not ultimately conventions of purpose, but instrumental conventions, or process conventions. Focusing on instrumentality makes Wochenklausur also fit in a world dominated by the Technological System and its demand for efficiency and formal-rational discourses and practices (while complexity would rather call forward a focus on open-ended process).

Going further, WochenKlausur recognizes art to be useful for a social purpose.

"Why art, then? First, with every successful project that is recognized as art, intervention in existing social circumstances wins increased significance. The word 'social' is then used more positively again […] Second, the mythos 'art' is of assistance when one is interested in helping realize an intention […] Third, the media reports less about the most exciting social work than about the dullest cultural events" (ibid., p. 131-132).

WochenKlausur's reflexive awareness on their own art conventions have been necessary to their ability to question some of these conventions and therefore to perform conventional entrepreneurship towards both the art world and targeted audiences. This possible way out of the entrepreneur's dilemma will now be presented: *Double Entrepreneurship*.

24 Cf. Abbing 2002, pp. 34-51.

Double Entrepreneurship

As should be clear to the reader by now, making changes to conventions in art worlds is likely to be both difficult and challenging. If we add to this the slowness of the traditional diffusion process for innovations in art (which takes place at first through 'peers' and through 'art worlds' and 'serious audience members') and its digestion and institutionalization into official "Art" reserved to cultural elites (including a few professional revolutionaries), we soon come to the preliminary conclusion that any artist who would want, here and now, to set social processes in motion among selected target groups, and not to be trapped in the internal strife of the art world, would have to set him- or herself as far as conventionally possible from the traditional distribution-system of his/her art world and would also have to directly involve the chosen social target group (using intensely participative forms of action, as I argued above).

Besides this strategic choice, the artist will have to trigger and to manage a double movement of entrepreneurship, as the only way out of the double-bind of the social context: His or her entrepreneurship in conventions towards the participants in a given project, will have to be combined with a similarly willfully constructed entrepreneurship in conventions towards his/her art world of origin. A double entrepreneurship means an internal entrepreneurship upon the entrepreneur's own conventions combined with an external entrepreneurship upon the conventions of outsiders. A single entrepreneurship, touching only a designated target-group, will be likely to be restrained by the conventions of the art-world. In another scenario, a single entrepreneurship touching only the art world itself, will not bring much 'social change' about, in the relationships with social groups outside the art world. It would rather develop complacency for the current status of social reality, as Gablik (1991) denounced (as discussed in chapter 5, section 4) – a very common situation in much of the contemporary visual art world, including in a large proportion of the practice of Institutional Critique.

To be aware of the necessity of double entrepreneurship, the artist would need to keep a critical distance to both the art world and other realms of socially constructed reality. In this perspective he or she should have developed a high degree of reflexivity, while retaining an openness to others (however "naive" they may seem to him or her). In this sense, the artist as double entrepreneur in conventions bears the comparison with the reflexive sociologist (as suggested by Schinkel 2004), but only as far as the reflexive sociologist engages with social reality.[25] In his *Critique de la Modernité*, Alain Touraine also advocated more generally for a social actor to constitute him- or herself who "fits into social relationships by transforming them, but

25 Besides, as Dieleman (2008) argued, the kind of reflexivity that is required here corresponds to 'hermeneutic reflexivity', i.e. a type of reflexivity that many social scientists too are lacking.

without ever becoming identified completely with any group, with any community" (Touraine 1992, p. 243).

In this process, the 'double' entrepreneur will also have to avoid becoming completely marginal and not sécant anywhere anymore (i.e. becoming unintelligible on either side). His/her double entrepreneurship will necessarily have to be convincing for both targets and it will allow both sides to consider the artist as a relative insider (i.e. 'one of us'). The issue with such double entrepreneurship is its intricacy in terms of relative strategic opportunities: The entrepreneur is compelled to play with double-sworded elements of suspicion and of conviction upon two different target-groups through one overall project. Each element is then indeed double-sworded in two ways: its connotation plays across the two targets, as does its implication (in terms of making new conventions more or less attractive).

To start with, the artist has to convince an inner circle of early adopters. On the side of the art world, this clearly means that the artist must shape (or join, or seize) a specific circle of recognition. However, if this is necessary, this step will not prevent this circle from being altogether marginalized or expelled from the art world.[26] Nor will it prevent the artist from getting trapped in the working conventions of one's own circle of recognition. The double entrepreneurship is a complex and fragile dynamics, and implies that working conventions be kept from crystallizing completely and that the art project be kept from institutionalizing beyond a certain level.[27]

Opportunities for such a peculiar process of double entrepreneurship in conventions will also not necessarily be present. They will especially depend on the artist's position in the art world and on the degree of relative polyarchy in the polity conventions of the art world (as will be analyzed in the next section). Is the artist already recognized as a marginal-sécant? How far can the contestation of conventions go in the art world, and how open can participation in the contestation be?

Depending on the polity setting of a specific art world at a specific historical moment, but also (outside the art world) depending on the specific institutional setting in the background of an intervention project affecting a specific social-political-economic-ecological (i.e. complex) situation, the opportunities for double entrepreneurship in conventions, and the best strategies to achieve such a double entrepreneurship, will differ widely.

---

26 The dynamic tensions between art worlds and circles of recognition will be described in the next section.

27 See next section.

Entrepreneurship in conventions from a Batesonian perspective: hard-programmed ideas and deutero-learning

The concept of conventions bears parallels with Gregory Bateson's systemic psychology, and his notions of "soft vs. hard programmed ideas" and of "deutero-learning" are especially relevant in the context of an analysis of conventions embedded in an inter- and trans-disciplinary research.

> "In mental evolution, there is also an economy of flexibility. Ideas which survive repeated use are actually handled in a special way which is different from the way in which the mind handles new ideas. The phenomenon of habit formation sorts out the ideas which survive repeated use and puts them in a more or less separate category. These trusted ideas then become available for immediate use without thoughtful inspection, while the more flexible parts of the mind can be saved for use on newer matters. In other words, the frequency of use of a given idea becomes a determinant of its survival in that ecology of ideas which we call Mind; and beyond that the survival of a frequently used idea is further promoted by the fact that habit formation tends to remove the idea from the field of critical inspection" (Bateson 1973, p. 477).

This is in line with the argument that conventions allow 'normal' information to be used trustfully and/or habitually without much consideration, while only 'abnormal' information is attracting attention.

Bateson adds that "the survival of an idea is also certainly determined by its relations with other ideas. Ideas may support or contradict each other; they may combine more or less readily" (Bateson 1973, p. 478). This focus is in line with the discussion on the internal coherence of a convention, and of the coherence between conventions, but highlights its psychological dimension, more than does the rather sociological and economic focus of the analysis of conventions.

Furthermore, Bateson distinguishes hard-programmed from soft-programmed ideas (a distinction which was absent from my discussion of conventions so far):

> "It is commonly the more generalized and abstract ideas that survive repeated use. The more generalized ideas thus tend to become premises upon which other ideas depend. These premises become relatively inflexible.
> In other words, in the ecology of ideas there is an evolutionary process, related to the economics of flexibility, and this process determines which ideas shall become hard programmed. The same process determines that these hard-programmed ideas become nuclear or nodal within constellations of other ideas, because the survival of these other ideas depends on how they fit with the hard-programmed ideas. It follows that any change in the hard-programmed ideas may involve change in the whole related constellation" (Bateson 1973, p. 478).

I only tangentially approached this focus, without considering it sufficiently, when mentioning 'orders of worth', principles and paradigms, i.e. a more abstract level (both in social institutions, and in the habitus / repertoire of dispositions of individual social agents). I can now come back to it, thanks to Bateson's notion of deutero-learning which allows to explore how even hard-programmed ideas can change.

The repetition and expansion of experience across contexts allows 'learning to learn', i.e. what Bateson defined as deutero-learning. Aesthetics of sustainability, or the sensibility to patterns that connect, also facilitates learning operating at the level of deutero-learning, gaining systemic understanding across different contexts, thanks to hermeneutic and ontological reflexivities. And vice versa, deutero-learning contributes to the cultivation of a sensibility to patterns that connect. Bateson described deutero-learning in his empirical analysis of the learning processes of dolphins, humans and dogs:

"[The subject] also, in some way, learns to learn. He not only solves the problems set him by the experimenter, where each solving is a piece of simple learning; but, more than this, he becomes more and more skilled in the solving of problems. In semi-gestalt or semi-anthropomorphic phraseology, we might say that the subject is learning to orient himself to certain types of contexts, or is acquiring 'insight' into the contexts of problem solving. In the jargon of this paper, we may say that the subject has acquired a habit of looking for contexts and sequences of one type rather than another, a habit of 'punctuating' the stream of events to give repetitions of a certain type of meaningful sequence" (Bateson 1973, p. 140).

Deutero-learning is to be understood both as a psychological and a sociological concept. To prevent any confusion, I have to point out that in Bateson's systemic view of the mind, psychological processes can only be understood if also analyzed as social (inter-subjective) processes. Bateson argues as follows:

"How do we learn those learnings or wisdoms (or follies) by which "we ourselves" - our ideas about self – seem to be changed? [...] *Learning the contexts of life* is a matter that has to be discussed, not internally, but as a matter of the external relationship between two creatures. And *relationship is always a product of double description.* [...] Relationship is not internal to the single person. It is nonsense to talk about 'dependency' or 'aggressiveness' or 'pride', and so on. All such words have their roots in what happens between persons, not in some something-or-other inside a person. No doubt there is a learning in the more particular sense. [...] But the relationship comes first; it *precedes.* Only if you hold on tight to the primacy and priority of relationship can you avoid dormitive explanations. The opium does not contain a dormitive principle, and the man does not contain an aggressive instinct. [...] The same is true for 'dependency,' 'courage,' 'passive-aggressive behavior,' 'fatalism' and the like. *All* characterological adjectives are to be reduced or expanded to derive their definitions from patterns of interchange, i.e. from combinations of double description" (Bateson 1979, p. 124-125).

Understood as both a sociological and psychological process, deutero-learning is thus pointing at an important dimension of the conditions under which entrepreneurship in inter-conventional interaction could be successful, not only superficially, but also at the level of hard-programmed ideas. Those two processes (i.e. deutero-learning and entrepreneurship in inter-conventional interaction) can become mutually reinforcing. And both processes can be fostered through the autoecopoïetic practicing of aesthetics of sustainability (as defined in chapter 4).

Only now can I bring forward a claim from Hildegard Kurt, that was unfounded in her argumentation (as Margolin noticed – cf chapter 6, section 2), but gains in weigh if re-interpreted in the current context: Hildegart Kurt argues indeed that "far from restricting itself to surfaces, art is involved in designing values, and increasingly becoming a medium for exploration, cognition and for changing the world. One characteristic of art as a form of knowledge is that it brings not only ratio, but also aesthetic competence into the cognitive process – which makes it different from science and at the same time its equal" (Kurt in eds. Strelow, Prigann and David 2004, pp. 239-240). I propose to understand what Kurt pointed at as "aesthetic competence", as the autoecopoïetic practicing of aesthetics of sustainability (as defined in chapter 4), and to understand the "exploration, cognition and changing the world" as an expression of deutero-learning and entrepreneurship in inter-conventional interaction.

Caveat: the bulk of the psyche's iceberg

In the preceding pages, I have proposed to elaborate an analysis of conventions, and of entrepreneurship in conventions, informed by the earlier exploration of the paradigm of complexity and of aesthetics of sustainability. I have also built a frail bridge between this analysis and the systemic psychology of deutero-learning. However, one important psychological dimension is still missing, which I will be unable to properly consider in this work (although I repeatedly pointed at it, for example in chapter 4, with Bateson's view of art as moving beyond the limitation of "purposive consciousness"): The *working of*, and *working with*, the subconscious (which is commonly referred to as the bulk of the psyche's iceberg). I suspect that a thorough analysis of the subconscious, delving into the interface of cognitive studies and philosophy, would reveal the uni-duality of emotions and reason, of analogy and logic, and the uni-plurality of reflexivities (and not only their diversity as explored in Dieleman 2008). To substantiate this suspicion, I would need to explore how reason is understood by Antonio Damasio as inseparable from emotions, even in the most seemingly linear, formal rational decision-making processes (cf. Damasio 1994, 2003; and critiques of Damasio: Gluck 2007). I would need to take into account Howard Gardner's famous claim for "multiple intelligences" (Gardner 1983, 1999). I would need to discuss Lakoff and Johnson (1999)'s concept of the "embodied mind" and to assess how they argue that even mathematics and logic derive from metaphorical thinking (Lakoff and Johnson 2001). I would need, through all this, to explore more closely how the subsconscious works as multiple open windows that keep opening when the autopoïetically-driven doors of consciousness shut out certain disturbances from the outside world, and thus fosters the 'eco'- in autoecopoïesis.[28] I am pointing at a few elements in this direction, in a speculative excursus to be found in appendix 1.

SECTION 2: POLITY CONVENTIONS: THE POLITICAL SETTING FOR DOUBLE ENTREPRENEURSHIP IN CONVENTIONS

An art world is also a space of political relationships. As will be discussed here, the political setting of an art world does matter, for the possibility of double entrepreneurship in conventions. Everyone does not have an equal say in the construction of the conventions and especially in the aesthetics of an art world, because an art world is not necessarily an even playing-field where all individuals and groups would be granted equitable chances to made their voice heard or even to elaborate their own perspectives. The question I will address now is how different configurations of authority in different art worlds determine *who* can contribute *what* to the construction of the aesthetic conventions and other major conventions of an art world.[29]

Howard Becker (1982) pointed out that what he calls the "entitlement" to make judgments that are recognized by most participants in an art world as legitimate, "depends on a variety of contingencies that arise from political shifts and struggles within the art world".[30] He did not analyze these internal political phenomena further. However, an understanding of art worlds' contrasted political settings is necessary, to identify the different ways in which these art worlds change, and especially how their conventions are susceptible to an internal entrepreneurship in conventions. In an art world as an arena of conflicts, certain circumstances enable occasional change, others almost continuous change, while some circumstances make change especially difficult for long periods. One major dimension of such circumstances lies in the polity conventions of the art world.

---

28 Edgar Morin also delved into this direction, in the 3rd volume of his method, *La connaissance de la connaissance* (Morin 1986).

29 This section is based largely on an unpublished conference paper (Kagan and Abbing 2006).

30 Cf. Becker 1982, p. 152.

A polity convention is a convention relating to the relationships of power, authority and contestation within an art world. In an art world, I will however not be speaking of just one single polity convention, because different groups in the art world will follow relatively variable sets of formal and informal rules and because different types of operations (e.g. distribution, aesthetic discourses, and employment relationships) may involve different sets of habits and beliefs about power, authority and contestation. Therefore I will rather speak of a bundle of polity conventions when assessing the overall political setting of an art world.

I will now introduce this concept of 'polity conventions' in the context of the art worlds and of 'circles of recognition' and together with notions of authority and legitimacy. This will allow, in a further step, to propose a number of conceptual tools to assess the polity settings of art worlds (in between the two extreme polarities of 'polyarchy' and 'hegemony') and to observe shifts in polity settings, in terms of moving contestation, participation and changing beliefs and habits making up polity conventions.

The art world as a field of routine cooperation and an order of legitimization

The definition of an art world and its borders does not rest on an arbitrary academic distinction, but on a socially constructed boundary, that is, a boundary that exist in people's heads and that is based on shared conventions. However, this does not mean that the 'art world' is a completely relative floating concept: As Becker (1982) argued, one can only start speaking of *one* art world when there is clearly "routine cooperation" taking place within the supposed art world. Different art world participants being more or less aware of such routine interactions, they form in their minds, relatively different opinions on the definition and borders of an art world, which are then tested in the interactions with other participants. They are aware of these boundaries, and sometimes, some participants want to cross them, while other participants try to prevent this, exercising exclusionary practices. Therefore the boundaries are also barriers.

I propose to establish the boundary zone of an art world where routine interactions, legitimacy and recognition end, i.e. where stand the most peripheral and adventurous recognized and legitimate artists, who produce art with aesthetic value according to accepted conventions in a given art world, and who are still being engaged in relatively routine social interactions with other participants in that art world. Across the boundary stand non-recognized and not legitimate artists, who produce art with no or little aesthetic value (according to the prevailing, or to generally accepted conventions in the art world) and who are engaged at most in non-routine, sparse social interactions with other participants in that art world.

An art world is thus as vast as the order of legitimization that it holds together. It is an arena offering a common identity, understanding and recognition around the definition of a type of activity/work deemed artistic. It has a network of collective action with organizations and shared conventions that move the art world beyond a local distinctiveness, and with institutions and an official history that place it beyond immediate and total control by a single group of people.[31] An art world, understood in social-systemic terms (following partly Luhmann 2000) tends to be self-referential with therefore a relative autonomy, path-dependency and resistance to fast-paced change in its underlying conventions. But if art worlds have boundaries, the social construction of these boundaries by art world participants can sometimes generate overlaps or voids between art worlds. In the case of overlaps, such situations can have a destabilizing impact on these art worlds. The destabilization will be all the more present if the overlapping art worlds hold diverging underlying conventions.

Circles of recognition

I am using, as did Hans Abbing (2002) the expression 'circles of recognition'.[32] It may seem more usual to refer to different art worlds, to sub art worlds or to scenes, schools or styles within one art world.[33] These subunits (under these different names) play a role in the definition of the overall art world. But unlike schools or styles within one art world, the space occupied by a circle of recognition does not necessarily always fit within the confines of the boundaries of a single art world. It can relate to groups not yet recognizable as art worlds or groups at the borders or crossing borders of existing art worlds. However it relates to groups that are recognizable as such (i.e. as a specific group of people). For example, the surrealists around André Breton formed a prominent circle of recognition operating across art worlds of their time such as the modern visual arts world and the art world of literature.

Circles of recognition are shaping around certain standards/values members usually agree on and they are explicitly or implicitly grouping around some key individual figures. Besides, at least some people outside the circle

---

31 In Becker (1982), this is what would be considered a fully blown art world, i.e. what he sees as the last stage in the rise of new art worlds (cf. chapter 10 in Becker 1982).

32 The expression "circles of recognition" has already been used in sociology by Pizzorno 1991, and in the context of the sociology of art by Nathalie Heinich (1997, p. 4), but with a different meaning. I am introducing it here as a notion with a specific meaning within an analysis of conventions in art worlds.

33 Within the confines of an art world, these circles of recognition do have some degree of relative autonomy; but the expression 'sub art world' would imply an a priori exaggeration of such autonomy, going against the notion of an art world as the institution that legitimizes art. The expression 'sub art world' is thus potentially self-contradictory, and I will abstain from using it.

are also aware of the standards that characterize the group and these outsiders can also play a role in setting up the boundaries and standards of the circle (even against the will and interests of the circle itself).

According to mainstream sociological understanding of art worlds and fields, people in the center of an art world — the 'establishment' as common speech characterizes them— have the largest say in the definition of art; people outside the center have a smaller say. People outside an art world, most of all those with little interest in an art world, have a much smaller or negligible say. But as soon as we introduce the strategies of smaller groups (circles of recognition) in the picture, the actual dynamics of power and authority in the art world have to be more closely observed, in order to consider how far the mainstream understanding is correct, and on the other hand, how soon and in what way peripheral agents and groups can play a constitutive role in the (re)definition of the art world's canon and other underlying conventions, through the dynamics of entrepreneurship in conventions.

The authority core of an art world

How much unified and uniform an art world is, will therefore have an influence on the opportunities for entrepreneurship in conventions within the art world. This question is related to the existence of a 'core' in the art world.[34] Metaphorically, I propose to think of the 'core' as a 'gravitational field' of authority.

Authority is often confused with power in daily speech. However in political sociology the two terms have distinctive meanings, although they most often work in pair.[35] While power expresses the effective possession of means to achieve certain ends, authority expresses the effective possession of attributes of legitimacy and rightfulness that authorize the pursuit of certain ends. Having authority implies having influence (because of one's

34 I speak of a 'core', instead of simply looking at the 'center' of an art-world. Apart from the various idea associations this term fosters (one can especially think of hard-cores vs. soft-cores), the 'core' as opposed to the 'center' is a 3-dimensional metaphor, pointing at the notion of gravity: Core gravity is not just more dynamic than a metaphor of pyramidal hierarchy. It is more telling than a two-dimensional metaphor about the interactive strength of authority in the art world as a sort of 'gravitational field'.

35 Authority actually originates from the Latin word 'auctoritas', as does the concept of 'authorship' (authority and author come from the same Latin root). While the 'author' is the beneficiary of conventions establishing the paternity in the production of something, authority benefits from conventions establishing the paternity of legitimate deed. Power on the other end, comes from the Latin word 'potestas': power of coercion.

competence, status or charisma).[36] The strength of authority lies in the beliefs of those that are subjected to its realm: If physical violence can enforce power in the short term, it is of limited effect in the absence of authority. Authority is doubly interactive because (1) it lies on a convention binding the powerful and the powerless and (2) it validates enactments of power (i.e. it doesn't celebrate its own existence: there would be no authority if it is not an authority *on* something else).

In art worlds organized around informal rules rather than around formal control and command structures, the political value of authority is especially relevant. The 'core' refers to a material space (an arena or platform), to organizations and to people occupying this space and these organizations. The 'core' is a relatively objectified notion: its socially constructed reality is not only relative to the judgement of participants but acquires an institutional consistency, the more so with the 'solidification' of the core (see below).

An art world may or may not actually have just one core, i.e. it may have one centre of authority and power or multiple centres of power. If the existence of an art world implies necessarily a degree of social organization and of institutionalization, this does not necessarily imply that every art world should a priori be conceived as following a unidimensional hierarchization around a single core. (Thinking as a political sociologist in this section, I do not a priori subsume all art worlds under a single institutional model.[37]) Especially in periods of important change, i.e. when an art world becomes relatively unstable, it becomes helpful to conceive of more than one core. But even if there is no single core around which the whole art world gravitates, there still exists at least a tacit agreement about the focus of the art world: the different groups involved share a convention informing adherents that the art world 'would have its core around here', with certain values and certain characteristics. Otherwise the art world would enter severe volatility and turmoil and ultimately there would be no art world left at all.

Degrees of institutionalization of an art world's core

The core may be more or less institutionalized, i.e. to various degrees.

In a first situation, at the lowest degree of institutionalization, no core has formed concretely and the focus of the art world is a matter of common understanding between a number of circles of recognition. Following

36 After Max Weber, political sociology traditionally distinguishes three forms of authority: traditional (based on history and status), charismatic (based on the gift of grace, on inspiration) and legal-rational (based on articulated discourses of rationalization and on explicit rules).

37 Ideally, one should analytically make a distinction between two factors: The dimensionality of power (one or several cores) and its centralization (there may be, next to a central centre of power, one or more sub-centres of power, i.e. comparable with central government and local governments).

Becker: "An art world is born when it brings together people who never co-operated before to produce art based on and using conventions previously unknown or not exploited in that way".[38] Members of the different circles of recognition feel tied together by a relative bond originating from the imagined core of the art world. The core exists *only* as a shared but hazy bundle of conventions. Several discourses are being developed that point at a vague common ground, setting in motion the collective construction of an imaginary linking with common symbols and heroes. Canon formation is at work with elements converging and a weak path-dependency. But one cannot speak of a 'recognized aesthetic canon' yet: There is not yet an operating common platform, a common arena with organizations allowing a sustained coordination of these discourses and imaginations.

Howard Becker describes this first stage in practical terms in his depiction of the birth of an art world.[39] He insists that the most difficult thing for a beginning art world is not for dispersed innovators to experiment but for them to find each other and for conversations to take shape that lead to canon-forming discourses. In such a stage, networks e.g. of suppliers, distributors, supporters only start clustering locally.[40] In this first situation, I will speak of the art world's *diluted core*. Such a situation is likely to correspond to a stage in the history of an art world when it is coming into existence or is in a major transition period (a revolutionary period).

In a second situation, a core has formed as a concrete common arena and members of one circle (or several overlapping circles) of recognition occupy the core. They try to take advantage of this position to draw the identity of the core closer to the identity of their own circle of recognition. Depending on many conditions, their strategies may know relative success, or the core may slip out of their hands (and they may then even slide out of the core). Members of the different circles of recognition situate themselves towards the current trend in the core.

One important player in this process already described by Becker is the successful local supplier outgrowing the local market and the subsequent development of distribution.[41] This process generates wider attention and establishes connections between local circles of recognition that happen to be taking part in similar working processes. The extension of a network of co-operation goes together with an integration of circles of recognition. The art world is then institutionalized enough to be endowed with widespread recognition as 'true art' in the rest of society.

38 Cf. Becker 1982, p. 310.

39 Cf. ibid., pp. 310-322.

40 In the contemporary context of Internet-based networks, this aspect may have to be reconsidered, i.e. speaking of trans-local clustering.

41 Cf. ibid., pp. 320-321, 325-329.

However, if a canon has emerged and path-dependency is established, the canon is still debatable. The coordination of discourses and imaginations entering the core is not locking them up into a completely rigid mould.

The core exists both as a shared bundle of conventions and as a substantial space (i.e. a concrete platform) where the shared conventions are being constituted. I will speak then of the art world's *alterable core*. Such a situation is likely to correspond to either an emerging and expanding art world or a flexible art world preventing/limiting excessive monopolization and hegemonization of the core: a near-polyarchic art world (see below).

For example, for the past century, the core of the art world of modern/contemporary visual art has been probably often an alterable core, with alternating leaders in the figures of avant-gardes, movements, schools or individual artists.[42]

In a third situation, the core has acquired its own identity and interests. It can no longer be identified to one or several circle(s) of recognition occupying its space, but becomes in and for itself the heart and soul of the art world. Even if members of the art world move in and out of the core, those currently in the core identify themselves to the continuity of the core at least as much as to their own circles of recognition. Even members farther from the core notice this situation, and they may also identify themselves at least as strongly to the core as to their own circles of recognition.

The art world has developed a canon that is rigid, almost unassailable and strongly path-dependent.

In other words, the art world, and especially its core, has then acquired very rigid institutions. I will speak then of the art world's *solid core*. Such a situation is likely to correspond to *a near-hegemonic art world* (see below) and may especially indicate the apogee of a near-hegemonic art world (which, thanks to its now established institutions, enjoys the social recognition that usually flows to legitimate art).[43]

One shall remark that an art world is not assumed to necessarily move in a measured way from the diluted, to the alterable, to the solid core and vice versa. For example, the move from diluted to solid may take place within a relatively short time span.

The core of an art world cannot be fully compared to the political institutions of a State. Unlike explicit political systems, the political arena of an art world (its polity) is mainly built around *implicit* conventions and relationships of power, decision and debate. Explicit formal power networks and in-

---

42 Some of the arguments put forward in this text about specific art worlds may be speculative and contestable by the reader, but even then, they advantageously illustrate how one can attempt at understanding the art world, looking at its polity conventions.

43 In my unpublished conference paper (Kagan and Abbing 2006) is discussed the case of classical music in the late 20th century, as an example of a near-hegemonic art world with a relatively solid core.

stitutions do exist, but their role is more secondary than in the political institutions of a State. This peculiarity stems from an ongoing "Romantic Order"[44] in the arts (ruling since the later 19th century) in which formal organizations have lost some of their legitimacy (although the arts do have numerous dealings with the institution and the formal organizations of the State). Formal organizations do not offer the main framework through which conflict is organized in arts worlds. Political interactions in the arts world rather take place primarily through informal configurations between circles of recognition, and only secondarily through formal configurations. For example, even the members of formal selection committees for official art prizes pay a great deal of attention to the influence of informal networks of gatekeepers and opinion-makers in their art world. They do not feel out of the grasp of an informal community of influential critics, artists, experts and other peers. The exact configuration of formal and informal 'rules of the game', in an art world, is however variable, from art world to art world.

I propose to analyze the range of possible political configurations in an art world along an axis, with at its two poles, 'hegemony' and 'polyarchy'.

The Art World: between Hegemony and Polyarchy[45]

In a lively art world on the move, there may well be a number of different important circles of recognition which partly govern the art world together and partly govern separately their own backyard (and thus have some relative autonomy). Important circles of recognition often expel people and groups (i.e. minor circles of recognition). There are continuous attempts by teams and networks of individual participants to socially monopolize one's own circle and the art world as a whole. Nevertheless, due to a number of factors I will now discuss, on the whole social monopolization of the art world is less successful in such a case than in the case of an art world locked up by a hegemonic setting.

---

44 Several art sociologists have repeatedly pointed at the influence of romantic values and conventions in the arts (e.g. Nathalie Heinich, Hans Abbing). On the "Romantic Order", cf. chapter 1, section 2.

45 In the following lines, the model used is largely inspired from Robert A. Dahl's *Polyarchy*: A 'polyarchy' is a political system which is both "highly inclusive and extensively open to public contestation" (Dahl 1971, p. 8 – page numbers refer to the 1975 reprint). In his work, Robert Dahl developed the concept of polyarchy in order to account for political regimes at the level of a country. However, he welcomed the possibility and the interest of an application of the concept to subnational and socially more specific levels of political space "such as [...] trade unions, firms, churches, and the like" (Dahl 1971, p. 12). *Remark*: The term "polyarchic" as used in the present text is directly inspired by Robert A. Dahl's *Polyarchy*. However, Dahl himself was using the adjective "polyarchal", when referring to a country's political system with the characteristics of Polyarchy.

Let's imagine two extreme cases of art worlds. I will later describe how actual art worlds can be positioned relatively to those 2 extreme poles along three dimensions relating to contestation and participation in the politics of the art world. But first, I point now at the extreme configurations to chart out the conceptual territory for the following expositions: In the first case (the hegemonic setting), the art world allows no contestation and offers no room for open debate. In the second case (the polyarchic setting), the art world works as a more 'democratic' arena.[46]

Readers accustomed to 'cultural studies' might be confused by my use of the term 'hegemony'. I am using the term following the tradition of American political science (and especially Robert Dahl), and not in the Marxist tradition of Antonio Gramsci (and after him, Stuart Hall), who speaks of "hegemony" to theorize the "hegemonic domination" of the ruling social class over the other social classes.[47] That meaning is absent from my use of the term.

I will first describe the extreme case of a fully hegemonic art world:

- In a *hegemonic setting*, the quality conventions for the definition of an art world are imposed by the circles of recognition that (at a given point in history) have entered the decisional core of the art world.[48] These circles have indeed acquired the ability to monopolize a number of conventions and especially conventions about quality.
- A hegemonic art world will resist and fight all forms of contestation from any individual, circle of recognition or other group at the periphery of the art world or outside the art world. A hegemonic art world will be more likely to exclude heterodox/critical/oppositional circles of recognition and individuals at its periphery than to include them, and will close its gates to newcomers coming from the outside world. In the process of maintaining power art world 'officials' rely on so-called 'rookies', not on 'rebels'. Rookies usually gradually become leaders.
- In a hegemonic art world, the groups and individuals who are closest to (or 'in') the power-core (the 'officials') will attempt at maintaining their power through segregation and exclusion (i.e. the old 'divide and rule' principle). But they will also resort to one of the following 2 possible forms of political processes (or alternatively each of the 2 forms): a process of organic 'consensus' (where everybody must agree, unless a

46 'Democratic' in terms of the common speech contemporary understanding of this word, i.e. relating to so-called 'representative democracy'. See Aron (1965) for an overview of what this meaning entails.

47 Cf. Hall 1996, pp. 42-45.

48 A hegemonic art world can well be understood in terms of fields of power, following Bourdieu. However, the more polyarchic an art world, the less rigid the field (and the more advantageous the conceptual tools of 'conventions' become, by comparison to Bourdieu's 'field').

minority blocks the decision, but with a high risk of being ostracised) and a process of absolute political competition where no circle of recognition trusts any other circle, where individual agents also distrust each other and where the 'winner takes all'.[49]

An art world relatively close to the hegemonic setting can be found in the case of classical music across the 20th century (as discussed in more details in Kagan and Abbing 2006): Most quality conventions of classical music have remained stable and largely uncontested across the last century, e.g. concerning conventions on the instruments used (the intrusion of electronic innovations being confined to the experimental practices of contemporary composers) or e.g. concerning the relatively low conventional tolerance for adaptations of earlier compositions.

The more hegemonic an art world, the less likely that an internal entrepreneurship in conventions be successful (unless it is carried out as a conservative entrepreneurship in conventions, aimed at reinforcing conventions considered as traditionally legitimate).

To highlight the maximum possible contrast between imaginable polity settings in an art world, I now turn to the other extreme: The polyarchic setting.

- In a *polyarchic setting*, a number of different circles of recognition are able to come to a negotiated and contestable agreement on quality conventions for the definition of the art world (of its collective values, artistic goals and evaluation criteria).
- A polyarchic art world will be open to contestation from the circles of recognition at the periphery of the art world. A polyarchic art world will also be more likely to try to include heterodox/critical/oppositional circles of recognition and individuals at its periphery. A polyarchic art world will also be more likely to open its gates even to newcomers coming directly from the outside worlds. In the process of maintaining power art world 'officials' not only rely on 'rookies' but also on 'rebels' who may in due time take their place. Some current leaders were former 'rebels' too who, in collaboration with others, successfully reshaped the conventions of their art world and who are now bound by their own success.
- In a polyarchic art world, the groups and individuals who are closest to the power-core (the elite or 'officials') will attempt at maintaining their power through integration and political processes of compromises that are known to be still contestable later on (i.e. each settlement is a con-

49 An absolute form of political competition implies that the losers of the day encounter a relatively high probability of losing any means to pursue the competition. The losing party runs a great risk of losing all means of action and of being expelled from the playing field.

testable give and take: the negotiation is never completely ended – see more details on compromise below).

An art world relatively close to the polyarchic setting can be found in the case of the visual arts across the 20th century (as discussed in more details in Kagan and Abbing 2006): Through recurrent self-referential mocking of art and through the currently widespread beliefs in a post-modern condition of art, quality conventions are ritually questioned and displaced. Contemporary art inherited from a century of raging debates and oppositions around quality conventions, with a fast-paced succession of competing leading circles of recognition (or "avant-gardes").

The more polyarchic an art world, the more likely that an internal entrepreneurship in conventions will be successful (but the higher the risk that the entrepreneurship in conventions be relativized, as the entrepreneurs are included within the art world and bound to make compromises).

Furthermore, in general political terms, the type of political setting of a polyarchic polity, is relatively closer both to the forms of democracy advocated by "social ecology" (after Murray Bookchin) and to the participative democracy that the Fowkes linked to a "sustainable art", than a hegemonic polity.

The political space for contestation: compromise vs. consensus

Polyarchy, the ideal of contestation and the political philosophy of Liberalism are about respecting 'the rights of minorities', i.e. allowing minorities to retain a space of contestation, and a space for their interests and values to exist even when these clash with majority values, with average values. In the ideal of polyarchy, every contestant must be guaranteed 'security', i.e. be sure that even when he or she loses the political contest and the contested seat of power to determine what is the commonly applied value, he or she will still be able to contest, i.e. to dispute and challenge (up to a certain level) the outcome of the political process. The loser is merely complied to accept a defeat as a temporary fact; to accept a compromise as a temporary configuration that one must live with, but not necessarily agree with. To put it shortly, there is no definitive 'winner' in a polyarchy, but a relative and temporary winner that still should seek dialogue with relative losers in establishing a temporary compromise (and if there is no compromise, i.e. if the winner imposes his views without due debate, the minority is allowed to contest vehemently and even exceptionally sabotage the decision process by means of 'direct action'/civil disobedience). Through its combined obligation to accept temporary defeat and right to go on contesting, the losing party keeps enough political resources to remain on the playing field.

Absolute political competition is inimical to polyarchy. To be viable in a polyarchy, the dynamics of political competition must be corrected by a sense of compromise from all sides. It becomes then regulated competition. As compared to political competition, political contestation occurs at a higher level than regulated competition because it also involves mapping and moving the terms of trade, rethinking the conventions under which competition is legitimized. Regulated political competition is expected to occur within given conventions whereas contestation is expected to work also upon conventions. (Contestation can also be present outside situations of competition, e.g. within relationships of cooperation.) The right to contest is necessary for entrepreneurship in conventions.

However the situation in art worlds is not often as straightforward as just described: Even without the emergence of formal structures and groups acting out in the open against the existence of a shared space for contestation, forms of informal social monopolization can take hold of an art world. Rules are indeed often implicit and vague and informal networks operate next to decision-making organizations, as is the case in most art worlds where for example diplomas have only limited significance as formal entrance barriers to arenas of competition. In art worlds, informal social monopolization will tend to starve off contestation, while effective contestation will tend to expose and denounce informal social monopolization.

The notion of compromise implies a 'give and take' in a temporary configuration: It does not mean that all compromising parties have to come to a full agreement but it means that they must have a relative tolerance for the reached agreement. Compromise is a specific modality of the negotiation of conventions and of decision-making processes in an art world. Compromise is by definition temporary because it does not a priori imply that it is itself (i.e. the attained solution) an expression of a higher truth. In making a compromise, contestants are aware that one is defending a particular interest against other particular interests and that the solution must be an arbitration, based on relative majorities while trying to minimize the harm eventually done to minorities. In compromises concessions are made and some disagreements subsist. Therefore compromise imposes a practical limited 'cease-fire' where contestation can still be allowed as long as the contestant respects the binding practical consequences of the compromise (i.e. temporary status quo). It is indeed often the case in a compromise that no party feels fully satisfied with the solution. They are not led to believe that they have attained truth. But they do believe that the compromise is relatively acceptable.[50]

I am using the political concept of 'compromise' in a way specific to the art worlds.[51] As already noted, art worlds involve more informal social con-

50 See Pellizzoni 2001. The praise of compromise vs. consensus as a guarantee of effective representative democracy (in State politics) can also be found in the conferences of legal and political theorist Slobodan Milacic.

51 Thereby I depart from the definition I would give to compromise in a political science analysis of very formal processes of decision-making in and around hou-

figurations than e.g. political parties, workers unions and houses of parliaments. Therefore negotiations take different forms. Formal collective meetings will be less frequent and less relevant to negotiations than ongoing bilateral meetings, private conversations and informal chatting. Even apparently formal and concrete discussions will be likely to indirectly and tacitly bring common understandings on shared conventions.

I am also using the political concept of consensus in a way specific to the art worlds.[52] With consensus, a broad agreement is being shaped on an issue and every party apparently agrees on the terms of the solution/decision. Consensus is not a priori temporary because it is supposed to solve the issue of particular interests by the shaping of a superior common interest. Consensus is indeed supposed to bring cohesion to the overall group in a process of collective thinking. Actually, the idea of consensus assumes that society has shared values which can outweigh particular interests.

The notion of consensus relates to the old distinction in sociology (since Ferdinand Tönnies) between community and society or between "solidarité contractuelle" and "solidarité fusionnelle" (as coined more recently by Guy Bajoit).[53] But consensus is not just some vestige from the old intensely close communities: It is community under a new disguise. Habermas for example claims that genuine consensus could allow the recognition of the truth of a indisputable public good.[54] The praise of consensus as universal reason giving way to truth by Habermas has been perceptively criticized as falling into essentialism and therefore ignoring the multiplicity of values and of reasons across different cultural spheres.[55] Essentialism indulges in ethnocentrism and the most elaborate defence of consensus, by Habermas, constitutes symbolic violence against minority groups with diverging understandings of good and evil.[56]

---

ses of parliament. In those cases compromise acquires a more delineated definition. Parties claim upfront that they have interests to defend (although they may reveal only some parts of their own interests in strategic manoeuvres) and gradually make concessions to each other. A relative winner is also more clearly emerging from the process.

52 In the usual formal political process, the definition of consensus is stricter. No solution can be proclaimed as long as a minority puts its veto on it. But after consensus has been reached, every party is restrained from further contesting the content of the consensus. If made into a systematic decision system, each consensus further restrains the scope of contestation and therefore restrains polyarchy (until the system eventually reaches the level of near-hegemony if all solutions are sought by consensus).

53 Cf.Bajoit 1992, p. 101.

54 Cf. Outhwaite 2003.

55 Cf. Heyting, Kruithof and Mulder 2002.

56 Cf. Baumeister 2003.

The attainment of consensus implies restraint from further contestation by oneself and lesser tolerance for contestation by others. It therefore most often moves an art world rather in the direction of a near-hegemonic polity setting than in the opposite direction. But if the context is one of severe antagonisms and extreme competition, replacing competition with attempts at consensus may contribute to a movement towards a relatively more polyarchic setting – therefore the political value of consensus is relatively contextual.

Compromise is also different from political absolute competition: With political absolute competition, the winner's solution/decision would directly become a reality and the loser's ability to contest would be severely undermined. Further contestation would be imperilled and rather come in a dialectic manner, from the reconstruction from scratch of the means of the loser or from new groups of contesters appearing in and around the art world. By contrast, compromises allow dynamics of regulated competition.

To illustrate these different modalities of negotiations in the production process of specific art works, I will point at the art world of contemporary dance (see boxed text below). For the remainder of this section, I will come back several times to the art worlds of dance, to illustrate my analysis of polity conventions in art worlds.

Compromise vs. consensus in dance companies

I focus more specifically on French contemporary dance choreography in the 1980's and 1990's. The procedures through which a dance company[57] constructs a given choreography reveal different political models of coordination.

In some cases the choreographer is recognised as the creative leader and there will be a consensus among dancers that the choreographer's choices are legitimate (although an external observer might see them as autocratic – which is also true to some extent). This basic scheme has been most dominant in the cases of choreographers like Angelin Preljocaj and Régine Chopinot.[58] In other cases, there is a collective process through which 'the collective' and 'the greater imperative of the work of art itself' express themselves with a relatively consistent production process: A consensus builds up that seems to legitimately outweigh personal visions. A choreographer and group that seems to have come clos-

57 As a collective, a dance company includes: the choreographer and dancers, and eventually also the other stakeholders including the stage designer, composer and musicians.

58 Cf. Guigou 2004, pp. 222-225, 228-230, and Kagan 2004b.

est to this scheme is the Groupe Emile Dubois around the choreographer Jean-Claude Gallotta.[59] In these first two scenarios, the dancers run the risk of ignoring their potential individual interests, strengths and values, which may be at stake in the production process. But in the scenario that follows, the situation is different: In some cases, there is a collective process through which different participants are encouraged to bring individual contributions even if they may be at odds with other contributions or with the 'unity of the work'. There is a higher tolerance for conflict and appreciation of relative confrontation. The role of the choreographer is not passive either but active and relatively confrontational, negotiating his or her own vision with the others. The most typical illustration of this scheme is to be found in the working process of the Compagnie Dominique Bagouet in the 1980's.[60] Here the model of compromise clearly emerges.

Distinguishing compromise from consensus at the theoretical level is highly valuable in an effort to differentiate the polity settings of different art worlds. Becker (1982) failed to do so as he claimed aesthetic systems of art worlds to be based on consensus (without any mention of an alternative basis for common judgments of value, such as compromise). Confronting this mechanism of consensus to the phenomenon of competition between schools and styles (i.e. circles of recognition) in the art worlds, Becker came to the logical a priori conclusion that all art worlds are conservative.[61] But if an art world bases its aesthetic system on a mechanism closer to compromise than to consensus, this art world will less easily be qualified a priori as conservative.

To take again an example from dance, one can contrast the negotiations between the artists and the audiences in modern and contemporary dance performances vs. classical dance ballet performances. In the first case, compromises are made on the basis of the tension between the audiences' needs for recognizable aesthetic conventions and the artists' drive for innovations and experiments. In the second case, a much stronger consensus exists that

59 Cf. Guigou 2004, pp. 185-186, 231-232. But Gallotta and the group he worked with only came *relatively* close to this scheme, as personalities of dancers (and tensions) never did completely fade away.

60 Cf. Guigou 2004, pp. 69-79.

61 "The conservatism of art worlds [...] means that changes will not find an easy reception. [...] Because of all these conservative pressures, innovators must make a strong argument in defence of any substantially new practice" (Becker 1982, p. 135).

binds audiences and performers.[62] The audiences have quite precise expectations about both the narrative streams and the technical characteristics of the dance. Relative tolerance is low for performers *compromising* the canons.

In some situations 'compromises' are even relatively relevant to the definition of 'art': In the practices of 'Intervention Art' (which is more a circle of recognition than an art world), already introduced with the case of Wochenklausur, the researcher may sometimes witness negotiations tending to compromises concerning the role of the artist in projects involving local populations as participants. This is for example the case with the group 'Creatief Beheer' around the artist Rini Biemans in Rotterdam.[63]

My advocacy for compromise, rather than consensus, and for the maintenance of political spaces of contestation, also meets the analysis conducted by Claire Bishop, after Laclau and Mouffe, in her critique of 'Relational Art' (as a movement within contemporary visual arts, which promoted an unreflected notion of consensus):

> "Laclau and Mouffe argue that a fully functioning democratic society is not one in which all antagonisms have disappeared, but one in which new political frontiers are constantly being drawn and brought into debate – in other words, a democratic society is one in which relations of conflict are sustained, not erased. Without antagonism there is only the imposed consensus of authoritarian order – a total suppression of debate and discussion, which is inimical to democracy" (Bishop 2004, pp. 65-66).

Shifts in polity setting along three dimensions of contestation and participation

Along which dimensions can an art world shift from a near-hegemonic to a near-polyarchic setting, and vice versa?

The nature of the art world's core (diluted, alterable or solid) does not by itself determine whether the setting will be hegemonic or polyarchic, although it does affect the likeliness of either setting. An alterable core is compatible with both near-hegemonic and near-polyarchic settings. The institutionalisation of the core reduces the likeliness of a polyarchic setting because, unlike explicitly political institutions in contemporary liberal democracies, art worlds do not put emphasis on mechanisms of representation of differences that would affect the institutionalised core: i.e. no explicit political organization is organizing the distribution of power and the negotiations over rules and values (such as for example parties or unions). Even if

62 The second case also applies to specific circles of recognition in the art world of contemporary dance, with some companies achieving a new academism not incomparable to the academism of classical dance. Here again Angelin Preljocaj in his late career can illustrate this point.

63 Cf. Kagan 2004, pp. 132-137, 157-159.

some formal organizations do exist, they do not matter to a very significant extent.[64] But, even if art worlds do not have formal political representation in place to regulate their specific polities, they do have numerous relationships of influence and power and of contestation of power.

The degree of polyarchy (or hegemony) in a given art world's polity (i.e. an art world's space of political relationships) depends on three dimensions that are connected and combined with each other but that can (and do historically) vary more or less independently. The first dimension deals with contestation, as discussed above. The second and third dimensions deal with the notion of 'participation'. These second and third dimensions should always be considered together in order to evaluate the *degree of participation* in the polity of an art world: A given degree of political participation in an art world is a combination of the proportion of active participants (relatively to the total number of participants of the art world) and of their intensity of participation.

These three dimensions are thus the following:

- The *degree of contestation* of power which is effectively possible and observed within the art world. This means: opportunities for debate and room for opposition. An increasing degree of contestation means that a process is under way, of *liberalization* of power in the political field of the art world, involving a heightened tolerance for *competition* and questioning.
- The proportional size of the *area of participation* in the polity of the art world that is effectively observed, in terms of numbers of groups and individuals involved in the interactions contributing to the polity. One should take into account all stakeholders, including different strata of audiences, as Becker (1982) did. An increasing area of participation means that a process is under way, of *opening up* of the polity to more players.[65] This does not mean that all stakeholders exert the same extent of participation: That is what the third dimension assesses:
- The relative *intensity of participation* in the polity of the art world for the different groups and individuals participating. This dimension should not be confused with the degree of contestation (although contestation is

64 Because of the informalization of the art worlds since the demise of formal academies by the turn of the 20th century, it is generally taboo for artists to organize themselves formally (except, in some cases, when interacting with heteronomous cultural political institutions).

65 "Opening up to more players" can mean: (1) a higher number of circles of recognition participating in the art world's polity (not including circles existing in the art world but being passive in terms of political interactions); (2) a higher number of individual stakeholders active in the art world's polity (not including individuals passively present in the art world); (3) a higher number of individual stakeholders active in the polity of circles of recognition.

of course a form of participation): Intensive participation can also take the form of intensive 'yeah-saying'.

An increasing degree of participation also involves an extension of *cooperation* between players. As already mentioned, regulated competition involves cooperation. At the political level, the participation of an increasing number of groups and individuals in the discussions dealing with the polity conventions, as well as the intensification of such an activity, imply the broadening of co-operations into and across different old and new combinations.

Shifts at the level of an art world, and at the level of a circle of recognition

Polity conventions in art worlds are not as straightforward as they may be in formally political settings. Especially the interplay between the level of circles of recognition and the level of the art world as a whole will be more subtle than the interplay between e.g. party systems and the Nation-State.

Along the dimension of the degree of contestation, opportunities for contestation vary along two lines: the liberalization or de-liberalization of the general art world, and the liberalization or de-liberalization of circles of recognition: If some art worlds may be moving as a whole in the direction of liberalization or in the direction of de-liberalization, other art worlds may be nearly hegemonic at the level of the art world while allowing contestation within circles of recognition or may allow contestation at the level of the art world while having near-hegemonies in circles of recognition.[66]

A relatively high degree of localized hegemony is not necessarily contradictory with near-polyarchy at the larger level of the art world; it is even relatively helpful for the constitution of a common polity arena with an alterable core (which is in the long run more stable than a diluted core). Emerging groups and ideas are fragile, can be easy targets for criticism, have not yet developed a body of conventions and are prone to destructive suspicion. Therefore an emerging art world with a diluted core is at risk. The pre-emptive response to that risk is often a relatively high degree of localized social monopolization and hegemony at the level of the circle of recognition. But localized hegemony is not the only possible pre-emptive response to destructive suspicion: Alternative strategies can be based on a circumvention of criticism through an appeal to new, yet uninvolved, players.

66 A further complication could come from the distinction of different types of circles of recognition. I already suggested implicitly, while discussing the 'core' of an art world, to make distinctions between different circles according to their proximity and relationships to the core of the art world (i.e. central circles vs. peripheral circles).

Within a circle of recognition, it is often difficult (though not impossible) to sustain long-term near-hegemony: Being in a circle of recognition even more than being in an art world, involves participation. Whether participants are aware of it or not, they not only profit from (or sometimes suffer from) the circle and its reputation, but also contribute to the circle. In this context the metaphor of a conversation is helpful. In a circle a conversation goes on which is the continuous outcome of the collective inputs of all participants and of the reactions to the inputs of outsiders (like outside critics). Therefore the conversation (and its outcome in the sense of a reputation) is a common good. Nobody can appropriate it exclusively with ease, and without the continuous inputs of the participants it would not exist.[67] But if all participants contribute at least something to the conversation, this does not imply that, in practice, the impact of each individual contribution to the conversation is equivalent. Again, as at the level of an overall art world, they do not all have an equal say. Circles of recognition also know some kind of 'establishment' which, depending on the context, can be more or less volatile than the 'establishment' of the overall art world. Attention, and thereafter influence, soon moves to the individual (or group) who manages to set up and convey a convincing discourse into the conversation. Here, rhetorical qualities, identification of opportunities in the ongoing conversation and a willful attitude (which can be labeled 'entrepreneurial' or 'political') are essential elements for (would-be) leaders in the circle of recognition. These qualities contribute to an ability to arouse 'affectual action' around oneself, i.e. to become relatively charismatic in the eyes of other participants in the circle of recognition.[68]

Security and stability

For an art world to be sustainably open to contestation, i.e. for mutual contestation to be legitimized in the social interactions of the art world, there needs to be a higher degree of *expectation of security*, both on the side of circles of recognition situated at the core of the art world and on the side of circles of recognition situated at its periphery. Expected security means that the legitimacy of the circle of recognition (and of strategic individuals in the circle) and its existence expectedly will not be endangered.

67 Here I follow the interpretation Klamer (2004) has given to the existing notion of the 'common good' with the exemplary illustration of the conversation.

68 On charismatic legitimacy, see Weber 1965. The charismatic leader need not be an artist. Besides, it is not necessarily a single individual, but can be a small alliance of individuals, such as e.g. a triumvirate.

A counter-example

Coming back to the art worlds of dance as an illustration, here is a very concrete counter-example concerning the art world of classical dance in France in the 1980's: Failing to build up mutually expected security, the experiment of the GRCOP[69], introducing a contemporary dance 'research group' into the major French institution of the classical ballet dance art world, was a failure. Between 1981 and 1989, mutual mistrust built up 'till the Opéra de Paris managed to get rid of this alienating appendix which contrasted too violently with both the entrenched practices of the 'old house' and with its dusty aesthetic canons.[70] Actually, the GRCOP disappeared only 6 months after the sudden death of its inceptor Jacques Garnier, who had been attempting to bring together the art worlds of contemporary dance and classical dance. Closing down this group, the Opéra did however continue to invite established contemporary choreographers to create pieces for the main corps de ballet of the Opéra. Contemporary dance would then remain an exotic element from another art world, with less risks of further destabilizing the art world of classical ballet. Although these two worlds have since then gotten used to regular exchanges and collaborations, classical and contemporary dance in France have not come close enough to each other for them to merge into one art world. A low expectation of security is one among the factors crowding out 'routine interactions'. Becker (1982) indeed argued that people can only start speaking of *one* art world when there is clearly "routine interaction" within the supposed art world.

Which circumstances influence the evolution of mutual expected security?

- Internal Context: If the art world is being divided and polarized into highly antagonistic groups, its existence in its current form is endangered (and it may then either split up into a number of diverging art worlds, or successfully repress antagonistic groups, or itself cease to exist). The recognition of this risk may or may not motivate the core of an art world to become more open to contestation: If it does, a movement towards more polyarchy will likely be observed (because of the voluntary attempt to enhance mutual expected security); if not, the situation will favour a relatively more hegemonic setting.

---

69 i.e. Groupe de Recherche Chorégraphique de l'Opéra de Paris, created in 1981 by Jacques Garnier and closed down in 1989.

70 I am aware that this case description contradicts the official hagiography of the GRCOP and of the Opéra de Paris. My claims are based on interviews that I conducted with former stakeholders in the Opera/GRCOP, who however wish to remain anonymous.

- External Context: If the art world is confronted to seemingly antagonistic groups in the rest of society, a threat to security may be perceived, especially if the hostility is thought to be growing among the general population[71]. The growth of indifference about the art world in the rest of society is another source of long-term risks for the security of the art world, which the art world may or may not perceive and react to. Confronted with hostility or indifference from the outside world, the core and central circles of the art world will be likely to attempt to disregard contestation from outside, until a certain stage, i.e. until vital positions become endangered like the status of officials in (parts of) society or until the economic prospects of art world officials and workers are seriously endangered. But if the setting of the art world is polyarchic, the tendency to try and disregard contestation from outside will be less strong (and therefore outside threats will likely be taken into account at an earlier stage, and maybe taken as opportunities for change rather than for conservatism).

Shifting beliefs and habits

Which sets of beliefs and habits, among the members of an art world, are more likely to allow a near-polyarchic setting?[72] And, by contrast, which beliefs and habits will rather be associated to a near-hegemonic setting? The power of conviction and the effectiveness of each belief about political relations in the art world, depend on the interaction of this belief with other beliefs of the same order and on its translation into habits. Only a relatively coherent network of beliefs and habits is convincing enough to support a given political setting.[73]

An art world's political setting does not depend equally on every art world member's beliefs and habits. Whether an art world turns towards a near-hegemonic or a near-polyarchic setting, what matters most is the beliefs and habits of those participants and circles who are involved in the political activities of the art world (i.e. the stakeholders with relatively high intensity of participation and degree of influence). As already described, in a near-hegemonic art world, these politically active members will for most of them be found close to the core (i.e. 'officials'); while in a near-polyarchic art world, the distribution of these active members will be more scattered

71 "Seemingly" antagonistic, because the degree of antagonism depends partly on how members of the art world a priori perceive the groups from outside worlds.

72 On the analytical advantage of distinguishing habits from beliefs, see Lahire 2003. In short, Lahire argues that the discrepancy between beliefs and actual behavior as embodied in habits, makes sociological interpretation at the individual level more complex than often assumed.

73 This conclusion was already drawn by Dahl in his analysis of political beliefs (Dahl 1971, p. 132), although he did not use the concept of convention.

and include a good number of players from the periphery of the art world.[74] Besides, individuals who profess more articulate beliefs are more likely to get involved in the conversations on the conventions of the art world, i.e. to be more active, than individuals with inarticulate beliefs.

In any case, for any art world to exist, there needs to be a shared belief that the art world ought to exist, next to other existing art worlds and instead of emerging/potential alternative art worlds. A polyarchic setting implies belief in the legitimacy (1) of contestation and (2) of open participation in the debate.[75] A hegemonic setting implies the belief that debate is unnecessary or should otherwise be restricted to the core of the art world and that contestation is a priori illegitimate. In a polyarchic setting, influential members of the art world will be likely to convey their respect for and make a habit of actual debate and participation. In the extreme case of a purely hegemonic setting, they would show contempt for any form of debate.

In a polyarchic setting, members will less easily accept to be associated with the art world if it turns away from contestation, because they will feel their own freedom and chances are threatened: They identify indeed their own interests to contestation. For example, in a typical near-polyarchic setting, many art world participants will be vociferously expressing their condemnation of 'any form of censorship' and will interpret the autonomy of art as implying the right to express almost any opinion in almost any "unconventional" way.

By contrast, in a hegemonic setting, it will be relatively easy for core-participants to impose the institutionalization of the art world to subdued and peripheral members, and especially to have them support a more rigid canon with forms of 'censorship', because:

- their political voice is less valued (individual expression may be even deemed irrelevant to the higher canon of the art world);
- they have the habit of tending to delegate to the core (i.e. to 'officials') the authority to decide how the art ought to be done (and, in behaving so, they become politically passive - and therefore their voice is getting even less valued).

If belief in certain influential (core) members of the art world as authority figures is very strong, and if players in the art world make a habit of looking up to core players exclusively, this will tend to hinder contestation and debate. But if belief in authority figures extends to peripheral, 'unconven-

74 The beliefs of the passive members of the art world matter much less, i.e. per capita. But being 'active' or 'passive' is a matter of degree, not a binary switch.

75 In the art world of contemporary visual art, belief in the legitimacy of contestation is strong and widespread. However, belief in the legitimacy of open participation in the debate is more ambiguous and limited (cf. Kagan and Abbing 2006 for an account of this phenomenon).

tional' or minor members of the art world or to outsiders to the art world, this will tend to foster contestation and debate in the art world. The reality is often somewhere in the middle, nowadays in contemporary art forms, with different circles of recognition recognizing different authority figures. For example, in the art worlds of contemporary visual art and of contemporary dance, almost officially, the belief in core authority figures is down, at least since the advent of so-called post-modern discourses. However, a set of authority figures, different in each circle of recognition, is still present for many members of the art world (e.g. it can be Joseph Beuys for some, the Harrisons or Hans Haacke for others).

Of crucial importance to a polity convention is the belief, among active members, about the relative effectiveness of the art world: How effectively can the art world (or could a more polyarchic or more hegemonic art world) address what they believe to be critical problems and issues? For example, if the critical and the peripheral active individuals believe that the art world is not able to deal effectively with problems and issues that they identify as critical, they will be likely to turn towards peripheral circles of recognition and away from the art world as a desirable arena for participation and contestation.

Another determinant belief/habit variable is trust.[76] To which extent do players in the art world have trust and confidence in specific other players in the art world? Also, how far do they believe one can trust individuals and groups in general in the art world? A sustained polyarchic setting requires social interactions involving a degree of trust. Furthermore, the mutual security of opponents would be endangered if contestation and oppositions would develop in a context of distrust.

Yet another important belief/habit variable concerns the prospects of cooperation among active members of the art world. Do they believe in (and practice) almost absolutely competitive relationships, almost absolutely cooperative relationships, or something in between?

(1) Absolutely competitive	(2) Cooperative competitive	(3) Absolutely cooperative
◄──────	──────	──────►
Irreconcilable differences	Compromise	(a) only win-win or lose-lose (b) special case: Consensus

If individuals were to believe in absolute political competition as the only way to deal with each other (1), then a near-polyarchic setting would be less likely to occur: When members interpret the setting of their political interactions as only and absolutely competitive, this will sweep out of their minds

76 As the discussion of competition showed, trust is also crucial in the institution and maintenance of regulated competition. Deep distrust in social relationships endangers regulated competition.

the opportunities for cooperation.[77] In a polity convention more attuned to polyarchy, beliefs and habits about cooperation facilitate both the manifestation of oppositions and the search for compromises (2). But "if compromises are disdained, cooperation is difficult and conflicts will be more likely to go unresolved" (Dahl 1971, p. 162). Then a move towards a near-hegemonic setting will be more likely to bring cooperation. However, given the largely informal character of relationships in art worlds, the notion of 'compromise' will be often more subtle and implicit than in the usual political negotiations Dahl was referring to. Therefore, an observer should not conclude that 'compromises are disdained' if he or she fails to observe explicit compromises established in a formal fashion as would occur in a traditional political negotiation.

If members were to believe in absolute cooperation as the only way to deal with each other (3), then too would a near-polyarchic setting be less likely to occur: When members intend to have only full cooperation among them, this very belief undermines their own legitimacy as an individual or as a subgroup. Heterodox circles of recognition will be especially looked at as 'unwilling to cooperate' and conflicts will be regarded as undesirable. Any development of contestation will be perceived as a cancer growing in a previously sane body: belief in absolute cooperation implies a strongly organic metaphorical understanding of social organization. Then, 2 options will come out: Either (a) the differences get largely dissolved into a unified stream and circles of recognition get integrated into a uniform art world (with a kind of organic social monopolization, not unlike what happens, at another level, in a totalitarian regime). Or (b) the circles behave themselves around the implicit constraint of consensus, so that a consensual art world shapes up (where contestation is under-toned). The later is a softer alternative and could be labelled 'pseudo-polyarchic'. But consensus is not compromise, as consensus stops the process of contestation: reaching a consensus implies the end of contestation by the minority, as we have already seen. Therefore, the correct term would not be 'pseudo-polyarchic' but rather 'soft-totalitarian'.[78]

Coming back to the case of dance: The development of a strong belief in absolute organic consensus, in an art world, and its highly nefarious political consequences, will be discussed further down with the case of German modern dance in the 1920's and under the Third Reich. By contrast, in post-modern dance (e.g. in the Judson group in the USA in the 1960's), team-

77 This observation does not apply to regulated economic competition. My analysis here deals clearly with the political level that lies beyond the instituted economic relationships. More generally, the reader should not confuse political competition with economic competition.

78 The Serbian political scientist Slobodan Milacic (in his conferences at Sciences Po Bordeaux) called contemporary forms of consensus and the rule of so-called public opinion, forms of a new 'soft totalitarianism'.

work and cooperative creativity did not involve strong consensus but rather mutual stimulations and provocations, therefore installing compromising attitudes rather than consensual communion.[79]

Polity conventions and romantic myths

The distinctiveness of beliefs and habits relating to the polity conventions of art worlds (compared to other social worlds) comes from their relationships with other categories of beliefs and habits that make up the specific conventions of the art worlds (which were already evoked in chapter 1, under the heading of a 'romantic order'). The specific beliefs and habits making up the different possible polity conventions of art worlds indeed also involve different interpretations of typical founding myths in the modern definitions of art, such as the 'autonomy of art' and the 'creative genius'.

I have already indirectly described above the relationship to 'autonomy' in a near-polyarchic setting: It connects to beliefs in (and habits of) contestation for its own sake (with its condemnation of censorship). And it ultimately advocates that every artist, or even every individual participant, should be autonomous. However, such a set of beliefs contributes to a double-morality as long as an art world claims to defends the freedom of speech coming from inside the art world, while muzzling, ignoring or denouncing contestation coming from outside the art world (accusing it of being 'censorship').

Autonomy in a near-hegemonic setting, is more an autonomy of the art world at large than an autonomy of every artist; therefore here the imperative of autonomy is not incompatible with a stricter canon, central authority figures and forms of internal 'censorship'.

In a near-polyarchic setting, the individualization of the myth of the creative genius tends to be widely put into practice: Ultimately every participant aims at being recognized as genuinely original (and a discourse of democratization of originality is developed).

In a near-hegemonic setting, it is not necessary for every participant to act out one's 'potential' as a fully original genius, because the core authority figures (e.g. the choreographer) monopolize this role, relegating others (e.g. the dancers) to a second order of originality and individuality.

Illustrations from dance

Starting in the first half of the 20th century and beyond, the rise of Modern and Contemporary Dance in Germany, the United States and later France offered a liberation of the individual expressiveness of the dancer, away from the straitjacket of the classical dance canon with its

79 See e.g. Banes 1993.

strict rules and stress on hierarchy. On average, modern and contemporary dance did away with the established hierarchy of dancers in classical dance, giving every dancer in the group the ability to improvise, experiment and contribute to the choreography (e.g. Trisha Brown's company is one of the most characteristic of this development in 'postmodern dance').

The reverse trend, towards a more pro-hegemonic art world, also occurred: With the move towards the institutionalization of contemporary dance in France in the 1990's under the established authority of some choreographers, a process took place of monopolization of the genius-figure and originality by these leaders (e.g. Angelin Preljocaj aiming to retain exclusive control over the 'creation' of his choreographies)[80]. However, this process was not uniform, and some leading figures worked against it, continuing to pursue shared creativity with the dancers (e.g. Mathilde Monnier).

Two examples of near-hegemonic and near-polyarchic polities

In the previous pages, I have repeatedly used the art form of dance as an illustration. Dance in the second half of the 20th century offers indeed an insightful case of two highly contrasted art worlds in terms of polity conventions, of which I will now draw a quick sketch:[81]

Classical dance, on the one hand, comes closer to the hegemonic setting and is characterized by submission to the central authority of the choreographer, standardization and codification of the canon (i.e. quality conventions) with a disciplinary appeal to technical virtuosity and a focus on the production of frontal and symmetrical performances for a traditionally passive and sitting audience. Classical dance imposes a hierarchy of dancers, strives for a standardized body, does everything to hide conflicts under an idyllic consensual image and moves as far as possible from any affiliation to the earthly body of everyday life in its search for purity and lightness.

Postmodern dance, on the other hand, comes clearly closer to the polyarchic setting and is characterized by horizontal exchanges without hierarchy (the dancers are at the same level as the choreographer). Postmodern dance accepts and values individual differences in the bodies of dancers as well as everyday movements (e.g. walking) and earthly imperfection (e.g. falling). It pursues improvisation, spontaneity and experimentation and explores conflicts (e.g. contact-improvisation). Moreover, post-modern dance interacts

---

80 Author's interview with the choreographer in November 2004. See also Kagan 2004b.

81 See also Guigou 2004, and on post-modern dance: Banes 1987.

with other art forms, integrates non-dancers and moves away from the halls, into new spaces (e.g. the streets) where interaction with the audience is sought.

Through the contrast of classical and post-modern dance, it becomes clear that the polity conventions of an art world have far reaching consequences both in terms of the art-historical and of the broader social significance of that art world: Near-polyarchic polities in the arts will not only be facilitating double entrepreneurship in conventions, but will also be advancing the value of participatory democratic practices, which also form an important dimension of cultures of sustainability: its political dimension (as suggested by Bookchin and noted by the Fowkes).[82]

Polyarchy and high complexity

However, a skeptical reader might still wonder, why is a polyarchy/'demo–cracy' relevant to cultures of sustainability? A number or arguments have been emerging from the preceding pages, demonstrating the advantages of polyarchy for social transformations and a cultural evolution. Besides the own value of polyarchy in terms of social justice – which should in no way be under-estimated as a primordial dimension of sustainable human societies (against the hypothetical threat of a 'green fascism'), polyarchic polities are also, thanks to their overall greater flexibility than hegemonic polities, better fit for the evolution of 'high complexity' (vs. 'low complexity').

Morin speaks of "hypercomplexity" in the historical evolution of human societies. Hypercomplexity points at a "capacity to organize diversity in increasingly disorganizing conditions [and to] tolerate, use and handle noise and disorders", at a correlative development of the autonomy of the subject and of communications and communities with the other and of the interactions with the environment, and at a development of the capacities to learn, elaborate strategies and create (Morin 1980, p. 434). Hypercomplexity is especially fragile and nonlinear, and can only sustain itself by movements between two polarities of 'low complexity' and 'high complexity': Where the polarity of low complexity means centralization, hierarchy, coercion, low individual autonomy, lesser interactions between individuals and between social groups, under- and/or over-specialization, repression of disorder and noise, dogma and faith, stability, and an overall low evolutionary capacity, the polarity of high complexity means a dynamic tension of polycentrism and decentralization, of polyarchy, heterarchy and anarchy, of individual freedoms/autonomy and multiple social interactions, of specializations and poly-competence, together with tolerance to disorder and non-conformity,

82 I am aware that the political practices in the arts, are only constituting a small part of an overall social system. Their relative significance will depend on the changing external context and constraints in societies (at national and international levels).

doubts, interrogations, instability, and an overall greater evolutionary capacity (ibid., p. 437).

Morin's analysis further highlights that high complexity also requires dimensions and moments of lower complexity, while low complexity must be, carefully, kept from taking over society. In terms of polity conventions, this means that a near-polyarchic political setting also requires and tolerates dimensions and moments of near-hegemony, while preventing hegemony from spreading further.

A cautionary historical case about how the near-hegemonic political setting of an art world can lead it to embrace and support a totalitarian political system: German Dance under Nazism

The relationships between wider political developments and the polity conventions in an art world, are a very complex matter. My aim here is not to attempt at describing causal relationships, but to point at: (1) the existence of such relationships, as an important issue, and (2) the fact that polity conventions in art worlds are neither innocent (or fully autonomous from polity conventions in society at large), nor necessarily always championing 'democratic' (i.e. polyarchic) values in society.

As a cautionary historical warning, I choose to focus on one of the most dramatic worst-case examples of how an art world came to embrace and support a totalitarian political system: the development of German modern dance in the 1920's and 1930's, with the support that its leading figures Mary Wigman and Rudolf von Laban, brought to Nazism.[83]

This historical example has another value as a warning: It shows how anti-modernity and holistic utopias of harmony with nature, in the arts, can degenerate into 'green fascism': a precedent that should be kept in mind, to keep us away from simplistic forms of holism and community-celebrating.

Historian Laure Guilbert (2000) describes the gradual transition of the utopias of early modern dance in Germany, towards the ideology of Nazism.

---

83 This analysis is mainly based on the groundbreaking dance-historical research of Laure Guilbert (2000). See also Karina and Kant 1996 and Müller and Stöckemann 1993. For an introduction to the relationships of other art forms to Nazism, see ed. Faulstich 2009 (which however does not cover the case of dance). Surprisingly, I continue to notice denial from a part of the dance community about the past of Wigman and Laban, in late 2009 (personal communications with different representatives of international dance-education associations, along side the Congress of the World Alliance for Art Education, in which I was taking part). Unfortunately, Guilbert's book is not yet translated into English. Ten years after the publication of Guilbert's work, several hagiographies of Laban continue to unashamedly ignore or deny historical research.

Modern dance was born in the cradle of 'romantic anti-capitalism', which expressed a critique of Industrialization and a longing for a mythic golden age of harmony with nature, mixing reactionary nostalgia, utopian sensibility and a critique of the excesses of Enlightenment, reason and materialism.[84] In their history of 'Ausdrucktanz' (dance of expression), Müller and Stöckemann (1993) further characterize this utopian background as an anti-intellectual "Aufbruchsstimmung" (spirit of action) aiming to base self-experience and knowledge on the human body and on nature.[85]

Inspired by the ideas of philosopher Ludwig Klages on reviving organic rhythm against the mechanized cadence of productivism, Rudolf Bode, the founder of 'Ausdrucksgymnastik' (gymnastics of expression), aimed at re-instating the holistic "rhythm of the cosmos" through the whole human body, for a cultural revolution. In his texts, Bode combined holistic and racial-nationalist arguments: "Every movement interferes on the movement of the whole body: the same way a planetary disturbance influences the whole system". But according to Bode, holistic community and harmony is lost because of "the intensification of internationalism [and] the loss of the feeling of racial, familial and national belonging" (Bode, quoted in Guilbert 2000, p. 29) Bode joined the NSDAP (the Nazi party) already in 1922.

The case of Rudolf von Laban, in his early career, was less obvious: He was not yet promoting a reactionary discourse, but longing too for a new social order reinstating the 'universal cosmic rhythm' under a pantheist conception of nature, lost to modernity.[86] Laban soon became the founder and leader of German modern dance, aiming to form the human being as a whole.[87] At the utopian Monte Verita community, during the first world war, Laban developed cult-like ceremonies aimed at overcoming social classes through archaic forms of occultism and spirituality, developing a form of unanimous collectivism through dance. In the early 1920's, turning away from mysticism, Laban moved on to create the 'dance choruses', amateur dance groups aiming to transcend individuality in the movements of a dancing community. In the choruses, a leader would be expected to emerge spontaneously, expressing not his individuality but the "group spirit" (as expressed by Mary Wigman).[88] With his choruses in the 1920's, Laban wanted to shape an organic mass, to reconcile the individual with the holistic community, as an alternative to the technical mass of industrial civilization.[89] He was also increasingly fascinated by the powerful impressions of hundreds of amateur dancers moving together, as a mass spectacle (a feat made possible thanks to his invention of cinetography – the notation of movement).

---

84 Cf. Guilbert 2000, pp. 24-26.

85 Cf. Müller and Stöckemann 1993, pp. 9-10.

86 Cf. Guilbert 2000, pp. 30-35.

87 Cf. Müller and Stöckemann 1993, p. 14.

88 Cf. Laban and Wigman, quoted in Müller and Stöckemann 1993, pp. 19-20.

89 Cf. Guilbert 2000, pp. 50-52.

A student of Laban for five years, and the other leading figure of German modern dance, Mary Wigman also envisioned a new world, liberated by "primitivist" movement. But in the early 1920's, she developed into a different direction, with a mystic cult of individualism. Nevertheless, her exploration of the tension between the individual soloist dancer and the mass of the other dancers, gradually moved along the 1920's towards a longing for the charismatic guide and leader in the person of the lead dancer, i.e. herself, e.g. in her choreography *Die Feier* (celebration, 1928).[90] Wigman also developed a notion of dance as "absolute" and dominating the human will.

At the turn of the 1930's, i.e. before the Nazis came to power, the leading figures of modern dance already came closer to them, in words and in practice. In the late 1920's, Laban moves closer and closer to the far-right 'Volkstum' movement. In 1930, he declares the "dance choruses" as "the new folkloric dance [of the] white race" (Laban quoted in Guilbert 2000, p. 55).

As to Wigman, with *Totenmahl* (Monument to the Dead) in 1930, she expressed an authoritarian/hegemonic tendency in a choreography dominated by a high priestess soloist surrounded by compact block-like groups of dancers, and finally sacrificing herself for the German nation. About that piece, Wigman herself commented: "The potential element of conflict must not be resolved anymore inside the group itself. What this is about, is the unification of a group of human beings into one single moving body that [...] moves towards a common goal, with the agreement of all, according to a single point of view" (Wigman, quoted in Guilbert 2000, p. 61). Most commentators of the time celebrated in this piece, the realization that only a charismatic leader could save the world, and noted the nationalist theme of the piece, literally the resurrection of the German nation through its dead soldiers (with, as Guilbert notes, a fascination for death and sacrifice to the body of the Nation/Race). But not all commentators did so: the Brechtian circles denounced the piece and the importance of "emotional fusion" in Wigman's work.[91] In *Der Weg* (The Way, 1933), Wigman celebrates ineluctable destiny under the order of nature: Nature is here no longer about harmony and life, but about an immense force which the individual cannot resist.[92]

A counter-example is Kurt Joos, who performs in 1932 *Der grüne Tisch* (The Green Table), a satire against the culture of war, where, noticeably, dancers's individualities were not erased (and where Joos appeals to individual responsibilities in the face of political crisis). Unlike his colleagues, Joos rejected the first antisemitic laws touching some of his dancers, and escaped the country in September 1933.

90 Cf. ibid., pp. 60-61.

91 Cf. ibid., pp. 61-65.

92 Cf. ibid., pp. 96-97.

But unlike Joos and a couple of others, the immense majority of his colleagues did immediately collaborate actively with the new power in place, apply antisemitic laws, and seek the support of the regime for modern dance. By 1934, they adopted discourses celebrating "Germanity" and posed as heroes of German identity, by contrast to the French-Russian influences of classical ballet. For example, in 1934 Laban writes that foreign influences in dance are "destructive", and that the role of modern dance is to elevate the German race.[93] Between 1933 and 1936, Laban knows a very successful carrier at the top of the Nazi cultural administration, thanks to the support of Goebbels, until his gradual demise in 1936-1937 (after the Olympic games representing an international apotheosis for several German modern dancers) which owes more to internal power-struggles than to any ideological opposition from his side. Laban also pushes away his rival Rudolf Bode (who was collaborating with the Nazi ideologue Alfred Rosenberg since 1929, and who eventually comes back to the forefront when Laban is himself pushed away in late 1937). Mary Wigman, who receives her biggest subsidies ever thanks to the Nazi regime, and will continue to work in its service until the 1940's, claims her willingness to carry out Nazi propaganda and her attachment to the Führer (in a documented correspondence), and writes in 1933: "The belonging to the species creates the common, forces into community. Who could dissolve himself from that? The human being is born into the community of the species, bound to community, obliged by community" (Wigman, quoted in Müller and Stöckemann 1993, pp. 127-128).

In 1933, Bode's publications elaborated a Nazi discourse on German modern dance, with principles of a new culture, based on totality, rhythm, the economy of efforts, with a revolutionary order for collective life. His new laws of movement claimed to regenerate the Aryan race: especially a law of rhythm (regenerating creative forces by liberating subconscious drives) and a principle of totality integrating the individual in the body of the nation:

"The organic totality of the individual is the condition for the totality of the State, the same way each leaf of the tree both serves the whole and represents an individual form. [...] The people unites itself in the dance, in the circular movement of the whole, without the existence of the individual self mattering [...] the individual only has existence in relation to the whole, and has no private existence, but defines himself by his civic responsibility" (Bode, quoted in Guilbert 2000, p. 154).

This quote from Bode is a summary of a holistic ideology which fits well in a totalitarian regime.[94]

93 Cf. ibid., p. 174.

94 I do not mention here, further ideologues of Nazi dance in the 1930's, such as Fritz Böhme. For a thoroughly detailed account, see Guilbert 2000.

It should be noted than when the Nazis came to power, dance was not a priority for them.[95] It was the activism from the leading figures of modern dance that attracted their interest and their support, especially between 1934 and 1940.

With few exceptions, the modern German dancers willingly eliminated polysemy in their work and promoted increasingly hegemonic values through their work, not only under the Nazi regime, but already in the years before. From that position, it was not very difficult for them to adapt to the new totalitarian order.

What also should raise critical awareness, from the insights of this historical case, is that a mythology of a holistic communion with nature is potentially very compatible with highly hegemonic polity conventions and with a totalitarian ideology.

CONCLUSION

Social change at the interface of art worlds and other worlds (i.e. other parts of society), requires a double entrepreneurship in conventions, transforming the art worlds to allow art to function autoecopoïetically as a field of experimentation (like the apprentice acrobat on the wire in Bateson's metaphor), and potentially transforming various social conventions held by diverse social agents. With entrepreneurship in inter-conventional interaction, and deutero-learning, even hard-programmed ideas and conventions may change.

But a successful double entrepreneurship in conventions also requires that the art world in question be sufficiently open to such an enterprise, which is more likely in art worlds with a more polyarchic political setting. Furthermore, only relatively polyarchic polity conventions in art worlds, may be compatible with entrepreneurship in conventions towards cultures of sustainability, while polity conventions tending to hegemony pose much greater challenges to such an attempt. In this concrete way, the artistic is eminently political, and the power/authority relationships inside an art world, are not innocent.

The extreme case of German modern dance in the 1920's and 1930's should work as a warning against the risks of degeneration into some form of 'green fascism', if discourses and practices claiming to advance ecology and sustainability, limit themselves to holistic ideals of consensus and organic unity. On the contrary, should be sought discourses and practices valuing pluralism and contestation, compromises and regulated competition, as the political dimension of cultures of sustainability based on complex uni-plurality and exploring a 'queer ecology' (and not a straightforwardly harmonious image of nature).

95 Cf. Guilbert 2000, pp. 202-203.

Conclusion

> We shall not cease from exploration
> And the end of all our exploring
> Will be to arrive where we started
> And know the place for the first time.
> T.S. ELIOT 1959, P. 59

The emergence of cultures of sustainability is not a process that can be implemented at the tip of a pen during a UN conference, even though institutional and political frameworks do play an important role in stifling or fostering cultural transformations. Everyone speaks nowadays of climate change, which is indeed one of the most pressing global ecological issues of the day (together with other ecological issues such as biodiversity and water, and with socio-economic and cultural issues). Given the increasingly tight 'window of opportunity' for mitigation of climate change (one or two decades at most), it might seem to some readers that cultures of sustainability are a longer-term concern while urgency calls for swift measures at the levels of regulations and technologies. However, the longer term concern of cultures of (un)sustainability is running deep and matters not only for short-term climate change mitigation measures, but also for adaptation to climate change (and other domains of crisis) in the coming decades. Cultural strategies across sectors of arts, science, education, urbanism, social policy and many others, if they may not perform a magical trick allowing to easily achieve the mitigation goals for 2020 that were discussed by NGOs alongside the failed COP15 conference, may well, however, engage a deeper if slower transformational process. The capacity of our civilization to transform itself in the next decades without collapsing or deteriorating radically, may depend at least partly on cultural transformations such as described in the present volume.

Furthermore, the developments of aesthetics of sustainability can develop stronger incentives for the preservation of diversity than merely utilitarian rationality (biodiversity in so far as it servcs our own species' survival), as argued by Loïc Fel after Aldo Leopold: The own aesthetic value of natural diversity as an end in itself forming the experiential frame for aesthetic appreciation, as fostered by "green aesthetics", can strengthen the plea for biodiversity in terms of ecosystemic resilience.[1]

Connecting Patterns: Culture and the arts towards sustainability

The Western development model is plagued by a culture of unsustainability constituted by several fundamental dimensions of modernity. The Western worldview and mode of knowing is in crisis. Ervin Laszlo characterized this worldview as atomistic, materialistic, individualistic and Eurocentric, and as based in the "classical scientific method", as opposed to a "systems view" emerging in the 20th century. Basarab Nicolescu further criticized the principles of classical modern physics, i.e. continuity, local causality, determinism and binary logic. Both authors deplored the resulting fragmentation of understanding, and this classical mode of knowing was further deconstructed by Edgar Morin as disjunctive simplification, which affects not only science but modern modes of knowing in general, with three modes of simplifying thought: "to idealize... to rationalize... to normalize." These critiques also echo Husserl's critique of the overrunning of, and estrangement from, life-world experience, and Gregory Bateson's warning against the empire gained by the narrow views of "purposive consciousness", which are both especially pregnant in modern societies. Since the second half of the 20th century, these issues are combined with a culture of hyper-consumption, which continues to grow despite the announced shift to service economy, knowledge/information economy and creative economy in most recent decades.

These dimensions of our culture of unsustainability are path-dependent and rooted in historical developments marked by patriarchalism, Cartesian method, Baconian faith in progress and by a progressive estrangement from nature, which may have its roots as far as the invention of phonetic alphabets according to David Abram.

Modern and contemporary further developments of this culture of unsustainability include the transformation of science and techniques into a techno-scientific system, or 'Technological System' in the analysis of Jacques Ellul, following a shift from enlightenment to positivism (as discussed by Horkheimer and Adorno), and the increasingly autopoïetic operation of social systems in modern societies through functional differentiation, as analyzed by Niklas Luhmann.

The autopoïetic development analyzed by Luhmann is found back in art as a social system in modern societies, where, as analyzed by art sociologists from Pierre Bourdieu and Howard Becker to Hans Abbing, the "high arts" institutionalize social segregations. In this world of high arts, a very peculiar combination has emerged, of a 'Romantic Order' embedded in a

1 Cf. Fel 2009, pp. 308-309.

'Technological System', the articulation of which I explored, after Jacques Ellul and Maarten Doorman. This complex combination (which I shortened as 'RO/TS') is also the first example of a macro-concept that I articulated in this volume, before introducing the very notion of macro-concepts. However, even if the situation is complex and open to change, overall in the past century, the romantic order in art mainly functioned as a safety valve for society, relieving some of the 'pressure' caused by our overall culture of unsustainability, but without achieving significant transformations beyond the sandbox of the art system. As analyzed by Luhmann, modern art as a social system, is incapable of communicating directly with other social systems (and vice versa), that is, as long as its operations are dominated by autopoïesis. This situation was denounced from within the art world by art historian Suzi Gablik, whose critique of modernism in art targets especially the aesthetics of detachment and the individualistic and antagonistic autonomy developed in the high arts, but also denounces the nihilistic self-consciousness of postmodernists.

How can we find alternatives to this culture of unsustainability and the accompanying art of unsustainability? A response to this challenge requires a difficult and wide-ranging transformation of the modern mode of knowing. But fortunately, this transformation has already started to emerge, as demonstrated by the rise of systems thinking and of transdisciplinarity in recent decades.

Systems thinking is however only a step, in the pathway toward a culture of complexity. With the case of systems thinker Fritjof Capra, I explored the insights and promises of systems thinking: Especially, the notions of networks, non-linearity, emergence and self-creativity opened up a first phase in the outlining of a paradigmatic shift, away from linear causality. Some principles have also come out of the systems perspective, most especially Capra's six principles of ecological literacy: the networking principle of living systems, the continual cycles of the web of life, the openness of the Earth system to solar energy, the importance of partnership and cooperation for evolution, the necessity of diversity for resilience, and the search for dynamic balance (rather than the attempt to optimize any single value or dimension). At the social level, the systems perspective highlighted how social organizations may be able to evolve, if understood and managed as living and learning organizations. Systems thinking also points out the importance of culture for both social reproduction and social change. But with the case of systems thinker (and 'integral studies' thinker) Ervin Laszlo, I also pointed at the risks of simplification by holism which are present in systems thinking (together with a risk of technocratization of cybernetics).

I have argued in this volume that a culture of sustainability would be a culture of complexity, following most especially the insights of Edgar Morin's 'method' and more generally of transdisciplinarity. With Morin was introduced the possibility to think unity and diversity alongside each other, and to think about any pair of terms, with a combination of unity, complementarity, competition and antagonism, altogether forming a complex relationship and calling forward a dia-logical thinking process, whereby macro-concepts can emerge. One such macro-concept developed by Morin is eco-auto-organization, which explores the complex organizational relationships between individual life forms and the ecosystems in which they co-evolve and eco-evolve. Eco-organization highlights the complementarity of diversity, complexity, spontaneity and organization in all life. At the social level, Morin introduced the macro-concept of socio-eco-organization, which aims to balance the organicist tendencies with the eco-organizational tendencies in contemporary human societies, in order to prevent the risks of both organicist totalitarianism (e.g. the totalitarian state) and eco-organizational conflictuality (e.g. class struggle under an unbridled market economy). Morin's insights thus also revealed, next to the centrality of culture in social evolution already highlighted with Capra, the issue of the State (pointing at the political dimension of sustainability).

More generally, Morin's insights pointed at the genesic force of chaos at the root of any creation, and highlighted the continuous presence of chaos (and of order from disorder) in generative endo-exo-causality and de-re-organization, two macro-concepts that, in turn, are at the root of phenomena of emergence, are fueling the poïesis of machine-beings (i.e. the creative evolution of the universe) and are imposing an uncertainty principle which disarms all forms of determinisms.

Furthermore, complexity thinking allows to redeem poïesis from its grim state under Luhmann's autopoïesis, rethinking it as autoecopoïesis, i.e. systems operating in ways creatively sensible to chaos (i.e. having a certain productive openness to disturbances) – the ecopoïetic element being understood as the tendency of psychic systems and social systems to construct themselves in open communications with their environments (implying a co-determination and co-evolution of both the system and its environment through the emergence of properties stemming from the communication between system and environment).

The aesthetic dimension of potential cultures of sustainability implies a sensibility to patterns that connect (after Gregory Bateson) grounding aesthetics of sustainability, as an alternative to modern, Kantian aesthetics of detached contemplation. Bateson characterized aesthetic experience as responsiveness to the meta-pattern uniting the living world, rather than an arrested perception, stumbling upon the relative differences and being limited, bounded by those differences. Aesthetics of sustainability also require a sensibility to complexity, non-content with fixed and holistic harmonies but, after the aesthetic insights of Heraclitus, revealing harmonies-in-oppositions. This further requires a transdisciplinary sensibility, beyond the pitfalls of both the simplifying clarity of modernism and the confusion and rootlessness of postmodernism.

Bateson's understanding of the aesthetic experience further points at the value of a more-than-conscious mode of knowing. Such a shift can and should be informed by some non-Western cultures which, as argued by David Abram, have developed a phenomenological and animistic sensibility to a more-than-human world. The recovery of the eco- in an autoecopoïetic sensibility requires an acknowledgment of the participatory nature of perception, as an animistic, integrated, synaesthetic mind-body experience (the very term 'mind-body' is to be dissolved, as dualism is to be abandoned for a monist immanent perspective à-la Bateson). Such a participatory perception would hopefully revive the aesthetic sensibility to the "grace" of the patterns that connect which, as argued by both Bateson and Abram, was lost under our civilizational development (with purposive consciousness and alphabetic culture, as already argued).[2]

The interplay of art and ecology became the focus of a specific artistic movement, with the emergence of 'environmental/ecological art' from the late 1960's onwards. Ecological art critically distinguished itself from the shortcomings of land art. I discussed the insights of a relatively small number of prominent artists who were especially active from the 1960's or 1970's onwards, and who developed areas of artistic work that opened up the field of environmental/ecological art:

- Hans Haacke, influenced by the "Systems Esthetics" of Jack Burnham, developed an increasingly complex work on ecological and social systems (and relations between them) from the 1960's onwards.
- Alan Sonfist started working on natural processes from the 1960's onwards, aiming to include the natural world in our concept of community.
- Joseph Beuys, in the 1970's and 1980's, explored the connection between humans and nature, and stressed the political and socially transformative dimension of ecology and art. He introduced the concept of "social sculpture" which inspired subsequent artists and aims at reshaping and restructuring society, considering social change processes as aesthetic processes.
- Helen Mayer Harrison and Newton Harrison, whose work at the intersection of art and natural sciences in the 1970's led them to the understanding of the complexity of ecological and cultural systems, developed an attention to relations, and a method of conversational drift allowing emergence in social interactions. The couple is still very active and their work addresses the co-evolution of biodiversity and cultural diversity in specific ecosystems, with the ambition to engage policies and planning into the process.
- Mierle Laderman Ukeles, since her "maintenance art" manifesto in 1969, celebrated in her interventions the human work of maintenance of dynamic balance in eco-systems, rather than the "death instinct" of development and progress. She developed rituals celebrating the importance of maintenance work, e.g. by sanitation workers.

2 On Bateson's use of the term "grace", see Charlton (2008).

I also discussed the work of other artists who explored further dimensions of environmental/ecological art from the 1970's onwards: Patricia Johanson (who was a pioneer at the crossroads of environmental art and urban planning), Herman Prigann (who coined the notion of 'metamorphic artworks' and realized numerous site reclamation projects), Aviva Rahmani (who developed a methodology of ecological restoration), Lynne Hull (who introduced "Trans-Species Art"), David Haley (who developed an understanding of "ecopoiesis"), Shelley Sacks (who further developed Joseph Beuys' concept and practice of 'social sculpture'), James Turrell (as an example of the interest of Earthworks) and Andy Goldsworthy (as an example of the interest of 'art in nature').

Other cultural actors accompanied the movement of ecological art, such as the art historian Suzi Gablik who celebrates the rise of a connective and responsible art. I discussed at length the insights brought by Gablik (such as the notions of "reconstructive postmodernism", of the "connective self") as well as the limits of her discourse (such as her holistic, over-harmonious view). With the case of the 2004 Monongahela conference "on Post-Industrial Community Development", I also evoked some of the themes and issues discussed within the movement of ecological art in recent years, such as the role of ecological art in interdisciplinary collaborations with communities, different levels and depths of dialogues with communities and the different methods and tools of eco-art, among other topics discussed by Tim Collins and Reiko Goto, the Harrisons, Suzi Gablik, Ann Rosenthal, the collective Temporary Services, Grant Kester, the London-based collective PLATFORM, A. Laurie Palmer, Stacy Levy, Stephanie Flom, Tom and Connie Merriman, Jackie Brookner and Erica Fielder.

The interest of ecological art, and more specifically of eco-art interventions, was highlighted especially in the exhibition *Ecovention* (2002). The notion of ecovention, pointing at "an artist-initiated project that employs an inventive strategy to physically transform a local ecology", is highlighted in the works of "Ocean Earth,"the "Center for Land Use Interpretation", George Steinmann, Basia Irland, Kathryn Miller, Tim Collins, Reiko Goto, Suzanne Leibovitz Steinman, Patricia Johanson, Betty Beaumont, Lynne Hull, Brandon Ballengée, Tera Galanti, Alan Sonfist, Mierle Laderman Ukeles, Buster Simpson, Viet Ngo, Laurie Lundquist, Superflex, Mel Chin, Betsy Damon, Jackie Brookner, Aviva Rahmani, Shai Zakai, Agnes Denes, Georg Dietzler and the group AMD&ART, as was discussed after Sue Spaid.

The historical emergence and development of environmental and ecological art marks an example in the way art worlds can play a constructive role in the construction of culture(s) of sustainability. With its diversity in approaches (from social sculpture to land reclamation) and with its common threads and principles (as discussed at the Monongahela conference), it seems that this movement fosters an advanced, elaborate form of aesthetics of 'the pattern which connects'. However, a simplistic form of holistic discourse is looming behind some of the discourses of eco-art. Assessing the degree of sensibility to complexity, would require elaborate qualitative empirical assessments on a project basis (i.e. case by case) and no general statement can be made.

The interplay of ecology, sustainability and art also came into the focus of attention of a diversity of other actors in the art world of contemporary art, within the past decade. Art exhibitions have been organized around the world, dealing with the themes of ecology, sustainability (and recently often with the specific topic of climate change). I discussed how some curators/writers constructed specific discourses in the past decade, including specifically Hildegard Kurt (with her definition of "aesthetics of sustainability"), Maja and Reuben Fowkes (with their notion of "sustainable art") and very lately, *Third Text* editor Rasheed Araeen's notion of "ecoaesthetics".

Curator and writer Linda Weintraub discussed how artists whose work usually is not associated to ecological or environmental art brought interesting perspectives on themes such as desire in the culture of hyper-consumption (Reverend Billy Talen), the culture-nature relation (Dave Burns and Matias Viegener), biodiversity (whereby she points at the work of Eduardo Kac – which I however discussed as very questionable), and the difficult relation of contemporary western societies to processes of death, decay and dirt (with the works of Catherine Chalmers, Damien Hirst and Joe Scanlan). Weintraub's analysis focuses for a great part on "art space" artists, i.e. artists who mainly produce work at art galleries, unlike many ecological artists, and her discourse demonstrates that they do bear some limited relevance, at least as contributors to a relative potential for change in perspectives among the international cultural elites that do pay some attention to contemporary art.

I further conducted a detailed critical discussion of the curatorial discourses of a selection of exhibitions that specifically addressed ecology and sustainability in the past decade:

- *Beyond Green: Towards a Sustainable Art* (SMART Museum, Chicago, 2005) curated by Stéphanie Smith, and with works by Jennifer Allora and Guillermo Calzadilla, Wochenklausur, Nils Norman, Frances Whitehead, Marjetica Potrc, Learning Group, Kevin Kaempf, Dan Peterman (whose work at the Kunstraum of the University of Lueneburg I also discussed), Temporary Services and Free Soil among others.
- *Auszeit. Kunst und Nachhaltigkeit (Timeout. Art and Sustainability* - Kunstmuseum Liechtenstein, 2007), curated by Friedemann Malsch and Christiane Meyer-Stoll, an exhibition which reduces the understanding of sustainability to 'slowing down' and offers little more than a detached white-box discussion on time and nature, pointing at works by Giovanni Anselmo, Piero Gilardi, Jochen Gerz and On Kawara among others.
- *Greenwashing* (Fondazione Sandretto Re Rebaugendo, Turin, 2008), curated by Ilaria Bonacossa and Latitudes (Max Andrews and Mariana

> Canepa Luna), an exhibition with the aim to highlight critical artistic perspectives on environmental issues, and with a peculiar, "post-environmental" curatorial discourse to which I have given a careful but non-complacent, heavily critical attention, pointing at its many flaws (especially in the curatorial discourse of Latitudes, more than Bonacossa's). The exhibition itself gathered works by Jennifer Allora and Guillermo Alzadilla, Lara Almarcegui, Cyprien Gaillard, Santiago Serra, Wilfredo Prieto, Maria Thereza Alves, Tomas Saraceno, Nikola Uzunovski and Ibon Aranberri, but also by Amy Balkin, The Bruce High Quality Foundation, A Constructed World, Minerva Cuevas, Tue Greenfort and Sergio Vega (works which do not fit with the curators' simplistic discourse), among others. Despite the flaws of the curatorial discourse, many of the exhibited artists do contribute valuable, and relatively subtle/complex critical insights, especially into the issue of 'greenwashing'.

Finally, I ended my exploration of discourses in contemporary art in the last decade, with a discussion of a short text by the French philosopher and psychoanalyst Félix Guattari (*Les Trois Ecologies*) because this text was often quoted by curators (including Stéphanie Smith and Max Andrews), and criticized by some eco-artists (including David Haley). Guattari's *Three Ecologies* offers an insightful, though very dense, introduction to thinking the ecology in complex terms. But, probably because of the very short length of this essay and of its lack of articulation of the complexity of understanding complexity, Guattari's work was maybe not appreciated by all its readers from the world of contemporary art, to its full value.

The transversal overview I conducted of some themes addressed in contemporary art, revealed the emergence of "eco-centric" perspectives and insights as analyzed by Linda Weintraub, even coming from unlikely sources. Furthermore, the keyword 'sustainability' is gradually integrating the vocabulary of a number of art curators, and generating discourses and practices exploring the interplay of social and ecological critique and contemporary art. The growing debate is lively and diversified, even though some of its elaborations reveal especially questionable discourses (such as post-environmentalism, or un-reflexively pro-technological practices and discourses).

The discussion of ecological art pointed at the many instances (or 'ecoventions') where the artists intervene directly in ecological and social reality, engage with communities, and aim at setting transformations into motion. Such ambitions beg a more careful attention to the potentials and challenges facing artists aiming to contribute to social change, not only within their art worlds but also directly in specific social contexts. This is why I re-introduced and expanded upon my analysis of artistic intervention in society as "double entrepreneurship in conventions" (already engaged since 2004), in the last chapter of the present volume (which I also replaced in the context of Batesonian "deutero-learning"). The possibility to perform this 'entrepreneurship' was shown to depend on the specific complexity of social systems, on specific qualities of social agents as 'entrepreneurs' and on the context of art worlds and more specifically the political setting of the relevant art world (i.e. on 'polity conventions' in art worlds).

The analysis of polity conventions in art worlds also gave me the opportunity to discuss the political aspect of cultures of sustainability as requiring that the political settings of social interactions be at least 'near-polyarchic' (rather than 'near-hegemonous'), in order to allow rather than stifle, processes of emergence, eco-evolution and what Morin characterized as "high complexity". Near-polyarchy is however not a self-evident 'given', neither in the arts, nor in environmentalist movements. I pointed at the necessity to prevent a degeneration of ecological art into green fascism, with the extreme example of modern dance under Nazism, where some organicist-holistic ideals, for lack of complexity, embedded themselves comfortably within probably the most monstruous totalitarian political system of the past century: With few exceptions, the modern German dancers willingly eliminated polysemy in their work and promoted increasingly hegemonic values through their work, not only under the Nazi regime, but already in the years before. This case showed that a mythology of a holistic communion with nature is potentially very compatible with highly hegemonic polity conventions and with a totalitarian ideology. Against this totalitarian drift, a sensibility to complexity is crucial.

My analysis suggested that, with entrepreneurship in inter-conventional interaction, and deutero-learning, even hard programmed ideas and social conventions may change. But a successful double entrepreneurship in conventions also requires that the art world in question be sufficiently open to such an enterprise, which is more likely in an art world with a more polyarchic political setting. Furthermore, only relatively polyarchic polity conventions in art worlds, may be compatible with entrepreneurship in conventions towards cultures of sustainability. In this very concrete way, the artistic is eminently political. The political dimension of aesthetics and cultures of sustainability should include discourses and practices valuing pluralism and contestation, compromises and regulated competition, fostering complex uni-plurality and exploring a 'queer ecology,' away from a straightforwardly harmonious image of nature.

At all the levels explored in the preceding chapters, not only the disjunctive bias of the modern scientific method and of autopoïesis in social systems is looming, but also the biases of unitary views maiming the *unitas multiplex* (in universalist and holistic forms) are constantly threatening, with their simplifying short-circuits, the fragile (but hopefully resilient enough) evolution of cultures of complexity.[3]

3 In the course of the present volume, I discussed the risk of holism more directly (in chapter 2) than the risk of universalism (which is connected to the belief in

What is to be done?

What, then, is to be done? As I come to the completion of this volume, I ask this question first of all to myself, and by extension, to the eventually like-minded reader.

At a practical level concerning the researcher in the social sciences (which was my disciplinary point of departure when I started this research), I shall point out two consequences of my analysis:

- The degree to which specific art projects/interventions can be transformative, and can contribute especially to a sensibility to complexity, is a difficult question which at best, can only be assessed on a case-by-case basis and qualitatively. However:
- A meaningful assessment can only be achieved if the qualitative observation is engaging the researcher as a full person, and beyond the limitations of purposive consciousness. The relevance of, and quest for aesthetics of sustainability does not only concern a specialized category of individuals (artists) or a specific social system (the arts), but all social actors, including the academic/scientific researcher.

Therefore, the insights unfolded in my analysis apply to the scientific researcher as much as to the other social actors involved. The call for transdisciplinarity applies of course across all disciplines. This does not mean that 'artists' and 'scientists' should altogether cease their more disciplinary activities as 'artists' or 'scientists', but it does mean that such activities are limited and autistic, thus insufficient, if not completed with interdisciplinary and especially transdisciplinary activities. In the domain of the search for cultures of sustainability, transdisciplinarity calls forward an "artscience" (Dieleman 2007, after Edwards 2008) i.e. the construction of a genuinely transdisciplinary methodology. This endeavor will require further elaborations. However, it is not a transformative process in which a solitary researcher can engage him- or herself. This is a collective, combined exploration that has to engage artists as well as (social and natural) scientists.

It is with this goal in mind that I initiated the transdisciplinary project of an "International Summer School of Arts and Sciences for Sustainability in Social Transformation" (ASSiST), organized by the international network Cultura21 together with partner organizations I3C (International Council of

---

universal natural laws, associated with the modern scientific method, discussed in chapter 1). For example, Laville and Leenhardt (1996) plead for universalism as well as for holism. The term "holism" is often uncritically accepted and used by many of the authors I cited earlier (including Fritjof Capra, Hans Dieleman, David Haley) and whose analyses are otherwise very insightful. (This may however characterize a lack of terminological precision on their part, rather than a lack of sensibility to complexity.)

Community Cultural Centers, based in Sofia), Red-LATS (Latin American Network of Art for Social Transformation, based in Buenos Aires), and the Center for the Study of Culture and Society (based in Bangalore).[4] In the vision statement of the school, which we wrote together, is stated:

"The Summer School addresses the HOW questions of art & science for sustainability: For example: How can academic research contribute to community artists and contemporary artists developing new social, aesthetic and community approaches? How can the experience of post-modern off-balance dance contribute to a rethinking of democracy? How can scientists enhance their reflexivity and their creativity thanks to insights from the arts? How can artists effectively work in communities and at other levels of social reality, for social transformation together with academics? In order to address these questions, the Summer School will explore: How do artists and scientists from a diversity of backgrounds, do their work today? Which methods and approaches do they use and how can these be transferred and transmuted to other artists and scientists around the world? And which cross-breedings between these different approaches should be further developed? The Summer School will also foster thinking at a paradigmatic and normative level and link the WHY and the HOW questions, in order to avoid mere instrumentality. That 'deeper' level will address issues of spirituality, epistemology, ontology, aesthetics and introspection, and critically reflect the orientations and values we vest in 'social transformation' and in 'sustainability'. Art and philosophy, conceived not only as vehicles, but also as possibilities to step outside ourselves, will contribute to this deeper, epistemological and 'more than rational' level of the Summer School's methodology" (Kagan et al. 2010).

The school aims to bridge the inter- to the transdisciplinary level, and thus enable "participants to develop varied ways to work transdisciplinarily, no longer only as determined 'artists', 'activists' or 'scientists', but, as artiscientists, artivists, intellectactors, i.e. shaping new identities for a new culture" (ibid.).

Other transdisciplinary laboratories can also serve as exemplary case studies for the social scientist and would-be-'artiscientist' (for which in-depth and participative forms of empirical research will be required). The organization "FoAM" in Brussels, may be a good example of this. In the online self-presentation text of FoAM, one can read:

"FoAM is a transdisciplinary laboratory and a research group committed to developing a holistic culture, by actively propagating resilient cultural forms. FoAM's members include artists, gardeners, cooks, technologists, designers, writers and scientists from all walks of life. [...] FoAM is dedicated to advancing and supporting a commu-

4 The first edition of this summer school took place in Gabrovo, Bulgaria, from August 21st to 27th 2010, with as the key theme of this year's edition "Walking and Places: building transformations", i.e. studying walking as an artscience, action-research methodology. See http://assist2010.ning.com

> nity of generalists, people who live and work in the interstitial spaces between professional & cultural boundaries, operating under the motto 'grow your own worlds'. [...] Ecologies of worlds should guide us in considering our future. We imagine this future to be responsive, adaptive and interconnected. We abandon the static and universal designs of the industrial era and move towards a world of malleable materials, objects and spaces. [...] By accepting that our actions and our futures are deeply entwined with many others, we can glimpse the larger patterns of these connections [...] we learn by doing. We have pieced together our education and constructed our collaborative setting, which meanders between different intellectual, economic and political models. We gather creative expression and cultural values where they are least expected - in science and engineering, traditional crafts, play and games, or the chores of everyday life [...] exercising our abilities to 'DIWO' (do-it-with-others) [...] we are looking at ways in which human systems and societies could adopt some characteristics of plants - valuing diversity and collaboration over monocultures of competition; approaching problem-solving through whole systems thinking; redesigning industry and economics to adopt more cyclical, regenerative processes" (http://fo.am).

These self-descriptive arguments glimpse at an *apparent* transdisciplinary quality of their discourse, which might also then be found back in their work.[5]

However, an artscience will not significantly emerge without an adequate institutional setting. Policy frameworks are thus of prime importance, in order to foster genuine practice of transdisciplinarity (and not only proclamations thereof). This is the second level of 'what has to be done', in parallel with the first level mentioned above (i.e. methodological elaboration): Lobbying on policy makers at the levels of cultural, scientific and educational policies worldwide, in order to rethink policies towards support for transdisciplinary artscience. But then comes forward the challenging question of a culture of complexity at the policy-level itself, vs. the tradition of planning: How can cultural policies allow the emergence of cultures of complexity, rather than design top-down cultural and scientific programs?

In other words, the challenge is to foster emergence rather than to design rigid frameworks, to facilitate iterative processes rather than linear developments. The iterative , i.e. not deciding/thinking and then implementing in

---

5 In its discourse, FoAM sounds promising. However, one issue of concern that I have with the concrete projects of FoAm (and examples they use and speakers they invite at their events) is that they often seem to be over-emphasizing technological experiments and design. This focus is less stressed in the discourse than it is practiced in the projects, but it is present: For example, one can read in a 2007 presentation by Maja Kuzmanovic (one of the leading organizers of FoAM): "I'd like to think that there are technological and design solutions that will help us out of the current situation into a luminous future" (http://libarynth.org/symposium 2007/maja_kuzmanovic).

a linear sequence, but learning- and thinking-while-doing in circular reflexive sequences and in parallel, overlapping, telescoping processes, is also relevant at the policy level itself. It appears then that, in the spiral of complexity, the policy level itself is to be engaged into a transdisciplinary art-ministration rather administration of separately considered policy sectors such as culture, education and science.

Finally, some readers might dismiss some of my arguments, on the account of technophobia. To illustrate how such a judgment would be 'besides the point', I will end my exploration with a short critical discussion of a small bit of Bruno Latour's discourse about similar topics, whereby Latour fails to move beyond the linear belief in what he labels himself as "the promises of science [and] technology".

In a recent discussion of Nordhaus and Shellenberger's post-environ–mentalism (which I already discussed in the context of the exhibition "Greenwashing" in chapter 6), the sociologist Bruno Latour (2008) points out in well thought-out terms that while the ideology of modernity was disjunctive and believed-to-be an emancipator (through the idea of progress), the reality of modern times has been one of increasing entanglement of humans and nonhumans (and increasing complexity).[6] So far so good, I can follow Latour. But then, Bruno Latour shares with the post-environmentalist a criticism against the "asceticism" of "Green politics". Latour skillfully argues that our "sin [is] not technology itself, but the absence of love for the technology we have created [...] the sin is not to wish to have dominion over nature but to believe that this dominion means emancipation and not attachment" (Latour 2008, pp. 11-12). An interesting argument, but not quite enough. Actually, Latour's discourse seems to stop short of complexity, although peeking into complexity.[7] It is stuck in dialectics, and in a binary alternative, as he states to conclude his article: "We want to develop, not withdraw" (ibid., p. 13). From a dialogical perspective however, my claim would rather be: We want to evolve, i.e. to emerge and to be-come resilient. Beyond merely developing or withdrawing linearly, we (re)connect to, and reconstruct a queer ecology. After all, the entanglements that Latour rightly points at, were far from being absent or negligible before modernity. The queer ecology predates modernity, and can survive modernity. Latour, like

6 Even though his article flirts with complexity thanks to its sensibility to entanglements (cf. Latour 2008, pp. 5-6), Latour does not quote Morin, instead repeatedly insisting on referring to Ulrich Beck's "Risk Society"; Beck with whom he shares the dream of "modernizing modernization". The limitations of Latour's Actor-Network-Theory in the context of turning to complexity theories in sociology, are shortly mentioned in Smith and Jenks (2006).

7 A similar observation is made by Loïc Fel, who comments that those who fail to fully appreciate the high level of qualitative complexity of natural systems, will tend to boast around with exaggeratedly optimistic (and often techno-naive) discourses (as seems to be happening in Latour's case): Cf. Fel 2009, pp. 290-291.

the so-called post-environmentalists, is also stuck in a binary alternative in his time narrative (i.e. either we move forward or backwards). Hence his preference for "development, stupid" and his plea, after Nordhaus and Shellenberger (and not unlike Michael Braungart too) for "the limits of the notion of limits", with the slogan "develop more, not less". However, in a nonlinear, queer ecology of co-evolving species, the linear obsession with moving forward (vs. leaning 'backwards' or being 'frozen' by 'limits') has become irrelevant. What matters much more is shifting, metamorphosing and adapting, or rather co-adapting, contextually and 'glocally' (trans-locally). And in the eco-evolving Earth system of co-evolving human & nonhuman systems, there are constantly mutually evolving limits. These limits teach us humility, but humility in front of complexity is not the same thing as a postmodern resignation following the loss of modernist illusions. There is no reason to be afraid of limits, when limits are no mere barriers of modern autopoïesis, but autoecopoïetic interfaces, that is, not obstacles to linearly conceived progress but challenges to dynamically conceived balancing.

Appendices

APPENDIX 1: THE SUBCONSCIOUS AND THE 'ECO-' IN AUTOECOPOÏESIS: A SPECULATIVE EXCURSUS

How can the subsconscious work as multiple open windows that keep opening when the autopoïetically-driven doors of consciousness shut out certain disturbances from the outside world, and thus foster the 'eco'- in autoecopoïesis? Whereas I do not pretend to offer any conclusive analysis of this question, I propose here a few speculative elements of reflection.[1]

The brain filters out almost all of the sensory/informational inputs from the outside world, only a tiny portion of which is accessible to consciousness. But some part of the filtered-out information still may reaches the subconscious.

Consciously, we only see what we believe is possible, because it matches patterns that already exist within ourselves. We always perceive something after reflection in the mirror of memory.

1 Given the largely speculative character of the following arguments, i.e. lacking the weigh of proper scientific references backing-up my claims, I decided to move this text into an appendix, but not to altogether remove it from this work, because it touches upon an important dimension for an understanding of autoecopoïesis. The reflection in this appendix is partly inspired by hasty reading of passages in authors like Damasio, and partly by the viewing of a largely unacademic, and questionable documentary film entitled What the Bleep do we know? (2005). I do not support all the claims made in this film, especially not its spiritualist/New Age claims and its severely biased and simplistic interpretation of quantum physics. However, I do recognize that some interviews of cognitive scientists in the film, are providing certain insights to the layman that I am concerning cognitive sciences, about the working of the subconscious. I also remark that I am not the only one to point only sketchily at the importance of the subconscious: e.g. Smith and Jenks (2006) also do so. They especially stress, after Damasio and some others, that consciousness is not a unitary phenomenon but "a distributed parallelism of interactive processing" (p. 63).

Besides, a correction to Luhmann's account of the brain needs to be made. Luhmann argued that the brain has no direct connection to the outside world (except through the mechanism of structural coupling and irritations), and that the mind's (i.e. in his terminology, "psychic system", emerging from, distinct from, the brain) representations are not representations of the outside world but own constructions.[2]

However, the brain can be potentially confused and incapable of making the difference between what is a representation of the outside, thanks to perception, and what is an own imaginary construction. Why is this so? Because seeing and remembering/imagining involve the same neural networks: The brain does not make the difference between what it sees and what it remembers, because the same neural nets are firing in both cases.

What comes out of this corrected perspective on the brain's operations, is less clear-cut than Luhmann's strict operative closure. We are not dealing here with a strict closure, but with a confusion about what is coming from inside and what is coming from the outside. The brain cannot distinguish whether information is generated from inside or outside itself.

In other words: We don't know the difference between what's taking place 'out there' and what's taking place 'in here'. Or do we?

Luhmann (2000) insisted a lot on this making of distinction, this prime imperative 'draw a distinction'. But one has to be aware that, when it comes to the mental constructions of a given human being, the brain cannot clearly establish the distinction between the inside and outside, the self and the environment. But the psychic system does decide, on a daily basis, whether an experience is a hallucination, a dream, a memory or an actual lived, sensory experience. But this decision is contextual and subject to doubts; it is not a kind of business that can be easily settled by an autopoïetic system: The autopoïetic system always needs clues from the outside, in order to reassure itself about where the inside is and where the outside is. For example, sensory deprivation beyond 24 to 48 hours has been repeatedly reported to lead to hallucinations and to severe, sustained psychic 'damage'.

The important aspect to note here, is that the construction of the self, the delimitation and distinction of the inside of the system, can only be trustworthy to the psychic system as long as there exists a genuine communication with the environment of the system (which in Luhmannian terms, would be understood as a 'structural coupling' rather than a 'communication'). *Even if operationally piloted from the inside, that communication necessarily includes inputs that are not controlled from the inside... and it is*

2 Luhmann based his claims on one stream of cognitive research, but cognitive research has evolved, and for example, the work of Maturana and Varela, that Luhmann took as his foundation rather uncritically, has been hotly debated and contested (and furthermore, Maturana and Varela have expressed diverging perspectives since then).

because they are not controlled from the inside, that they reassure the system about its distinctions of inside and outside.

This also explains why these distinctions might be changed through the operation of the subconscious: Beyond the reach of consciousness, the subconscious may have stored sensory inputs and memories and may assemble them with own constructions, so that the hallucinated experience may appear as a real experience to consciousness, given that it includes elements that seem to be non-controlled by the self (but are more or less 'controlled', so to say, at subconscious level). This additional point adds to the general rule of confusion as to the inside/outside distinction.

But doubts and confusions allow an evolution in the thresholds of conscious perception. Such doubts can be stimulated in different ways (as will be described further down) but I focus now on only one of them:

Surprise, informational discrepancies, as soon as perceived i.e. recognized as such, instigate doubts, suspicions, and thereby constitute a crucial ecopoïetic tendency in the evolution of the 'psychic system'. As mentioned above, the subconscious does receive some of the informational inputs from the environment that are filtered and out of reach for purposive consciousness. Therefore, the subconscious is playing a role in the ecopoïetic tendency of suspicions, i.e. in the breaking open of habitual patterns of thought (on the role of suspicion in conventions, see chapter 7, section 1). In other words, the subconscious allows the random to enter into the ordered psychic system.[3] It is the ecopoïetic irruption of the new that allows the system to be creative, in spite of its strong autopoïetic tendencies! This line of thought echoes Bateson (1979) in his developments on randomness and order, chaos and selection, *and suggests creativity as one of the 'great stochastic processes' of evolution* that Bateson evokes.

So, we do build our models of reality in an autopoïetic operational way; so far can I follow Luhmann. But the more information we receive from the environment (allowing distinctions to be trustworthy and thus allowing distinguish-abilities), the more we can refine those models autopoïetically, or, in cases of discrepancies that get 'irritating' enough to get noticed, the more we modify them auto-ecopoïetically.

The human brain has a relatively high postnatal plasticity. The structures of the brain, the neural nets, are, to some extent, constantly modified by experiences. Nerve cells that fire together wire together. Therefore, through repeated practices/experiences (e.g. through the introduction of new social conventions), certain neural cells establish long-term relationships. Also, nerve cells that no longer fire together, no longer wire together. Every time we interrupt a thought-process that produces chemical responses in our body, those nerve cells start breaking their long-term relationships.

3 Luhmann comes closest to this with the notion of "irritation", operating through structural couplings. But I find that line of argumentation insufficiently helpful, to understand the creativity allowed by the subconscious.

There is however no absolute postnatal plasticity, and no "blank slate" onto which society can imprint anything, but some structures, some "attractors" that limit the possibilities of plasticity, as argued at length by Smith and Jenks (2006) in their criticism of post-modernism.

It is possible to forget, to lose habits, to replace old habits with new habits. To most people this sounds obvious, but to some social scientists it may come as a surprise to realize that an individual's habitus established in early socialization, despite its stability and resilience to change, cannot totally escape change, given that the rewiring process of neural nets is continuing to operate in adult age (but this argument is less surprising, from the modified habitus-concept of Lahire, i.e. the repertoire of dispositions – as described in chapter 7, section 1). What does this all mean? Two things:

- Social change that affects the individual (micro-social level) literally can change the structure of the brain, in short it 'rewires it'! (But this is possible only up to a certain extent, depending on the only-relative postnatal plasticity of the human brain...)
- The keyword here is 'interrupt': Change happens when given autopoïetic thought processes are interrupted. The next question then becomes: How can there be such a thing as this interruption of the thought process?

This *interruption can come in several ways*:

An effort of *self-observation* by the individual, i.e. by a psychic system, at the level of 'second-order observation', can allow breaking certain thought processes, and thereby breaking certain neural nets. The development of reflexivities in social agency can foster this capacity of self-interruption (which may remain operationally autopoïetic).

Next to self-observation, the constraints imposed by the environment on the psychic system do also provoke interruptions in the thought processes. *Environmental constraints* do not determine directly how the system will respond to the interruptions, but they *do impose interruptions* upon the system. Here, I do not follow Luhmann who argued that the psychic system can altogether determine, from the inside, what will 'irritate' it (i.e. which interruptions will occur and which not, ignoring the environmental constraints). The autopoïetic capacity of the system to ignore environmental constraints is only relative.

The third important way in which the interruptions occur, comes from the subconscious part of the human mind. Indeed, *the subconscious mind ignores its environment much less than consciousness does*: much more of the sensory information coming from the environment is being received at subconscious level than at conscious level (where, as already noted, most information has been filtered out). Therefore, when the subconscious invites itself unexpectedly into our consciously monitored trains of thoughts, it introduces *an 'auto-ecopoïetic' trigger* that can substantially disrupt established thought processes.

These three modes of interruption of thought processes are not to be seen separately from each other, as three different options, but are to be seen as interacting and possibly reinforcing/complementing/opposing each other.

Three modes of interruption of autopoïetic inertia

(1) *From the inside*: Self-observation interrupts a certain thought process (cf. second-order observation as discussed in system theories, including in Luhmann; and cf. some of the reflexivities as discussed by Dieleman 2008). Although Luhmann would argue that this too is an autopoïetic process (and it can be), it can still affect and change some other autopoïetic processes; i.e. 2nd-order self-observation can change first-order observations. And, at the level of neural networks, such an observation, that interrupts habitual thought processes, literally changes the structure of the brain because, as soon as it occurs, some neural links disconnect themselves (and thereby, some long-term relationships between neurons can gradually get lost).

(2) *The constraints from the environment of the system*: Even from the autopoïetic perspective coming from Maturana and Varela, the environmental constraints are active. For example, just before sitting down to write this specific piece of text, I wanted to sit on a bench in a Lüneburg park. But as all the benches were taken, I had to resort to an alternative recourse, and I am now sitting in a nearby café, having to order drinks, among other constraints. Here the consciousness can have recourse to its own internally programmed set of options and thus the adaptive process can also still be said by the Luhmannians to be autopoïetic. But there is here too an interruption of thought processes, i.e. a cutting loose of neural links, a minimal change in the neural structures. Thus both this and the previous case, i.e. both the environmental constraints and the self-observation, do co-determine the structuring of the system, even though the process is here possibly still 'piloted' from the inside i.e. autopoïetically. But the pilot does not control everything in the environment: He cannot choose the constraints. I could not ignore those people already sitting on he benches, and sit on their laps as if they were part of the bench (although I could have chosen to sit on the grass). Furthermore, *the pilot is not 'one'*, i.e. the second-order and first -order observations can collide, contradict each other (or second-order observation can reveal contradictions between different first-order observations), and this can facilitate suspicions, doubts, questions about one's autopoïeticallly constructed reality and thus allow consciousness to open itself to alternative options from the realms of perceptions, outside of the established frames of the autopoïesis-fueling existing mental patterns, i.e. there is a bridge to the third mode of interruption of the autopoïetic closure:

(3) *Introduction of the new from the outside*: I argue that ecopoïesis, i.e. the determination of the mind system from the relationship with the

outside, can and does occur simultaneously to autopoïesis, but not directly at the conscious level. Here the subconscious is of the utmost importance. Although the brain filters out almost all perceptive information (so that they do not reach consciousness), some reach the subconscious. As Bateson argued, the conscious, "purposive" part of mind is very limited. The new, present at subconscious level, can under some circumstances, start invading consciousness (i.e. it may burst in from the windows even if the doors are closed, sealed, shut off from disturbing subconscious information). *Most especially, the news from the subconscious can become a reality when suspicion, doubt, self-questioning, surprise, etc. is present within the system of consciousness, i.e. autopoïesis has to collaborate with ecopoïesis.* But, as the brain is not always able to make the difference between what's coming from the outside and what from the inside (what I see and what I remember), and as the conscious cannot manage to control what's happening at the subconscious level, the conscious "I" cannot always know clearly how, when and where these subconscious news are inviting themselves and are let in (by consciousness). Neither – in the case when the subject realizes that subconscious news are entering and modifying the conscious autopoïetic system – can one be absolutely sure whether these subconscious experiences are the fruits of ecopoïetically induced experiences – from the outside – or are imagined i.e. internal constructions at subconscious level. Thus there is a *great deal of confusion, within the 'self-reflexive' subject, as to what is autopoïetic and what is ecopoïetic in the psychic system.* Not only do autopoïesis and ecopoïesis depend on each other. They are interacting 'beneath the surface' of perception, most often within the blind spots of observations/perceptions/thoughts.

Why is it helpful to know that the noise from the outside comes in, unexpected, in the psychic system? Because such an event is necessary as a prerequisite for an evolutionary process. "Every random fluctuation is a possibility of new organization. [... If a]mplified by positive feedback, any fluctuation [may] thus constitute a random generator of variety, at the base of any evolution." (De Rosnay 1975, p. 258).[4]

But the importance of the autopoïetic element in the evolutionary process should not be under-estimated either: The random fluctuations most often do not reorganize reality; only sometimes do they invade their environment, i.e. when emerges an 'autocatalysis' (which may eventually lead to the self-constitution of a new autopoïetic system – e.g. eventually even maybe one new sub-system within the distributed parallelism of the brain which constitutes the impression of an united consciousness). And an al-

4 Translation from the original French text by the author.

ready established autopoïetic system has some degree of control over processes of autocatalysis emerging within itself (but only *some* degree)...

As already mentioned, the preceding paragraphs are offered merely as speculative open propositions, prone to tentatively complement my main text while stimulating a discussion and inviting clarifications from better-qualified authors in this domain.

APPENDIX 2: DEUTSCHE ZUSAMMENFASSUNG: KUNST UND (NICHT-)NACHHALTIGKEIT

Diese Arbeit macht es sich zur Aufgabe, geistes-, sozial- und kulturwissenschaftliche Ansätze des systemischen Denkens, der Komplexität und der Transdisziplinarität als zentrale Bausteine des aktuellen Nachhaltigkeitsdiskurses zu deuten und eine solchermaßen erörterte „Kultur der Nachhaltigkeit" mit der (Nicht-)Nachhaltigkeit zeitgenössischer Kunstausprägungen zu konfrontieren.

Die Kultur und die Kunst der Nicht-Nachhaltigkeit

Das westliche Entwicklungsmodell ist von einer Kultur der Nicht-Nachhaltigkeit geprägt, die sich aus verschiedenen fundamentalen Dimensionen der Moderne ergibt. Die westliche Weltsicht und ihr modus cognoscendi sind in der Krise. Ervin Laszlo (1996) hat diese Weltsicht als atomistisch, materialistisch, individualistisch und eurozentrisch, und als auf der „klassischen wissenschaftlichen Methode" beruhend charakterisiert – im Gegensatz zu einem „systemischen Blick" der aus dem 20. Jahrhundert erwachsen sei. Basarab Nicolescu (2002) hat darüber hinaus die Prinzipien der klassischen modernen Physik, also Kontinuität, lokale Kausalität, Determinismus und binäre Logik kritisiert. Beide Autoren bedauern die daraus entstehende Fragmentierung des gemeinen Verständnisses, und dieser klassische modus cognoscendi wurde von Edgar Morin (1977) weiter dekonstruiert als eine disjunktive Vereinfachung, die nicht nur die Wissenschaft sondern generell die modernen modi cognoscendi durch drei Simplifikationen des Denkens beeinflussten: „Idealisieren, Rationalisieren, Normalisieren". Diese Kritik gibt sowohl Edmund Husserls (1970) Kritik am Übergehen und sich Entfremden von der Lebenswelterfahrung, wie auch Gregory Batesons (1973) Warnung vor dem Reich, das aus den engen Blickwinkeln des „zweckbestimmten Bewusstseins" erwachse, wieder, da beide Probleme in modernen Gesellschaften besonders ausgeprägt seien. Seit der zweiten Hälfte des zwanzigsten Jahrhunderts sind diese Probleme mit einer Kultur des Hyperkonsums verbunden, der trotz der für die letzten Jahrzehnten angekündigten Bewegung in Richtung einer Dienstleistungs- Wissens- oder Informationsgesellschaft bzw. Kreativwirtschaft immer noch wächst.

Diese Dimensionen unserer Kultur der Nicht-Nachhaltigkeit sind wegabhängig und verwurzelt in geschichtlichen Entwicklungen, die von Patriarchalismus, Kartesischer Methode, Baconschem Fortschrittsglauben und von einer progressiven Entfremdung von der Natur, die David Abram (1996) zu folge ihre Wurzeln in der Erfindung des phonetischen Alphabetes haben, markiert.

Moderne und zeitgenössische Weiterentwicklungen dieser Kultur der Nicht-Nachhaltigkeit beinhalten die Umwandlung von Wissenschaft und Technik in ein techno-wissenschaftliches System, in ein „Technisches System" (Jacques Ellul 1977), die der Bewegung von der Aufklärung zum Positivismus folgt (Horkheimer und Adorno 2002), und die zunehmend autopoïetische Operationen von sozialen Systemen in modernen Gesellschaften durch funktionale Ausdifferenzierung präferiert (Niklas Luhmann 1984, 1999, 2000).

Die autopoïetische Entwicklung nach Luhmann findet sich in der Kunst als sozialem System wieder, wo, wie die Kunstsoziologen von Pierre Bourdieu, Howard Becker bis Hans Abbing analysieren, die „hochkulturellen Künste" („high arts") gesellschaftliche Segregationsprozesse institutionalisieren. In dieser Welt der hohen Kultur hat sich eine eigentümliche Kombination aus einer „Romantischen Ordnung", eingebunden in ein „Technisches System", entwickelt, deren Interaktionen in dieser Arbeit nach den Vorgaben von Jacques Ellul (1980) und Maarten Doorman (2004) erkundet werden. Diese komplexe Verbindung ist auch das erste Beispiel für ein Makro-Konzept, das ich in diesem Werk beschreibe. Auch wenn die Situation komplex und offen für Veränderung ist, hat insgesamt im vergangenen Jahrhundert die romantische Ordnung in der Kunst hauptsächlich als Sicherheitsventil für die Gesellschaft funktioniert, um etwas von dem Druck, den unsere allumfassende Kultur der Nicht-Nachhaltigkeit verursacht, herauszulassen, aber ohne dass es signifikante Transformationen über den „Sandkasten" der Kunst hinaus gegeben hätte. Wie Luhmann analysiert, ist die Kunst der Moderne als soziales System unfähig mit anderen sozialen Systemen zu kommunizieren (und umgekehrt), jedenfalls so lange seine Operationen von Autopoïesis dominiert werden. Diese Situation wurde von der Kunsthistorikerin Suzi Gablik (1984, 1991) aus der Kunstwelt heraus angeprangert. Die Kritik des Modernismus in der Kunst zielt hierbei besonders auf die Ästhetik der Loslösung, auf die individualistische und antagonistische Autonomie, die sich in der „hohen Kultur" entwickelt habe und auf die nihilistische Selbstbezogenheit der Postmodernisten.

Systemisches Denken

Wie können wir also Alternativen zu dieser Kultur der Nicht-Nachhaltigkeit und die damit verbundene Kunst der Nicht-Nachhaltigkeit finden? Eine Antwort auf diese Herausforderung benötigt eine weitreichende Transformation des modernen modus cognoscendi. Glücklicherweise jedoch hat diese Transformation schon begonnen, wie der Aufstieg des systemischen Denkens und der Transdisziplinarität in den jüngsten Jahrzehnten aufzeigt.

Systemisches Denken ist allerdings nur ein Schritt auf dem Weg zu einer Kultur der Komplexität. Anhand des Beispiels des systemischen Denkers Fritjof Capra (1996, 2002), erforsche ich die Einsichten und Versprechen des Systemischen Denkens. Besonders die Begriffe von Netzwerken, Nichtlinearität, Emergenz und Selbst-Kreativität haben die erste Phase eines Paradigmenwechsels weg von linearer Kausalität eingeläutet. Einige Prinzipien haben sich auch aus der systemischen Perspektive entwickelt, insbesondere Capras sechs Prinzipien der ökologischen Grundbildung (ecological literacy): Das Netzwerkprinzip der lebenden Systeme, die kontinuierlichen Zyklen des Netzwerks des Lebens, die Offenheit des Systems der Erde für Sonnenenergie, die Wichtigkeit von Partnerschaft und Kooperation für die Evolution, die Notwendigkeit von Diversität für Resilienz und das Streben nach dynamischer Balance (im Gegensatz zum Versuch einzelne Werte oder Dimensionen zu optimieren). Auf der sozialen Ebene hat die systemische Perspektive herausgestellt, wie soziale Organisationen sich entwickeln könnten, wenn sie als lebende und lernende Organisationen wahrgenommen und behandelt werden. Das systemische Denken stellt auch die Wichtigkeit der Kultur, sowohl für die soziale Reproduktion, als auch für den sozialen Wandel heraus. Am Beispiel des systemischen Denkers (und „integral studies“ Denkers) Ervin Laszlo (1996, 2007) weise ich aber auch auf die Risiken von Vereinfachungen durch den Holismus hin, die (zusammen mit der Technokratisierung der Kybernetik) im systemischen Denken gegenwärtig sind.

Die Kultur der Komplexität

Der *Méthode* Edgar Morins (1977, 1980, 1992, 2008) und genereller der Transdisziplinarität (Nicolescu 2002) folgend muss eine Kultur der Nachhaltigkeit eine Kultur der Komplexität sein. Durch Morins Begriff der „Komplexität“ eröffnete sich die Möglichkeit, sich Einheit und Vielfalt auch als zusammengehörig und nicht nur als Gegensätze zu denken. Jede Dichotomie kann über Einheit, Komplementarität, Konkurrenz und Antagonismus gedacht werden und somit komplexe Beziehungen (Makro-Konzepte) schaffen und dia-logische Denkprozesse initialisieren. Ein solches Makro-Konzept wird nach Morin als „Öko-auto-Organisation“ etikettiert, was recht gut die komplexen organisatorischen Verhältnisse zwischen individuellen Lebensformen und den Ökosystemen, in denen sie ko-evolvieren und öko-evolvieren, beschreibt. Das Konzept der „Öko-Organisation“ beleuchtet die Komplementarität von Vielfalt, Komplexität, Spontaneität und Organisation des Lebens schlechthin.

Auf der sozialen Ebene hat Morin das Makro-Konzept der Sozio-Öko-Organisation eingeführt, das darauf abzielt die organizistischen Tendenzen mit den öko-organisatorischen Tendenzen in zeitgenössischen menschlichen

Gesellschaften auszubalancieren, um den Risiken sowohl von Totalitarismus (also zum Beispiel dem totalitären Staat), als auch öko-organisatorischem Konflikt (zum Beispiel Klassenkampf in einer ungezügelten Marktwirtschaft) vorzubeugen. Morins auf Nachhaltigkeit anwendbare Definition der Komplexität verweist auf die Zentralität der Kultur (in der sozialen Evolution) und auf die politischen Dimensionen der Nachhaltigkeit (bezüglich den Strukturen und Funktionen staatlicher Organisation).

Allgemeiner hat Morins die genesische Kraft aufgezeigtund die kontinuierliche Präsenz des Chaos (der Schaffung von Ordnung aus Unordnung) betont. Die schöpferische Endo-Exo-Kausalität bzw. ihre Triebkraft zur Des-Um-Organisation (de-re-organization) sind dabei zwei Makrokonzepte als Wurzeln verschiedener Erscheinungsformen von „Emergenz", also auch als Grundlage der kreativen Evolution des Universums (Poïesis). Morin postuliert hier das Unbestimmtheitsprinzip, das alle Formen des Determinismus ablehnt.

Weiterhin erlauben Komplexitätstheorien die Läuterung der Poïesis aus ihrem trostlosen Zustand unter Luhmanns Autopoïesis, in dem sie sie als Autoökopoïesis umdenken, also als ein System, das mit kreativer Sensibilität der chaotischen Umwelt gegenüber operiert, eine produktive Offenheit für Störungen aufweisen. Ökopoïesis wird dabei als die Tendenz psychischer und sozialer Systeme verstanden, sich ständig in offener Kommunikation mit ihren Umwelten zu befinden. Dabei wird eine Kodetermination und Koevolution des Systems und seiner Umwelt aufgrund der Emergenz von Eigenschaften angenommen, die aus dieser Kommunikation zwischen System und Umwelt stammen.

Eine Ästhetik der Nachhaltigkeit

Die ästhetische Dimension von möglichen Kulturen der Nachhaltigkeit impliziert eine Empfänglichkeit für „Muster, die Verbinden" (nach Gregory Batesons „pattern which connects" – Bateson 1979). Diese Muster und ihre Erfahrung begründen eine Ästhetik der Nachhaltigkeit als Alternative zur modernen Kantischen Ästhetik der losgelösten Kontemplation. Bateson hat dieses ästhetische Erleben als Empfänglichkeit für ein Meta-Muster verstanden, das die lebendige Welt als etwas Ganzes versteht, im Gegensatz zu einer eingefrorenen Wahrnehmung, die sich an relativen Unterschieden festmacht und sich durch diese Unterschiede selbst limitiert und einschränkt. Diese Ästhetik der Nachhaltigkeit benötigt auch eine Empfänglichkeit für Komplexität, die sich allerdings nicht mit holistischer Harmonie zufrieden gibt, sondern (nach den ästhetischen Einsichten des Heraklit) die Harmonie in Gegensätze aufbricht. Diese Ästhetik benötigt weiterhin ein transdisziplinäres Verständnis, das die Irrwege sowohl der vereinfachenden Klarheit des Modernismus, als auch der Verwirrung und Entwurzeltheit des Postmodernismus überkommt.

Batesons Verständnis des ästhetischen Erlebens deutet auf den Wert eines modus cognoscendi hin, der mehr als nur rational-kognitiv bewusst sei. Eine solche Verschiebung könne inspiriert sein von einigen nicht–westlichen Kulturen, die, wie David Abram (1996) argumentiert, eine phänomenologische und animistische Empfänglichkeit für eine mehr-als–nur-menschliche Welt (*„more-than-human world“*) haben. Die Überwindung der autopoïetischen Abkapselung hin zur autoökopoïetischen Empfänglichkeit bedarf der Anerkennung der partizipatorischen Natur der Wahrnehmung als eine animistische, integrierte, synästhetische und geistig–körperliche Erfahrung;der Dualismus aus „Geist und Körper“ müsse hierbei, die monistisch-immanente Perspektive Batesons folgend, zugunsten eines einheitlichen Begriffes aufgegeben werden. Eine solche umfassend natürlich-partizipatorische Wahrnehmung würde hoffentlich die ästhetische Empfänglichkeit für die Anmut und Umfassendheit der *„patterns that connect“* (verbindende Muster) erhöhen. Diese Empfänglichkeit ist, wie sowohl Bateson als auch Abram beklagen, in unserer zivilisatorischen Entwicklung, die zweckgerichtetes Bewusstsein und alphabetische Kulturen betonen, verlorengegangen.

Auf Grundlage der theoretischen Erkenntnisse, die erarbeitet wurden, erörtert diese Arbeit weiterhin wie die Erforschung der Themen Ökologie und Nachhaltigkeit sich in den vier zurückliegenden Jahrzehnten in der Kunstwelt entwickelt hat, zunächst durch den Aufstieg einer Bewegung der „ecological art“ und in jüngerer Zeit durch ein erweitertes Interesse an diesen Themen quer durch die Kunstwelt.

Öko-Kunst

Das Zusammenspiel von Kunst und Ökologie rückte mit dem Aufkommen der „environmental/ecological art“ seit den späten 1960er Jahren auch in Abgrenzung zur „land art“ in den Fokus einer spezifischen künstlerischen Bewegung Ich diskutiere die theoretische und künstlerische Relevanz dieses neuen Zusammenspiels anhand einer ausgesuchten Anzahl von prominenten Künstlern, die seit den 1960er und 1970er Jahren besonders aktiv waren, und die künstlerischen Arbeitsfelder entwickelt haben, die diesen Bereich der environmental/ecological art eröffnet haben.

Hans Haacke entwickelte seit den 1960er Jahren, beeinflusst durch die „systemische Ästhetik“ (Systems Esthetics) von Jack Burnham, eine zunehmend komplexe Arbeit zu ökologischen und sozialen Systeme und die Beziehungen zwischen diesen.

Alan Sonfist begann in den 1960er Jahren über natürliche Prozesse zu arbeiten, wobei er darauf abzielte die natürliche Welt in unser Konzept der Gemeinschaft einzubeziehen.

Joseph Beuys erforschte die Verbindung zwischen Menschen und Natur und betonte die politisch und sozial transformative Dimension von Ökologie und Kunst.

Helen Mayer Harrison und Newton Harrison, deren Arbeit an der Schnittstelle von Kunst und Naturwissenschaften sie in den 1970er Jahren zu einem Verständnis der Komplexität von ökologischen und kulturellen Systemen führte, entwickelten eine Methode des konversationellen Abdriftens (conversational drift), die Emergenz in sozialen Interaktionen erlaubt. Das Paar wendet sich in seiner künstlerischen Arbeit noch heute vor allem der Koevolution von Biodiversität und kultureller Diversität in spezifischen Ökosystemen zu, wobei sie politische Grundsatzentscheidungen und Planungen als Teil ihrer Arbeiten begleiten.

Mierle Laderman Ukeles fordert seit ihrem „maintenance art"-Manifest von 1969 in ihren Interventionen die menschliche „Erhaltung" (maintenance) der dynamischen Balance in Ökosystemen, die im Gegensatz zum „Todesinstinkt" der Entwicklung und des Fortschritts steht. Sie hat Rituale entwickelt, die die Wichtigkeit der „Erhaltungsarbeit", beispielsweise durch Müllwerker, feiern.

Ich setze mich weiterhin mit der Arbeit anderer Künstler auseinander, die weitere Dimensionen der environmental/ecological art seit den 1970ern entwickelt haben: Patricia Johanson (eine Pionierin der Kreuzungsbereiche von environmental art und Stadtplanung), Herman Prigann (der den Begriff der „metamorphic artworks" prägte und viele Projekte zur Wiederurbarmachung realisierte), Aviva Rahmani (die ebenfalls eine Methode der ökologischen Wiederherstellung entwickelt hat), Lynne Hull (die „Trans-Species-Art" einführte), David Haley (der ein Verständnis für „Ökopoïesis" entwickelt hat), Shelley Sacks (die Joseph Beuys' Konzept und Praxis der sozialen Skulptur weiterentwickelte), James Turrell (zum Beispiel sein Interesse an „Earthworks") und Andy Goldsworthy (als Beispiel für ein Interesse an der „art in nature").

Andere Kulturakteure haben die ökologische Kunstbewegung begleitet, wie die Kunsthistorikerin Suzi Gablik (1991, 2004), die den Aufstieg einer verbindenden und verantwortungsbewussten Kunst feiert. Ich diskutiere die Einsichten, die Gablik beigetragen hat, ausführlich (wie etwa die Begriffe des „rekonstruktiven Postmodernismus", oder des „verbindenden Selbst"), wie auch die Grenzen ihres Diskurses (etwa ihre holistische, überharmonisierende Sicht). Im Fall der Monongahela Konferenz von 2004 über die „Post-industrielle Gemeinschaftsentwicklung", rufe ich auch einige der Themen und Problemstellungen wach, die in der ökologischen Kunstbewegung in den zurückliegenden Jahren diskutiert wurden, wie die Rolle der ökologischen Kunst in der interdisziplinären Zusammenarbeit mit lokalen Gemeinschaften, die verschiedenen Ebenen und Tiefen des Dialogs mit lokalen Gemeinschaften und die unterschiedlichen Methoden und Werkzeuge der eco-art. Diese und ähnliche Themen werden auch von Tim Collins und Reiko Goto, den Harrisons, Suzi Gablik, Ann Rosenthal, der Gruppe „Temporary Services", Grant Kester, die Londoner Gruppe „Platform", A. Laurie Palmer, Stacy Levy, Stephanie Flom, Tom und Connie Merriman, Jackie Brookner und Erica Fielder besprochen.

Das Interesse der ökologischen Kunst, und spezifischer der „Eco-art interventions", wurde in der Ausstellung *Ecovention* von 2002 besonders herausgestellt. Der Begriff der „ecovention", der auf „ein künstler-iniziiertes Projekt, das eine erfinderische Strategie einsetzt, um eine örtliche Ökologie physisch zu transformieren", hindeutet, wird in den Werken von „Ocean Earth", dem „Center for Land Use Interpretation", George Steinmann, Basia Irland, Kathryn Miller, Tim Collins, Reiko Goto, Suzanne Leibovitz Steinman, Patricia Johanson, Betty Beaumont, Lynne Hull, Brandon Ballengée, Tera Galanti, Alan Sonfist, Mierle Laderman Ukeles, Buster Simpson, Viet Ngo, Laurie Lundquist, Superflex, Mel Chin, Betsy Damon, Jackie Brookner, Aviva Rahmani, Shai Zakai, Agnes Denes, Georg Dietzler und der Gruppe AMD&ART herausgestellt.

Die historische Emergenz und Entwicklung einer "environmental and ecological art" stellt ein Beispiel dafür da, wie Kunstwelten (art worlds) eine konstruktive Rolle in der Konstruktion von Kultur(en) der Nachhaltigkeit spielen können. Mit der Vielfalt der Herangehensweisen (von sozialer Skulptur bis zur Wiederurbarmachung von Land) und mit den gemeinsamen roten Fäden und Prinzipien (wie in der Monongahela Konferenz diskutiert), scheint diese Bewegung auf dem Weg zu einer entwickelten, ausgearbeiteten Form der Ästhetik des „verbindenden Musters" zu sein. Allerdings scheint hinter einigen Diskursen der „eco-art" eine simplistische Form des holistischen Diskurses unter Vernachlässigung der oben erörterten „Komplexität" zu stehen.

Kunst und Nachhaltigkeit im vergangenen Jahrzehnt

Das Zusammenspiel von Ökologie, Nachhaltigkeit und Kunst ist im vergangenen Jahrzehnt auch in den Fokus der Aufmerksamkeit einer Vielfalt von anderen Akteuren der zeitgenössischen Kunstwelt geraten. Kunstausstellungen, die sich mit den Themen Ökologie und Nachhaltigkeit und in letzter Zeit häufig mit dem Klimawandel befassen, werden auf der ganzen Welt organisiert. Einige Kuratoren und Autoren haben im vergangenen Jahrzehnt dazu besondere Diskurse konstruiert, insbesondere Hildegard Kurt (mit ihren Definitionen der „Ästhetik der Nachhaltigkeit"), Maja und Reuben Fowkes (mit ihrem Begriff der „nachhaltigen Kunst") und erst kürzlich der *Third Text-* Herausgeber Rasheed Araeen (mit dem Begriff der "ecoaesthetics").

Kuratorin und Autorin Linda Weintraub (2006) hat erörtert, wie Künstler, deren Arbeit normalerweise nicht mit „ecological oder environmental art" assoziiert werden, interessante Perspektiven in diese Themenstellungen eingebringen, wie zum Beispiel das Begehren in der Kultur des Hyper-Konsums (Reverend Billy Talen), die Kultur-Natur Beziehung (Dave Burns und Matias Viegener), Biodiversität (womit sie auf die Arbeit von Eduardo Kac – die ich allerdings als sehr fragwürdig behandle, hinweist), und die schwierigen Beziehungen der zeitgenössischen westlichen Gesellschaften zu den Prozessen Tod, Verwesung und Erde (Catherine Chalmers, Damien Hirst und Joe Scanlan). Weintraubs Analyse legt weitgehend den Schwerpunkt auf „art space" Künstler, also Künstler, die, im Gegensatz zu vielen ökologischen Künstlern, hauptsächlich Arbeiten in Kunstgalerien produzieren. Ihr Diskurs zeigt, dass diese Künstler tatsächlich begrenzte Relevanz aufweisen, da sie (zu einem geringen Ausmaße) zu einem Perspektivenwechsel der internationalen kulturellen Eliten beitragen, die deshalb dieser zeitgenössischen Kunst etwas mehr Aufmerksamkeit zuwenden.

Ich nehme weiter eine detaillierte kritische Erörterung der kuratorischen Diskurse zu Ausstellungen vor, die im vergangenen Jahrzehnt spezifisch Ökologie und Nachhaltigkeit angesprochen haben:

Beyond Green: Towards a Sustainable Art (SMART Museum, Chicago, 2005) kuratiert von Stéphanie Smith, und mit Arbeiten von Jennifer Allora und Guillermo Calzadilla, Wochenklausur, Nils Norman, Frances Whitehead, Marjetica Potrc, Learning Group, Kevin Kaempf, Dan Peterman, Temporary Services und Free Soil u.a..

Auszeit. Kunst und Nachhaltigkeit (Timeout. Art and Sustainability - Kunstmuseum Liechtenstein, 2007), kuratiert von Friedemann Malsch und Christiane Meyer-Stoll, eine Ausstellung die das Verständnis von Nachhaltigkeit auf 'verlangsamen' reduziert und wenig mehr als eine abgehobene „white-box discussion" über Zeit und Natur anbietet, und auf Werke von Giovanni Anselmo, Piero Gilardi, Jochen Gerz and On Kawara u.a. hinweist.

Greenwashing (Fondazione Sandretto Re Rebaugendo, Turin, 2008), kuratiert von Ilaria Bonacossa und Latitudes (Max Andrews und Mariana Canepa Luna), ist eine Ausstellung mit dem Ziel kritische künstlerische Perspektiven auf Umweltproblematiken herauszustellen. Diese Ausstellung beinhalteteinen eigentümlichen kuratorischen "post-environmental" Diskurs, dem ich deshalb kritische Aufmerksamkeit widme, weil er vielen Fehler aufweist. Die Ausstellung beinhaltet Arbeiten von Jennifer Allora and Guillermo Alzadilla, Lara Almarcegui, Cyprien Gaillard, Santiago Serra, Wilfredo Prieto, Maria Thereza Alves, Tomas Saraceno, Nikola Uzunovski und Ibon Aranberri, aber auch von Amy Balkin, The Bruce High Quality Foundation, A Constructed World, Minerva Cuevas, Tue Greenfort und Sergio Vega. Trotz der Fehler des kuratorischen Diskurses tragen viele der ausgestellten Künstler wertvolle und relativ subtile/komplexe kritische Einsichten, besonders zur Problemstellung des „Greenwashing" bei.

Schließlich beende ich meine Erforschung der Diskurse der zeitgenössischen Kunst im letzten Jahrzehnt, in dem ich einen kurzen Text des französischen Philosophen und Psychoanalytikers Félix Guattari (*Les Trois Ecologies*, 1989) erörtere, denn er wird oft von Kuratoren (wie etwa Stéphanie Smith 2005 und Max Andrews 2006) zitiert und von einigen eco-artists (wie David Haley 2008) kritisiert. Guattari's *Die Drei Ökologien* bietet eine einsichtige und dichtgewobene Einführung in das komplexe Denken von Ökologie. Guattaris Arbeit wurde nicht von all seinen Lesern aus der Welt der zeitgenössischen Kunst geschätzt, wahrscheinlich wegen der Dichte seines Essays und wegen einer mangelnden Artikulierung des Verstehens von Komplexität.

Dieser Überblick, den ich quer über Themen der zeitgenössische Kunst- vorgenommen habe, zeigt das vielfache und vielseitige Aufkommen von „eco-zentrischen" Perspektiven und Einsichten. Das Schlüsselwort „Nachhaltigkeit" wird Stück für Stück in das Vokabular von mehr und mehr Kuratoren übernommen und generiert Diskurse und Praktiken, die die Verbindung von sozialer und ökologischer Kritik und zeitgenössischer Kunst stärken. Die wachsende Debatte ist lebendig und vielfältig, auch wenn einige ihrer Ausarbeitungen fragwürdige Diskurse hervorbringen (wie „post-environmentalism" oder unreflektierte pro-technologische Praktiken und Diskurse).

Über soziale und politische Transformationsprozesse: „Double Entrepreneurship in Conventions"

Die Diskussion der ecological art weist auf die vielen Gelegenheiten (oder "ecoventions") hin, bei denen die Künstler in der ökologischen und sozialen Realität direkt zusammenkommen, sich mit lokalen Gemeinschaften auseinandersetzen, und darauf hinarbeiten Transformationsprozesse in Gang zu bringen. Solche Ambitionen verlangen nach einer umsichtigen Aufmerksamkeit für die Potentiale und Herausforderungen, die Künstlern begegnen, die auf soziale Veränderungen, nicht nur in ihren „Kunstwelten" (art worlds) sondern direkt in spezifischen sozialen Kontexten, abzielen. Deshalb analysiere ich im folgenden Teil der Arbeit künstlerische Interventionen in der Gesellschaft als „double entrepreneurship in conventions" und ergänze meine bisherigen Erörterungen zu diesem Thema (Kagan 2004, 2008a) um Batesons Konzept des „deutero-learning". Die Möglichkeit zur „double entrepreneurship" hängt dabei von der Komplexität sozialer Systeme, von den Qualitäten sozialer Akteure als „künstlerische Unternehmer" und vom Kontext der Kunstwelten und spezifischer von der politischen Aufstellung der relevanten Kunstwelt („polity conventions" in „art worlds") ab.

Die Analyse der „polity conventions" in „art worlds"gibt mir zum Schluß die Möglichkeit politische Aspekte der Kulturen der Nachhaltigkeit zu diskutieren. Vorausgesetzt wird dabei, dass die politischen Rahmenbedingungen der sozialen Interaktionen „nahezu-polyarchisch" (nach Dahl 1971) im Gegensatz zu „nahezu-hegemonistisch" sind, um Emergenzprozesse, Koevolution und „Hochkomplexität" zu ermöglichen statt zu unterdrücken. Nahezu-Polyarchie ist allerdings keine Selbstverständlichkeit, weder in der Kunst noch in Umweltbewegungen. Ich weise zum Beispiel auf die Notwendigkeit hin, eine Degeneration von ecological art in einen grünen Faschismus zu verhindern. Ein Extremfall einer solchen Verbindungt von Kunst und Politik war der moderne Tanz unter dem Nazismus, wo sich einige organistisch-holistische Ideale aus Mangel an Komplexität bequem mit dem wahrscheinlich monströsesten totalitären System des vergangenen Jahrhunderts eingerichtet hatten. Mit wenigen Ausnahmen hatten die modernen deutschen Tänzer willig die Mehrdeutigkeit in ihrer Arbeit eliminiert und zunehmend hegemonistische Werte durch ihre Arbeit gefördert, und zwar nicht nur unter dem Naziregime, sondern bereits in den Jahren davor. Dieser Fall (nach Guilbert 2000) zeigt, dass eine Mythologie der holistischen Vereinigung mit der Natur potentiell mit in hohem Maße hegemonistischen polity conventions und mit einer totalitären Ideologie zusammenpasst. Gegen diese totalitäre Strömung ist deshalb eine Empfänglichkeit für Komplexität elementar wichtig.

Meine Analyse suggeriert, dass mit „Unternehmertum" (Entrepreneurship) bei „interkonventionellen" Interaktionen und „deutoro-learning" sich sogar die eingefahrensten Ideen und sozialen Konventionen ändern können. Ein erfolgreiches „Doppelunternehmertum" in Konventionen setzt voraus, dass die in Frage stehende Kunstwelt für ein solches Unternehmen offen genug ist, was in einer Kunstwelt mit einer polyarchischeren politischen Aufstellung wahrscheinlicher ist als in einer hegemonistischen politischen Konstellation. In Kunstwelten können nur relativ polyarchische polity conventions den notwendigen Kontext bilden für ein künstlerisches Unternehmertum, das auf Kulturen der Nachhaltigkeit gerichtet ist. Auf diese sehr konkrete Weise ist der Künstler also eminent politisch. Die politische Dimension der Ästhetik und Kultur der Nachhaltigkeit sollte Diskurse und Praktiken beinhalten, die Pluralismus und Streitfragen, Kompromisse und regulierten Wettbewerb wertschätzen und komplexe Uni-pluralität sowie eine „Queer-Ecology" weg vom harmonisierenden Bild der Natur fördern.

Bibliography

Hans Abbing. *Why are artists poor? The exceptional economy of the arts.* Amsterdam: Amsterdam University Press, 2002.

—— *From High Art to New Art.* Amsterdam: Vossiupers UvA, 2006.

—— *Van hoge naar nieuwe kunst.* Groningen: ihistorische uitgerevij, 2009.

Hans Abbing and Sacha Kagan. "Social Price and Social Monopolization : Combining Economic Structure with Social Analysis." Paper presented at ACEI (Association for Cultural Economics International) conference, Vienna, July 2006.

David Abram. *The Spell of the Sensuous.* New York: Random House, 1996.

Tim Adams. "Natural talent." *The Observer*, 11-03-2007. Available online at: http://observer.guardian.co.uk/review/story/0,,2031027,00.html [last accessed: 07/02/2010].

Theodor Adorno. "On the Logic of the Social Sciences," In Theodor Adorno, Hans Albert, Ralf Dahrendorf, Jürgen Habermas, Harald Pilot and Karl Popper. *The Positivist Dispute in German Sociology.* Translated by Glyn Adey and David Frisby. London: Heinemann, 1976, pp. 105-122.

Victoria D. Alexander. *Sociology of the Arts. Exploring fine and popular forms.* Malden: Blackwell, 2003.

Günter Altner. "Ethical and Methodical Aspects of the Concept of Sustainability for Academia's Contribution to Sustainability Research – Seven Theses." In Maik Adomssent, Jasmin Godemann, Alexander Leicht and Anne Busch (eds.), *Higher Education for Sustainability: New Challenges from a Global Perspective.* Waldkirchen: VAS, 2006, pp. 67-78.

Virginia Anderson and Lauren Johnson. *Systems thinking basics : from concepts to causal loops.* Waltham: Pegasus Communications, 1997.

Max Andrews, Ed. *LAND, ART: A Cultural Ecology Handbook.* London: Royal Society for the encouragement of Arts, Manufactures and Commerce (RSA), 2006.

Andras Angyal. *Foundations for a science of personality.* Cambridge: Harvard University Press, 1941.

Rasheed Araeen. "Ecoaesthetics: A Manifesto for the Twenty-First Century." *Third Text: critical perspectives on contemporary art & culture*, 23, 5, September 2009, pp. 679-684.
Chris Argyris and Donald Schön. *Organizational learning: A theory of action perspective*. Reading: Addison Wesley, 1978.
Raymond Aron. *Démocratie et Totalitarisme*. Paris: Gallimard, 1965.
William Ross Ashby. "General systems as a new discipline." *General Systems Yearbook*, 3, 1958, pp. 3-6.
—— "Principles of the self-organizing system." In Heinz von Foerster and G.W. Zopf, Eds., *Principles of Self-Organization*. New York: Pergamon Press, 1962.
Henri Atlan. "Rôle positif du bruit en théorie de l'information appliquée à une définition de l'organisation biologique." *Annales de physiologie biologique et médicale*, 1, 1970, pp. 15-33.
—— "Du bruit comme principe d'auto-organisation." *Communications*, 18, 1972a, pp. 21-35.
—— *L'Organisation biologique et la Théorie de l'information*. Paris: Hermann, 1972b.
—— "On a formal definition of organization." *Journal of Theoretical Biology*, 45, 1974, pp. 1-9.
Francis Bacon. *Novum Organum. Or true directions concerning the interpretation of nature*. 1620. 1863 English translation available at: http://www.constitution.org/bacon/nov_org.htm [last accessed: 07.07.2010].
—— *New Atlantis: A Work Unfinished.* [Publisher unknown] 1627. Available online at Project Gutenberg: http://www.gutenberg.org/etext/2434 [last accessed: 07.07.2010].
Guy Bajoit. *Pour une Sociologie Relationnelle*. Paris: Presses universitaires de France, 1992.
Sally Banes. *Terpsichore in Sneakers : Post-Modern Dance*. Middletown: Wesleyan University Press, 1987.
—— *Democracy's Body: Judson Dance Theater, 1962-1964*. Durham: Duke University Press, 1993.
Dominique Baqué. *Pour un nouvel art politique. De l'art contemporain au documentaire*. Paris: Flammarion, 2004.
Milford Bateman. *Why Doesn't Microfinance Work? The Destructive Rise of Local Neoliberalism*. London: Zed Books, 2010.
Gregory Bateson. "Cybernetic Explanation." *The American Behavioral Scientist*, April 1967, pp. 29-32.
—— *Steps to an Ecology of Mind: Collected Essays in Anthropology, Psychiatry, Evolution, and Epistemology*. New York: Paladin, 1973.
—— *Mind and nature: A necessary unity*. New York: Dutton, 1979.
Pierre Batifoulier, Ed. *Théorie des conventions*. Paris: Economica, 2001.
Jean Baudrillard. *La société de consommation*. Paris: Denoël, 1970.
Andrea Baumeister. "Habermas: Discourse and Cultural Diversity." *Political Studies*, 51, 2003, pp. 740-58.

Ulrich Beck. *Gegengifte. Die organisierte Unverantwortlichkeit.* Frankfurt am Main: Suhrkamp, 1988.

Carol Becker, Ed. *The Subversive imagination : artists, society, and social responsibility*. London: Routledge, 1994.

Howard S. Becker. "Art As Collective Action." *American Sociological Review*, 39, 6, Dec., 1974, pp. 767-776.

—— *Art Worlds*. Berkeley: University of California Press, 1982.

Hubert Beckmann and Jesus Garzon Heydt. "Transhumance as a Tool of Species Conservation in Times of Climate Change." In David Knaute and Sacha Kagan, Eds, *Sustainability in Karamoja? Rethinking the terms of global sustainability in a crisis region of Africa*. Köln: Rüdiger Köppe Verlag, 2009, pp. 235-262.

Janine Benyus. *Biomimicry*. New York: Morrow, 1997.

Peter L. Berger and Thomas Luckmann. *The Social Construction of Reality: A Treatise in the Sociology of Knowledge*. Garden City: Anchor Books, 1966.

Henri Bergson. *Essai sur les données immédiates de la conscience*. Paris: Alcan, 1889.

—— *L'Évolution créatrice*. 1907. Available online at: http://classiques.uqac.ca/classiques/bergson_henri/evolution_creatrice/evolution_creatrice.html [last accessed: 07.07.2010].

Arnold Berleant. *Art and Engagement*. Philadelphia: Temple University Press, 1991.

—— *The aesthetics of environment*. Philadelphia: Temple University Press, 1992.

—— "The Aesthetics of Art and Nature." In Salim Kemal and Ivan Gaskell, eds., *Landscape, Natural Beauty and the Arts*. Cambridge: Cambridge University Press, 1993, pp. 228-243.

Ludwig von Bertalanffy. "The Theory of Open Systems." *General System yearbook*, 1956.

—— *General systems theory. Essays on its foundation and development*. New York: Braziller, 1968.

Nicole Woolsey Biggart and Thomas D. Beamish. "The economic sociology of conventions: Habit, custom, practice and routine in market order." *Annual Review of Sociology*, 29, 2003.

Marga Bijvoet. *Art as Inquiry: Toward New Collaborations between Art, Science and Technology*. Rotterdam: University of Rotterdam, 1994.

—— *Greening of Art*. Online publication, Stichting MAI, nd. Available at: http://www.stichting-mai.de/hwg/amb/goa/greening_of_art_06.htm#07 [last accessed: 10.07.2010].

Claire Bishop. "Antagonism and Relational Aesthetics." *October,* 110, 1, 2004, pp. 51-80.

Beatrice von Bismarck, Diethelm Stoller and Ulf Wuggenig, Eds. *Games Fights Collaborations: Das Spiel von Grenze und Überschreitung.*

Kunst und Cultural Studies in den 90er Jahren. Ostfildern-Ruit bei Stuttgart: Cantz, 1996.

Maurice Blondel. *L'Action. Le problème des causes secondes et le pur agir*, vol. 1. Paris: Presses Universitaires de France, 1949 (1893).

David Bohm. *On Dialogue*. London: Routledge, 2006.

Luc Boltanski and Laurent Thévenot. *De la justification: les économies de la grandeur*. Paris: Gallimard, 1991.

Luc Boltanski and Eve Chiapello. *The new spirit of capitalism.* London: Verso, 2005 (original French version: 1999).

Ilaria Bonacossa and Latitudes. *Greenwashing. Environment: Perils, Promises and Perplexities.* Turin : The Bookmakers, 2008.

Linda Booth Sweeney and Dennis Meadows. *The Systems Thinking Playbook: Exercises to Stretch and Build Learning and Systems Thinking Capabilities*. Turning Point Foundation, 1995.

Pierre Bourdieu. *La Distinction. Critique sociale du jugement*. Paris: Minuit, 1979.

—— *Les règles de l'art. Genèse et structure du champ littéraire*. Paris: Seuil, 1992.

—— *The Field of Cultural Production*. Cambridge: Polity Press, 1993.

Pierre Bourdieu and Loïc Wacquant. *An Invitation to Reflexive Sociology*. Chicago: The University of Chicago Press, 1992.

Dominique Bourg. *Nature et technique: Essai sur l'idée de progrès*. Paris: Hatier, 1997.

Nicolas Bourriaud. *Esthétique relationnelle*. Dijon: les presses du réel,1998.

Sam Bower. "A Profusion of Terms." *greenmuseum.org*. 2009. Accessible online at: http://www.greenmuseum.org/generic_content.php?ct_id=306 [last accessed: 24.01.2010].

Alan Bowness. *The conditions of success. How the modern artist rises to fame*. London: Thames and Hudson, 1989.

Hans J. van de Braak. *Ontsnappingskunst: evolutie van de creatieve geest.* Amsterdam: Amsterdam University Press, 2002.

Emily Brady. *Aesthetics of the Natural Environment.* Edinburgh: Edinburgh University Press, 2003.

Davide Brocchi. "The Cultural Dimension of Sustainability." In Sacha Kagan and Volker Kirchberg, Eds., *Sustainability: a new frontier for the arts and cultures*. Waldkirchen: VAS – Verlag für akademische Schriften, 2008, pp. 26-58.

Jackie Brookner. Position Paper. The Monongahela Conference on Post-Industrial Community Development, Carnegie Mellon University, Pittsburgh, 2004. Available online at: http://moncon.greenmuseum.org/papers/brookner1.html [last accessed: 10.07.2010].

Lester Brown. *Building a Sustainable Society*. Washington: Worldwatch Institute, 1981.

Larissa Buchholz and Ulf Wuggenig. "Nomadische Treibhäuser: Dan Petermans Projekt ‚Greenhouse'." In Hildegard Kurt and Bernd Wagner,

Eds., *Kultur – Kunst – Nachhaltigkeit: Die Bedeutung von Kultur für das Leitbild Nachhaltige Entwicklung*. Bonn: Klartext Verlag, 2002, pp. 155-174.

David Buckland et al. *Burning ice: art & climate change*. London: Cape Farewell, 2006.

Malcolm Budd. *The Aesthetic Appreciation of Nature*. New York: Oxford University Press, 2002.

Fritjof Capra. *The Web of Life: A new synthesis of mind and matter*. London: HarperCollins, 1996.

—— *The Hidden Connections: A Science for Sustainable Living*. New York: Random House, 2002.

Manuel Castells. *The Information Age, vol. 2: The Power of Identity*. London: Blackwell, 1997.

—— *The Information Age, vol. 3: End of Millennium*. London: Blackwell, 1998.

Cornelius Castoriadis. *L'institution imaginaire de la société*. Paris: Seui, 1975.

Noel G. Charlton. *Understanding Gregory Bateson: mind, beauty and the sacred earth*. Albany: The State University of New York Press, 2008.

Noam Chomsky. *Language and Mind*. Harcourt, Brace & World, 1968.

Jean Clam. *Droit et société chez Niklas Luhmann: fondés en contingence*. Paris: Presses Universitaires de France, 1997.

Tim Collins. "Thoughts on Strategic Action." Position Paper. The Monongahela Conference on Post-Industrial Community Development, Carnegie Mellon University, Pittsburgh, 2004a. Available online at: http://moncon.greenmuseum.org/papers/collins1.html [last accessed: 10.07.2010].

—— "Reconsidering the Monongahela Conference." The Monongahela Conference on Post-Industrial Community Development, Carnegie Mellon University, Pittsburgh, 2004b. Available online at: http://moncon.greenmuseum.org/recap.htm [last accessed: 10.07.2010].

Paul Colomy. "Neofunctionalism and Neoinstitutionalism: Human Agency and Interest in Institutional Change." *Sociological Forum* 13, 2, 1998, pp. 265-300.

Thomas Comiotto, Eric Kluitenberg, David Garcia and Menno Grootveld, Eds. *Next 5 Minutes 4 Reader*. Amsterdam: Next 5 Minutes International Festival for Tactical Media, 2003.

Ian Cook et al. "Social sculpture and connective aesthetics: Shelley Sacks' 'Exchange Values'." *Ecumene*, 3, 2000, pp. 338-344.

Joel M. Cooper. *Cognitive Dissonance: 50 Years of a Classic Theory*. London: Sage Publications Ltd, 2007.

Leda Cosmides and John Tooby. "Evolutionary Psychology: A Primer." Center for Evolutionary Psychology, University of California, Santa Barbara, 1997. Online document available at: http://www.psych.ucsb.edu/research/cep/primer.html [last accessed 09/08/2010].

Tyler Cowen. *Creative Destruction: How Globalization is Changing the World's Cultures*. Princeton: Princeton University Press, 2002.

Michel Crozier and Erhard Friedberg. *L'acteur et le système: Les contraintes de l'action collective*. Paris: Seuil, 1977.

Robert Dahl. *Polyarchy*. New Haven: Yale University Press, 1971.

Antonio Damasio. *Descartes' Error, Emotion, reason and the human brain*. New York: Putnam Books, 1994.

—— *Looking for Spinoza: Joy, Sorrow, and the Feeling Brain*. Orlando: Houghton Mifflin Harcourt, 2003.

James A. Davis. *The Logic of Causal Order*. Beverly Hills: Sage, 1985.

Richard Dawkins. *The Selfish Gene*. Oxford: Oxford University Press, 1976.

Guy Debord. *La Société du Spectacle*. Paris: Buchet-Chastel, 1967.

Jules Delvaille. *Essai sur l'histoire de l'idée de progrès jusqu'à la fin du XVIIIe siècle*. Paris: Félix Alcan, 1910.

Richard Demarco. "Beyond Sensationalism: Social sculpture - an expanded concept of art." nd. Available online at www.exchange-values.org [last accessed: 02.03.2010].

Jacques Derrida. *Of Grammatology*. Baltimore: Johns Hopkins University Press, 1976.

René Descartes. *Discours de la méthode: Pour bien conduire sa raison, et chercher la vérité dans les sciences*. Leyde: Ian Maire, 1637. Available online at Project Gutenberg: http://www.gutenberg.org/etext/13846 [last accessed: 07.07.2010].

—— *Les Principes de la Philosophie*. Paris: Paléo, 2000 (original edition: 1644).

—— *Les Méditations Métaphysiques de René Descartes Touchant la Première Philosophie*. 1647 (1641: original in Latin). Available online at: http://www.wright.edu/cola/descartes/medf.html [last accessed: 07.07.2010].

John Dewey (with James Tufts). "Ethics" (1932). In Jo-Ann Boydston, Ed., *The Collected Works of John Dewey, 1882-1953*. Carbondale: Southern Illinois University Press, 1991.

John Dewey. *Art as Experience*. New York: Perigee, 2005 (1934).

Jared Diamond. *Collapse: How societies choose to fail or succeed*. New York: Viking Adult, 2004.

Hans Dieleman. *Art & Sustainability: a research, educational and art-production project*. Rotterdam: Erasmus University Rotterdam, 2001.

—— "Art & Sustainability." Unpublished lecture at the Erasmus Universiteit Rotterdam, 2002.

—— *Nieuwe Kunst in Oude Noorden: Een onderzoek naar de werking van kunst als social-artistieke interventie in de openbare ruimte in Rotterdam-Noord*. Rotterdam: Erasmus University Rotterdam: Erasmus studiecentrum voor Milieukunde, 2006.

—— "Art and Science." *Cultura21 Webmagazin*, 01.08.2007. Available online at: http://magazin.cultura21.de/piazza/english/art-and-science.html [last accessed: 01.07.2010].

—— "Sustainability, Art and Reflexivity: why artists and designers may become key change agents in sustainability." In Sacha Kagan and Volker Kirchberg, Eds., *Sustainability: a new frontier for the arts and cultures.* Waldkirchen: VAS – Verlag für akademische Schriften, 2008, pp. 108-146.

Paul DiMaggio. "Classification in Art." *American Sociological Review* 52, 1987, pp. 440-455.

—— "Cultural boundaries and structural change: The extension of the high-culture model to theatre, opera, and the dance, 1900-1940." In Michèle Lamont and Marcel Fournier, Eds., *Cultivating differences: Symbolic boundaries and the making of inequality.* Chicago: The University of Chicago Press, 1992. 1992.

David Diringer. *The Alphabet.* New York: Philosophical Library, 1948.

Maarten Doorman. *De Romantische Orde.* Amsterdam: Uitgeverij Bert Bakker, 2004.

Sarah Douglas. "In Their Words: James Turrell and Andy Goldsworthy." *Artinfo*, October 24th 2005. Available online at: http://www.artinfo.com/news/story/1365/in-their-words-james-turrell-and-andy-goldsworthy/ [last accessed: 06/02/2010].

Elizabeth Drew. *T.S. Eliot: The Design of his Poetry.* New York: Charles Scribner's Sons. 1949.

Daniel Durand. *La Systémique.* 10th ed. Paris: Presses Universitaires de France, 2006.

Nancy Duxbury, Eileen Gillette and Kaija Pepper. "Exploring the Cultural Dimensions of Sustainability". Newsletter, Special Edition 4, 2007. Vancouver: Creative City Network of Canada. Retrieved January 25, 2010, from http://www.creativecity.ca/se-newsletters/special-edition-4/index.html [last accessed: 07.07.2010].

David Edwards. *Artscience: Creativity in the Post-Google Generation.* Cambridge: Harvard University Press, 2008.

Riane Eisler. *The Chalice and the Blade: Our History, Our Future.* Sand Francisco: Harper & Row, 1987.

Mircea Eliade, *The Myth of the Eternal Return.* New York: Harper & Row, 1959.

Thomas Stearns Eliot. *Four Quartets: Little Gidding.* London: Faber and Faber, 1959.

Jacques Ellul. *L'illusion politique.* Paris: Robert Laffont, 1965.

—— *Métamorphose du bourgeois.* Paris: Calmann-Lévy, 1967.

—— *Les nouveaux possédés.* Paris: Fayard, 1973.

—— *Le système technicien.* Paris: Calmann-Lévy, 1977.

—— *L'empire du non-sens - L'art et la société technicienne.* Paris: Presses Universitaires de France, 1980.

Jale Erzen. "Ecology, Art, Ecological Aesthetics." In Heike Strelow, Herman Prigann and Vera David, Eds., *Ecological Aesthetics: Art in Environmental Design: Theory and Practice*. Berlin: Birkhäuser Verlag für Architektur, 2004, pp. 22-49.

Antoine Faivre, *Access to Western Esotericism,* State University of New York Press, 1994.

Werner Faulstich, Ed. *Die Kultur der 30er und 40er Jahre*. München: Wilhelm Fink Verlag, 2009.

Olivier Favereau and Emmanuel Lazega, Eds. *Conventions and Structures in Economic Organization: Markets, Networks and Hierarchies*. Cheltenham: Edward Elgar, 2002.

Loïc Fel. *L'esthétique verte: de la représentation à la présentation de la nature*. Seyssel: Champ Vallon, 2009.

Estelle Ferrarese. *Niklas Luhmann, une introduction.* Paris: Pocket / La Découverte, 2007.

Paul Feyerabend. *Against Method*. Brooklyn: Verso, 1993.

Richard Field. "Dewey, John." *The Internet Encyclopedia of Philosophy*, 2005. Available online at: http://www.iep.utm.edu/dewey/ [last accessed September 11, 2009].

Erica Fielder. Position Paper. The Monongahela Conference on Post-Industrial Community Development, Carnegie Mellon University, Pittsburgh, 2004. Available online at: http://moncon.greenmuseum.org/papers/fielder1.html [last accessed: 10.07.2010].

Stephanie Flom. "The Persephone Project." Position Paper. The Monongahela Conference on Post-Industrial Community Development, Carnegie Mellon University, Pittsburgh, 2004. Available online at: http://moncon.greenmuseum.org/papers/persephone1.html [last accessed: 10.07.2010].

Heinz von Foerster. "Communication amongst automata." *American Journal of Psychiatry*, 118, 1962, pp. 865-871.

Roger Fouts. *Next of Kin*. New York: William Morrow, 1997.

Maja Fowkes and Reuben Fowkes. "Human/Nature." Unpublished text. Budapest: Traffo Gallery, ca. 2002. Available online at: http://www.translocal.org/shows/humannature.htm [last accessed: 08.03.2010].

—— "Principles of Sustainability in Contemporary Art." *Praesens: central European contemporary art review*, 1, 2006, pp. 5-12.

—— "Planetary Forecast: The Roots of Sustainability in the Radical Art of the 1970s." *Third Text: critical perspectives on contemporary art & culture*, 23, 5, September 2009.

Suzi Gablik. *Has Modernism Failed?* London: Thames and Hudson, 1984.

—— *The Reenchantment of Art*. London: Thames and Hudson, 1991.

—— "Beyond the Disciplines: Art without Borders." Position Paper. The Monongahela Conference on Post-Industrial Community Development, Carnegie Mellon University, Pittsburgh, 2004.

Moacir Gadotti. *Education for Sustainability: A Contribution to the Decade of Education for Sustainable Development.* Sao Paulo : Editora e Livraria Instituto Paulo Freire, 2009.

Howard Gardner. *Frames of Mind: The theory of multiple intelligences.* New York: Basic Books, 1983

—— *Intelligence Reframed: Multiple Intelligences for the 21st Century.* New York: Basic Books, 2000.

Jean-Yves Gauthier, Jean-Claude Lefeuvre, Gaston Richard and Paul Trehen. *Eco-éthologie.* Paris: Masson, 1978.

Kirsten Gerdes, Ed. *Weather Report: Art and Climate Change.* Boulder: Boulder Museum of contemporary art, 2007.

Arie de Geus. "The Living Company." *Harvard Business Review*, March-April 1997.

Michael Gibbons et al., *The new production of knowledge: the dynamics of science and research in contemporary societies*, London: Sage, 1994.

Anthony Giddens. *The constitution of society: outline of the theory of structuration.* Cambridge: Polity Press, 1984.

—— *Conversations with Anthony Giddens: Making Sense of Modernity.* Stanford : Stanford University Press, 1998.

Jennifer Gidley. "The Evolution of Consciousness as a Planetary Imperative: An Integration of Integral Views." *Integral Review: A Transdisciplinary and Transcultural Journal for New Thought, Research and Praxis*, 5, 2007, pp. 5-226.

René Girard. *Mensonge romantique et vérité romanesque.* Paris: B. Grasset, 1961.

Andrew L. Gluck. *Damasio's Error and Descartes' Truth: An Inquiry into Consciousness, Metaphysics, and Epistemology.* Scranton: University of Scranton Press, 2007.

Reiko Goto. "My artistic subject matter is the life of nature. By contemplating nature, I seek to renew my own identity." Position Paper. The Monongahela Conference on Post-Industrial Community Development, Carnegie Mellon University, Pittsburgh, 2004. Available online at: http://moncon.greenmuseum.org/papers/collins1.html [last accessed: 10. 07.2010].

Pierre Yves Gomez. *Le gouvernement de l'entreprise, modèles économiques de l'entreprise et pratiques de gestion.* Paris: InterEditions, 1996.

Lisa Grabe. *Das „Projekt Nachhaltigkeit". Zu den Grenzen des Nachhaltigkeitskonzepts aus kultureller Perspektive.* Cultura21 eBooks Series on Culture and Sustainability, 1, 2010. Available online at: http://magazin. cultura21.de/piazza/texte [last accessed:10.12.2010].

Brian Greene. *The Elegant Universe: Superstrings, Hidden Dimensions, and the Quest for the Ultimate Theory.* London: Vintage Books, 2000.

Ulrich Grober. "Modewort mit tiefen Wurzeln - Kleine Begriffsgeschichte von ‚sustainability' und ‚Nachhaltigkeit'", in: *Jahrbuch Ökologie 2003.* München: Beck, 2002, pp. 167-175.

Morton M. Grodzins. *The Metropolitan Area as a Racial Problem.* Pittsburgh: University of Pittsburgh Press, 1958.

Félix Guattari. *Les Trois Ecologies.* Paris: Galilée, 1989.

Jorge Guerra Gonzalez. "Interculturality as Background for Sustainability." In David Knaute and Sacha Kagan, Eds., *Sustainability in Karamoja? Rethinking the terms of global sustainability in a crisis region of Africa.* Köln : Rüdiger Köppe Verlag, 2009, pp. 263-270.

Muriel Guigou. *La Nouvelle Danse Française.* Paris: L'Harmattan, 2004.

Laure Guilbert. *Danser avec le troisième Reich: Les danseurs modernes sous le nazisme.* Paris: Editions Complexe, 2000.

Ted Gurr. *Why men rebel.* Princeton: Princeton University Press, 1970.

Anke Haarmann and Harald Lemke, Eds. *Culture|Nature: Art and Philosophy in the Context of Urban Development.* Illustrated Volume. Berlin: jovis Verlag, 2009.

Jacques Hadamard. *The Psychology of Invention in the Mathematical Field.* Mineola: Dover Publications, 1954.

David Haley. "The Limits of Sustainability: the art of ecology." In Sacha Kagan and Volker Kirchberg, Eds., *Sustainability: a new frontier for the arts and cultures.* Waldkirchen: VAS – Verlag für akademische Schriften, 2008, pp. 194-208.

David Haley and Jaya Iyer. "The Art of Sustainable Living: A Creative Approach to Global Social and Environmental Crises." Discussion Paper commissioned by The Asia-Europe Foundation (ASEF), for the British-Council & ASEF workshop "Climate leaders: release your creative powers / How can art enhance our ability to think and act differently?", Bangkok, Thailand, October 2009.

Stuart Hall. "The problem of ideology: Marxism without guarantees." In David Morley, and Kuan-Hsing Chen, Eds., *Stuart Hall: Critical Dialogues in Cultural Studies.* London: Routledge, 1996, pp. 24-45..

Volker Harlan. *Was ist Kunst? Werkstattgespräch mit Beuys.* Stuttgart: Urachhaus, 1986.

Eddie Harmon-Jones and Judson Mills, Eds. *Cognitive dissonance: Progress on a pivotal theory in social psychology.* Washington: American Psychological Association, 1999.

Jean-Marie Harribey, Ed. *Le développement a-t-il un avenir? Pour une société économe et solidaire.* Paris: Mille et Une Nuits / ATTAC, 2004.

Vaclav Havel. *Disturbing the Peace.* London: Faber and Faber, 1990.

Eric Havelock, *The Muse Learns to Write: Reflections on Orality and Literacy from Antiquity to the Present*, New Haven: Yale University Press, 1986.

N. Katherine Hayles, Ed. *Nanoculture: implications of the new technoscience.* Bristol: Intellect, 2004.

Martin Heidegger. *Being and Time.* Oxford: Basil Blackwell, 1967.

—— *On Time and Being.* New York: Harper & Row, 1972.

Nathalie Heinich. *The glory of Van Gogh: An Anthropology of Admiration.* Princeton: Princeton University Press, 1997.

Frieda Heyting, Bernard Kruithof, and Ernst Mulder. "Education and Social Integration: On Basic Consensus and the Cohesion of Society." *Educational Theory,* 52, 4, 2002, pp. 381-96.

Pekka Himanen. *The Hacker Ethic and the Spirit of the Information Age.* New York: Random House, 2001.

Albert O. Hirschman. *Exit, voice, and loyalty; responses to decline in firms, organizations, and states.* Cambridge: Harvard University Press, 1970.

Max Horkheimer. *Eclipse of Reason.* New York: Oxford University Press, 1947.

Max Horkheimer and Theodor Adorno. *Dialectic of Enlightenment. Philosophical Fragments.* Palo Alto: Stanford University Press, 2002.

Patrick Hossay. *Unsustainable: A primer for global environmental and social justice.* London: Zed Books, 2006.

Edmund Husserl. *The Crisis of the European Sciences and Transcendental Phenomenology.* Evanston (Ill.): Northwestern University Press, 1970.

—— "Foundational Investigations of the Phenomenological Origin of the Spatiality of Nature." In Peter McCormick and Frederick A. Elliston, Eds., *Husserl: Shorter Works.* Brighton: Harvester Press, 1981.

Lewis Hyde. "Making It." Sunday Book Review, *The New York Times*, 06. 04.2008. Available online: http://www.nytimes.com/2008/04/06/books/review/Hyde-t.html?pagewanted=1&_r=1 [last accessed: 13.02.2010].

Roman Jacobson and Linda R. Waugh. *The Sound Shape of Language.* Bloomington: Indiana University Press, 1979.

David Jaffee. *Organization Theory: Tension and Change.* New York: McGraw-Hill, 2001.

Sacha Kagan. *ENT³, Art Projects for intercultural interaction: Entrepreneurship in Conventions.* M.A. Thesis. Rotterdam: Erasmus Universiteit Rotterdam, 2004. Available online at: http://sachakagan.chez.tiscali.fr/docs/ent3.doc [last accessed: 13.02.2010].

—— "Le Spectateur Symbolique a rencontré Angelin Preljocaj." *Le GT – quotidien des grandes traversées,* 18.11.2004 [2004b].

—— "Art effectuating social change: double entrepreneurship in conventions." In Sacha Kagan and Volker Kirchberg, Eds., *Sustainability: a new frontier for the arts and cultures.* Waldkirchen: VAS – Verlag für akademische Schriften, 2008a, pp. 147-193.

—— "Introduction: Sustainability as a new frontier for the arts and cultures." In Sacha Kagan and Volker Kirchberg, Eds., *Sustainability: a new frontier for the arts and cultures.* Waldkirchen: VAS – Verlag für akademische Schriften, 2008b, pp. 14-24.

—— "Qué TAL ? - Transdisciplinary Action-Research & Literacy." Paper presented at the 2009 Summit of the World Alliance for Arts Education. Available online at http://sachakagan.wordpress.com/writings/ [last accessed: 07.07.2010]. [2009a]

—— "Land of Thorns: a filmic attempt to mediate issues of (un-) sustainable development." In David Knaute and Sacha Kagan, Eds, *Sustainability in Karamoja? Rethinking the terms of global sustainability in a crisis region of Africa*. Köln: Rüdiger Köppe Verlag, 2009, pp. 515-526. [2009b]

—— "Cultures of Sustainability and the aesthetics of the pattern that connects." *Futures: The journal of policy, planning and futures studies*, 42 (10), 2010.

Sacha Kagan and Hans Abbing. "Polity Conventions in Art Worlds: the constitution of authority configurations." Paper presented at the STP&A (Social Theory, Politics and Arts) 32nd annual conference, Vienna, July 2006. Available online at: http://www.culture.info/images/stories/kagan _abbing.doc [last accessed: 11.07.2010].

Sacha Kagan and Bettina Steinbrügge, Eds. *Everything will be fine / alles wird gut*. Lüneburg: Leuphana Universität Lüneburg, 2007.

Sacha Kagan, Liv Pedersen and Sally Ollech. "The Karamoja Syndrome: transdisciplinary systems research informing policy and advocacy." In David Knaute and Sacha Kagan, Eds, *Sustainability in Karamoja? Rethinking the terms of global sustainability in a crisis region of Africa*. Köln: Rüdiger Köppe Verlag, 2009, pp. 143-211.

Sacha Kagan et al. "Vision : a breeding place for transdisciplinary and transformative art&science." Summer School of Arts and Sciences for Sustainability in Social Transformation (ASSiST) website. Available online at http://assist2010.ning.com [last accessed: 01.07.2010].

Stephen Kalberg. "Max Weber's Types of Rationality: Cornerstones for the Analysis of Rationalization Processes in History." *American Journal of Sociology* 85, no. 5, 1980.

Lilian Karina and Marion Kant. *Tanz unterm Hakenkreuz*. Berlin: Henschel Verlag, 1996.

Tim Kasser. *The high price of materialism*. Cambridge: MIT Press, 2002.

Michael Keating. *Sommet de la Terre 1992: Un programme d'action: Version pour le grand public de l'Agenda 21 et des autres accords de Rio*. Genève: Centre pour notre Avenir à Tous, 1993.

Evelyn Fox Keller. *The century of the Gene*. Cambridge: Harvard University Press, 2000.

Caffyn Kelley. *Art and Survival: Patricia Johanson's environmental projects*. Salt Spring Island: Islands Institute, 2006.

—— *Culturing sustainability: A cookbook for artists and educators*. Athabasca: Athabasca University, 2008. Available online at: http://islandsinstitute.ning.com/forum/topic/show?id=1163070%3ATopic%3A16092 [last accessed: 05.03.2010].

Grant Kester. Position Paper. The Monongahela Conference on Post-Industrial Community Development, Carnegie Mellon University, Pittsburgh, 2004. Available online at: http://moncon.greenmuseum.org/papers/kester1.html [last accessed: 10.07.2010].

Volker Kirchberg. "Cultural consumption analysis: Beyond Structure and Agency." *Cultural Sociology* 1, no. 1, 2007, pp. 115-135.

—— "Zur Transdisziplinarität des Thema ‚Stadt und Kultur' in den Kulturwissenschaften – Eine Einleitung." Inaugural lecture of the lectures series "Transdisziplinarität der Kulturwissenschaften: Stadt und Kultur, Kulturwissenschaftliche Perspektiven", Leuphana Universität Lüneburg, October 15th 2009.

Volker Kirchberg, Sacha Kagan and Christoph Behnke. *Final Report: ESA Research Network for the Sociology of the Arts Conference in Lüneburg, 27 March until 1 April 2007*. Lüneburg: Leuphana Universität Lüneburg, 2007.

Arjo Klamer. "The economics of attention." Unpublished paper. Rotterdam: Erasmus Universiteit Rotterdam, 2002.

—— "Art as Common Good." Paper presented at the Association for Cultural Economics International (ACEI) Conference, Chicago, 2004.

Julie Thompson Klein. "Prospects for transdisciplinarity." *Futures: The journal of policy, planning and futures studies*, 36, 2004, pp. 515-526.

Naomi Klein. *No Logo – Taking Aim at the Brand Bullies*. Toronto: Vintage Canada, 2000.

David Knaute and Sacha Kagan, Eds. *Sustainability in Karamoja? Rethinking the terms of global sustainability in a crisis region of Africa*. Köln: Rüdiger Köppe Verlag, 2009.

Julien Knebusch. "Art and Climate (Change) Perception: outline of a phenomenology of climate." In Sacha Kagan and Volker Kirchberg, Eds., *Sustainability: a new frontier for the arts and cultures*. Waldkirchen: VAS – Verlag für akademische Schriften, 2008, pp. 242-262.

Oleg Koefoed. "Zones of Sustension: an exploration of eventality, culturality, and collective intuition in life and work." In Sacha Kagan and Volker Kirchberg, Eds., *Sustainability: a new frontier for the arts and cultures*. Waldkirchen: VAS – Verlag für akademische Schriften, 2008, pp. 59-92.

Arthur Koestler. *The ghost in the machine*. London: Hutchinson, 1967.

Melvin Kranzberg and Carroll Purcell, *Eds. Technology in Western Civilization.* 2 vols. New York: Oxford University Press, 1967.

Thomas Kuhn. *The Structure of Scientific Revolutions*. Chicago: University of Chicago Press, 1962.

Hildegard Kurt. "Kultur und Kunst in der Agenda21? Anmerkungen zu einem schwierigen Verhältnis." In Mathias Fechter and Margret Krannich, Eds., *Gesellschaftliche Perspektiven. Arbeit, Geschlecht, Natur, Neue Medien*. Essen: .Klartext Verlag, 2000, pp. 112-118.

—— "Nachhaltigkeit – Eine Herausforderung an die Kunst?" *Kulturpolitische Mitteilungen*, 97, 2, 2002, pp. 46-49.

Hildegard Kurt and Michael Wehrspaun. "Kultur: der verdrängte Schwerpunkt des Nachhaltigkeits-Leitbildes." In Tina Jerman, Ed., *Zukunfts-*

Formen: Kultur und Agenda 21. Bonn: Kulturpolitische Gesellschaft, 2001, pp. 79-93.

Hildegard Kurt and Bernd Wagner, Eds. *Kultur – Kunst – Nachhaltigkeit: Die Bedeutung von Kultur für das Leitbild Nachhaltige Entwicklung*. Bonn: Klartext Verlag, 2002.

Mierle Laderman Ukeles. "Maintenance Art—Proposal for an Exhibition." 1969. Available on the website of Ronald Feldman gallery, at: http://www.feldmangallery.com/media/pdfs/Ukeles_MANIFESTO.pdf [last accessed: 07.07.2010].

Bernard Lahire. *L'homme pluriel: les ressorts de l'action*. Paris: Nathan, 1998.

—— "From the habitus to an individual heritage of dispositions. Towards a sociology at the level of the individual." *Poetics*, 31, 5, October 2003, pp. 329-355.

—— *La culture des individus: dissonances culturelles et distinction de soi*. Paris: La découverte, 2004.

George Lakoff and Mark Johnson. *Philosophy in the flesh*. New York: Basic Books, 1999.

—— *Where Mathematics Comes From: How the Embodied Mind Brings Mathematics into Being*. New York: Basic Books, 2001.

Michèle Lamont and Marcel Fournier, Eds. *Cultivating differences: Symbolic boundaries and the making of inequality*. Chicago : The University of Chicago Press, 1992.

Klaus Landwehr, Ed. *Ecological Perception Research, Visual Communication, and Aesthetics*. Berlin: Springer-Verlag, 1990.

Friederike Lang. *Kunst im Diskurs um Kultur und Nachhaltigkeit: Eine empirische Untersuchung des doppelten Defizits zwischen Akteuren aus den Bereichen der Kunst und der Nachhaltigkeit*. Diplomarbeit. Lüneburg: Leuphana Universität Lüneburg, 2007.

John Lane. *A snake's tail full of ants: Art, Ecology & Consciousness*. Bideford: Resurgence, 1996.

Kalle Lasn. *Culture Jam: The Uncooling of America*. New York : Eagle Brook, 1999.

Ervin Laszlo. *Goals for Mankind : a report to the Club of Rome on the new horizons of global community*. New York : Dutton, 1977

—— *The Creative Cosmos: A Unified Science of Matter, Life and Mind*. Edinburgh: Floris Books, 1993.

—— *The systems view of the world : a holistic vision for our time*. Cresskill : Hampton Press, 1996.

—— *Science and the akashic field : an integral theory of everything*. Rochester : Inner Traditions, 2007 (updated 2nd ed.).

Pierre de Latil. *La pensée artificielle. Introduction à la cybernétique*. Paris: Gallimard, 1953.

Serge Latouche. *Survivre au développement*. Paris : Mille et Une Nuits, 2004.

Bruno Latour. *Nous n'avons jamais été modernes: Essai d'anthropologie symétrique*. Paris: La Découverte, 1991.

—— *Reassembling the social: an introduction to Actor-Network-Theory*. Oxford: Oxford University Press, 2005.

—— "'It's Development, Stupid!' or How to Modernize Modernization." 2008. Available online at: http://www.bruno-latour.fr/articles/article/107-NORDHAUS&SHELLENBERGER.pdf [last accessed: 07. 07.2010]. [NB: Was marked by Latour as "for a book of essays edited by Jim Proctor *Postenvironmentalism* to be published by the MIT Press, 2008."]

Bettina Laville and Jacques Leenhardt. *Villette-Amazone: Manifeste pour l'Environnement au XXIe Siècle*. Arles: Actes Sud, 1996.

Henri Lefebvre. *Position: contre les technocrates*. Paris: Gonthier, 1967.

Enrique Leff. *Saber Ambiental: sustentabilidad, racionalidad, complejidad, poder*. Mexico: Siglo xxi editores, 1998.

Timothy M. Lenton, Hermann Held, Elmar Kriegler, Jim W. Hall, Wolfgang Lucht, Stefan Rahmstorf and Hans Joachim Schellnhuber. "Tipping elements in the Earth's climate system." *Proceedings of the National Academy of Sciences of the United States of America*, 105, 6, February 2008, pp. 1786-1793. Available online at: http://www.pnas.org/content/105/6/1786.long [last accessed: 07.07.2010].

André Lepecki. "For a Sensorial Manifesto (On Dances that Failed)." In Chantal Pontbriand, Ed., *Dance: Distinct Language and Cross-Cultural Influences*. Montréal: Parachute, 2001, pp. 161-168.

Lawrence Levine. *Highbrow, Lowbrow: The emergence of cultural hierarchy in America*. Cambridge: Harvard University Press, 1988.

Stacy Levy. Position Paper. The Monongahela Conference on Post-Industrial Community Development, Carnegie Mellon University, Pittsburgh, 2004. Available online at: http://moncon.greenmuseum.org/papers/levy1.html [last accessed: 10.07.2010].

David K.Lewis. *Convention: a philosophical study*. Cambridge: Harvard University Press, 1969.

Lucy Lippard and Kirsten Gerdes, Eds. *Weather Report: Art and Climate Change*. Boulder: Boulder Museum of Contemporary Art, 2007.

James Lovelock. *Gaia: A New Look at Life on Earth*. Oxford: Oxford University Press, 1979.

Jean-Claude Lugan, *La systémique sociale*. 4th ed. Presses Universitaires de France, Paris, 2005.

Niklas Luhmann. *Politische Theorie im Wohlfahrtsstaat*. München: Olzog, 1981.

—— *Soziale Systeme: Grundriß einer allgemeinen Theorie*. Frankfurt am Main: Suhrkamp, 1984.

—— *Ökologische Kommunikation:Kann die moderne Gesellschaft sich auf ökologische Gefährdungen einstellen?*. Opladen: Westdeutscher Verlag, 1986.

—— *Die Wirtschaft der Gesellschaft.* Frankfurt am Main: Suhrkamp, 1988.
—— *L'amour comme passion.* Paris: Aubier, 1990a.
—— *Die Wissenschaft der Gesellschaft.* Frankfurt am Main: Suhrkamp, 1990b.
—— *Beobachtung der Moderne.* Opladen: Westdeutscher Verlag, 1992.
—— "La malice du sujet et la question de l'homme." *Sociétés*, 43, 1994, pp. 3-15.
—— *Soziologische Aufklärung*, vol. 6. Opladen: Westdeutscher Verlag, 1995.
—— *Politique et complexité. Les contributions de la théorie générale des systèmes*. Paris: Cerf, 1999.
—— *Art as a Social System*. Palo Alto: Stanford University Press, 2000.
Stéphane Lupasco. *Le Principe d'antagonisme et la Logique de l'énergie. Prolégomènes à une science de la contradiction.* Paris: Hermann, 1951.
—— *L'énergie et la matière vivante. Antagonisme constructeur et logique de l'hétérogène.* Paris: Julliard, 1962.
Michel Maffesoli. *Eloge de la raison sensible*. Paris: La Table Ronde, 2005.
Friedemann Malsch, Ed. *Auszeit. Kunst und Nachhaltigkeit / Timeout. Art and Sustainability.* Nürnberg: Verlag für moderne Kunst, 2007.
Karl Mannheim. *Man and society in an age of reconstruction*. London: Routledge and Kegan Paul, 1940.
Lynn Margulis. *Symbiotic Planet.* New York: Basic Books, 1998.
Filippo Tommaso Marinetti. "Manifeste du futurisme." *Le Figaro*, 20.02 1909. Available online at: http://zinclafriche.org/mef/?page_id=386 [last accessed: 07.07.2010].
Herbert Marcuse. "Some Implications of Modern Technology." *Studies in Philosophy and Social Science*, 9, 1941, pp. 414-439.
—— *The aesthetic dimension: toward a critique of Marxist aesthetics*. Boston: Beacon Press, 1978.
Magoroh Maruyama. "The Second Cybernetics: Deviation-amplifying Mutual Causal Processes." *American Scientist*, 51, 1963, pp. 164-179, 250-256.
Magoroh Maruyama. "Paradigmatology and its Application to Cross-disciplinary, Cross-professional and Cross-cultural Communication." *Cybernetica*, 17, 1974, pp. 136-156, 237-281.
Humberto Maturana and Francisco Varela. *Autopoietic systems.* Santiago de Chile: Facultad de Ciencias, Universidad de Santiago, 1972.
—— *Autopoïesis and Cognition.* Boston: D. Reidel, 1980.
Helen Mayer Harrison and Newton Harrison. *The Lagoon Cycle*. Ithaca: Herbert F. Johnson Museum of Art, Cornell University, 1985.
—— Position Paper. The Monongahela Conference on Post-Industrial Community Development, Carnegie Mellon University, Pittsburgh, 2004. Available online at: http://moncon.greenmuseum.org/papers/harrison1.html [last accessed: 10.07.1980].

Patricia A. McAnany and Norman Yoffee, Eds. *Questioning collapse: Human resilience, ecological vulnerability, and the aftermath of empire*. Cambridge: Cambridge University Press, 2009.

William McDonough and Michael Braungart. *Cradle to Cradle: Remaking the Way We Make Things*. New York: North Point Press, 2002.

Pierre-Michel Menger. *Portrait de l'artiste en travailleur: Métamorphoses du capitalisme*. Paris: Seuil, 2002.

Colin Mercer. *Towards Cultural Citizenship: Tools for Cultural Policy and Development*. Stockholm: The Bank of Sweden Tercentenary Foundation & Gildlunds Förlag, 2002.

Maurice Merleau-Ponty. *Phenomenology of Perception*. London: Routledge & Kegan Paul, 1962.

—— *The Visible and the Invisible*. Evanston: Northwestern University Press, 1968.

Tom Merriman and Connie Merriman. Position Paper. The Monongahela Conference on Post-Industrial Community Development, Carnegie Mellon University, Pittsburgh, 2004. Available online at: http://moncon.greenmuseum.org/papers/merriman1.html [last accessed: 10.07.2010].

Max Miller. "Intersystemic Discourse and Co-Ordinated Dissent: A Critique of Luhmann's Concept of Ecological Communication". *Theory, Culture & Society*, 11, 1994, pp 101-121.

Piet Mondrian. *Le Néoplasticisme*. Paris: Éditions de l'Effort Moderne, 1921.

Janet Moore. "Is higher education ready for transformative learning? A question explored in the study of sustainability." *Journal of Transformative Education*, 3(5), 2005, pp. 76-91.

Gareth Morgan. *Images of Organization*. Thousand Oaks: Sage, 1997.

Robert C. Morgan. "Touch Sanitation: Mierle Laderman Ukeles." In Linda Frye Burnham and Steven Durland, Eds., *The Citizen Artist: 20 Years of Art in the Public Arena. An Anthology from High Performance Magazine 1978-1998*. Critical Press, 1998. (Article originally published in *High Performance Magazine*, Fall 1982.) Available Online at Community Arts Network Reading Room: http://www.communityarts.net/readingroom/archivefiles/2002/09/touch_sanitatio.php [last accessed 06.02.2010].

Edgar Morin. *Le Paradigme perdu: la nature humaine*. Paris: Seuil, 1973.

—— *La méthode, volume 1: La nature de la nature*. Paris: Seuil, 1977.

—— *La méthode, volume 2: La vie de la vie*. Paris: Seuil, 1980.

—— *La méthode, volume 3: La connaissance de la connaissance*. Paris: Seuil, 1986.

—— *La méthode, volume 4: Les idées: leur habitat, leur vie, leurs moeurs, leur organisation*. Paris: Seuil, 1991.

—— *Method. Towards a Study of Humankind. Volume 1: The Nature of Nature*. New York: Peter Lang, 1992.

—— *L'an I de l'ère écologique. Et dialogue avec Nicolas Hulot*. Paris: Tallandier, 2007.

—— *La méthode*. [Joint edition of the 6 volumes.] Paris: Seuil, 2008.

—— "Changer le rapport de l'homme à la nature n'est qu'un début." *Le Monde*, June 12th 2009.

Harold Morowitz. *Beginnings of cellular life*. New Haven: Yale University Press, 1992.

Raymonde Moulin. "La genèse de la rareté artistique. " *Revue d'ethnologie Française*, 8, 1978, pp. 241-258.

Hedwig Müller and Patricia Stöckemann. *...jeder Mensch ist ein Tänzer: Ausdruckstanz in Deutschland zwischen 1900 und 1945.* Giessen: Anabas-Publishing, 1993.

Lewis Mumford. *The Myth of the Machine. The Pentagon of Power*. London: Secker and Warburg, 1971.

Basarab Nicolescu. *Manifesto of Transdisciplinarity*. Albany : State University of New York Press, 2002.

—— Ed. *Transdisciplinarity: theory and practice*. Cresskill : Hampton Press, 2008.

Wouter de Nooy et al. *Exploratory social network analysis with Pajek*. Cambridge: Cambridge University Press, 2005.

Observatoire de la Mondialisation. *Lumière sur l'AMI. Le test de Dracula*. Paris: Esprit Frappeur, 1998.

André Orléan, Ed. *Analyse économique des conventions*. Paris: Presses Universitaires de France, 1994.

William Outhwaite. "Jurgen Habermas." In George Ritzer, Ed., *The Blackwell Companion to Major Contemporary Social Theorists*. Malden: Blackwell Publishing, 2003.

A. Laurie Palmer. Position Paper. The Monongahela Conference on Post-Industrial Community Development, Carnegie Mellon University, Pittsburgh, 2004. Available online at: http://moncon.greenmuseum.org/papers/palmer1.html [last accessed: 10.07.2010].

Talcott Parsons. *Theories of Society: foundations of modern sociological theory*. New York: Free Press, 1961.

René Passet. *L'économique et le vivant*. Paris: Payot, 1979.

Luigi Pellizzoni. "The Myth of the Best Argument: Power, Deliberation, and Reason." *British Journal of Sociology*, 52, 1, 2001, pp. 59-86.

Alessandro Pizzorno. "On the individualistic theory of social order." In Pierre Bourdieu and James S. Coleman, Eds., *Social Theory for a Changing Society*. Boulder: Westview, 1991, pp. 209-231.

PLATFORM. Position Paper. The Monongahela Conference on Post-Industrial Community Development, Carnegie Mellon University, Pittsburgh, 2004. Available online at: http://moncon.greenmuseum.org/papers/platform1.html [last accessed: 10.07.2010].

Plato, *Phaedrus*. In Edith Hamilton and Huntington Cairns, Eds., *Plato: The collected Dialogues*. Princeton: Princeton University Press, 1982.

Chantal Pontbriand, Ed. *Dance: Distinct Language and Cross-Cultural Influences*. Montréal: Parachute, 2001.

Karen Press. Exhibition opening talk. Cape Town: 2001. Available online at www.exchange-values.org [last accessed: 02.03.2010].

Herman Prigann. "Thoughts about 'Art in Nature'." *IO. The Internet Magazine of Applied Aesthetics*, Summer 1998. Available online at: http://www.helsinki.fi/jarj/iiaa/io1998/prigann.html [last accessed: 09.03.2010].

Ilya Prigogine. "La thermodynamique de la vie." *La Recherche*, 3, 24, June 1972, pp. 547-562.

Thomas Princen et al., Eds. *Confronting consumption*. Cambridge: MIT Press, 2002.

Aviva Rahmani. "The Butterfly Effect of Hummingbirds: environmental triage: disturbance theory, triggerpoints, and virtual analogs for physical sites." In Sacha Kagan and Volker Kirchberg, Eds., *Sustainability: a new frontier for the arts and cultures*. Waldkirchen: VAS – Verlag für akademische Schriften, 2008, pp. 264-289.

—— *What the World Needs is a Good Housekeeper*. Charleston: BookSurge Publishing, 2009.

Paul H. Ray and Sherry Ruth Anderson. *The Cultural Creatives*. New York: Harmony Books, 2000.

Richard Register. *Ecocities*. Berkeley: Berkeley Hills Books, 2001.

George Ritzer. *Enchanting a Disenchanted World: Revolutionizing the Means of Consumption*. Thousand Oaks: Pine Forge Press, 1999.

John Robinson. "Squaring the circle? Some thoughts on the idea of sustainable development." *Ecological Economics*, 48, 2004, pp. 369-384.

Pierre Rosenstiehl and Jean Petitot. "Automate asocial et systèmes acentrés." *Communications*, 22, 1974, pp. 45-62.

Ann Rosenthal. Position Paper. The Monongahela Conference on Post-Industrial Community Development, Carnegie Mellon University, Pittsburgh, 2004. Available online at: http://moncon.greenmuseum.org/papers/rosenthal1.html [last accessed: 10.07.2010].

Joël De Rosnay. *Le Macroscope: Vers une vision globale*. Paris: Seuil, 1975.

Rosemary Ross Johnston, "On Connection and Community: Transdisciplinarity and the Arts." In Basarab Nicolescu, Ed., *Transdisciplinarity : theory and practice*. Cresskill: Hampton Press, 2008.

Shelley Sacks. "Exchange Values Six Years On." 2002. Available online at www.exchange-values.org [last accessed: 02.03.2010].

—— "Banana is not an Easy Thing." nd. Available online at: www.exchange-values.org [last accessed: 02.03.2010].

Laurien Saraber. *Tussen rasa en redoble: werelddans in de Nederlandse kunsteducatie*. Utrecht: LOKV, 2000.

Ferdinand de Saussure. *Cours de linguistique générale*. Genève: Payot, 1931.

Willem Schinkel. "The autopoiesis of the artworld after the end of art." Paper prepared for the workshop "Collapsing cultural canons: elite culture, popular culture, and politics in late modernity." Paris: EHESS, October 2004.

Maximilian Schlosshauer. "Decoherence, the measurement problem, and interpretations of quantum mechanics." *Review of Modern Physics* 76, 4, 2005, pp. 1267-1305.

Donald Schön. *The Reflective Practitioner: How Professionals Think In Action*. London: Maurice Temple Smith Ltd, 1983.

Juliet Schor and Douglas Holt, Eds. *The consumer society reader*. New York: New Press, 2000.

William Richard Scott. *Organizations: Rational, Natural, and Open Systems*. Fifth Edition. Upper Saddle River: Prentice Hall, 2003.

Peter M. Senge. *The Fifth Discipline: The Art and Practice of The Learning Organization*. London: Random House, 1999 (original publication: 1990).

Richard Sennett. *The Corrosion of Character: The Personal Consequences of Work in the New Capitalism*. New York: W.W. Norton, 1998.

—— *The Culture of the New Capitalism*. New Haven: Yale University Press, 2005.

—— *The craftsman*. New Haven: Yale University Press, 2008.

Michel Serres. "Ce qui est écrit dans le code: 2. Vie, information, deuxième principe." *Critique*, 290, 1971, pp. 579-606.

James Shapiro. "Genome System Architecture and Natural Genetic Engineering in Evolution." In Lynn Helena Caporale, Ed., *Molecular Strategies in Biological Evolution, Annals of the New York Academy of Science*, vol. 870, 1999.

Simon Sheikh. "Notes on Institutional Critique". 2006. Available online at: http://eipcp.net/transversal/0106/sheikh/en [last accessed: 07.07.2010].

Michael Shellenberger and Ted Nordhaus. "The Death of Environmentalism: Global Warming in a Post Environmental World." Available online at http://www.thebreakthrough.org/images/Death_of_Environmentalism.pdf [last accessed: 06.03.2010].

Richard Shusterman, *Pragmatist Aesthetics: Living Beauty, Rethinking Art*, Cambridge: Blackwell Publishers, 1992.

Herbert A. Simon. "From substantive to procedural rationality." In Spiro J. Latsis, Ed., *Method and appraisal in economics*. Cambridge: Cambridge University Press, 1976.

Derek Simons and Katie Warfield. "The Biocentric and Culture-centric Orientations of Cultural Ecology: A Conceptual Review with a View to Cultural Policy and Planning." Working Paper no. 2. Vancouver: Creative City Network of Canada – Centre of expertise on culture and communities, 2007.

Joost Smiers. *Arts under Pressure: promoting cultural diversity in the age of globalization*. London: Zed Books, 2003.

John Smith and Chris Jenks. *Qualitative Complexity: Ecology, cognitive processes and the re-emergence of structures in post-humanist social theory*. London: Routledge, 2006.

Stephanie Smith. *Ecologies : Mark Dion, Peter Fend, Dan Peterman*. Chicago: Smart Museum of Art, 2001.

——, Ed. *Beyond green: toward a sustainable art*. Chicago: Smart Museum of Art, 2005.

Alastair Sooke. "He's got the whole world in his hands." *The Daily Telegraph*, 2007-03-24. Available online at: http://www.telegraph.co.uk/arts/main.jhtml?xml=/arts/2007/03/24/bagoldsworthy124.xml [last accessed: 07/02/2010].

Sue Spaid. *Ecovention: Current Art to Transform Ecologies*. Cincinnati: Contemporary Arts Center, 2002.

Charlene Spretnak, Ed. *The Politics of Women's Spirituality*. New York: Anchor/Doubleday, 1981

Heike Strelow, Herman Prigann and Vera David, Eds. *Ecological Aesthetics: Art in Environmental Design: Theory and Practice*. Basel: Birkhäuser Verlag für Architektur, 2004.

Arthur G. Tansley. "The use and abuse of vegetational concepts and terms." *Ecology*, 16, 1935, pp. 284-307.

Eric Tariant. "Il est possible de ressusciter l'espérance." Interview of Edgar Morin. *TGV Magazine*, 116, July-August 2009, pp. 96-97.

Richard Tarna, *The Passion of the Western Mind*. New York: Ballantine Books, 1991.

Wladyslaw Tatarkiewicz, Jean Harrell, Cyril Barrett and Danuta Petsch. *History of Aesthetics*. Bristol: Thoemmes Press, 2006.

Temporary Services. "As we live, so we work.." Position Paper. The Monongahela Conference on Post-Industrial Community Development, Carnegie Mellon University, Pittsburgh, 2004. Available online at: http://moncon.greenmuseum.org/papers/temp1.html [last accessed: 10.07.2010].

Gunther Teubner. "Economy of Gift – Positivity of Justice. The mutual paranoia of Jacques Derrida and Niklas Luhmann". *Theory, Culture and Society*, 18, 1, 2001.

René Thom. *Modèles mathématiques de la morphogénèse: recueil de textes sur la théorie des catastrophes et ses applications*. Paris: Union générale d'éditions, 1974.

Sue Thornton. "Lynne Hull." *High Performance*, 62, Summer 1993. Available online at Community Arts Network Reading Room: http://www.communityarts.net/readingroom/archivefiles/1999/12/lynne_hull.php [last accessed: 07/02/2010].

Alain Touraine. *Critique de la Modernité*. Paris: Fayard, 1992.

Jörg Tremmel. *Nachhaltigkeit als politische und analytische Kategorie. Der deutsche Diskurs um nachhaltige Entwicklung im Spiegel der Interessen der Akteure*. München: Ökom-Verlag, 2003.

Thorstein Veblen. *Theory of the leisure class. An Economic Study in the Evolution of Institutions.* New York: Macmillan, 1899. eBook available at Project Gutenberg: http://www.gutenberg.org/etext/833 [last accessed: 07.07.2010].

Christopher Vecsey, *Imagine Ourselves Richly: Mythic Narratives of North American Indians.* San Francisco: HarperCollins, 1991.

Laura Verdi. "From Art & Fashion to Homo Civicus." In Sacha Kagan and Volker Kirchberg, Eds., *Sustainability: a new frontier for the arts and cultures*. Waldkirchen: VAS – Verlag für akademische Schriften, 2008, pp. 498-528.

Franz de Waal. *The ape and the sushi-master: Cultural Reflections of a Primatologist*. New York: Basic Books, 2001.

Brian H. Walker et al. "A handful of heuristics and some propositions for understanding resilience in social-ecological systems." *Ecology and Society*, 11, n°1, 2006. Available online at: http://www.ecologyandsociety.org/vol11/iss1/art13 [last accessed: 07.07.2010].

Paul Watzlawick, Ed. *L'invention de la réalité - - contributions au constructivisme*. Paris: Seuil, 1996.

Max Weber. *Politics as a Vocation*. Philadelphia: Fortress Press, 1965.

—— *The Protestant Ethic and the Spirit of Capitalism.* Penguin Books, 2002. Available online at: http://www.ne.jp/asahi/moriyuki/abukuma/weber/world/ethic/pro_eth_frame.html [last accessed: 07.07.2010].

Linda Weintraub. *Ecocentric Topics: Pioneering themes for eco-art*. Rhinebeck: Artnow Publications, 2006.

John C. Welchman, Ed. *Institutional Critique and After*. Zurich: JRP|Ringier, 2006.

Etienne Wenger. *Communities of Practice*. Cambridge: Cambridge University Press, 1998.

Etienne Wenger, Richard McDermott, and William Snyder. *Cultivating communities of practice: a guide to managing knowledge*. Boston: Harvard Business School Press, 2002.

Robert Wesson. *Beyond Natural Selection.* Cambridge: The MIT Press, 1991.

Lynn White Jr. "The historical roots of our ecologic crisis". *Science* 155, 1967, pp. 1203-1207.

Ken Wilber. *A brief history of everything*. Boston : Shambhala, 2007.

Helmut Wilke. *Systemtheorie III: Steuerungstheorie.* Stuttgart: Lucius & Lucius, 1995.

Edward O.Wilson. *On Human Nature*. Cambridge: Harvard University Press, 1978.

Ulf Wuggenig. "Nachhaltige Entwicklung, Ethik und zeitgenössische Kunst." In Günther Altner and Gerd Michelsen, Eds., *Ethik und Nachhaltigkeit: Grundsatzfragen und Handlungsperspektiven im universitären Agendaprozess*. Frankfurt am Main: VAS – Verlag für akademische Schriften, 2001, pp. 240-264.

Jacqueline Zanoni de los Santos. *Tracing the Experiential Path: Large Scale Environmental Sculpture by Patricia Johanson*. Master Thesis. Norman: University of Oklahoma Graduate College, 1997.

Wolfgang Zinggl, Ed. *Wochenklausur: Sociopolitical Activism in Art*. Vienna: Springer, 2001.

Wolfgang Zumdick. "Beuys." Website of the Social Sculpture Research Unit, Oxford Brookes University, 2008. Accessible online at: http://www.social-sculpture.org/influences-and-underpinnings/beuys.htm [last accessed: 02.03.2010].